PRACTICAL SAMPLING TECHNIQUES

STATISTICS: Textbooks and Monographs

A Series Edited by

D. B. Owen, Founding Editor, 1972–1991

1. The Generalized Jackknife Statistic, *H. L. Gray and W. R. Schucany*
2. Multivariate Analysis, *Anant M. Kshirsagar*
3. Statistics and Society, *Walter T. Federer*
4. Multivariate Analysis: A Selected and Abstracted Bibliography, 1957–1972, *Kocherlakota Subrahmaniam and Kathleen Subrahmaniam*
5. Design of Experiments: A Realistic Approach, *Virgil L. Anderson and Robert A. McLean*
6. Statistical and Mathematical Aspects of Pollution Problems, *John W. Pratt*
7. Introduction to Probability and Statistics (in two parts), Part I: Probability; Part II: Statistics, *Narayan C. Giri*
8. Statistical Theory of the Analysis of Experimental Designs, *J. Ogawa*
9. Statistical Techniques in Simulation (in two parts), *Jack P. C. Kleijnen*
10. Data Quality Control and Editing, *Joseph I. Naus*
11. Cost of Living Index Numbers: Practice, Precision, and Theory, *Kali S. Banerjee*
12. Weighing Designs: For Chemistry, Medicine, Economics, Operations Research, Statistics, *Kali S. Banerjee*
13. The Search for Oil: Some Statistical Methods and Techniques, *edited by D. B. Owen*
14. Sample Size Choice: Charts for Experiments with Linear Models, *Robert E. Odeh and Martin Fox*
15. Statistical Methods for Engineers and Scientists, *Robert M. Bethea, Benjamin S. Duran, and Thomas L. Boullion*
16. Statistical Quality Control Methods, *Irving W. Burr*
17. On the History of Statistics and Probability, *edited by D. B. Owen*
18. Econometrics, *Peter Schmidt*

19. Sufficient Statistics: Selected Contributions, *Vasant S. Huzurbazar (edited by Anant M. Kshirsagar)*
20. Handbook of Statistical Distributions, *Jagdish K. Patel, C. H. Kapadia, and D. B. Owen*
21. Case Studies in Sample Design, *A. C. Rosander*
22. Pocket Book of Statistical Tables, *compiled by R. E. Odeh, D. B. Owen, Z. W. Birnbaum, and L. Fisher*
23. The Information in Contingency Tables, *D. V. Gokhale and Solomon Kullback*
24. Statistical Analysis of Reliability and Life-Testing Models: Theory and Methods, *Lee J. Bain*
25. Elementary Statistical Quality Control, *Irving W. Burr*
26. An Introduction to Probability and Statistics Using BASIC, *Richard A. Groeneveld*
27. Basic Applied Statistics, *B. L. Raktoe and J. J. Hubert*
28. A Primer in Probability, *Kathleen Subrahmaniam*
29. Random Processes: A First Look, *R. Syski*
30. Regression Methods: A Tool for Data Analysis, *Rudolf J. Freund and Paul D. Minton*
31. Randomization Tests, *Eugene S. Edgington*
32. Tables for Normal Tolerance Limits, Sampling Plans and Screening, *Robert E. Odeh and D. B. Owen*
33. Statistical Computing, *William J. Kennedy, Jr., and James E. Gentle*
34. Regression Analysis and Its Application: A Data-Oriented Approach, *Richard F. Gunst and Robert L. Mason*
35. Scientific Strategies to Save Your Life, *I. D. J. Bross*
36. Statistics in the Pharmaceutical Industry, *edited by C. Ralph Buncher and Jia-Yeong Tsay*
37. Sampling from a Finite Population, *J. Hajek*
38. Statistical Modeling Techniques, *S. S. Shapiro and A. J. Gross*
39. Statistical Theory and Inference in Research, *T. A. Bancroft and C.-P. Han*
40. Handbook of the Normal Distribution, *Jagdish K. Patel and Campbell B. Read*
41. Recent Advances in Regression Methods, *Hrishikesh D. Vinod and Aman Ullah*
42. Acceptance Sampling in Quality Control, *Edward G. Schilling*
43. The Randomized Clinical Trial and Therapeutic Decisions, *edited by Niels Tygstrup, John M Lachin, and Erik Juhl*
44. Regression Analysis of Survival Data in Cancer Chemotherapy, *Walter H. Carter, Jr., Galen L. Wampler, and Donald M. Stablein*
45. A Course in Linear Models, *Anant M. Kshirsagar*
46. Clinical Trials: Issues and Approaches, *edited by Stanley H. Shapiro and Thomas H. Louis*
47. Statistical Analysis of DNA Sequence Data, *edited by B. S. Weir*
48. Nonlinear Regression Modeling: A Unified Practical Approach, *David A. Ratkowsky*
49. Attribute Sampling Plans, Tables of Tests and Confidence Limits for Proportions, *Robert E. Odeh and D. B. Owen*
50. Experimental Design, Statistical Models, and Genetic Statistics, *edited by Klaus Hinkelmann*
51. Statistical Methods for Cancer Studies, *edited by Richard G. Cornell*
52. Practical Statistical Sampling for Auditors, *Arthur J. Wilburn*
53. Statistical Methods for Cancer Studies, *edited by Edward J. Wegman and James G. Smith*

54. Self-Organizing Methods in Modeling: GMDH Type Algorithms, *edited by Stanley J. Farlow*
55. Applied Factorial and Fractional Designs, *Robert A. McLean and Virgil L. Anderson*
56. Design of Experiments: Ranking and Selection, *edited by Thomas J. Santner and Ajit C. Tamhane*
57. Statistical Methods for Engineers and Scientists: Second Edition, Revised and Expanded, *Robert M. Bethea, Benjamin S. Duran, and Thomas L. Boullion*
58. Ensemble Modeling: Inference from Small-Scale Properties to Large-Scale Systems, *Alan E. Gelfand and Crayton C. Walker*
59. Computer Modeling for Business and Industry, *Bruce L. Bowerman and Richard T. O'Connell*
60. Bayesian Analysis of Linear Models, *Lyle D. Broemeling*
61. Methodological Issues for Health Care Surveys, *Brenda Cox and Steven Cohen*
62. Applied Regression Analysis and Experimental Design, *Richard J. Brook and Gregory C. Arnold*
63. Statpal: A Statistical Package for Microcomputers—PC-DOS Version for the IBM PC and Compatibles, *Bruce J. Chalmer and David G. Whitmore*
64. Statpal: A Statistical Package for Microcomputers—Apple Version for the II, II+, and IIe, *David G. Whitmore and Bruce J. Chalmer*
65. Nonparametric Statistical Inference: Second Edition, Revised and Expanded, *Jean Dickinson Gibbons*
66. Design and Analysis of Experiments, *Roger G. Petersen*
67. Statistical Methods for Pharmaceutical Research Planning, *Sten W. Bergman and John C. Gittins*
68. Goodness-of-Fit Techniques, *edited by Ralph B. D'Agostino and Michael A. Stephens*
69. Statistical Methods in Discrimination Litigation, *edited by D. H. Kaye and Mikel Aickin*
70. Truncated and Censored Samples from Normal Populations, *Helmut Schneider*
71. Robust Inference, *M. L. Tiku, W. Y. Tan, and N. Balakrishnan*
72. Statistical Image Processing and Graphics, *edited by Edward J. Wegman and Douglas J. DePriest*
73. Assignment Methods in Combinatorial Data Analysis, *Lawrence J. Hubert*
74. Econometrics and Structural Change, *Lyle D. Broemeling and Hiroki Tsurumi*
75. Multivariate Interpretation of Clinical Laboratory Data, *Adelin Albert and Eugene K. Harris*
76. Statistical Tools for Simulation Practitioners, *Jack P. C. Kleijnen*
77. Randomization Tests: Second Edition, *Eugene S. Edgington*
78. A Folio of Distributions: A Collection of Theoretical Quantile-Quantile Plots, *Edward B. Fowlkes*
79. Applied Categorical Data Analysis, *Daniel H. Freeman, Jr.*
80. Seemingly Unrelated Regression Equations Models: Estimation and Inference, *Virendra K. Srivastava and David E. A. Giles*
81. Response Surfaces: Designs and Analyses, *Andre I. Khuri and John A. Cornell*
82. Nonlinear Parameter Estimation: An Integrated System in BASIC, *John C. Nash and Mary Walker-Smith*
83. Cancer Modeling, *edited by James R. Thompson and Barry W. Brown*
84. Mixture Models: Inference and Applications to Clustering, *Geoffrey J. McLachlan and Kaye E. Basford*

85. Randomized Response: Theory and Techniques, *Arijit Chaudhuri and Rahul Mukerjee*
86. Biopharmaceutical Statistics for Drug Development, *edited by Karl E. Peace*
87. Parts per Million Values for Estimating Quality Levels, *Robert E. Odeh and D. B. Owen*
88. Lognormal Distributions: Theory and Applications, *edited by Edwin L. Crow and Kunio Shimizu*
89. Properties of Estimators for the Gamma Distribution, *K. O. Bowman and L. R. Shenton*
90. Spline Smoothing and Nonparametric Regression, *Randall L. Eubank*
91. Linear Least Squares Computations, *R. W. Farebrother*
92. Exploring Statistics, *Damaraju Raghavarao*
93. Applied Time Series Analysis for Business and Economic Forecasting, *Sufi M. Nazem*
94. Bayesian Analysis of Time Series and Dynamic Models, *edited by James C. Spall*
95. The Inverse Gaussian Distribution: Theory, Methodology, and Applications, *Raj S. Chhikara and J. Leroy Folks*
96. Parameter Estimation in Reliability and Life Span Models, *A. Clifford Cohen and Betty Jones Whitten*
97. Pooled Cross-Sectional and Time Series Data Analysis, *Terry E. Dielman*
98. Random Processes: A First Look, Second Edition, Revised and Expanded, *R. Syski*
99. Generalized Poisson Distributions: Properties and Applications, *P. C. Consul*
100. Nonlinear L_p-Norm Estimation, *Rene Gonin and Arthur H. Money*
101. Model Discrimination for Nonlinear Regression Models, *Dale S. Borowiak*
102. Applied Regression Analysis in Econometrics, *Howard E. Doran*
103. Continued Fractions in Statistical Applications, *K. O. Bowman and L. R. Shenton*
104. Statistical Methodology in the Pharmaceutical Sciences, *Donald A. Berry*
105. Experimental Design in Biotechnology, *Perry D. Haaland*
106. Statistical Issues in Drug Research and Development, *edited by Karl E. Peace*
107. Handbook of Nonlinear Regression Models, *David A. Ratkowsky*
108. Robust Regression: Analysis and Applications, *edited by Kenneth D. Lawrence and Jeffrey L. Arthur*
109. Statistical Design and Analysis of Industrial Experiments, *edited by Subir Ghosh*
110. *U*-Statistics: Theory and Practice, *A. J. Lee*
111. A Primer in Probability: Second Edition, Revised and Expanded, *Kathleen Subrahmaniam*
112. Data Quality Control: Theory and Pragmatics, *edited by Gunar E. Liepins and V. R. R. Uppuluri*
113. Engineering Quality by Design: Interpreting the Taguchi Approach, *Thomas B. Barker*
114. Survivorship Analysis for Clinical Studies, *Eugene K. Harris and Adelin Albert*
115. Statistical Analysis of Reliability and Life-Testing Models: Second Edition, *Lee J. Bain and Max Engelhardt*
116. Stochastic Models of Carcinogenesis, *Wai-Yuan Tan*
117. Statistics and Society: Data Collection and Interpretation: Second Edition, Revised and Expanded, *Walter T. Federer*
118. Handbook of Sequential Analysis, *B. K. Ghosh and P. K. Sen*
119. Truncated and Censored Samples: Theory and Applications, *A. Clifford Cohen*
120. Survey Sampling Principles, *E. K. Foreman*

121. Applied Engineering Statistics, *Robert M. Bethea and R. Russell Rhinehart*
122. Sample Size Choice: Charts for Experiments with Linear Models: Second Edition, *Robert E. Odeh and Martin Fox*
123. Handbook of the Logistic Distribution, *edited by N. Balakrishnan*
124. Fundamentals of Biostatistical Inference, *Chap T. Le*
125. Correspondence Analysis Handbook, *J.-P. Benzécri*
126. Quadratic Forms in Random Variables: Theory and Applications, *A. M. Mathai and Serge B. Provost*
127. Confidence Intervals on Variance Components, *Richard K. Burdick and Franklin A. Graybill*
128. Biopharmaceutical Sequential Statistical Applications, *edited by Karl E. Peace*
129. Item Response Theory: Parameter Estimation Techniques, *Frank B. Baker*
130. Survey Sampling: Theory and Methods, *Arijit Chaudhuri and Horst Stenger*
131. Nonparametric Statistical Inference: Third Edition, Revised and Expanded, *Jean Dickinson Gibbons and Subhabrata Chakraborti*
132. Bivariate Discrete Distribution, *Subrahmaniam Kocherlakota and Kathleen Kocherlakota*
133. Design and Analysis of Bioavailability and Bioequivalence Studies, *Shein-Chung Chow and Jen-pei Liu*
134. Multiple Comparisons, Selection, and Applications in Biometry, *edited by Fred M. Hoppe*
135. Cross-Over Experiments: Design, Analysis, and Application, *David A. Ratkowsky, Marc A. Evans, and J. Richard Alldredge*
136. Introduction to Probability and Statistics: Second Edition, Revised and Expanded, *Narayan C. Giri*
137. Applied Analysis of Variance in Behavioral Science, *edited by Lynne K. Edwards*
138. Drug Safety Assessment in Clinical Trials, *edited by Gene S. Gilbert*
139. Design of Experiments: A No-Name Approach, *Thomas J. Lorenzen and Virgil L. Anderson*
140. Statistics in the Pharmaceutical Industry: Second Edition, Revised and Expanded, *edited by C. Ralph Buncher and Jia-Yeong Tsay*
141. Advanced Linear Models: Theory and Applications, *Song-Gui Wang and Shein-Chung Chow*
142. Multistage Selection and Ranking Procedures: Second-Order Asymptotics, *Nitis Mukhopadhyay and Tumulesh K. S. Solanky*
143. Statistical Design and Analysis in Pharmaceutical Science: Validation, Process Controls, and Stability, *Shein-Chung Chow and Jen-pei Liu*
144. Statistical Methods for Engineers and Scientists: Third Edition, Revised and Expanded, *Robert M. Bethea, Benjamin S. Duran, and Thomas L. Boullion*
145. Growth Curves, *Anant M. Kshirsagar and William Boyce Smith*
146. Statistical Bases of Reference Values in Laboratory Medicine, *Eugene K. Harris and James C. Boyd*
147. Randomization Tests: Third Edition, Revised and Expanded, *Eugene S. Edgington*
148. Practical Sampling Techniques: Second Edition, Revised and Expanded, *Ranjan K. Som*

Additional Volumes in Preparation

Multivariate Statistical Analysis, *Narayan C. Giri*
Bayesian Biostatistics, *edited by Donald A. Berry and Dalene Stangl*

PRACTICAL SAMPLING TECHNIQUES

SECOND EDITION, REVISED AND EXPANDED

RANJAN K. SOM
Consultant
New York, New York

CRC Press
Taylor & Francis Group
Boca Raton London New York

CRC Press is an imprint of the
Taylor & Francis Group, an **informa** business

Library of Congress Cataloging-in-Publication Data

Som, Ranjan Kumar.
Practical sampling techniques / Ranjan K. Som — 2nd ed., rev. and expanded.
p. cm. — (Statistics, textbooks and monographs; v. 148)
Rev. ed. of: A manual of sampling techniques. 1973.
Includes bibliographical references and index.
ISBN 0-8247-9676-4 (hc: alk. paper)
1. Sampling (Statistics) I. Som, Ranjan Kumar. Manual of sampling techniques. II. Title. III. Series.
HA31.2.S65 1995
001.4'33—dc20 95-35436
CIP

First edition, *A Manual of Sampling Techniques*, ©1973 Heinemann Educational Books, London.

The publisher offers discounts on this book when ordered in bulk quantities. For more information, write to Special Sales/Professional Marketing at the address below.

MARCEL DEKKER, INC.
270 Madison Avenue, New York, New York 10016

To

KANIKA, SUJATO, and BISHAKH

Foreword

A sample survey properly conducted, with an optimal survey design, can supply information of sufficient accuracy useful for planning and policy purposes with great speed and low cost. It is especially useful in countries where official statistical systems are not well developed and enough resources are not available for a complete enumeration. It is also claimed that estimates based on sample data collected by well-trained investigators under proper supervision, though subject to sampling errors, are more accurate than those based on complete enumeration, which would be subject to non-sampling errors usually of a larger magnitude than the sampling errors. It is with these ideas that the methodology of sample surveys was developed during the forties.

A major contribution to survey methodology was made by P.C. Mahalanobis and his colleagues at the Indian Statistical Institute in designing and instituting a multipurpose National Sample Survey to collect data periodically, at the national level, on the socio-economic-demographic characteristics of the people. In this program, Dr. R.K. Som played an important role as a chief executive in charge of designing and conducting large-scale demographic surveys and analyzing data. Later he became the Chief of the Population Program Center of the United Nations Economic Commission for Africa and then a Special Technical Adviser on Population at the United Nations Headquarters: he thus acquired considerable international experience in demographic data collection and analysis.

Practical Sampling Techniques is the result of his rich and varied experience in conducting large scale sample surveys combined with his expertise in statistical theory.

A sample survey involves a variety of operations, starting with a statement of objectives, identification of the target universe, construction of sampling frame, choice of sampling plan, preparation of schedules, and training of investigators and ending up with the collection and analysis of data, and reporting on survey results. Dr. Som's manual covers all these aspects with immense detail and lively illustrations. It is, indeed, a valuable and indispensable guide to the practitioners of sample surveys.

Dr. Som also tries to bring an academic touch to his book by discussing some of the theoretical aspects of sample surveys, which should be of interest to students and others studying sampling, as well as to professional workers from other fields who want to acquire a deeper understanding of survey methodology.

C.R. Rao
Eberly Professor of Statistics
Pennsylvania State University

Preface

Using only simple mathematics and statistics, the book aims to present the principles of scientific sampling and to show in a practical manner how to design a sample survey and analyze the resulting data. It seeks a *via media* between, on the one hand, monographs on sampling and special chapters in textbooks on statistics that set forth the basic ideas of sampling, and, on the other hand, advanced sampling texts with detailed proofs and theory. It is more in the nature of a manual or a handbook which attempts to systematize ideas and practices and could also be used as a textbook for a one- or two-semester course. For an understanding of the book, the basic requirement is a knowledge of elementary statistics (including the use of multiple subscripts and summation and product notations) and algebra, but not of calculus.

The approach is essentially practical and the book is primarily aimed at those who are not sampling specialists but are interested in the subject, namely undergraduate, graduate and continuing education students of statistics and professional and research workers from other fields such as sociology, economics, agriculture, forestry, public administration, public health, family planning, demography, psychology, and public opinion and marketing research. For them, the book presents the basic knowledge required to allow, at a minimum, intelligent consultation with professional sampling statisticians, and, at best, the design, implementation, and analysis of sampling inquiries when such expert services are not easily available. The latter applies particularly in quite a few developing countries where many professionals or researchers must work in their own subject fields without the ready benefit of consultation with sampling experts. It is hoped that survey practitioners would find the book useful and would know when to seek expert advice, e.g., for complex topics, such as varying probability sampling without replacement, that are not covered here in detail.

For both types of readers, viz., students and practicing professionals, the book tries to be as complete as possible to permit self-study. The requirements of students have been considered by including proofs of some theorems of sampling and by following pedagogical methods: for the latter purpose, some repetition has been allowed.

After an introduction to "with and without replacement" schemes for simple random sampling, attention has been confined mainly to sampling with replacement, which results in considerable simplicity for the presentation and understanding of principles. The theoretical results have generally been approached from simple principles and derived from examples, intuitive discussions and some fundamental theorems. Proofs have, however, been supplied only for those theorems which do not require a knowledge of calculus or advanced probability, and even these have been placed in an appendix in order not to interrupt the flow of discussions.

While remaining reasonably rigorous at a low mathematical level, the main emphasis has been not so much on providing proofs of all the theorems and statements as on imparting an understanding of why these are true, what makes them work and how they are applied. An effort has been made to avoid misleading generalizations, and the theories have been tested by their usefulness and relevance in application. References to texts with advanced theory and applications in specific fields have been given and the basis has been provided for at least a nodding acquaintance with topics such as inverse sampling and the bootstrap and jackknife methods for estimating sampling variances. At the same time, an attempt has been made to shield beginners from an over-exposure to terminologies and concepts: theories have not been identified with authors, except when it simplifies references. Inevitably, some results are more or less pulled out of the hat.

After an introductory chapter, the book is divided into five parts dealing respectively with: single-stage sampling; stratified single-stage sampling; multi-stage sampling; stratified multi-stage sampling; and miscellaneous topics, including errors in data and estimates; planning, execution, and analysis of surveys; and use of computers in survey sampling.

The arrangement of the topics within each of the first four parts is such that all the aspects relating to any part are covered. For example, Part I, "Single-stage Sampling", covers simple random sampling (with and without replacement), ratio and regression estimators, systematic sampling, unequal probability sampling (sampling with probability proportional to size), choice of sampling units (cluster sampling), size of sample (cost and error considerations), and self-weighting designs.

For each type of sampling, estimating procedures are shown not only for the total and the mean of a study variable, but also for the ratio of the totals of two study variables.

It is intended that the first four parts of the book be read in sequence. The entire book should be read for understanding the principles and their intelligent application; at a later stage, the reader who is concerned, for example, with a stratified single-stage design need refer only to Parts I and II for estimation purposes.

Appendix I gives the list of notations and symbols, Appendix II elements of probability and proofs of some theorems in sampling, Appendix III some mathematical and statistical tables, Appendix IV a hypothetical universe from which samples are drawn, Appendix V a case study of the Indian National Sample Survey, 1964-65, and Appendix VI a review of the major multi-country survey programs in the socio-economic field, including population, consumer expenditure, family planning, and health.

Drawing from different fields, numerous examples have been provided, as well as exercises with hints for solution. Working out non-trivial exercises should be as important as following the textual material. A specially constructed hypothetical universe has been used for selecting samples, which have been followed through in order to illustrate different sampling designs. Real-life examples from large-scale inquiries have also been provided.

The book has evolved from my association with the Indian Statistical Institute and the National Sample Survey, since its inception in 1950, and with the United Nations (first at its regional office in Africa and then at its headquarters in New York), and from lectures delivered at national and international statistical and demographic courses.

A previous edition of the book was published in 1973 by Heinemann Educational Books, London, under the title *A Manual of Sampling Techniques*. I was particularly touched by the gracious comments on the book by pioneers in the field of sampling who themselves have authored seminal books on the subject such as W. Edwards Deming, P.V. Sukhatme, and Leslie Kish, by survey directors such as Sir Claus Moser and W.R. Buckland, by reviewers and sampling experts, and by students and survey professionals.

I owe another debt, a basic one, to Professor P.C. Mahalanobis, Director of the Indian Statistical Institute, who had pioneered the development of statistics, including survey sampling in India since 1922 and, as Honorary Statistical Adviser to the Cabinet of the Government of India, was instrumental in launching the Indian National Survey in 1950. Murthy (1984) and C.R. Rao (1993) have reviewed Mahalanobis's contribution to statistical development in India; I myself have written on Mahalanobis's contributions to the statistical activities of the United Nations and the International Statistical Institute (Som, 1994). At the Indian Statistical Institute, I was fortunate to have as my colleagues specialists in sampling theory and practice, such as A. Bhattacharyya, I.M. Chakravarty, Nanjamma Chinnappa, A.C. Das, A.N. Ganguly, M.N. Murthy, A.S. Roy, J. Roy and V.K. Sethi, in addition to preeminent statisticians such as C.R. Rao, M. Ganguli and D.B. Lahiri for consultation and advice.

The text for the present edition of the book has been revised and expanded with the inclusion of techniques that are either new or have recently gained a certain currency, such as the jackknife and bootstrap methods of variance estimation and the use of focus groups in data collection, and subjects of topical interest such as the use of personal computers in sample surveys, internationally sponsored multi-country survey programs, and survey considerations for some of today's concerns, such as the nutritional status of children, disability, status of women, and HIV/AIDS infection.

Computers, particularly personal computers, have become indispensable in survey data processing, analysis and publication and, increasingly in developed countries, in the data collection process itself. This topic is covered in Chapter 27 and is a special feature of the book.

Internationally sponsored multi-country survey programs are in operation in a large number of developing countries. While providing substantive results, these multi-country survey programs are also assisting in the transfer of technology in survey sampling and microcomputer use. A review of these programs, even of the major ones, is not available in one place and I thought that survey professionals and students would benefit from the inclusion of such a review here.

I have not covered the general topic of foundations of survey sampling, as I consider it beyond the scope of this book. However, for interested readers, I have given some references under "Further reading" at the end of Chapter 1.

For readers in French and Spanish, I have given references to sampling texts in these languages.

Although most computer and software brand names have trademarks or registered trademarks, the individual trademarks have not been listed here.

Ranjan K. Som

Acknowledgments

My grateful thanks are due to Dr. I.M. Chakravarti, Dr. R.G. Laha, and Dr. J. Roy for permission to quote six examples/exercises from their book, *Handbook of Methods of Applied Statistics, Vol. II, Planning of Surveys and Experiments* (Wiley, 1967); to John Wiley & Sons Inc., New York for permission to quote five examples/exercises from *Sampling Techniques* by W.G. Cochran (1977); an extract from the article, "Boundaries of statistical inference," by W. Edwards Deming in *New Developments in Survey Sampling* (1969), edited by Norman L. Johnson and Harry Smith, Jr.; and two exercises from *Sampling of Populations: Methods and Applications* by Paul S. Levy and Stanley Lemeshow (1991); to Dover Publications, New York and the Deming Institute for permission to quote an extract from W. Edwards Deming's *Some Theory of Sampling* (Wiley, 1950; Dover, 1966); to John Wiley & Sons Inc., New York, and the Deming Institute for permission to quote an extract from W. Edwards Deming's *Sample Design in Business Research* (1960); to the Government of Ethiopia for permission to use the data of Example 15.1; to the American Statistical Association to quote an extract and Fig. I.1 from its publication, *What Is a Survey?* by R. Ferber, P. Sheatsley, A. Turner, and J. Waksberg; to Longman & Green on behalf of the Literary Executor of the late R.A. Fisher and Frank Yates for permission to reprint two tables from their book, *Statistical Tables for the Biological, Agricultural and Medical Research* (6th ed., 1973); to the Food and Agriculture Organization of the United Nations for permission to reproduce (in an adapted form) a figure from the book, *Quality of Statistical Data*, by S.S. Zarkovich); to the Academic Press for permission to quote an exercise from *Probability in Science and Engineering* (1967) by J. Hájek and V. Dupac; to the Scarecrow Press for permission to quote an example and a table from *The Mathematical Theory of Sampling* (1956) by W.A. Hendricks; to the German Statistical Office for permission to quote as an exercise and an example from the article, "Organization and functioning of the Micro-Census in the Federal Republic of Germany," by L. Herbergger in *Population Data and Use of Computers with Special Reference to Population Research* (1971), German Foundation for Developing

Countries, Berlin, and Federal Statistical Office, Wiesbaden; to the Government of India for two tables from National Sample Survey reports; to Professor Leslie Kish for permission to quote extracts from his book, *Survey Sampling* (Wiley, 1965), and two articles; to the Statistical Publishing Society, Calcutta, for permission to reproduce three exercises from *Sampling Theory and Methods* (1967) by M.N. Murthy; to Dr. C.R. Rao for permission to reproduce an exercise from his book, *Advanced Statistical Methods in Biometric Research* (Hafner, 1973); to the Iowa State University for permission to reproduce an example from *Statistical Methods* (3rd ed., 1967) by G.W. Snedecor and W.G. Cochran; to the Literary Executor of the late M.R. Sampford for permission to reproduce an example from his book *An Introduction to Sampling Theory* (Oliver & Boyd, 1962); to Dr. P.V. Sukhatme, Dr. Sashikala Sukhatme, and Dr. C. Asok for permission to reproduce three examples from the book *Sampling Theory of Surveys with Applications* (3rd ed., 1984) by P.V. Sukhatme, B.V. Sukhatme, Sashikala Sukhatme, and C. Asok, Iowa State University Press, Ames, Iowa, and Indian Society of Agricultural Statistics, New Delhi; to the United Nations for permission to reproduce (in adapted forms) some examples from *A Short Manual on Sampling, Volume I,* and extensive extracts from *Preparation of Sample Survey Reports, Provisional Issue*; to the U.S. Bureau of the Census for permission to quote a table from its publication, *Technical Paper No. 2*; and to Dr. A. Weber for permission to quote two examples from his mimeographed lecture notes *Les Méthods de Sondage* (1967); and to Dr. Frank Yates and Longman Green for permission to cite some examples from *Sampling Methods for Censuses and Surveys* (4th ed., 1981) by Frank Yates (London, Edwin Arnold, and New York, Oxford University Press.)

I benefitted greatly from the comments and suggestions of the following on some draft chapters of the first edition of the book: Prof. B.C. Brookes (Reader in Information Studies, University of London); Dr. W.A. Ericson (Professor and Chairman, Department of Statistics, University of Michigan); Mr. Benjamin Gura (Deputy Chief, International Statistical Programs Center), Mr. Leo Solomon (Chief, Research & Analysis Branch), Dr. B.J. Tepping (Chief, Center for Research for Measurement Methods), Dr. Peter Robert Ohs (Mathematical Statistician), and Mr. David Bateman (Sampling Statistician) at the U.S. Bureau of the Census; Mr. C.M.H. Morojele (Regional Statistician for Africa, U.N. Food and Agriculture Organization); Sir Claus A. Moser (Director, Central Statistical Office, U.K.); Dr. M.N. Murthy (Lecturer in Statistical Methods, United Nation Statistical Institute for Asia and the Pacific); Mr. M.D. Palekar (Chief, Special Projects Section, U.N. Statistical Office); Mr. John C. Rumford (Demographic Survey Adviser to the Government of Liberia) at the U.S. Agency for International Development; Mr. Jacques Royer (Director, Projections and Programming Branch, U.N. Economic Commission for Europe); Dr. H. Schubnell (Director, Statistical Office, Federal Republic of Germany); Dr. S.R. Sen (Executive Director for India, World Bank); Dr. Egon Szabady (President, Demographic Committee of the Hungarian Academy of Sciences); Dr. R. Zasepa (Professor, Central School of Planning and Statistics, Poland); and Mr. A.S. Roy (Sampling Statistician, Indian Statistical Institute), who had read the final draft of the first edition and offered many constructive suggestions. The late Patrick J. Loftus, Director of the U.N. Statistical Office (1963-1972), encouraged me in the study and wrote a foreword to the first edition: I remain deeply grateful to him.

For providing material and helpful discussions on the new material of the present edition, I am grateful to: Mr. William P. Butz, Associate Director, U.S. Bureau of the Census, and his colleagues, especially Dr. Eduardo Arriaga, Dr. J. Howard Bryant, Mr. Bob Bush, Mr. Lawrence S. Cahoon, Ms. Cathleen Chamberlain, Mr. Evan Davey, Dr. Peter D. Johnson, Dr. David Megill, Mr. Stan Rolark, Ms. Karen Stanecki, Mr. Michael Stroot, and Mr. Bob Tiari at the U.S. Bureau of the Census; Mr. Richard M. Cornelius (Deputy Coordinator for Population, U.S. Agency for International Development); Mr. William Seltzer (Director), Mr. Parmeet Singh and Mr. Anthony G. Turner (respectively Coordinator and Sampling Specialist, National Household Survey Capability Programme), Ms. Mary Chamie (Statistician), Mr. Ramesh Chanda (Special Technical Advisor on Computer Methods and Applications), Mr. Carlos Ellis (Population Statistics Data Processing Specialist), Mr. Patrick Gerland (Demographic Software Specialist), Dr. Anis Maitra (Inter-regional Adviser), Mr. James T. Maxwell (Programme Officer), Mr. Sam Suharto (Population Census Methods Specialist), Dr. Sirageldin Suliman (Civil Registration and Vital Statistics Specialist), and Ms. Joann Vanek (Statistician) at the U.N. Statistical Division; Dr. Brian Williams (Head, Department of Public Health Medicine, University of Sheffield Medical School, U.K.; Consultant to the U.N. Statistical Division); Dr. M. Nizamuddin (Chief, Population Data, Policy and Research Branch), Mr. Abdul-Muneim Abu Nuwar (Country Director to Syria), Mr. Hedi Jemiai (Deputy Chief, Division of Arab States and Europe), Dr. Atef Khalifa (Director, Country Support Team in Amman, Jordan) and Dr. K.V. Ramachandran (Demographic Specialist, Country Support Team in Addis Ababa, Ethiopia) at the U.N. Population Fund; Ms. Birgitta Bucht (Assistant Director) and Mr. C. Stephen Baldwin (Chief, Population Projects Section), U.N. Population Division; Dr. Beverley A. Carlson (Senior Advisor on Population and Statistics, UNICEF); Mr. James Cheyne (Programme Officer) and Dr. Gilles Desvé (Consultant) at the World Health Organization; Mr. Timothy Merchant (Senior Economist/Statistician) at the World Bank; Dr. Jeremiah Sullivan and Ms. Thanh N. Le at the Demographic &

Health Surveys; Mr. W. Parker Mauldin, Professor Emeritus, Rockefeller University; Dr. Moni Nag and Dr. Louise Kantrow of the Population Council; Prof. P.K. Bannerjee (Secretary, Calcutta Statistical Association); Dr. S.K. Chakrabarty (Retired Director of the Bureau of Applied Economics and Statistics, Government of West Bengal), Mr. B.R. Panesar (Retired Statistician, Indian Statistical Institute), Mr. S.K. Nath (Joint Director, National Sample Survey, Government of India), Dr. Sambhu Nath Sen (Retired Secretary, Ministry of Agriculture, Government of Bihar, India), and Prof. N. Balakrishnan (Associate Professor of Mathematics & Statistics, McMaster University, Hamilton, Canada).

For supplying references to recent sampling books in French and Spanish, I am grateful to Mr. J.L. Bodin, Chief of the Department of International Relations and Cooperation, National Institute of Statistics and Economic Studies (INSEE), France, and to Prof. J.L. Sanchez-Crespo, University of Madrid.

My special thanks to Ms. Debbie Iscoe for preparing an excellent camera-ready copy of the manuscript with all the equations and tables.

For reviewing selected new chapters or sections in the second edition, I am indebted to the following: Dr. Toré Dalenius (Division of Applied Mathematics, Brown University); Dr. A.S. Hedayat (Professor of Mathematical Statistics and Computer Science, University of Illinois); Dr. John F. Kantner (Professor Emeritus, Johns Hopkins University); Dr. Richard Maisel (Professor of Sociology, New York University); Dr. Zvia Napthali (Adjunct Professor of Statistics, New York University); Dr. Robert K. Merton (Professor Emeritus, Columbia University); and Dr. Thomas N. Pullum (Professor, Department of Sociology, University of Texas at Austin). My sister, Dr. Prantika Som, Scientist at the Medical Department of the Brookhaven National Laboratory, advised me on the medical aspects of epidemiological surveys. I am also grateful to the reviewers of Marcel Dekker, Inc., for suggesting various improvements to the draft of the book. The responsibility for any errors of omission or commission in the book is of course entirely mine.

For writing the Foreword to this edition of the book, I am most grateful to Dr. C.R. Rao, Sc.D., F.R.S., Past President of the International Statistical Institute, former Director of the Indian Statistical Institute, and currently the Eberly Professor of Statistics, Pennsylvania State University.

Finally, I have great pleasure in expressing my thanks to Mr. Russell Dekker, Vice-President and Editor in Chief, Marcel Dekker, Inc., for encouraging me in updating and expanding the second edition of the book, and Mr. Brad Benedict, Assistant Production Editor, for his help and guidance in the book's production.

A textbook, it has been said, should be dedicated to one's children. My wife, Kanika, along with my sons, Sujato and Bishakh, provided continuing support. To them, I once again dedicate this book.

Contents

Guide to Readers

Topics which may be considered too advanced or too detailed may be deferred at the first reading. Some **advanced topics** are mentioned below:

> pps systematic sampling (section 5.4.3); **Stratified simple random sampling** - Ratio of ratio estimators of two study variables (section 10.5), Gain due to stratification (section 10.6), and Stratification after sampling (section 10.7); **Stratified varying probability sampling** - Ratio of ratio estimators of two study variables (section 11.6) and Gain due to stratification (section 11.7); **Multi-stage simple random sampling** - Three-stage srs - estimation of totals, means, ratios, and their variances (section 15.6) and Three-stage srs: estimation of sub-universes (section 15.7); **Multi-stage varying probability sampling** - Three-stage design with pps, srs, and srs (section 16.4) and Special cases of crop surveys (section 16.5); **Size of sample and allocations to different stages in multi-stage sampling** - Three-stage designs (section 17.3); Stratified three-stage srs (section 20.3); Inverse sampling (section 24.5); Estimation of rare events (section 24.6); Simplified methods of variance computation (section 24.8); Other methods of estimating sampling variances: Bootstrap and Jackknife (section 24.9); Methods of collapsed strata (section 24.10); Controlled Selection (section 24.11); Use of Bayes' theorem in sampling (section 24.12); Randomized response technique (section 24.14); Line and lattice sampling (section 24.15); More complex models (section 25.5); Elements of probability and proofs of some theorems in sampling (Appendix II); and Case Study: Indian National Sample Survey, 1964-5 (Appendix V).

Other topics which may be of interest to some readers only, such as agricultural surveys and opinion and marketing research, may be skipped by other readers.

In classroom teaching, the choice of topics will of course depend on the views of the teacher and the discipline under which sampling is being taught; topics such as the extension of two-stage sampling to three-stage sampling may be indicated briefly and the miscellaneous topics (Part V) either omitted (with the exception of chapter 25 on errors and biases in data and estimates, which may be condensed) or covered in an introductory course. Furthermore, the teacher might wish to supplement the examples and exercises with those relevant to the discipline: it would also be instructive for students, under the guidance of the teacher, to design a sample survey and analyze the resulting data.

The following chapters and sections could comprise a one-semester course on sampling techniques:

> Chapters 1-2 and 4-9; sections 10.1 10.3, 12.1 12.3.6, 12.4, 12.6; Chapters 13-14; sections 15.1-15.5; sections 17.1-17.2.3; Chapters 18-19; sections 20.1-20.2.9; Chapters 22-23; sections 24.1-24.5; Chapters 25 and 27; Appendices I, III and IV.

Use of Dedicated Software, Spreadsheet Software, and Scientific Calculators

A large set of actual survey data should best be processed with **dedicated software for estimation of totals and their variances in stratified multi-stage designs**, such as CENVAR of the U.S. Bureau of the Census, whose purpose is to provide estimates and their variances (see section 27.10.1).

Smaller sets of data, including those given in the book, could be worked with **spreadsheet software**, such as Lotus 1-2-3, Quattro Pro, Excel or As-Easy-As, and very small sets of data could be handled by handheld scientific calculators.

In using spreadsheet software, after entering the y values, the mean could be obtained by the function AVG. The function STDS in Quattro Pro and the function STDEV in Excel give the sample standard deviation *per unit*, which must be multiplied by $\sqrt{(1-f)/n}$ in simple random sampling without replacement (srswor) and by $\sqrt{(1/n)}$ in simple random sampling with replacement (srswr) to obtain an estimate of the standard error of the mean, where n is the size of the sample, N that of the "universe", and f the sampling fraction $= n/N$.

A still smaller set of data such as in Exercises 2.1 and 2.2 could also be analyzed quicker on a handheld scientific calculator with statistical functions. Note again that the value of σ_{n-1} obtained there is the same as that obtained by the STDS function in Quattro Pro and STDEV in Excel and therefore should be multiplied by $\sqrt{(1-f)/n)}$ in srswor and by $\sqrt{(1/n)}$

in srswr to obtain the standard error of the mean. In using these software programs or the scientific calculator, note the definition of the functions used. The table below shows the functions in four spreadsheet software programs and Radio Shack scientific calculators:

Function/Comm	Lotus 1-2-3 v5/WIN	Quattro Pro v5/DOS or WIN	Excel for v5/WIN	As-easy-As v5/DOS	Radio Shack scientific calculator
Number of items	COUNT	COUNT	COUNT	n	
Sum (or total) [Eq. (2.1)]	SUM	SUM	SUM	SUM	$\sum x$
Arithmetic Mean [Eq. (2.2)]	AVG	AVG	AVERAGE	AVG	$\bar{x}$
Raw sum of squares*	SUMSQ		SUMSQ		$\sum x^2$
Sum of products of two variables†	SUMPRODUCT	SUMPRODUCT	SUMPRODUCT		
Population variance per unit [Eq. (2.3)]	VAR	VAR	VARP	VAR	
Sample estimate of variance per unit [Eq. (2.6)]	VARS	VARS	VAR		
Population S.D per unit‡	STD	STD	STDEVP		σ_n
Sample S.D. per unit§	STDS	STDS	STDEV		σ_{n-1}
Standard error of sample mean [STDS/$\sqrt{n}$]	SEMEAN				

Equation numbers in square brackets refer to those in the text.

* $\sum Y_i^2$ or $\sum y_i^2$.

† $\sum Y_i X_i$ or $\sum y_i x_i$.

‡ The (positive) square root of VAR (Lotus 1-2-3 or Quattro Pro) or VARP (Excel).

§ The (positive) square root of VARS (Lotus 1-2-3 or Quattro Pro) or VAR (Excel).

Journals

AMS	Annals of Mathematical Statistics
BISI	Bulletin of the International Statistical Institute/Bulletin de l'Institut Internationale de Statistique
IMSB	Institute for Mathematical Statistics Bulletin
ISR	International Statistical Review
JASA	Journal of the American Statistical Association
JMR	Journal of Marketing Research
JOS	Journal of Official Statistics, Stockholm
JRSS	Journal of the Royal Statistical Society, London
PHR	Public Health Reports
POQ	Public Opinion Quarterly
RISI	Review of the International Statistical Institute
SFP	Studies in Family Planning
WHSQ	World Health Statistics Quarterly

Organizations

ASA	American Statistical Association
ECA	U.N. Economic Commission for Africa
ESCAP	U.N. Economic Commission for Asia and the Pacific
FAO	Food and Agriculture Organization of the United Nations
HMSO	Her Majesty's Stationery Office, U.K.
IASS	International Association of Survey Statisticians
ILO	International Labour Organisation
INSEE	Institut National de la Statistique et des Etudes Economiques, Paris
IRD	Institute for Resource Development
ISI	International Statistical Institute
IUSSP	International Union for the Scientific Study of Population
NCBS	Netherlands Central Bureau of Statistics
NIDI	Netherlands Inter-university Demographic Institute
NCHS	National Center for Health Statistics (U.S.)
U.N. (or UN)	United Nations
UNDP	United Nations Development Programme
UNFPA	United Nations Population Fund
UNICEF	United Nations Children's Fund
USAID	United States Agency for International Development
WHO	World Health Organization

Programs

AGFUND	Arab Gulf Programme for the United Nations Development Organizations
CPS	Contraceptive Prevalence Surveys
DHS	Demographic and Health Survey(s)
EPI	Expanded Programme on Immunization, WHO/UNICEF
FPS	Family Planning Surveys
GCHS	Gulf Child Health Survey
LSMS	Living Standards Measurement Study, World Bank
NHSCP	National Household Survey Capability Programme, U.N.
NSS	The National Sample Survey, India
PAPCHILD	Pan-Arab Project for Child Development
SDA	Social Dimensions of Adjustment Program, World Bank
WFS	World Fertility Survey

CHAPTER 1

Basic Concepts of Sampling

1.1 Introduction

Sampling is the process by which inference is made to the whole by examining only a part. It is woven into the fabric of our personal and public lives. In some cultures, a couple enters into a marriage partnership on the basis of a short courtship. With a single grain of rice, an Asian village housewife tests if all the rice in the pot has boiled; from a cup of tea, a tea-taster determines the quality of the brand of tea; medicine dosages are set on sampling investigations; and a sample of moon rocks provides scientists with information on the origin of the moon.

As a part of the information-collection and decision-making process, sample surveys are conducted on different aspects of life, culture, and science. The purpose of sampling is to provide various types of statistical information of a qualitative or quantitative nature about the whole by examining a few selected units; the sampling method is the scientific procedure of selecting those sampling units which would provide the required estimates with associated margins of uncertainty, arising from examining only a part and not the whole.

The monograph, *What Is A Survey?*, published by the American Statistical Association, lists some surveys in the United States; the list illustrates the wide variety of issues with which surveys deal (Ferber, Sheatsley, Turner, and Waksberg):

1. The U.S. Department of Agriculture conducted a survey to find out how poor people use food stamps.

2. Major TV networks rely on surveys to tell them how many and what types of people are watching their programs.

3. Auto manufacturers use surveys to find out how satisfied people are

with their cars.

4. The U.S. Bureau of the Census conducts a survey every month to obtain information on employment and unemployment in the nation.

5. The National Center for Health Statistics sponsors a survey every year to determine how much money people are spending for different types of medical care.

6. Local housing authorities make surveys to ascertain satisfaction of people in public housing with their living accommodations.

7. The Illinois Board of Higher Education surveys the interest of Illinois residents in adult education.

8. Local transportation authorities conduct surveys to acquire information on people's commuting and travel habits.

9. Magazines and trade journals utilize surveys to find out what their subscribers are reading.

10. Surveys are used to ascertain what sort of people use our national parks and other recreation facilities.

World-wide, an indication of the scope of sampling in different countries may be obtained from the following partial listing from the latest (1982) issue of the United Nations report, *Sample Surveys of Current Interest*, relating to surveys conducted during 1977-79: a survey of crime victimization in Hong Kong; surveys of industries in Ethiopia, India, Indonesia, and the U.S.A.; family/household budget/income and expenditure surveys in Belgium, Brunei, Canada, Chile, Hungary, Indonesia, Malawi, Norway, Sri Lanka, Sweden, the U.S.A., and Zimbabwe; health survey in Australia, France, Netherlands, Thailand, and the U.S.A.; fisheries surveys in Indonesia, Sri Lanka, Tufalu, the U.S.A., and Zimbabwe; population/demographic/fertility and family planning surveys in Cyprus, Egypt, Indonesia, Iran, Morocco, Nepal, Pakistan, Panama, Papua and New Guinea, Peru, Philippines, Singapore, Spain, Sri Lanka, Thailand, Trinidad and Tobago, the U.S.A., Yugoslavia, and Zimbabwe.

1.2 Complete enumeration or sample

Leaving aside the rôle of sampling in personal lives and the design of experimentation under controlled laboratory or field conditions, there arises in every country a demand for statistical information in two ways: first, when the data obtained from routine administrative and other sources have to be analyzed speedily, and second, and more frequently, when accurate data are not available from conventional sources. Consider, for example, the problem of how to conduct an inquiry into family budgets by employing field interviewers. Three alternatives seem possible for collecting data:

1. a complete enumeration of all the units of observation (commonly known as the census), such as all the families in the area of the inquiry, using the best available enumerators;

2. a complete enumeration of all the units by relatively less competent enumerators; and

3. a sample of a selected number of units by the best available enumerators.

In reality, especially for large parts of the world for even a limited number of inquiries, and for all the countries for an inquiry into every facet of community life, an adequate number of competent staff are simply not available. In practice, therefore, to enumerate fully all the units and still make use of the best available staff, the enumeration has to be spread over a long period. But the situation may change over time: we might then be measuring different things.

In general, therefore, only two alternatives remain, namely, a complete enumeration of all the units by less efficient enumerators and a sample of units with more efficient staff. The advantages of sampling (see Section 1.3) are such that the latter method is usually chosen, although errors can arise as a result of generalizing about the whole from the surveyed part.

There are situations where a complete enumeration would be essential, and if situations do not permit this, efforts should be made to establish it as soon as possible. Examples are the counting of population for census purposes, income tax returns and a voters' list.

Conversely, sampling is the only recourse in destructive tests such as testing of the life of electric bulbs, the effectiveness of explosives, and hematological testing.

In practice, sampling is often used to evaluate or augment the census process (see section 1.5 below). Post-enumeration sample surveys are one means of assessing the coverage and content errors of population censuses, almost universally used in all countries of the world; for the evaluation, particularly through post-enumeration field surveys, of the 1990 population censuses in France, see Coëffic (1993), and in the U.S., see Hogan (1993), and of the 1991 population census of China, see Zongming and Weimin (1993).

1.3 Advantages of sampling

Sample surveys have these potential advantages over a complete enumeration – **greater economy, shorter time-lag, greater scope, higher quality of work, and actual appraisal of reliability**.

A sample requires relatively better resources for designing and executing it adequately and so the cost per unit of observation is higher in a sample than in a complete enumeration. But the total cost of a sample will be much less than that of a complete enumeration covering the same items of inquiry. One airline found, for example, that inter-company settlements for the preparation of tickets due to other air carriers could be estimated quite accurately, within 0.07 per cent, for a 10 per cent sample of the total of about 100,000 documents per month and at a considerable saving of funds (Slonim, chapter XVII). In 1994 about 115 million income tax returns were expected to be received by the U.S. Internal Revenue Service: it would be unpractical to audit all these. The IRS now audits fewer than one in 100, selected by computer programs designed to spot likely problems, with a "progressive" sampling fraction ranging from one half of one per cent for returns in the range $25,000-$50,000 to 5 per cent for returns reporting more than $100,000.

With a smaller number of observations, it is also possible to provide results much faster than in a complete enumeration.

Sampling has a greater scope than a complete enumeration regarding the variety of information by virtue of its flexibility and adaptability and the possibility of studying the interrelations of various factors.

Data obtained from a complete enumeration or a sample are subject to various types of errors and biases, the magnitude of which depends on the particular survey procedure: this aspect is covered in greater detail in Chapter 25. However, if the same survey procedure is followed in both a sample and a complete enumeration, the accuracy of a single observation will be the same; but with a comparatively small scale of operations, a sample survey makes it possible to adopt a superior survey procedure by exercising better control over the collection and processing of data, by employing better staff and providing them with intensive training and better equipment, and in interview surveys by employing in-depth interviews. Under suitable conditions – and this is one of the main objectives of sampling – more accurate data can be provided by a sample than by a complete enumeration.

For example, the Current Population Survey, conducted by the U.S. Bureau of the Census, produced, in April 1950, a more accurate count of the labor force in the U.S.A. than did the Population Census of April 1950. Comparisons made through a case-by-case matched study with the sample returns, and individual comparisons made with the special sample interviews taken by the survey enumerators soon after the date of the census enumeration, revealed significant net differences in the labor force participation rates and the number unemployed between the census and the sample. The sample indicated about $2\frac{1}{2}$ million more persons in the labor force than the census and half a million more unemployed. These differences appeared to occur because the marginal classifications were more adequately identified in the sample than in the census, mainly because of the type of enumerators and their training and supervision; the census enumerators, with their necessarily limited training on labor force problems, were apparently less effective in identifying marginal labor force groups (Hansen and Pritzker, 1956). Further studies made at the U.S. Bureau of the Census have shown that because of the response biases of a substantial order in difficult items such as occupation, industry, work status, income, and education, the amount lost by collecting such items for a sample of 25 per cent instead of a complete census was far less than one might assume, even for very small areas. These considerations, plus those of economies and timeliness of results, led to the adoption, for the first time, in the 1960 census of a 25 per cent sample of households as the basic procedure for such information (Hansen and Tepping, 1969).

To be useful, a sample must be scientifically designed and conducted. A classic example of a biased sample is the ill-fitted *Literary Digest* poll of the 1936 U.S. presidential election, where a mammoth sample of ten million individuals was taken. But the sample was obtained from automobile registration lists, telephone directories and similar sources and only 20 per cent of the mail ballots were returned; both these factors resulted in a final sample biased heavily in favor of more literate and affluent individuals, with political affiliations generally different from the rest: the poll erred by the huge margin of 19 per cent in predicting votes for Franklin D. Roosevelt, when he won by a 20 per cent majority. The failure of the 1948 polls for the U.S. presidential election, in which Harry S. Truman was elected, is mainly explained by changes in the preferences of the people polled between the time they were interviewed and the time they had voted, but could be partly attributed to the quota system of sampling that was deficient in organized labor. In the 1970 British General Election also, pollsters performed very poorly when they predicted the return of the Labour Party.

1.4 Limitations of sampling

When basic information is required for every unit, obviously a complete enumeration has to be undertaken. Errors due to sampling also tend to be high for small administrative areas (but on this point see the preceding section) or for some cross-tabulations where the number of sample observations falling in a certain cell may be very small.

1.5 Relationship between a complete enumeration and a sample

Almost invariably a complete enumeration of population, housing and farms provides the frame (such as set of enumeration areas) from which the sample can be drawn in the first instance. In addition, the information obtained about the units from a complete enumeration, even if out of date, can be used to provide supplementary information with a view to improving the efficiency of sample designs.

Sampling is also used as an integral part of a complete enumeration, e.g., of population, housing, and farms in the following operations: tests of census procedures; enumeration of items in addition to those for which universal coverage is required; evaluation through post-enumerative field checks; quality control of data processing; advance tabulation of selected topics; extending the scope of analysis, such as that of the interrelations of various factors, for which results are required only for large areas and for a country as a whole; and for enumerating difficult items which may require the use of special techniques that are costly or time-consuming. For such purposes, sampling is often used in conjunction with the census of population in most countries of the world.

Sampling and complete enumeration are thus complementary and, in general, not competitive. For example, in areas where routine systems of data collection and analysis do not exist or are defective and early establishment and operation of such systems are not feasible, the required data and analysis for population programming could be obtained from integrated sample surveys where the interrelations between the demographic, social and economic factors are covered. For such programming at the local level, rough estimates could often be adequate; but at the stage of implementation, detailed data at these local levels (obtained only through a complete enumeration) would be required and would necessarily be secured.

1.6 Probability versus purposive samples

When we talk of a random sample, most often we mean a haphazard sample, the selection of the constituent units of which is not underpinned by probability theory. Scientifically, a survey is called a *probability sample* (and also sometimes, a *random sample*) when the method of its selection is based on the theory of probability. In order to provide a valid estimate of an unknown value along with a measure of its reliability, a probability sample must be used. The advantage of sampling referred to in the preceding section accrue only from a probability sample.

Only a probability sample, in addition to providing valid estimates, can provide measures of the reliability of these estimates by indicating the extent of error due to sampling a part, and not the whole, and can also set limits within which the unknown value that is being estimated from the sample data is expected to lie with a given probability, were all the units to be completely enumerated using the same survey procedure. With the required information, it is also possible to design a sample that would yield data of maximum reliability at a given cost or vice versa.

Note that we are talking of the ***reliability***, and not the ***accuracy*** of estimates: the distinction between the two terms is explained in Chapter 25 on errors and biases in data and estimates.

In addition, a probability sample, especially designed and using specific techniques, can provide estimates of errors and biases other than those due to samples, that affect both a sample and a complete enumeration, such as the response or observational errors, the differential bias of enumerators, errors arising from incomplete samples, processing errors, etc. These are detailed in Chapter 25.

A sample selected by a non-random method is known as a ***non-random sample***; with such a sample it is not possible to measure the degree of reliability of the sample results. Examples are ***purposive samples*** (selected by some purposive method and therefore subject to biases of personal selection), and ***quota samples*** (usually of human beings, in which each enumerator is instructed to collect information from the "quota" of the assigned number of individuals of either sex, and belonging to a certain age group, social class, etc. but the selection of the sample of individuals is left to the enumerator's choice); in quota sampling, errors of frame and non-response are not recognized. There still remain, especially in opinion polls and marketing research, the vestige of purposive and quota sampling, but their use is decreasing and is to be discouraged. This is not to say that non-random samples cannot be used for some tests of procedure, such as the pre-testing of questionnaires; one particular type of non-probability sample – focus groups – has recently gained certain currency (see section 24.13). But wherever quantitative measures are required that can also provide scientific and continuing basis for planning inquiries of a related nature, a probability sample must be used.

In what follows, we consider only probability samples, except for "focus groups" (in section 24.13).

1.7 Terms and definitions

An ***elementary unit***, or simply a unit, is an element or group of elements on which information is required. Thus persons, families, households, farms, etc. are examples of units. A ***recording unit*** is an element or group of elements for which information is required. In an inquiry, if information is required on sex and age of individuals, then individuals constitute the recording unit; but if information is required on family size, the family is the recording unit. A ***unit of analysis*** is the unit for which analysis is made. In a family budget inquiry, the units of analysis might be the persons in the family or all the families in a certain expenditure group.

The ***universe*** (also called ***population***, but we shall not use this term in order to avoid a possible confusion with the population of human beings) is the collection of all the units of a specified type defined over a given space and time. The universe may be divided into a number of ***sub-universes***. Thus in a country-wide inquiry into family budgets, all the families in the country, defined in a certain manner, would constitute the universe. The sub-universe might consist of the totality of families living in rural or urban areas or of families with incomes below and above a certain level.

The universe may be ***finite***, that is, it may consist of a finite number of elements. Almost all sample surveys deal with finite universes, but there are some advantages in considering a sample as having been obtained from an ***infinite*** universe. This is so when the universe is very large in relation to the sample size, or when the sampling plan is such that a finite universe is turned into an infinite one: the latter procedure is ensured by ***sampling with replacement*** that will be explained later.

A characteristic of the units which may take any of the values of a specified set with a specific relative frequency or probability for the different units is called a ***random variable*** (or a ***variate***). If such a characteristic is to be studies on the basis of information obtained through an inquiry, the variable is termed ***study variable***. Of course, in an inquiry there will be more than one study variable. In a family budget survey, per family and per capita expenditure on food and other items are examples of two study variables. An ***ancillary variable*** is a variable which, although not a subject of study of a particular inquiry, is one on which information can be collected that may help in providing, for a study variable, estimators with better approximation to the unknown true value. Thus, in a demographic inquiry, the ancillary information might be the previous census population for the sample areas.

1.8 Sampling frames

For drawing a sample from the universe, a frame of all the units in the universe with proper identification should be available. This frame may consist either of a list of the units or of a map of areas, in case a sample of areas is to be selected. The frame should be accurate, free from omissions and duplications, adequate and up-to-date, and the units should be identified without ambiguity. Supplementary information available for the field covered by the frame may also be of value in improving the sample design.

Two obvious examples of frames are lists of households (and persons) enumerated in a population census, and a map of areas of a country showing boundaries of area units.

The first, however attractive, suffers from two limitations, namely of possible under- or over-enumeration of households (and population) in censuses and of changes in the population due to births, deaths, and migration, unless the sample is conducted simultaneously with the census. There are remedial methods for these difficulties. One, for example, consists of selecting a sample of area units, such as villages and urban blocks, from the frame available from a population census and listing anew the households residing in these sample areas at the time of survey-enumeration: these households may then be completely enumerated with regard to the study variables or a sub-sample of households might be drawn for the inquiry.

1.9 Types of probability sampling

If, from the universe, a sample is taken in such a manner that the sample selected has the same a priori chance of selection as any other sample of the same size, it is known as ***simple random sampling***. This is the most direct method of sampling but has most often to be modified in order to make better approximations to the unknown universe values, or to obviate the lack of accurate and up-to-date frames, to simplify the selection procedure.

One variant is ***systematic sampling***. In this procedure, if we are to draw, say, a five per cent sample from a universe, a random number is first chosen between 1 and $(1/0.05 =)$ 20; suppose that the selected random number is 12. Then the 12th, $(12 + 20 =)$ 32nd, $(32 + 20 =)$ 52nd etc. units would constitute the systematic sample. This method is often used because of its simplicity.

Another method is to use the available information on an ancillary variable in drawing the sample units. If the sampling frame does not contain any information on the universe units except their identification, there will be no reason to prefer one unit to another for inclusion in the sample, i.e., all the units would have the same chance of selection; and this is what happens with simple random and systematic samples. However, if information is available on an ancillary variable that is known to have a high degree of positive correlation with the study variable, the information on the ancillary variable may be used to draw the units in such a manner that different units get different probabilities of selection with a view to obtaining more reliable estimates. This is known as ***sampling with varying probability*** (of which ***sampling with probability proportional to size*** is the most common case). In a survey of commercial establishments, for example, if, while preparing the list of all the commercial establishments, information is also collected on their size (either number of persons employed or the value of production), then a sample of establishments could be drawn such that each establishment has an unequal, but known, *a priori* probability of selection which is proportional to size.

Before drawing a sample, the units of the universe could be sub-divided into different sub-universes or strata and a sample drawn from each stratum according to one or more of the three procedures – simple random, systematic, and varying probability sampling. This is known as ***stratified sampling***. Thus in a family budge inquiry, the families constituting the universe could be sub-divided into a number of strata depending on residence in different areas, or into a number of strata according to the size of the family if such information is available prior to sampling; different systems of stratification might be used in a sampling plan. Stratified sampling is used when separate estimates are required for the different strata or to obtain results with less variability.

So far we have considered ***single-stage sampling***. The sample might, however, be drawn in a number of stages. Thus in a household inquiry where a list of the existing households is not available, a sample of geographical areas, such as urban blocks or villages, for which lists are available (first-stage units), may first be chosen, then a list made of all the households currently residing in the selected sample (urban blocks or villages), and, at the second-stage, a sample of households selected from the lists of current households (second-stage units); this is an example of a two-stage sample design. In practice, ***multi-stage samples*** are frequently used because an accurate and up-to-date frame is not often available.

The above types of sampling may be used singly or in combination. For example, there may be **unstratified single-stage sampling, stratified single-stage sampling, unstratified multi-stage sampling**, and, the most common of all, **stratified multi-stage sampling**. Other types of sampling, such as multi-phase sampling, will be discussed later.

We shall use the term "***sampling plan***" to mean the set of rules or specifications for selecting the sample, the term "***sample design***" to cover in addition the method of estimation, and "***survey design***" cover sample design and other aspects of the survey, e.g., choice and training of enumerators, tabulation plans, etc.

1.10 Universe parameters

Let the finite universe comprise N units, and let us denote by Y_i the value of a study variable for the ith unit ($i = 1, 2, \ldots, N$). For example, in a family budget inquiry, where the unit is the family, one study variable may be the total family size, when Y_i would denote the size of the ith family; another study variable may be the total family income in a specified reference period (e.g., the thirty days preceding the date of inquiry), when Y_i would denote the value of the income of the ith family; and so on.

A *universe parameter* (or simply a *parameter*) is a function of the frequency values of the study variable. Some important parameters are described below.

In general, we would be interested in the *universe total* and the *universe mean* of the values of the study variable. The total of the values Y_i is denoted by Y,

$$Y_1 + Y_2 + \cdots + Y_N = \sum_{i=1}^{N} Y_i = Y \tag{1.1}$$

and the universe mean is the universe total divided by the number of universe units and denoted by $\overline{Y}$,

$$\overline{Y} = Y / N \tag{1.2}$$

In addition, interest centers also on the variability of the units in the universe, for the variability of any measure computed from the sample data depends on it. This variability of the values of the study variable in the universe is measured by the mean of the squared deviations of the values from the mean, and is called the *universe variance per unit*; it is denoted by σ_Y^2,

$$\sigma_Y^2 = \sum_{i=1}^{N} (Y_i - \overline{Y})^2 / N \tag{1.3}$$

The positive square root of σ_Y^2 is termed the *universe standard deviation per unit*.

Note: In the notation for the universe variance per unit, σ_Y^2, the subscript Y is included to denote that the universe variance refers to a particular study variable; in later sections, the subscripts are used in a somewhat different way.

Note that both the universe mean and the universe standard deviation are in the same unit of measurement; for example, in a family budget inquiry, if the study variable is the total family income, then the universe mean and the universe standard deviation (per unit) would be expressed in the same currency, be it dollars, pounds, pesetas, or rupees.

To obtain a measure of the universe variability independent of the unit of measurement, the universe standard deviation is divided by the universe mean; the ratio is called the ***universe coefficient of variation*** and denoted by CV,

$$CV_Y = \sigma_Y/\overline{Y} \tag{1.4}$$

It is often expressed as a percentage. The square of the CV is called the ***relative variance***. With the CV it becomes possible to compare the variability of different items, e.g., the variability of family consumption in different countries and time.

Another universe measure of interest is the ratio of the totals or means of the values of two study variables, e.g., the proportion of income that is spent on food in a family budget inquiry; it is denoted by R,

$$R = Y/X = \overline{Y}/\overline{X} \tag{1.5}$$

where X is the universe total and $\overline{X}$ the universe mean of another study variable, similarly defined as Y and $\overline{Y}$. Let the universe standard variable of the second study variable be denoted by σ_X.

The ***universe covariance*** between two study variables is obtained on taking the mean of the products of deviations from their respective means, and is denoted by σ_{YX},

$$\sigma_{YX} = \sum_{i=1}^{n}(Y_i - \overline{Y})(X_i - \overline{X})/N \tag{1.6}$$

The ***universe linear (product-moment) correlation coefficient*** between two study variables is obtained on dividing the product of the two respective standard deviations into the covariance and is denoted by ρ_{YX},

$$\rho_{YX} = \sigma_{YX}/\sigma_Y\sigma_X \tag{1.7}$$

Note that the correlation coefficient is a ***pure number***. It varies from -1 (perfect negative linear correlation) through zero (no linear correlation) to $+1$ (perfect positive linear correlation).

If N' of the total N units possess a certain attribute or belong to a certain category (such as number of households with cable TV), the universe proportion P of the number of such units is

$$P = N'/N \tag{1.8}$$

In addition to the parameters already defined, there are other types of parameters. The values of the universe parameters are generally unknown and the primary objective of a sampling inquiry is to obtain estimates of these parameters along with measures of the reliability of the estimates obtained from the data of a sample.

1.11 Sample estimators

An *estimator* is a rule or method of estimating a universe parameter. Usually expressed as a function of the sample values, it is called a ***sample estimator*** (or simply, an ***estimator***). Note that there can be more than one estimator for the same universe parameter. The particular value yielded by the sample estimator for a given sample is called a ***sample estimate*** (or simply, an ***estimate***).

Let a simple random sample of n units be selected from the universe of N units, the value of a study variable for the ith sample unit being denoted by y_i $(i = 1, 2, \ldots, n)$. The ith sample unit may be any of the N universe units. The sample mean $\overline{y}$, defined by

$$\overline{y} = (y_1 + y_2 + \cdots + y_n)/n = \sum_{i=1}^{n} y/n \tag{1.9}$$

is an estimator of the universe mean $\overline{Y}$.

A sample estimator of the universe total Y, defined in equation (1.1), is

$$y = N\,\overline{y} \tag{1.10}$$

In general, the estimates obtained from different samples of the same size and taken from the same universe, following the same survey procedure, will vary among themselves and will only by accident coincide with the universe value being estimated, even if the same survey procedure is followed in both the sample and the complete enumeration. This is simply because a part, and not the whole, of the universe is covered in a sample. This variability due to sampling is measured by the sampling variance of the estimator. For example, the sampling variance of the sample mean in sampling with replacement is

$$\sigma^2_{\overline{y}} = \sigma^2_Y/n \tag{1.11}$$

In general, of course, σ^2_Y will not be known, and the sampling variance of the sample estimator has to be estimated from the sample data. For a simple random sample of n units, the sample estimator of the universe variance per unit is

$$s_y^2 = \sum_{i=1}^{n}(y_i - \overline{y})^2/(n-1) \tag{1.12}$$

If sampling is made with replacement, so that the same universe unit can occur more than once in the sample, or if the universe is considered infinite, then an estimator of the variance of the sample mean $\overline{y}$ in simple random sampling is

$$s_{\overline{y}}^2 = s_y^2/n \tag{1.13}$$

Sample estimators for other universe parameters can also be defined. The sample estimator of the universe covariance of two study variables is simple random sample of n units is

$$s_{yx} = \sum_{i=1}^{n}(y_1 - \overline{y})(x_i - \overline{x})/(n-1) \tag{1.14}$$

An estimator of the universe ratio R of the totals of two study variable, defined in equation (1.5), is the ratio of the estimators of the two respective totals, namely

$$r = y/x = \overline{y}/\overline{x} \tag{1.15}$$

and an estimator of the universe correlation coefficient ρ_{YX} is the sample correlation coefficient

$$\hat{\rho}_{yx} = s_{yx}/s_y s_x \tag{1.16}$$

1.12 Criteria of estimators

Several criteria have been established for the sample estimators. One is **unbiasedness**. A sample estimator is said to be unbiased if the average value of the sample estimates for all possible samples of the same size is mathematically identical with the value of the universe parameter; this average over all possible samples of the same size is also known as the *mathematical expectation* of, or the (mathematically) *expected survey value*. In the examples above, the sample estimators $\overline{y}$ and $\overline{x}$ are unbiased estimators of the respective universe parameters $\overline{Y}$ and $\overline{X}$ respectively; this will be shown in Chapter 2. That a sample estimator is unbiased does not, of course, mean that the estimate obtained from a particular sample will necessarily be the same as the universe value.

In general, a sample estimator which is obtained by taking a ratio of two sample estimators (which themselves might be unbiased) is not unbiased. Thus the sample ratio of the totals of two study variables, defined in equation (1.15), is generally biased.

The second criterion is that of **consistency**. An estimator is said to be consistent if it tends to the universe value with increasing sample size; an alternative definition for a finite universe is whether the universe value is obtained, when the universe frequencies are substituted in the formula for the sample estimator. The sample mean, sample total, the sample ratio, the sample correlation coefficients etc., defined above, are all consistent estimators. An example of an inconsistent estimator is the following: suppose that from the data of a simple random sample of size n, the ratios of the values of two study variables are computed, namely,

$$r_i = y_i/x_i \tag{1.17}$$

and a mean is taken of these ratios, namely,

$$\bar{r} = \sum_{i=1}^{n} r_i/n \tag{1.18}$$

The mean of the ratios, $\bar{r}$, is not a consistent estimator of the universe ratio R, defined in equation (1.5), for in sampling without replacement, if the sample is ultimately enlarged to include all the universe units, the sample estimator $\bar{r}$ will take the value

$$\sum_{i=1}^{N}(Y_i/X_i)/N$$

which will not be the same as the universe value, namely,

$$R = Y/X = \sum_{i=1}^{N} Y_i \Big/ \sum_{i=1}^{n} X_i \tag{1.5}$$

except in the trivial case of all the ratios Y_i/X_i being the same.

An unbiased estimator is not necessarily consistent or vice versa; but in the general cases that we shall consider, an unbiased estimator can be taken to consistent. Also, if a consistent estimator is biased, its bias will decrease with sample size.

The third criterion is that of **precision** or **efficiency**. We shall consider here the concept of **relative efficiency** and **relative precision**. Of two estimators, based on the same sample size, for the same universe parameter, one is said to be more efficient than the other when its sampling variance is smaller than the other's. The **precision** of the estimator t relative to that of t' is defined as

$$Precision\ (t,t') = \sigma_{t'}^2/\sigma_t^2 \tag{1.19}$$

The **efficiency** of the estimator t relative to that of t' is defined as

$$Efficiency\ (t,t') = MSE_{t'}/MSE_t \tag{1.20}$$

MSE_t denoting the mean square error defined in the next section. For unbiased estimators, precision and efficiency are equivalent.

When, as is common, the universe values of the variances are not available, the sample estimators are substituted in the above expressions.

It can be shown that in large samples (simple random), the variance of the median as an estimator of the universe mean $\bar{Y}$ is $\frac{1}{2}\pi\sigma_Y^2/n$ (where π is the ratio of the circumference of a circle to that of its diameter = 3.14159 ..., or 3.1416, approximately), so that the efficiency of the mean relative to the median is, from equations (1.11) and (1.20), is

$$\left(\frac{1}{2}\ \pi\,\sigma_Y^2/n\right) / \left(\sigma_Y^2/n\right) = \frac{1}{2}\ \pi = 1.5708\ldots$$

or 1.57 approximately, i.e., for simple random samples of the same size, the sample mean is 57 per cent more efficient than the sample median in estimating the universe mean.

A relative concept is that of a *minimum variance estimator*. A well-known inequality in mathematical statistics (*Cramér-Rao inequality*) states that the variance of an estimator of $\bar{Y}$ cannot be smaller than σ_Y^2/n. But as the sample mean itself has sampling variance σ_Y^2/n in simple random sampling (with replacement), we can say that under this sampling plan, the sample mean is an unbiased, consistent, minimum variance estimator of the universe mean.

1.13 Mean square error

If the sample estimator t has a mathematical expectation T', different from the universal parameter T which it (the sample estimator t) seeks to measure, the sampling variability of the estimator t around its mathematical expectation T' is given by the sampling variance σ_t^2; but the variability of the estimator t around the true value T is given by the mean square error (defined in section 25.4.3). The sampling variance and the mean square error of t are connected by the relation:

$$\begin{aligned} Mean\ square\ error &= Sampling\ variance + (Bias)^2 \\ MSE_t &= \sigma_t^2 + B_t^2 \end{aligned} \tag{1.21}$$

where $B_t = T' - T$ is the bias in the estimator t.

1.14 Principal steps in a sample survey

The principal steps that should be taken in planning and executing a sample survey and analyzing and publishing the results could be grouped under the following topics (the broad topics have been illustrated in Figure 1.1, from the American Statistical Association's booklet, *What Is A Survey?*):

1. planning and preparatory work – define major objectives; establish legal basis; create steering and technical committees; prepare first operational plans; estimate budget and equipment requirements; determine sponsoring agencies (national and external); obtain funds; order equipment; finalize sampling plans; conduct pre-tests; develop and conduct publicity campaigns.

2. questionnaire design and tabulation and analysis plans – develop questionnaire design and tabulation and analysis plans; print questionnaires, training documents and control sheets.

3. sampling frames – update or prepare census field maps and listing of housing units in a household survey.

4. training – arrange national and external fellowships in survey sampling, data processing, analysis, and (desktop) publishing.

5. sampling design – prepare sampling plan.

6. pre-test and pilot studies – plan, conduct and analyze pre-tests and pilot studies to determine suitable topics for investigation, respondent types, suitability of sampling plan and estimation of sampling estimates and variances to help in finalizing sampling plan.

7. enumeration – plan procedure, develop materials; develop training programs; establish field offices; distribute materials to field; recruit and train enumerators and supervisors.

8. conduct sample enumeration.

9. prepare preliminary estimates of major items.

10. survey data processing – plan and test electronic data processing procedures and programs; acquire and test equipment; receive and check in competed questionnaires; data editing an entry; data tabulation.

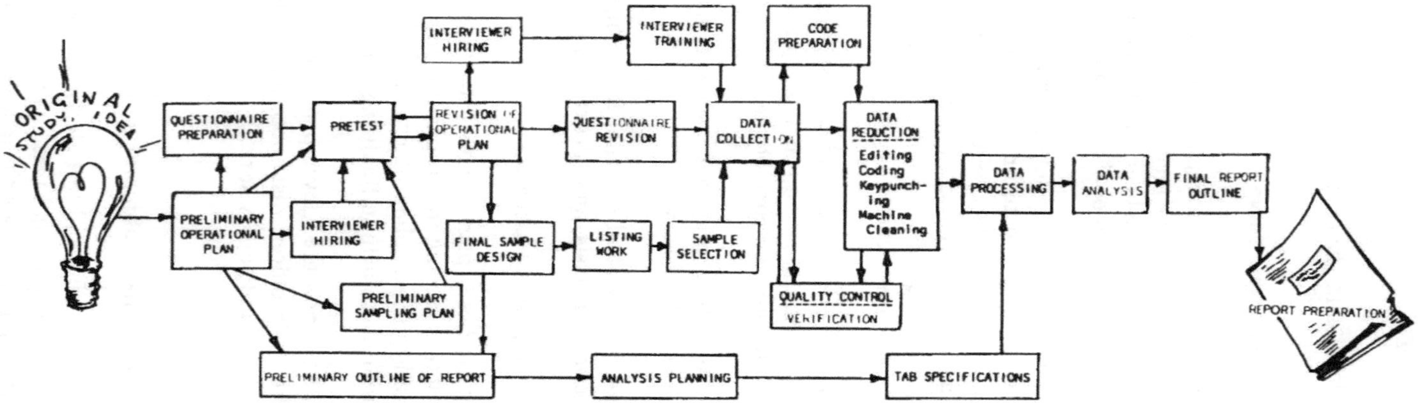

Figure 1.1: Stages of a sample survey (from Ferber *et al.*).

11. analysis and publication – analysis of data; preparation of technical and administrative reports; computation of sampling variances and indicators of non-sampling errors.

12. dissemination – seminars to discuss finding and their economic and social implications; recommend follow-up action.

The above-mentioned tasks are not all sequential (i.e., they do not necessarily follow one another): some, such as the preparation of the questionnaire (and its testing) and the recruitment of enumerators (and their training) are fully linked; others are partly linked; and several others, such as the construction of sampling frames and the preparation of training material for the enumerators may follow simultaneously. It would be most helpful to show these tasks in a **Gantt or Program Evaluation and Review Technique (PERT) Chart** so that the tasks could be monitored, evaluated, and corrective action taken. A project management computer software program, such as TimeLine (by Symantec Corporation) or Project (by Microsoft Corporation), should be installed to facilitate this.

The design and conduct of a sample survey require more technical expertise than that for a census. One principal recommendation of the United Nations should be heeded (U.N. 1947):

> ***"A sample survey should be carried out only under the technical guidance of professional statisticians not only with adequate knowledge of sampling theory but also with actual experience in sampling practice, and with the help of a properly trained filed and computing staff."***

1.15 Symbols and notations

Capital (and sometimes small) letters are used to denote the universe values of the study and ancillary variables (such as $Y_1, Y_2, \ldots, Y_N$; $X_1, X_2, \ldots, X_N$), and small letters those in the sample (such as $y_1, y_2, \ldots, y_n$; *and* $x_1, x_2, \ldots,$ x_n). N is the universe size (i.e., the total number of units in the universe), and n the sample size. Universe parameters are denoted by capital or Greek letters (such as $\overline{Y}$, the universe mean; σ^2, the universe variance per unit; ρ, the universe correlation coefficient of two variables), and the sample estimators by small letters or with circumflex or the hat ($\hat{}$) on the corresponding universe parameters (such as $\overline{y}$, the sample mean, as estimator of $\overline{Y}$; s^2, sample estimator for σ^2; and $\hat{\rho}$, sample estimator for ρ).

$\sum$ stands for summation (either for the universe or for the sample), and the number of terms to be summed is indicated by the letters at the bottom and the top of the summation sign. Thus,

$$Y_1 + Y_2 + \cdots + Y_N = \sum_{i=1}^{N} Y_i = \sum_{1}^{N} Y_i = \sum_{i}^{N} Y_i = \sum^{N} Y_i = \sum_{i} Y_i = \sum Y_i$$

the last five terms being used when there is no risk of ambiguity.

Double and triple summations may often be required. Thus, consider N (universe) first-stage units, the ith first-stage unit ($i = 1, 2, \ldots, N$) containing M_i second-stage units. Let the value of the study variable for the jth second-stage unit ($j = 1, 2, \ldots, M_i$) in the ith first-stage unit be denoted by Y_{ij}. Then the sum of the values of the study variable for any first-stage and for all first stage units can be represented as follows:

First-stage units	Sum of the values of the study variable of the second-stage units
1	$Y_{11} + Y_{12} + \cdots + Y_{1M_1} = \sum_{j=1}^{m_1} Y_{ij}$
2	$Y_{21} + Y_{21} + \cdots + Y_{2M_2} = \sum_{j=1}^{M_2} Y_{2j}$
$\vdots$	$\vdots$
i	$Y_{i1} + Y_{i2} + \cdots + Y_{iM_i} = \sum_{j=1}^{M_i} Y_{ij}$
$\vdots$	$\vdots$
N	$Y_{N1} + Y_{N2} + \cdots + Y_{NM_N} = \sum_{j=1}^{M_N} Y_{Nj}$
Grand Total	$Y_{11} + \cdots + Y_{1M_1} + Y_{21} + \cdots + Y_{2M_2} + \cdots +$ $+ Y_{N1} + \cdots + Y_{NM_N} = \sum_{i=1}^{N} \sum_{j=1}^{M_i} Y_{ij}$

In addition, the following notations will often be used for sums of squares and products. $\sum^n y_i^2$ will be called the "*raw*" or "*crude*" *sum of squares* of the y_i values and the sum of squares of deviations from the mean

$$\begin{aligned} SSy_i = \sum^{n}(y_1 - \bar{y})^2 &= \sum^{n} y_i^2 - \left(\sum^{n} y_i\right)^2 / n = \sum^{n} y_i^2 - n\bar{y}^2 \\ &= \sum^{n} y_i^2 - \bar{y}\left(\sum^{n} y\right) \end{aligned} \tag{1.22}$$

will be termed the *corrected sum of squares*. The choice of the particular expression for SSy_i in (1.22) will depend on computational convenience.

Similarly, $\sum^n y_i x_i$ will be called the "*raw*" or "*crude*" sum of products of the y_i and x_i values and the sum of products of deviations from the respective means

$$\begin{aligned} SPy_ix_i &= \sum^n (y_i - \bar{y})(x_i - \bar{x}) \\ &= \sum^n y_ix_i - \left(\sum^n y_i\right)\left(\sum^n x_i\right)/n \\ &= \sum^n y_ix_i - n\bar{y}\bar{x} = \sum^n y_ix_i - \bar{y}\left(\sum^n x_i\right) \\ &= \sum^n y_ix_i - \bar{x}\left(\sum^n y_i\right) \qquad (1.23) \end{aligned}$$

the *corrected sum of products*. Thus, the sample variance is by

$$s_y^2 = \sum^n (y_i - \bar{y})^2/(n-1) = SSy_i/(n-1) \qquad (1.24)$$

and the sample covariance given by

$$\begin{aligned} s_{yx} &= \sum (y_i - \bar{y})(x_i - \bar{x})/(n-1) \\ &= SPy_ix_i/(n-1) \qquad (1.25) \end{aligned}$$

$\prod$ (capital pi) stands for product. Thus,

$$Y_1 \times Y_2 \times \cdots \times Y_N = \prod_{i=1}^{N} Y_i \text{ or simply, } \prod^{N} Y_i.$$

Sub- and super-scripts are added, as necessary. A list of notations and symbols used in this book is given in Appendix I.

Further reading

References have been made only to the names of the authors (and to the year of publication when the author has more than one publication referenced); complete references are given at the end of the book. Some of the books mentioned in the first edition, as well as some published since then, are now out of print; however, I have retained references to the important ones among them, in the hope that they may still be available in university libraries.

For a **history of sampling**, see Bellhouse (1988a). For an **introduction to the ideas of scientific sampling**, see the books by Kalton (1983); Scheaffer,

Mendenhall, and Ott; Slonim (a popular account); Stuart; Sudman (who gives in chapter 1 of his book examples of some interesting sample surveys conducted in the U.S.A.); Warwick and Linninger; and special chapters on sampling in the book by Moser and Kalton and in textbooks on statistics (with generally the same mathematical level as this book), such as those by Blalock; Chakravarty, Laha and Roy; Hájek and Dupač; P.O. Johnson; Snedecor and Cochran; Tippet; and Yule and Kendall, and an article by Dalenius (1988).

A Short Manual on Sampling (1972) by the U.N., *Sampling Lectures* (1968) by the U.S. Bureau of the Census, and Toré Dalenius's *Elements of Survey Sampling* (1985) would also be valuable introductory reading. Barnett's *Sample Survey Principles and Methods* (1991) contains basic sampling techniques and provides many instructive examples and exercises: it would be a good preliminary read to this book.

For readers interested in **sampling theory**, texts by the following authors are recommended as follow-up:

> Chaudhuri and Stenger; Hájek; Hansen et al. (1953), vol. II; Hedayat and Sinha; Kish (1965); Murthy (1967); Sukhatme et al. (1984); and Yates.

For readers interested in the general topic of **foundations of survey sampling**, including **statistical inference in sampling from finite populations**, see articles, starting with Godambe (1955) and other relevant articles, in *New Developments in Survey Sampling*, edited by Norman L. Johnson and Harry Smith, Jr. (1969), *Foundations of Statistical Inference*, edited by V.P. Godambe and D.A. Sprott (1971), and *Handbook of Statistics, Vol. 6, Sampling*, edited by P.R. Krishnaiah and C.R. Rao (1988) and by T.M.F. Smith (1984). These collections and *Contributions to Survey Sampling and Applied Statistics*, edited by H.A. David (1978), and *Survey Sampling and Measurement*, edited by K. Namboodiri (1978), contain a number of important articles on sampling theory.

For relevant issues in statistical designs for research, see Kish (1987).

For definition of statistical terms, see *A Dictionary of Statistical Terms* by Kendall and Buckland: I have relied on it heavily.

For readers interested in **sampling methods and their application in specific subject fields**, the following are recommended:

Sampling methods (general): Cochran; Hansen *et al.* (1953), vol. I (especially relating to sample surveys undertaken by U.S. Government agencies); Kish (1965).

Agriculture, forestry and fishery: FAO (1989); Hendricks (1956); Jessen; Panse (1954); Sampford; Singh and Chaudhary; Sukhatme *et al.* (1984); Yates.

Auditing and accounting: T.F.M. Smith (1976); Trueblood and Cyert; Vance; Vance and Neter; Welburn.

Biological and geological studies: S.K. Thompson.

Business research: Deming (1960).

Demography surveys: Marks *et al.*; Som (1973); UN (1971b) and reports of the World Fertility Survey and Demographic and Health Surveys (see Appendix VI).

Health and morbidity: Levy and Lemeshaw; U.S. National Center for Health Statistics (1970a, 1970b, and 1976).

Household and social surveys: Ackoff; P.G. Gray and Corlett; Jones; Moser and Kalton; U.N. (1984b, 1993a).

Opinion surveys and marketing research: Blankenship; Bochtler and Löwenbein; Cantril; Dutka and Frankel; Gallup; Parten; Payne; Stephan and McCarthy; Velu and Naidu.

Social research: Fienberg and Tanur; Moser and Kalton (the British experience); Sudman.

Telephone surveys: Collins; Groves *et al.*; Waksberg (1978); Velu and Naidu.

Traffic surveys: Kish (1965); Rosander.

Rosander has also given the design and implementation of probability samples in the U.S.A. in the following fields: clerical operations; work sampling in a statistics and data processing division, an office and large work area, and in communications; post office parcel posts; freight car shortages; household goods shipments; shipment routing practices; truck terminal operators; and a legal proceeding.

For readers of **French**, books by the following authors are available as further reading: Ardilly (1994), Deroo and Dussaix (1980), Desabie (1962), Dussaix and Grosbras (1993), Gourieuroux *et al.* (1987), Grosbras (1987), Morin (1993), Thionet (1953), Thionet (1958), and Yates (an earlier edition translated into French).

Readers of **Spanish** are referred to the books by Sanchez-Crespo (1979) and Azorin and Sanchez-Crespo (1986) and the Spanish translation of the first edition of P. V. Sukhatme's book, *Sampling Theory of Surveys with Applications* (1954).

Note: In following the literature on sampling, one difficulty stems from the differences in concepts and notations followed by different authors. It is suggested that the reader first gets acquainted with the concepts and notations used in this book, and then, while consulting other references, notes the differences in concepts and notations. For example, some authors (including myself) use the y-notation for the (first) study variable while others use the x-notation. In dealing with single-stage sampling, all the authors cited use N and n to denote respectively the number of units in the universe and in the sample; for two-stage sampling, some authors (including myself) employ M_i and m_i to denote respectively the universe and the sample number of second-stage units in the ith first-stage unit ($i = 1, 2, \ldots, N$ for the universe, and $i = 1, 2, \ldots, n$ for the sample), but some others denote by M and m the universe and the sample number of first-stage units and by N_i and n_i the universe and sample numbers of second-stage units in the ith first stage unit. Again, in two-stage sampling, some authors use the term "size" to mean the number of second-stage units it (the first-stage unit) contains, but I and others have used the term "size" in a more general sense of the value of an ancillary variable for selection of units with probability proportional to "size" (pps), which can then be used also in single-stage pps sampling; a special case of "size" is, of course, the number of second-stage (or for that matter any subsequent-stage) units. Note also that some authors introduce multi-stage sampling under "cluster sampling."

As further reading to Chapter 1, the following are recommended:

Cochran, chapter 1; Deming (1950), chapter 1 (1960), chapter 1; Hansen *et al.* (1953), Vols. I and II, chapters 1-3; Kish (1965), chapters 1 and 2; Murthy (1967), chapters 1 and 2; Sukhatme *et al.* (1984), chapter 1; Yates, chapters 1-4.

Exercises

1. In which of the following situations would you prefer a sample survey to a complete enumeration?

    (a) Determination of the average number of match sticks in a box, and the proportion that actually light, from a carton containing a gross of match-boxes.

    (b) A survey of tuberculosis in the islands of Seychelles (estimated 1990 population of 70,000).

    (c) A longitudinal fertility inquiry in the U.S.A. (In a *longitudinal survey*, used especially in sociological and medical investigation, a sample of individuals or other units are observed at intervals over a period of time, so as to study their development as individuals. It may be contrasted with a *cross-sectional* survey, in which a sample of individuals in various stages of development, e.g., children of different ages, are observed at one point in time.)

    (d) Preparation of voters lists.

    (e) Output of steel in a country.

    (f) Hand-loom cloth production in India.

    (g) Production of coffee in Ethiopia. (*Note:* coffee grows in a wild state in some parts of Ethiopia.)

    (h) A survey of leisure time use by students in your university.

2. In family living and budget surveys, it is possible to use the questionnaire and interview method or the account book method (the "diary method"), or a combination of the two. What are the main advantages and disadvantages of each method?

(*Hint*: The "diary" method is not feasible in pre-literate or semi-literate societies. Although, recall errors should not be present if records are kept daily, the very fact that records are being kept may also change habits and patterns.)

3. A survey is required to determine the sex ratio of children (under fifteen years). Four procedures are suggested:

   (a) each boy in a sample of boys is asked to state the number of brothers and sisters he has;

   (b) each girl is a sample of girls is asked to state the number of sisters and brothers she has;

   (c) a sample of children is taken, and the number of boys and girls in the sample counted;

   (d) a sample of families are asked to give the number of boys and girls.

   Why would procedures (a) and (b) give biased results, over-representation of boys in one, and the over-representation of girls in the other? (C.R. Rao, 1952, Misc. Problem 20, adapted.)

   (*Hint*: In procedures (a) and (b) a single child and siblings consisting of the opposite sex would not be represented at all. Note that children in institutions are to be sampled too. Account has also to be taken of multiple reporting by siblings.)

4. Comment on the following extract from the proceedings of the Seminar on African Demography, Paris, 1959 (International Union for the Scientific Study of Population, 1960):

   "... suggested that under some conditions the principles of probability sampling might be sacrificed in the interest of obtaining properly qualified and reliable information in the villages to be included in the program [of study of African Demography]. These villages would then provide merely a "purpose sample"; but the extent to which they are representative of large population could be measured, at least partially, by comparing their characteristics in other respects on which information can be obtained with those of the whole population of the region."

   (*Hints*: "The suggestion is a hazardous one. ...Gini and Galvani (1929) did precisely what... suggested and they published their results to tell statisticians that the suggestion is a hazard. They selected and rejected counties in Italy until they found a sample that matched seven characteristics for the whole country, only to find that the sample broke down completely on all other characteristics": personal communication from Dr. W.E. Deming in 1970. Note also the observations by Yates (1981, section 3.13): "Gini and Galvani (1929) selected a sample from the Italian census data of 1921 which consisted of all the returns of

29 out of the 214 circondari into which the country is divided, using seven control characters. Agreement of the average values of the other characters in the sample and population was poor, and that of the frequency-distribution of such characters was even worse. The real weakness here is the use of excessively large units, though even with smaller units the use of purposive selection without rigorous rules of selection is always liable to give unsatisfactory results. There is, moreover, no means of judging its reliability. For these reasons, purposive selection has ceased to be extensively used and in modern sampling work it has largely been replaced by more thorough application of the principles of stratification, etc.").

PART I

SINGLE-STAGE SAMPLING

CHAPTER 2

Simple Random Sampling

2.1 Introduction

In this chapter we shall consider simple random sampling without stratification of the universe and without the selection of the sample units in different stages; unstratified single-stage simple random sampling is sometimes known as ***unrestricted simple random sampling***. In the next chapter will be shown the use of ancillary information in improving the efficiency of estimators obtained from a simple random sample. We shall often use the abbreviation "srs" for "simple random sample" or "simple random sampling".

We shall first consider the problem of sampling the universe of six households a, b, c, d, e, and f, with respective sizes (number of persons) 8, 6, 3, 5, 4, and 4. This is a hypothetical example, for, with such a small universe, it would be normal to enumerate all the units completely rather than take a sample. However, we shall use it to illustrate the principles.

2.2 Characteristics of the universe

We note first some characteristics of the universe. The total number of units, $N = 6$; if by Y_i we denote the size of the ith household ($i = 1, 2, \ldots, N$), the total number of persons in the universe is

$$Y = \sum_{i=1}^{N} Y_i = 8 + 6 + 3 + 5 + 4 + 4 = 30 \tag{2.1}$$

The average household size (number of persons per household) is

$$\overline{Y} = Y/N = \sum^{N} Y_i/N = 30/6 = 5 \tag{2.2}$$

Table 2.1: Sizes (numbers of persons) of a hypothetical universe of six households

Household i	Size Y_i	$Y_i - \overline{Y}$	$(Y_i - \overline{Y})^2$
a	8	3	9
b	6	1	1
c	3	-2	4
d	5	0	0
e	4	-1	1
f	4	-1	1
Total	30	0	16
Mean	$\overline{Y} = 5$	0	$\sigma^2 = 2.\dot{6}$

The variance per unit of the universe is defined as

$$\sigma^2 = \sum^{N}(Y_i - \overline{Y})^2/N = 16/6 = 2.\dot{6} \tag{2.3}$$

The computations are shown in Table 2.1. The standard deviation per unit of the universe is

$$\sigma = \sqrt{2.\dot{6}} = 1.6330$$

The coefficient of variation of the universe is defined as

$$\begin{aligned} CV &= \sigma/\overline{Y} = 1.6330/5 \\ &= 0.326\dot{6} \text{ or } 32.6\dot{6} \text{ per cent} \end{aligned} \tag{2.4}$$

Note: The characteristics of the universe will not be known in general, and the main objective in sample surveys is to estimate these characteristics.

2.3 Simple random sampling

If the sampling frame consists of an identifiable list of all the units in the universe, but without any information on the value (or magnitude) of the variable under study (the "study variable") or of any ancillary variable, and if we are sampling only one unit from the universe, then *a priori* there would be no reason to choose one unit over another, i.e. all the units should have the sample chance of being selected. Similarly, if we are selecting n (> 1) units from the universe of N units, then each combination of the n units should have the same chance of selection as every other combination: such a sample is termed a simple random sample of n units.

Note: In simple random sampling, the probability that a universe unit will be selected at any given draw is the same as that at the first draw, namely, $1/N$, and the probability that the specific unit is included in the sample of n units is n/N; see Appendix II, section A2.3.2, notes 1 and 2.

If we draw a simple random sample such that no unit occurs more than once in the sample, sampling is said to be *without replacement*, if a unit can occur more than once in the sample, it is said to be *with replacement*. We shall illustrate these with the hypothetical universe of six households given in Table 2.1, the objective being to estimate the average size of household from samples of different sizes.

2.4 Sampling without replacement

2.4.1 Mean and variance for the hypothetical universe

If we sample only one unit from the universe of six households without replacement, the six possible samples of size 1 are the six households in the universe. The estimator of the household size of the universe from each sample will be the household size of the sample unit itself; the average of the size estimators is 5, the same value as the universe household size. But we cannot compute the sampling variance from one sample in this case, as the sample consists of one unit only.

If we draw samples of two units each from the universe, such that no unit occurs more than once in the sample, and if we disregard the order of drawing so that the same combination of two units such as (ab, ba) is not repeated, we shall have ${}^6C_2 = 15$ combinations: these are written down in Table 2.2. For each of the 15 combinations (samples) we compute the following measures, as estimators of the corresponding universe values:

$$\text{Estimator of the universe mean: sample mean } \bar{y} = \sum_{i=1}^{n} y_i/n \qquad (2.5)$$

$$\text{Estimator of variance per unit: } s_y^2 = \sum^{n}(y_i - \bar{y})^2/(n-1) \qquad (2.6)$$

Table 2.2: All possible samples of size 2 drawn without replacement from the universe of Table 2.1

Sample no.	Sample units	Total size $y_1 + y_2$*	Mean size $(\bar{y})$†	s_y^2‡	$s_{\bar{y}}^2$§
1	ab	14	7.0	2.0	0.6667
2	ac	11	5.5	12.5	4.1667
3	ad	13	6.5	4.5	1.5000
4	ae	12	6.0	8.0	2.6667
5	af	12	6.0	8.0	2.6667
6	bc	9	4.5	4.5	1.5000
7	bd	11	5.5	0.5	0.1667
8	be	10	5.0	2.0	0.6667
9	bf	10	5.0	2.0	0.6667
10	cd	8	4.0	2.0	0.6667
11	ce	7	3.5	0.5	0.1667
12	cf	7	3.5	0.5	0.1667
13	de	9	4.5	0.5	0.1667
14	df	9	4.5	0.5	0.1667
15	ef	8	4.0	0	0
Total	-	-	75.0	48.0	16.0000
Mean	-	-	5.0	3.2	1.06

* y_1 and y_2 are the values of the two units included in the sample.
† $\bar{y} = \frac{1}{2}(y_1 + y_2)$ from equation (2.5).
‡ $s_y^2 = \frac{1}{2}(y_1 - y_2)^2$ from equation (2.8).
§ $s_{\bar{y}}^2 = (1 - f)s_y^2/n = \frac{1}{3}\, s_y^2$ from equation (2.7), as $n = 2$, $N = 6$, and $n/N = \frac{1}{3}$.

$$\text{Estimator of variance of sample mean: } s_{\bar{y}}^2 = s_y^2(1 - f)/n \qquad (2.7)$$

where n is the sample size, y_i $(i = 1, 2, \ldots, n)$ is the value of the ith unit of the study variable included in the sample, and $f = n/N$ is the sampling fraction.

Note the following points:

1. For $n = 2$, the formula for s_y^2 is simplified to

$$s_y^2 = \frac{1}{2}(y_1 - y_2)^2 \qquad (2.8)$$

where y_1 and y_2 are the values of the two sample units included in the sample.

2. In the general case, a workable formula for s_y^2 may be chosen from among the following, depending on the convenience of computation:

$$s_y^2 = \frac{SSy_i}{n-1} = \frac{\sum^n (y_i - \bar{y})^2}{n-1} = \frac{\sum^n y_i^2 - \left(\sum^n y_i\right)^2/n}{n-1} = \frac{\sum^n y_i^2 - n\bar{y}^2}{n-1} = \frac{\sum^n y_i^2 - \bar{y}\left(\sum^n y_i\right)}{n-1} \quad (2.9)$$

The results of all possible samples of size 3 drawn without replacement from the same universe are given in Table 2.3. The computations of the results of all possible samples of sizes 4, 5, and 6 drawn without replacement are not shown separately.

2.4.2 Distribution of sample mean

The distribution of the sample mean in all possible samples of different sizes are given in Table 2.4. Two points may be noted:

1. The average of the estimates of all possible samples for any sample size (this is known as the *mathematical expectation*) is the true universe value, namely 5. If the mathematical expectation of a sample estimator is the universe parameter, the estimator is said to be *unbiased* for the universe parameter; if not, the estimator is *biased*. We have seen that the sample mean, obtained from a simple random sample without replacement, is an unbiased estimator of the universe mean (see also section 1.12).

2. As the size of the sample increases, the sample estimates concentrate around the universe value; thus 73 per cent of the sample estimates of size two, 90 per cent of size three, and all from four and larger sizes fall within the range of household size 4-6 (see Figure 2.1). This characteristic of a sample estimator is known as *consistency* (see also section 1.12).

Table 2.3: All possible samples of size 3 drawn without replacement from the universe of Table 2.1

Sample no.	Sample units	Total sizes	Mean size ($\overline{y}$)*	s_y^2†	$s_{\overline{y}}^2$‡
1	abc	17	5.67	6.3333	1.0556
2	abd	19	6.33	2.3333	0.3889
3	abe	18	6.00	4.0000	0.6667
4	abf	18	56.00	4.0000	0.6667
5	acd	16	5.33	6.3333	1.0556
6	ace	15	5.00	7.0000	1.1667
7	acf	15	5.00	7.0000	1.1667
8	ade	17	5.67	4.3333	0.7222
9	adf	17	5.67	4.3333	0.7222
10	aef	16	5.33	5.333	0.8889
11	bcd	14	4.67	2.3333	0.3889
12	bcd	13	4.33	2.3333	0.3889
13	bcf	13	4.33	2.3333	0.3889
14	bde	15	5.00	1.0000	0.1667
15	bdf	15	5.00	1.0000	0.1667
16	bef	14	4.67	1.3333	0.2222
17	cde	12	4.00	1.0000	0.6667
18	cdf	12	4.00	1.0000	0.1667
19	cef	11	3.67	0.3333	0.0556
20	def	13	4.33	0.3333	0.0556
Total	-	-	100.00	64.0000	10.6667
Mean	-	-	5.00	3.2000	0.5333

* $\overline{y} = \frac{1}{3} \sum_1^3 \overline{y}_i$;

† $s_y^2 = SSy_i/(n-1) = \frac{1}{2}\, SSy_i$;

‡ $s_{\overline{y}}^2 = s_y^2(1-f)/n = \frac{1}{6}\, s_y^2$, as $n = 3$, $N = 6$ and $f = n/N = \frac{1}{2}$.

Table 2.4: Estimates of average household size from all possible samples of different sizes drawn without replacement from the universe of Table 2.1

Average size of household	Number of sample with estimated average size of household for samples of size n					
	$n=1$	$n=2$	$n=3$	$n=4$	$n=5$	$n=6$
3.00-3.24	1					
3.25-3.49						
3.50-3.74		2	1			
3.75-3.99						
4.00-4.24	2	2	2	1		
4.25-4.49			3	1	1	
4.50-4.74		3	2	2		
4.75-4.99				2	1	
5.00-5.24	1	2	4	2	3	1
5.25-5.49			2	3	1	
5.50-5.74		2	3	2		
5.75-5.99				2		
6.00-6.24	1	2	2			
6.25-6.49			1			
6.50-6.74		1				
6.75-6.99						
7.00-7.24		1				
7.25-7.49						
7.50-7.74						
7.75-7.99						
8.00-8.24	1					
Number of samples	6	15	20	15	6	1
Mean	5	5	5	5	5	5
Variance of mean	2.6667	1.0667	0.5333	0.2667	0.1067	0

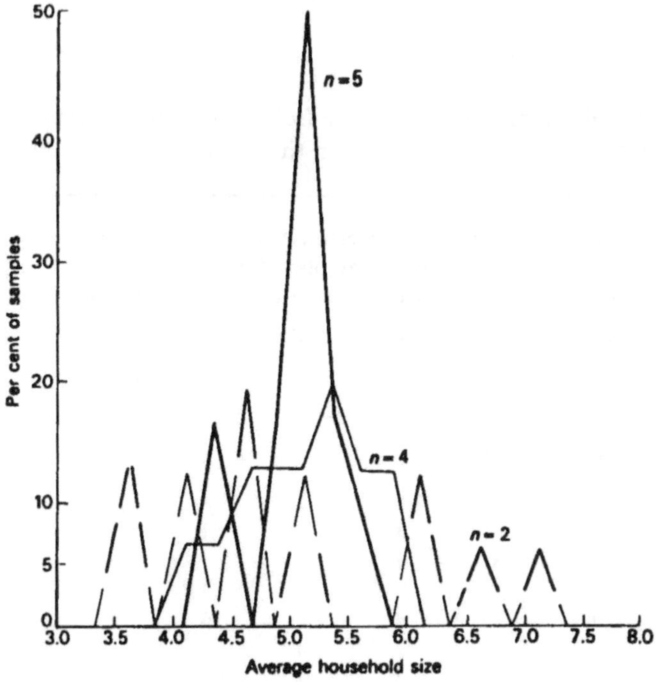

Figure 2.1: Distribution of estimates of average household size in simple random sampling without replacement from the universe of Table 2.1 (*Source* Table 2.4).

2.4.3 Variance estimators

A measure of the degree by which the estimate $\overline{y}$ obtained from a sample differs from the true value $\overline{Y}$ is given by $s^2_{\overline{y}}$. The values of $s^2_{\overline{y}}$ for the different sample sizes from the universe of Table 2.1 are given in Table 2.5. The following points may be noted from the data:

1. The average of s^2_y for all the sample sizes is a constant and is 3.2, whereas the universe variance is $\sigma^2 = 2.\dot{6}$ (Table 2.1). Thus s^2_y is not an unbiased estimator of the universe variance σ^2. The expected value of s^2_y is:

$$E(s^2_y) = \frac{N}{N-1}\,\sigma^2 = \frac{\sum^N (Y_i - \overline{Y})^2}{N-1} \tag{2.10}$$

In some textbooks, the expression on the right-hand side is denoted by S^2 and in others the universe variance is defined by this expression.

Table 2.5: Results of all possible samples of different sizes drawn without replacement from the universe of Table 2.1

Sample size n	Number of possible samples	Mean of sample means	Mean of s_y^2	Mean of $s_{\bar{y}}^2$
1	6	5	3.2	2.6667
2	15	5	3.2	1.0667
3	20	5	3.2	0.5333
4	15	5	3.2	0.2667
5	6	5	3.2	0.1067
6	1	5	3.2	0

In our example,

$$S^2 = \frac{\sum^N (Y_i - \overline{Y})^2}{N-1} = \frac{16}{5} = 3.2$$

which is the average of s_y^2 for all the sample sizes.

2. The universe variance of the sample mean for srs without replacement is

$$\sigma_{\bar{y}}^2 = \frac{\sigma^2}{n} \cdot \frac{(N-n)}{(N-1)} = \frac{S^2(1-f)}{n} \qquad (2.11)$$

Note that if $n = 1$, $\sigma_{\bar{y}}^2 = \sigma^2$, the universe variance per unit, as it should be.

3. The value of the estimator of the variance of the sample average $s_{\bar{y}}^2$ decreases with increasing sample size and becomes zero when $n = N$ (see Figure 2.2)

4. Tables 2.4 and 2.5 give the results obtained from the theoretical probabilities of the different samples. The same results are approximated when actual samples are drawn a large number of times from a universe.

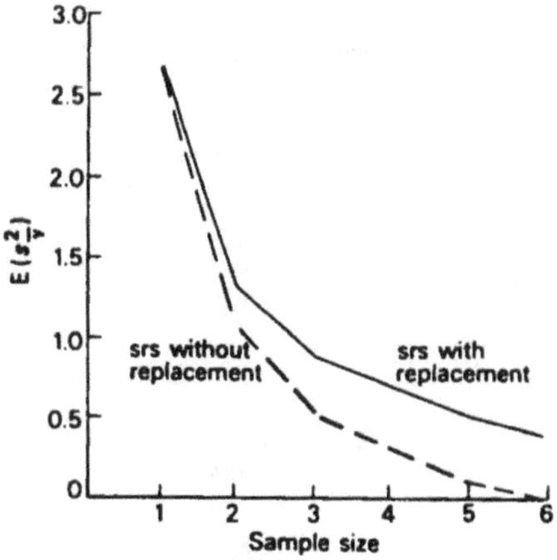

Figure 2.2: Expectation of the sampling variance of average household size drawn in simple random sampling with and without replacement from the universe of Table 2.1 (*Source:* Table 2.5 and equation (2.20)).

2.5 Sampling with replacement

2.5.1 Estimators of mean and variance

We have seen that in simple random sampling without replacement of n units out of a total of N units, there may be ${}^N C_n$ combinations (samples) if no unit is repeated in the sample and the order of selection is not taken into account; the probability of the occurrence of any combination is $1/{}^N C_n$, the total of all the probabilities for the ${}^N C_n$ combinations being 1. In practice, however, the above procedure of writing out all the ${}^N C_n$ combinations and then selecting one at random is not followed as this would be a needless, tedious procedure. It would also be unpractical: for example, there are ${}^{100}C_5 = 75, 287, 520$ combinations of 5 units to be selected without replacement out of 100 units and obviously it is out of question to write down all these combinations and then select one: furthermore, to use a computer to do the selection would be a waste of resources in this case. Actually, the n units of a sample are selected one by one with the help of random numbers (or in some other random way), and if a unit is once selected in the sample, it is not replaced.

In simple random sampling with replacement a unit can appear more than once in the sample. In Table 2.6 are shown the 36 possible samples of size 2 drawn with replacement from the hypothetical universe of Table 2.1: the probability of occurrence of any sample is $\frac{1}{36}$. note the repetition of units in the sample such as aa, bb etc. and ab, ba etc. For each of the 36 combinations (samples) we compute the following measures, as estimators of the universe parameters:

$$\text{Estimator of the universe mean: sample mean } \bar{y} = \sum^{n} y_i/n \qquad (2.12)$$

$$\text{Estimator of variance per unit: } s_y^2 = \sum^{n}(y_i - \bar{y})^2/(n-1) \qquad (2.13)$$

$$\text{Estimator of variance of sample mean: } s_{\bar{y}}^2 = s_y^2/n \qquad (2.14)$$

Note that the first two estimators are the same as those for srs without replacement (equations (2.5) and (2.6) respectively), but the third estimator is different.

The results of Table 2.6 show the following:

1. The mean of the estimates of all possible samples is the universe value, namely 5. Thus the sample mean is an unbiased estimator of the universe mean.

2. The mean of all the possible sample estimates of variance s_y^2 is 2.6, the universe variance, showing that s_y^2 is an unbiased estimator of σ^2.

2.5.2 Consistency of mean and sampling variance

In general, in srs with replacement of n units out of a total of N units, there will be N^n combinations, and the probability of selection of any combination is $1/N^n$. The results noted for srs with replacement of 2 units from the universe of 6 units will also hold for samples of other sizes. In addition, the following may be noted:

1. As the size of the sample increases, the sample estimates concentrate around the universe value. The sample mean is again a consistent estimator of the universe average.

2. The sampling variance of the sample mean for srs with replacement is

$$\sigma_{\bar{y}}^2 = \sigma^2/n \qquad (2.15)$$

   and its unbiased estimator is

$$s_{\bar{y}}^2 = s_y^2/n \qquad (2.16)$$

   This decreases with increasing sample size, but does not reach zero (Figure 2.2).

Table 2.6: All possible samples of size 2 drawn with replacement from the universe of Table 2.1

Sample no.	Sample units	Total size $y_1 + y_2$*	Average size $(\bar{y})$†	s_y^2‡	$s_{\bar{y}}^2$§
1	aa	16	8.0	0	0
2	ab	14	7.0	2.0	1.00
3	ac	11	5.5	12.5	6.25
4	ad	13	6.5	4.5	2.25
5	ae	12	6.0	8.0	4.00
6	af	12	6.0	8.0	4.00
7	ba	14	7.0	2.0	1.00
8	bb	12	6.0	0	0
9	bc	9	4.5	4.5	2.25
10	bd	11	5.5	0.5	0.25
11	be	10	5.0	2.0	1.00
12	bf	10	5.0	2.0	1.00
13	ca	11	5.5	12.5	6.25
14	cb	9	4.5	4.5	2.25
15	cc	6	3.0	0	0
16	cd	8	4.0	2.0	1.00
17	ce	7	3.5	0.5	0.25
18	cf	7	3.5	0.5	0.25
19	da	13	6.5	4.5	2.25
20	db	11	5.5	0.5	0.25
21	dc	8	4.0	2.0	1.00
22	dd	10	5.0	0	0
23	de	9	4.5	0.5	0.25
24	df	9	4.5	0.5	0.25
25	ea	12	6.0	8.0	4.00
26	eb	10	5.0	2.0	1.00
27	ec	7	3.5	0.5	0.25
28	ed	9	4.5	0.5	0.25
29	ee	8	4.0	0	0
30	ef	8	4.0	0	0
31	fa	12	6.0	8.0	4.00
32	fb	10	5.0	2.0	1.00
33	fc	7	3.5	0.5	0.25
34	fd	9	4.5	0.5	0.25
35	fe	8	4.0	0	0
36	ff	8	4.0	0	0
Total	-	-	180.00	96.00	48.00
Mean	-	-	5.00	$2.\dot{6}$	1.3

* y_1 and y_2 are the values of the two units included in the sample.

† $\bar{y} = \frac{1}{2}(y_1 + y_2)$ from equation (2.5).

‡ $s_y^2 = \frac{1}{2}(y_1 - y_2)^2$ from equations (2.13) and (2.8).

§ $s_{\bar{y}}^2 = \frac{1}{2} s_y^2$ from equation (2.14).

2.6 Important results

2.6.1 For simple random sampling

We shall summarize the important results for simple random sampling. We shall often refer to srs with replacement as "srswr" and srs without replacement as "srswor".

1. The sample mean $\overline{y}$ is an unbiased estimator of the universe mean $\overline{Y}$.

2. The variance of the sample mean is

$$\sigma^2_{\overline{y}} = \frac{\sigma^2}{n} \text{ in srswr} \tag{2.17}$$

$$= \frac{\sigma^2}{n}\frac{(N-n)}{(N-1)}$$

$$= \frac{S^2}{n}(1-f) \text{ in srswor} \tag{2.18}$$

where

$$S^2 = \frac{N}{N-1}\sigma^2 \tag{2.19}$$

and $f = n/N$ is the sampling fraction.

3. The sample estimator of variance per unit

$$s^2 = \frac{\sum^n (y_i - \overline{y})^2}{n-1}$$

is an unbiased estimator of σ^2 in srswr and of S^2 in srswor.

4. Unbiased variance estimators of the sample mean $\overline{y}$ are

$$s^2_{\overline{y}} = \frac{s^2}{n} \text{ in srswr} \tag{2.20}$$

$$s^2_{\overline{y}} = \frac{s^2}{n}(1-f) \text{ in srswor} \tag{2.21}$$

From the numerical examples, these results have been seen to hold for the universe of six households. Theoretical proofs are given in Appendix II that they hold for a universe of any size (sections A2.3.2 - A2.3.4).

2.6.2 Variance of the ratio of random variables

Another theorem of wide generality relates to the ratio of two random variables. If y and x are two random variables with expectations Y and X, variances σ_Y^2 and σ_X^2, and covariance $\sigma_{YX} = \rho\sigma_Y\sigma_X$ (where ρ is the correlation coefficient), then the variance of the ratio of two random variables $r = y/x$ is given approximately by

$$\sigma_r^2 = \frac{1}{X^2}(\sigma_Y^2 + R^2\sigma_X^2 - 2R\sigma_{YX}) \quad (2.22)$$

$$= R^2(CV_Y^2 + CV_X^2 - 2\rho CV_Y CV_X) \quad (2.23)$$

where $R = Y/X$. Also,

$$CV_r^2 = CV_Y^2 + CV_X^2 - 2\rho CV_Y CV_X \quad (2.24)$$

Theoretical proofs for simple random sampling are given in Appendix II, section A2.3.6. Estimators of σ_r^2 and CV_r are obtained on substituting the sample estimators for the universe parameters in the respective expressions.

Notes

1. In effect, sampling with replacement is equivalent to drawing samples from an infinitely large universe, and in the estimating formulae for the variance of the sample mean (and other related measures), the factor

$$1 - f = 1 - \frac{n}{N}$$

known variously as the *finite sampling correction, finite multiplier*, or *finite population correction*, does not enter.

2. As the finite sampling correction $(1 - f)$ is less than 1, the estimator of the sample mean (and other related measures) will, on an average, have less variance (i.e. be more efficient) for sampling without replacement than with replacement. However, sampling with replacement greatly simplifies the computations required for estimating universe parameters; for this reason, even if sampling is made without replacement, the sample may be treated as if it was obtained with replacement if the sample size is small relative to that of the universe. A useful rule of thumb for the latter is that the sampling fraction $f = n/N$ should be less than 10 per cent, and preferably less than 5 per cent.

3. *Behavior of sampling variance with increasing sampling size:* In sampling without replacement, the expression for the estimator of the variance of the sample mean, $s_y^2(1 - f)/n$, becomes zero when the universe is completely

enumerated, for then $n = N$ and $(1-f) = 0$. In sampling with replacement, however, the expression for the estimator of the variance of the sample mean, namely s_y^2/n, will not necessarily become zero even when the sample size n is made equal to the universe size N. This is because in sampling with replacement, a unit can be selected more than once and hence a sample of N units may not necessarily include all the distinct N universe units. We have already noted that sampling with replacement is equivalent to sampling an infinite universe, which can never be exhausted by sampling, as the units are replaced after each draw.

The coefficient of variation of the sample mean is

$$CV(\bar{y}) = \sigma_{\bar{y}}/\bar{Y} \tag{2.25}$$

$$= \sigma/\bar{Y}\sqrt{n}$$

$$= \text{universe } CV \text{ per unit}/\sqrt{n} \text{ in srswr} \tag{2.26}$$

$$= \frac{\sigma}{\bar{Y}\sqrt{n}}\sqrt{\frac{N-n}{N-1}}$$

$$= \frac{\text{universe } CV \text{ per unit}}{\sqrt{n}}\sqrt{\frac{N-n}{N-1}} \text{ in srswor} \tag{2.27}$$

The behavior of the coefficient of variation of the sample mean with increasing sample size in simple random sampling both with and without replacement is illustrated in Figure 2.3 for selected values of N and n, assuming the universe CV per unit to be 50 per cent.

Figure 2.3 illustrates the fact that when the sampling fraction is relatively large, sampling without replacement has substantial superiority over sampling with replacement; with a small sampling fraction, however, the gain is marginal.

In sampling with replacement, and approximately also in sampling without replacement when the sampling fraction is relatively small, the sampling variance of the sample mean is inversely related to the sample size, and does not depend on the size of the universe.

4. Although the variance estimators per unit in the universe s_y^2 and of the sample average $s_{\bar{y}}^2$ are unbiased estimators of the corresponding universe values, the sample estimators of the standard deviations, obtained on taking the positive square root of the variance estimators, are not unbiased estimators of the corresponding universe standard deviations: the reader may verify this for the numerical example of section 2.4. The bias, however, is negligible for large samples.

5. The sample CV is also not an unbiased estimator of the universe CV.

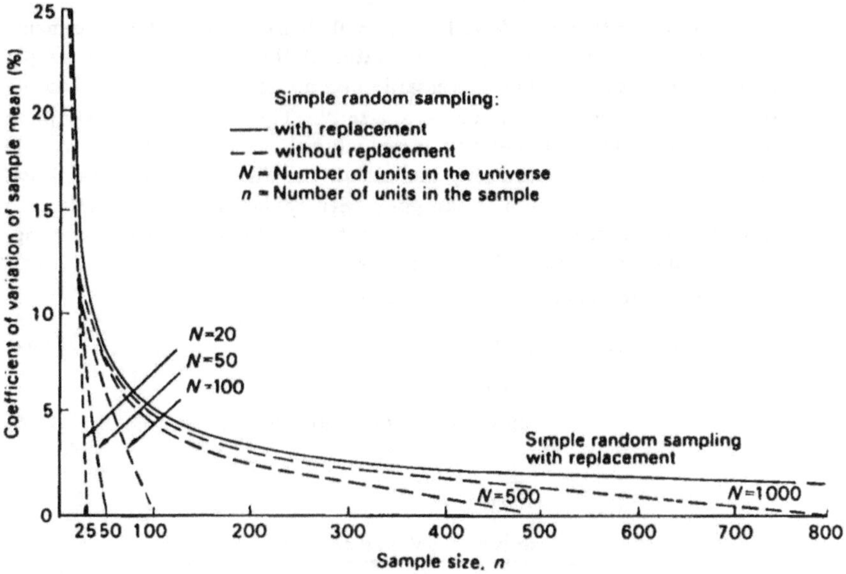

Figure 2.3: Behavior of the coefficient of variation of sample mean with increasing sample size in simple random sampling with and without replacement when the universe coefficient of variation per unit is 50%.

2.7 Fundamental theorems

We now introduce some theorems of fundamental importance in sampling.

1. If t_i $(i = 1, 2, \ldots, n)$ are n independent and unbiased estimators of the universe parameter T, then a combined unbiased estimator of T is the arithmetic mean

$$\bar{t} = \frac{1}{n} \sum^{n} t_i \tag{2.28}$$

and an unbiased estimator of the variance of $\bar{t}$ is

$$s_{\bar{t}}^2 = \frac{\sum^n (t_i - \bar{t})^2}{n(n-1)} = \frac{SSt_i}{n(n-1)} \tag{2.29}$$

If u_i $(i = 1, 2, \ldots, n)$ are n independent and unbiased estimators of another universe parameter U, then a combined unbiased estimator of U is similarly

$$\bar{u} = \frac{1}{n} \sum^{n} u_i \tag{2.30}$$

and an unbiased estimator of the variance of $\overline{u}$ is similarly

$$s_{\overline{u}}^2 = \frac{\sum^n (u_i - \overline{u})^2}{n(n-1)} = \frac{SSu_i}{n(n-1)} \tag{2.31}$$

2. An unbiased estimator of the covariance of $\overline{t}$ and $\overline{u}$ is the estimator

$$s_{\overline{t}\,\overline{u}} = \frac{\sum^n (t_i - \overline{t})(u_i - \overline{u})}{n(n-1)} = \frac{SPt_iu_i}{n(n-1)} \tag{2.32}$$

3. A consistent, but generally biased, estimator of the ratio of two universe parameters

$$R = T/U$$

is the ratio of the sample estimators $\overline{t}$ and $\overline{u}$

$$r = \overline{t}/\overline{u} \tag{2.33}$$

and an estimator of the variance of r is

$$s_r^2 = (s_{\overline{t}}^2 + r^2 s_{\overline{u}}^2 - 2rs_{\overline{t}\,\overline{u}})/\overline{u}^2 \tag{2.34}$$

4. A consistent, but generally biased, estimator of the universe correlation coefficient ρ between the two study variables is the sample correlation coefficient

$$\widehat{\rho}_{\overline{t}\,\overline{u}} = s_{\overline{t}\,\overline{u}}/s_{\overline{t}}s_{\overline{u}} \tag{2.35}$$

Notes

1. The above theorems have a very wide range of application in sampling. These can be generalized to stratified and multi-stage designs, with sample units being selected with equal or varying probabilities. The only requirement is that for the universe parameters, a number of independent and unbiased estimators be available from the sample. The independence of the sample estimators is ensured by selecting the sample units with replacement (at least at the first stage in a multi-stage sample design and in each stratum separately in a stratified design); in practice, when sampling is carried out without replacement this is approximated by making the (first-stage) sampling fraction fairly small (10 per cent or less, preferably 5 per cent or less: see note 2, section 2.6).

2. The estimators $\overline{t}$ and $\overline{u}$ generally refer to the totals of the characteristics; the advantage of this will be see later.

3. the estimator r is unbiased if the linear regression of $\bar{t}$ on $\bar{u}$ passes through the origin $(0, 0)$. An estimator of the bias (this estimator itself is generally biased) is

$$\frac{r s_{\bar{t}}^2 - s_{\bar{t}\bar{u}}}{\bar{u}^2} \tag{2.36}$$

4. Whenever a universe ratio or its sample estimator is considered, it is assumed that the denominator is not zero.

2.8 Standard errors of estimators and setting of confidence limits

The *standard error* of an estimator is given by the positive square root of the sampling variance of the estimator. Expressed as a ratio of the expected value of the estimator, it is called the *relative standard error* or the *Coefficient of Variation* (CV) and is conveniently expressed as a percentage measure. The square of the CV is sometimes called the *relative variance* of the estimator. When the universe values are not available, sample estimators are substituted in computing the CV.

The standard error of the estimator helps in setting probability limits to the estimator by assigning limits which are expected to include the (unknown) universe value with a certain probability: these limits are known as *confidence limits*, and the interval set by the limits, the *confidence interval.*

If $\bar{t}$ is an unbiased estimator of a universe parameter T (using the notations of section 2.7), and its sampling distribution moderately normal (which will happen when the sample size is not too small), and $s_{\bar{t}}$, the standard error of $\bar{t}$, is estimated from the sample data, then the statistic

$$t' = \frac{\bar{t} - \overline{T}}{s_{\bar{t}}} = \frac{(\bar{t} - T)\sqrt{n}}{s_t} \tag{2.37}$$

is distributed as *Student's t-distribution* with $(n-1)$ *degrees of freedom* on which $s_{\bar{t}}^2$, the sample estimator of the variance of t, is based; here s_t^2 is defined as

$$s_t^2 = \frac{\sum^n (t_i - \bar{t})^2}{n-1} \tag{2.38}$$

Denoting by $t'_{\alpha,n-1}$ the 100α percentage point of the t-distribution corresponding to $(n-1)$ degrees of freedom, we see that the inequality

$$|\bar{t} - T|\sqrt{n}/s_t \le t'_{\alpha,n-1} \tag{2.39}$$

is expected to occur on an average with probability $(1-\alpha)$. The chances are, therefore, $100(1-\alpha)$ per cent that the universe value T will be contained by the limits

$$\bar{t} - (t'_{\alpha,n-1} s_t)/\sqrt{n}, \qquad \bar{t} + (t'_{\alpha,n-1} s_t)/\sqrt{n} \tag{2.40}$$

Some illustrative values of the t-distribution are given in Appendix III, Table 2. For example, for $n = 10$, and $\alpha = 0.05$, $t'_{\alpha,n-1} = 2.26$. For large n, the normal distribution may be used, where for $\alpha = 0.05$, $t'_{\alpha,\infty} = 1.96$ or 2 approximately.

The following points should be noted:

1. When the universe standard error of $\bar{t}$, namely $\sigma_{\bar{t}} = \sigma/\sqrt{n}$, is known, then the statistic

$$(\bar{t} - T)/\sigma_{\bar{t}} \tag{2.41}$$

follows the normal distribution.

2. When the sample size is small, the t-distribution may still be used in setting probability limits to the universe values, but with the additional assumption that the sample itself comes from a normal distribution.

2.9 Estimation of totals, means, ratios and their variances

2.9.1 Simple random sampling with replacement

We have seen in sections 2.5 and 2.6 that if a simple random sample of n units is selected with replacement from the universe of N units, the sample mean $\bar{y}$ is an unbiased estimator of the universe mean $\bar{Y}$, and an unbiased estimator of the variance of $\bar{y}$ is

$$s^2_{\bar{y}} = \frac{s^2_y}{n} = \frac{\sum^n (y_i - \bar{y})^2}{n(n-1)} \tag{2.16}$$

Let us see, however, how the above and other results follow from the fundamental theorems of section 2.7.

With a simple random sample of n units selected with replacement, let the sample values of the study variable be $y_1, y_2, \ldots, y_n$. Each of these y_i values ($i = 1, 2, \ldots, n$), when multiplied by N, the total number of units in the universe, gives an independent, unbiased estimator of the universe total Y

$$y^*_i = N y_i \tag{2.42}$$

The combined unbiased estimator of Y is (from equation (2.28))

$$y_0^* = \frac{1}{n}\sum^n y_i^* = \frac{N}{n}\sum^n y_i = N\overline{y} \tag{2.43}$$

An unbiased estimator of the variance of the estimator y_0^* is, from equation (2.29),

$$\begin{aligned} s^2_{y_0^*} &= \frac{\sum^n (y_i^* - y_0^*)^2}{n(n-1)} = \frac{SSy_i^*}{n(n-1)} \\ &= \frac{N^2\sum^n (y_i - \overline{y})^2}{n(n-1)} = \frac{N^2 s^2 y}{n} \\ &= \frac{N^2 SSy_i}{n(n-1)} \end{aligned} \tag{2.44}$$

An unbiased estimator of the universe mean $\overline{Y}$ is obtained on dividing the unbiased estimator of the universe total by the number of universe units N, namely,

$$y_0^*/N = \overline{y} \tag{2.45}$$

i.e. the sample mean, and an unbiased estimator of the variance of $\overline{y}$ on dividing the unbiased estimator of variance of the estimator of the universe total by N^2, namely,

$$s^2_{y_0^*}/N^2 = s_y^2/n = s_{\overline{y}}^2 \tag{2.46}$$

From the same sample, unbiased estimators of the total and the mean of another study variable may be computed from estimation formulae of the above types. An unbiased estimator of the covariance between the estimators y_0^* and x_0^* of the two universe total Y and X is, from equation (2.32),

$$\begin{aligned} s_{y_0^* x_0^*} &= \frac{\sum^n (y_i^* - y_0^*)(x_i^* - x_0^*)}{n(n-1)} = \frac{SPy_i^* x_i^*}{n(n-1)} \\ &= \frac{N^2\sum^n (y_i - \overline{y})(x_i - \overline{x})}{n(n-1)} = \frac{N^2 s_{yx}}{n} \\ &= \frac{N^2 SPy_i x_i}{n(n-1)} \end{aligned} \tag{2.47}$$

A consistent, but generally biased, estimator of the universe ratio $R = Y/X$ of two universe totals Y and X is the ratio of the sample estimators from equation (2.33), namely,

$$r = y_0^*/x_0^* \tag{2.48}$$

and an estimator of the variance of r is (from equation (2.34))

$$\begin{aligned} s_r^2 &= \frac{s_{y_0^*}^2 + r^2 s_{x_0^*}^2 - 2r s_{y_0^* x_0^*}}{x_0^{*2}} = \frac{s_y^2 + r^2 s_x^2 - 2r s_{yx}}{n\overline{x}^2} \\ &= \frac{\sum^n (y_i - \overline{y})^2 + r^2 \sum^n (x_i - \overline{x})^2 - 2r \sum^n (y_i - \overline{y})(x_i - \overline{x})}{n(n-1)\overline{x}^2} \\ &= \frac{SSy_i + r^2 SSx_i - 2r\, SPy_i x_i}{n(n-1)\overline{x}^2} \\ &= \frac{\sum^n y_i^2 + r^2 \sum^n x_i^2 - 2r \sum^n y_i x_i}{n(n-1)\overline{x}^2} \end{aligned} \tag{2.49}$$

2.9.2 Simple random sampling without replacement

Here the unbiased estimators of universe totals and means, and the estimator of the universe ratio of two universe totals (or means) remain the same as for srs with replacement, but the finite sampling correction, $(1 - f) = (1 - n/N)$, is applied to the variance and covariance estimators. Thus the unbiased variance estimator of $\overline{y}$ is

$$s_{\overline{y}}^2 = (1 - f)\frac{s_y^2}{n} \tag{2.21}$$

$$= (1 - f)\frac{SSy_i}{n(n-1)} \tag{2.50}$$

An unbiased variance estimator of y_0^* is

$$\begin{aligned} s_{y_0^*}^2 = N^2 s_{\overline{y}}^2 &= (1 - f)N^2 \frac{s_y^2}{n} \\ &= (1 - f)\frac{N^2 SSy_i}{n(n-1)} \end{aligned} \tag{2.51}$$

An unbiased estimator of covariance of y_0^* and x_0^* is

$$\begin{aligned} s_{y_0^* x_0^*} &= (1 - f)N^2 \frac{s_{yx}}{n} \\ &= (1 - f)\frac{N^2 SPy_i x_i}{n(n-1)} \end{aligned} \tag{2.52}$$

A variance estimator of r is

$$
\begin{aligned}
s_r^2 &= \frac{1}{\bar{x}^2}\left(s_{\bar{y}}^2 + r^2 s_{\bar{x}}^2 - 2r s_{\bar{y}\bar{x}}\right) \\
&= \frac{(1-f)}{n\bar{x}^2}\left(s_y^2 + r^2 s_x^2 - 2r s_{yx}\right) \\
&= \frac{(1-f)(SSy_i + r^2 SSx_i - 2r\, SPy_ix_i)}{n(n-1)\bar{x}^2} \qquad (2.53)
\end{aligned}
$$

2.10 Use of random sample numbers

The first two plates of random sample numbers from *Statistical Tables for Biological, Agricultural and Medical Research* by R.A. Fisher and F. Yates are reproduced in Appendix III, Table 1. The use of these tables in selecting probability samples will be illustrated with examples. Random number tables are also provided by Kendall and Smith and by the Rand Corporation.

Notes

1. When a number of plates of random numbers is available, the student should select one plate by any manner and follow through the columns of random numbers from one exercise to the next until the end of the plates and then start again from the first plate.
2. When drawing a sample from a small number of units, such as selecting one adult from the total number of adults in a household, the "lottery method" may be followed by selecting one out of a number of plastic cards after proper shuffling.

2.11 Examples

In Appendix IV are given for thirty villages located in three states, the data on the current total population and also the current size of the households and the population obtaining during a census conducted five years ago. For some sample villages additional information on items such as the monthly income of the households will also be given. These will constitute our universe, from which samples will be drawn and the principles of sample selection and procedures of estimation of universe values illustrated.

Some universe values follow; normally, of course, not all these values will be known and will have to be estimated from a probability sample:

total number of states = 3
total number of villages = 30
total number of households = 600
total current number of persons = 3042
previous census number of persons = 2815
total area = $270km^2$
average number of households per village = 20
average number of persons per village = 101.42
average number of persons per household = 5.07

This is, of course, an artificial, hypothetical universe, as can be seen from the small number of households in a village. Examples from live data will, however, be given later.

Example 2.1

From village number 8 in zone 1 (Appendix IV), select a simple random sample of 5 households without replacement, and on the basis of the data on size and the monthly income of these 5 sample households, estimate for the 24 households in the village the total number of persons, total daily income, average household size, and average monthly income per household and per person, together with their standard errors.

Of the two plates of random numbers given in Appendix III (Table 1), we choose one plate at random, say, by tossing a coin; suppose the second plate is chosen. To obtain a simple random sample of 5 households without replacement from the present universe of 24 households in the village, we start from the top left-hand corner of the plate, and read down the first two columns (as the total number of households is of two digits).

One method of procedure is to select any random number that lies between 01 and 24, and not to consider all numbers between 25 and 99 as also 00; the five numbers thus selected would constitute our required sample of 5 households. (As selection without replacement, if a random number is repeated in the draws we would reject it, and continue drawing until five different numbers are selected.)

However, this procedure will lead to a large number of rejections of the random numbers, on an average $100 - 24 = 76$ per cent. A modified procedure is to divide all two-digit random numbers greater than 24 by 24 and take the remainder as the random number selected. Obviously, we should not consider numbers 97 to 99 and also 00 in order to give all the 24 households equal chances of selection.

The random numbers as they appear (and after division by 24 and taking the remainder, if the random number is larger than 24) are: 53 (remainder 5); 97 (rejected); 35 (remainder 11); 63 (remainder 15); 02; 98 (rejected); 64 (remainder 16).

The five households that constitute our sample are then households with serial numbers 5, 11, 15, 2, and 16. The data on the size and daily income of these 5

Table 2.7: Size and daily income of the 5 simple random samples of households selected from 24 households in village no. 8 in zone 1

Sample household no.	Household serial no.	Number of persons y_i	Total daily income in \$ x_i	y_i^2	x_i^2	$y_i x_i$
1	2	3	33	9	1089	99
2	5	4	40	16	1600	160
3	11	3	34	9	1156	102
4	15	5	68	25	4624	340
5	16	6	61	36	3721	366
Total	-	21	236	95	12190	1067
Mean	-	4.2 ($\overline{y}$)	47.2 ($\overline{x}$)			

sample households are given in Table 2.7, along with the required computations. The household size is denoted by y_i and the daily household income by x_i.

Here $N = 24$; $n = 5$; $f = n/N = 0.2083$; and $1 - f = 0.7916$. The corrected sums of squares and products are

$$SSy_i = \sum^n y_i^2 - \left(\sum^n y_i\right)^2 / n$$

$$= 95 - 21^2/5 = 95 - 88.2 = 6.8$$

$$SSx_i = \sum^n x_i^2 - \left(\sum^n x_i\right)^2 / n$$

$$= 12190 - 236^2/5 = 12190 - 11139.2 = 1050.8$$

$$SPy_ix_i = \sum^n y_ix_i - \left(\sum^n y_i\right)\left(\sum^n x_i\right) / n$$

$$= 1067 - 21 \times 236/5 = 1067 - 991.2 = 75.8$$

An unbiased estimate of the average household size is (from equation (2.45)) $\overline{y} = \sum^n y_i/n = 4.2$.

An unbiased variance estimate of $\bar{y}$ is (from equation (2.50)) $s_{\bar{y}}^2 = (1-f) \times SSy_i/n(n-1) = (0.79167 \times 6.8)/(5 \times 4) = 0.269178$, so that the estimated standard error of $\bar{y}$ is

$$s_{\bar{y}} = \sqrt{(0.269178)} = 0.519$$

and the estimated CV of $\bar{y}$ is

$$s_{\bar{y}} = 0.519/4.2 = 0.1236 \text{ or } 12.36\%$$

An unbiased estimates of the total number of persons in the village is (from equation (2.43))

$$y_0^* = N\bar{y} = 24 \times 4.2 = 100.8 \text{ or } 101$$

with an unbiased variance estimate (from equation (2.51))

$$s_{y_0^*}^2 = Ns_{\bar{y}} = N^2 s_{\bar{y}}^2 = 24^2 \times 0.269178$$

so that the estimated standard error of y_0^* is

$$s_{y_0^*} = Ns_{\bar{y}} = 24 \times 0.519 = 12.46$$

An unbiased estimate of the average daily income per household is $\bar{x} = \$47.20$, with an unbiased variance estimate $s_{\bar{x}}^2 = (1-f)SSx_i/n(n-1) = 41.5943$; estimated standard error $s_{\bar{x}} = \$6.45$; and estimated CV $s_{\bar{x}}/\bar{x} = \$6.45/\$47.20 = 0.1367$ or 13.67%.

An unbiased estimate of total daily income of all the 24 households in the village is $x_0^* = n\bar{x} = \$1132.80$ or \$1133, with an unbiased variance estimate $s_{x_0^*}^2 = 24^2 \times 4159.43$, so that the estimated standard error of x_0^* is

$$s_{x_0^*} = 24 \times \$6.45 = \$154.80$$

The estimated daily income per person is, from equation (2.48), the estimated total income divided by the estimated total number of persons, or

$$r = x_0^*/y_0^* = \sum^n x_i / \sum^n y_i = \$236/21 = \$11.24$$

The estimated variance of r is, from equation (2.53),

$$s_r^2 = (1-f)(SSx_i + r^2 SSy_i - 2r\, SPy_i x_i)/n(n-1)\bar{y}^2 = 0.46207$$

Therefore, the estimated standard error of r is $\sqrt{0.46207} = \$0.68$ and the estimated CV of r is

$$s_r/r = \$0.68/\$11.24 = 0.0605 \text{ or } 6.05\%$$

much smaller than those for the numerator and the denominator, namely, total income and total number of persons.

Note: The standard errors of the totals and the means are rather large: the standard errors could be decreased by increasing the size of the sample and/or by adopting more appropriate sampling techniques, detailed in later chapters.

Example 2.2

From the list of 600 households residing in 30 villages (Appendix IV), select a simple random sample of 20 households with replacement, and, on the basis of the data on the size of these 20 sample households, estimate for the whole universe the total number of persons the and average household size, together with their standard errors, and the 95 per cent confidence limits of the universe values.

Before the required sample can be drawn, we have to obtain a serial list of all the 600 households. The households in the 30 villages need not actually be re-numbered, but the purpose would be served by taking the cumulative totals of the number of households from the 30 villages, given in Table 2.8. Starting with the 3-digit numbers from where we had left off, and rejecting numbers between 601 and 999 as also 000, we get the following 20 acceptable random numbers:

585	348	39	84	70	18	451	433	504	226
317	366	72	101	551	538	518	359	377	29

Random number 585 refers to the first ($585 - 584 = 1$) household in the ninth village is state III; random number 348 refers to the seventh ($348 - 341 = 7$) household in the eighth village in state II; and so on. The selected sample of 20 households and the data on household size are given in Table 2.9.

Here $N = 600$ and $n = 20$. The reader should verify the following sums, sums of squares and products:

$$\sum y_i = 102 \qquad \sum y_i^2 = 594 \qquad SSy_i = 73.8$$

Therefore,

$$\overline{y} = \frac{1}{n} \sum y_i = \frac{102}{20} = 5.1$$

The estimated total number of persons in the 600 households in the universe is, from equation (2.43),

$$y_0^* = N\overline{y} = 600 \times 5.1 = 3060$$

The estimates variance of y_0^* is, from equation (2.44),

$$s_{y_0^*}^2 = 600^2 \times 73.8/380 = 600^2 \times 0.1942$$

Therefore the estimated standard error y_0^* is

$$s_{y_0^*} = 600 \times 0.4407 = 264.42$$

and the estimated CV of y_0^* is

$$s_{y_0^*}/y_0^* = 264.42/3060 = 0.0863 \text{ or } 8.63\%$$

From Appendix III, Table 2, the value of t for $n - 1 = 19$ degrees of freedom and 0.05 probability is 2.09. Therefore, from equation (2.40), the 95 per cent confidence limits of the total number of persons in the 30 villages are $3060 \pm 2.09 \times 264.42$ or 2508 and 3612.

Table 2.8: Total and cumulative total number of households in the 30 villages: data of Appendix IV

State no.	Village no.	Number of households	Cumulative number of households
I	1	17	17
	2	18	35
	3	26	61
	4	18	79
	5	24	103
	6	17	120
	7	20	140
	8	24	164
	9	24	188
	10	22	210
II	1	15	225
	2	22	247
	3	17	264
	4	19	283
	5	20	303
	6	19	322
	7	19	341
	8	23	364
	9	25	389
	10	23	412
	11	18	430
III	1	15	445
	2	21	466
	3	18	484
	4	21	505
	5	20	525
	6	21	546
	7	21	567
	8	17	584
	9	16	600

Table 2.9: Size of the srs of 20 households selected from 600 households in 30 villages

	State no. I							State no. II						State no. III						
Village no.	2	2	3	4	4	5	5	2	6	8	8	9	9	1	2	4	5	6	7	9
Household no.	1	12	4	9	11	5	22	1	14	7	18	2	13	3	6	20	13	13	5	1
Size y_i	6	4	5	4	4	4	3	9	5	5	5	4	8	7	3	4	7	8	1	6

The estimated average household size is $\overline{y} = 5.1$ and the estimated standard error of $\overline{y}$ is

$$s_{\overline{y}} = \sqrt{\frac{SSy_i}{n(n-1)}} = 0.4407$$

The 95 per cent confidence limits of the universe average household size are $5.1 \pm 2.09 \times 0.4407$ or 4.2 and 6.0.

Note: The total number and the serial numbering of households in each village will not in general be known beforehand; also, in the above sampling scheme, not all the villages will necessarily be represented. We shall see later how these limitations can be overcome by using stratified multi-sample designs.

Example 2.3

From the list of 30 villages (Appendix IV), select a simple random sample of 4 villages with replacement, and from the data on the numbers of households and persons, estimate for the universe the total numbers of persons and households, the average numbers of households and persons per village, and the average household size, together with their standard errors.

With the 30 villages in the 3 zones, it is easy to number these villages serially from 1 to 30. Starting from where we had left off, and using two-digit random numbers, we get the following: 19, 24, 18, 19, 15. In a "with replacement" sample, we would have to take the first four indicated villages, with village number 19 occurring twice; however, although the sampling fraction $4/30 = 1/7.5$ is over 10 per cent, we shall, for the sake of illustration, reject the repeated random number 19, and select the next non-repetitive random number, 15. The sample then becomes "without replacement", but we shall treat it as if it were selected with replacement. The required data for these sample villages are given in Table 2.10.

Table 2.10: Number of households and number of persons of the srs of 4 samples villages, selected from the 30 villages in Appendix IV

Serial no. i	Village serial no.	No. of households h_i	No. of persons y_i
1	19	25	127
2	24	18	105
3	18	23	114
4	15	20	105
Total	-	86	451
Mean	-	$\overline{h} = 21.5$	$\overline{y} = 112.75$

Here $N = 30$, $n = 4$. The reader should verify the following sums of squares and products: $\sum h_i^2 = 1878$; $SSh_i = 29$; $\sum y_i^2 = 51175$; $SSy_i = 324.75$; $\sum h_i y_i = 9787$; $SPh_i y_i = 90.5$.

From equations of the type (2.42) to (2.49), the estimated total number of households in the 30 villages is

$$h_0^* = N\overline{h} = 30 \times 21.5 = 645$$

The estimated variance of h_0^* is

$$s_{h_0^*}^2 = 30^2 \times 2.4167$$

so that the estimated standard error of h_0^* is

$$s_{h_0^*} = 30 \times 1.5055 = 45.16$$

and the estimated CV of h_0^* is

$$45.16/645 = 0.0700 \text{ or } 7.00\%$$

The estimated total number of persons in the 30 villages is

$$y_0^* = N\overline{y} = 30 \times 112.75 = 3382.5$$

The estimated variance of y_0^* is

$$s_{y_0^*}^2 = 30^2 \times 27.0625$$

so that the estimated standard error of y_0^* is

$$s_{y_0^*} = 30 \times 5.2022 = 156.066$$

and the estimated CV of y_0^* is 156.066/3382.5 = 0.0461 or 4.61 per cent.

The estimated average number of households per village is $\overline{h} = 21.5$, with estimated standard error of $s_{\overline{h}} = 1.5055$. The estimated average number of persons per village is $\overline{y} = 112.75$, with the estimated standard error of $s_{\overline{y}} = 5.2022$.

The estimated average household size is given by the estimated total number of persons divided by the estimated total number of households, or

$$r = y_0^*/h_0^* = 451/86 = 5.24$$

The estimated variance of r is

$$\begin{aligned} s_r^2 &= (SSy_i + r^2 SSh_i - 2r\, SPy_ih_i)/n(n-1)\overline{h}^2 \\ &= 172.5775/5547 = 0.03111, \end{aligned}$$

so that the estimated standard error of r is $s_r = 0.1706$ and the estimated CV of r is 0.1706/5.24 = 0.0326 or 3.26 per cent.

2.12 Estimation for sub-universes

Estimates of totals, means, ratios, etc. are often required for the study variables for sub-divisions of the universe or *domains of study*, e.g. for households in different occupation or income groups in household budget surveys, or for different family structures (such as nuclear and extended families) in social investigations. The general estimating equations for a simple random sample apply equally here, the only difference being that the study variable will take the sample value for units belonging to the particular sub-universe, and will be considered to have a zero value otherwise.

Thus, the sub-universe total is given by

$$Y' = \sum^{N} Y_i' \tag{2.54}$$

where $Y_i' = Y_i$ for units belonging to the sub-universe, and zero otherwise.

The sample values are y_i $(i = 1, 2, \ldots, n)$ for the study variable, and we define $y_i' = y_i$ if the sample unit belongs to the sub-universe, and $y_i' = 0$ otherwise.

Then if $y_i'^*$ is an unbiased estimator of the total of the sub-universe Y' obtained from the ith unit, a combined unbiased estimator of Y' is the arithmetic mean

$$y_0'^* = \frac{1}{n} \sum^{n} y_i'^* \tag{2.55}$$

and an unbiased estimator of the variance of $y_0'^*$ is

$$s^2_{y_0'^*} = \frac{\sum^n (y_i'^* - y_0'^*)^2}{n(n-1)} \tag{2.56}$$

The covariance and ratio of two sample estimators $y_0'^*$ and $x_0'^*$ are similarly defined, as also is the estimated variance of the ratio $y_0'^*/x_0'^*$.

From equation (2.43), the combined unbiased estimator of Y' from *a simple random sample* (with replacement) is

$$y_0'^* = \frac{N}{n} \sum^n y_i' = N\bar{y}' \tag{2.57}$$

where

$$\bar{y}' = \frac{1}{n} \sum^n y_i' \tag{2.58}$$

is the sample mean of the y_i' values.

An unbiased estimator of the variance of $y_0'^*$ is, from equation (2.44),

$$s^2_{y_0'^*} = \frac{N^2 \sum^n (y_i' - \bar{y}')^2}{n(n-1)} \tag{2.59}$$

If the number of units in the sub-universe N' is known, then an unbiased estimator of the mean of the sub-universe

$$\bar{Y}' = Y'/N' \tag{2.60}$$

is

$$\bar{y}_0'^* = y_0'^*/N' \tag{2.61}$$

An unbiased estimator of the variance of $\bar{y}_0'^*$ is

$$s^2_{\bar{y}_0'^*} = s^2_{y_0'^*}/N'^2 \tag{2.62}$$

If N' is not known, then an estimator of $\bar{Y}'$ is

$$\bar{y}_0'^* = y_0'^*/n_0'^* \tag{2.63}$$

where $n_0'^*$ is the combined unbiased estimator of N'. For a simple random sample, $n_0'^*$ is obtained on putting $y_i' = 1$ for the sample units which belong to the sub-universe, namely,

$$n_0'^* = Nn'/n \tag{2.64}$$

where n' is the number of sample units that belong to the sub-universe. From equations (2.57) and (2.64)

$$\overline{y}_0'^* = \sum^{n'} y_i/n'$$

the sample average (the divisor being n' rather than n).

An estimator of the variance of $\overline{y}'$ is

$$s^2_{\overline{y}'} = \frac{\sum^n (y_i' - \overline{y}')^2}{n'(n'-1)} \tag{2.65}$$

For an example, see Example 2.4.

2.13 Estimation of proportion of units

Estimates are often required of the total number or the proportion of the *units* that possess a certain qualitative characteristic or attribute or fall into a certain class. Thus, in a household survey, we might wish to known the proportion that live in houses they own.

If N' is the number of units in the universe possessing the attribute, then the universe proportion of the number with the attribute is

$$P = N'/N \tag{2.66}$$

where N is the total number of units in the universe. If we define our study variable so that it takes the value

$Y_i = 1$, if the unit has the attribute; and
$Y_i = 0$, otherwise,

then the universe number N' with the attribute is

$$N' = \sum_1^N Y_i$$

and the mean of the Y_is is

$$\overline{Y} = \sum^N Y_i/N = N'/N = P \tag{2.67}$$

Since Y_i can either be 1 or 0, the universe variance of Y_i is

$$\sigma^2 = \sum^N (Y_i - \overline{Y})^2/N = \sum^N Y_i^2/N - \overline{Y}^2 = P - P^2 = P(1-P) \tag{2.68}$$

If in the simple random sample of n units (drawn with replacement), r units possess the attribute, then the sample mean

$$\overline{y} = \sum^{n} y_i/n = r/n = p \tag{2.69}$$

is an unbiased estimator of the universe proportion P, where $y_i = 1$ if the sample unit has the attribute; and $y_i = 0$ otherwise.

The sampling variance of p is

$$\sigma_p^2 = \sigma_{\overline{y}}^2 = \frac{\sigma^2}{n} = \frac{P(1-P)}{n} \tag{2.70}$$

the coefficient of variation per unit is

$$\frac{\sigma}{P} = \frac{\sqrt{[P(1-P)]}}{P} = \sqrt{\frac{1-P}{P}} \tag{2.71}$$

and the CV of p is

$$\frac{\sigma_p}{P} = \frac{\sigma_{\overline{y}}}{P} = \sqrt{\frac{1-P}{nP}} \tag{2.72}$$

An unbiased estimator of σ^2 is

$$s^2 = \frac{\sum^2(y_i - \overline{y})^2}{n-1} = \frac{(np - np^2)}{n-1} = \frac{np(1-p)}{n-1} \tag{2.73}$$

and an unbiased estimator of the variance of p is

$$s_p^2 = s_{\overline{y}}^2 = \frac{s^2}{n} = \frac{p(1-p)}{n-1} \tag{2.74}$$

An estimator of the universe CV is

$$\frac{s}{p} = \sqrt{\left[\frac{(1-p)}{p} \cdot \frac{n}{(n-1)}\right]} \tag{2.75}$$

and an estimator of the CV of p is

$$\frac{s_p}{p} = \frac{s_{\overline{y}}}{p} = \sqrt{\left[\frac{1-p}{p(n-1)}\right]} \tag{2.76}$$

An unbiased estimator of N', the number of units in the universe with the attribute, is

$$n'^{*} = Np \tag{2.77}$$

Table 2.11: Size of sample households with TV satellite dishes in the srs of 20 households selected in Example 2.2

	Zone no. I		II		III
Village no.	2	5	6	8	6
Household no.	1	5	14	18	13
Size y_i	6	4	5	5	8

and unbiased estimator of whose variance is

$$s^2_{n'.} = N^2 s^2_p = N^2 p(1-p)/(n-1) \tag{2.78}$$

In large samples, the 100(1−α) per cent probability limits of the universe proportion P are provided by

$$p \pm t'_\alpha s_p \tag{2.79}$$

where t'_α is the α percentage point of the normal distribution.

Note: The results of this section could be derived from those of the preceding section by setting $y'_i = 1$, if the sample unit has the attribute, and zero otherwise. Theoretical proofs are given in Appendix II, section A2.3.5.

Example 2.4

In the sample of 20 households selected from 600 households given in Example 2.2, the data on the households which were found to have TV satellite dishes are given in Table 2.11. Estimate the proportion and the total number of households having TV satellite dishes and the total number of persons in these households, together with their standard errors.

Here $N = 600$, $n = 20$, and r (the sample number of households possessing the attribute) = 5. From equation (2.69), an unbiased estimates of the universe proportion of households with TV satellite dishes is $p = r/n = 5/20 = 0.25$, and an unbiased estimate of the variance of p is, from equation (2.74),

$$s^2_p = p(1-p)/(n-1) = 0.25 \times -.75/19 = 0.00986842$$

so that the standard error of p is $s_p = 0.09934$.

The unbiased estimate of the number of households with TV satellite dishes is, from equation (2.77),

$$n'^{*} = Np = 600 \times 0.25 = 150$$

An unbiased estimator of the variance of n'^{*} is, from equation (2.78)

$$s_{n'^{*}} = Ns_p = 600 \times 0.09934 = 39.74$$

The estimated number of persons in these households is obtained from equation (2.57) on putting $y'_i = y_i$ for households with TV satellite dishes, and $y_i = 0$ for other households in the sample. Then

$$y_0'^{*} = N \sum y'_i/n = 600 \times 28/20 = 840$$

An unbiased estimate of the variance of $y_0'^{*}$ is obtained similarly from equation (2.59),

$$s^2_{y_0'^{*}} = 600^2 = 0.333684$$

so that an estimated standard error of $y_0'^{*}$ is $600 \times 0.5776 = 346.56$.

Example 2.5

In the sample of 20 households selected from the 600 households (example 2.2), the numbers of males and females are given in Table 2.12. Estimate the universe proportion of males to total population and its standard error.

The procedure of section 2.13 will be inappropriate in this case, for the sample persons were not selected directly out of the universe of persons. However, we shall illustrate both the appropriate and the inappropriate procedures.

Assuming that the 102 sample persons were selected directly from the universe of persons, we apply the procedure of section 2.13, noting that $n = 102$, and r (the number of males) = 52. Hence from equation (2.69) the estimated proportion of males in the universe is $p = r/n = 52/102 = 0.5098$.

From equation (2.74) the estimated variance of the sample proportion p is $s_p^2 = p(1-p)/(n-1) = 0.00206545$, so that $s_p = 0.04545$.

Table 2.12: Numbers of males and females in the srs of 20 households selected in Example 2.2

																					Total
Male	3	2	3	2	2	2	1	5	3	2	2	2	5	4	2	2	4	3	1	2	52
Female	3	2	2	2	2	2	2	4	2	3	3	2	3	3	1	2	3	5	0	4	50
Total	6	4	5	4	4	4	3	9	5	5	5	4	8	7	3	4	7	8	1	6	102

The above is an inappropriate procedure. The appropriate procedure is that of section 2.9, and the estimated proportion of males in the universe is equal to the estimated total number of males divided by the estimated total number of persons. Note that $N = 600$ households and $n = 20$ households. In Example 2.2, we have already obtained the following for the estimated total number of persons:

$$y_0^* = N\overline{y} = 600 \times 5.1 = 3060$$

with corrected sum of squares

$$SSy_i = 73.8$$

Denoting by x_i the number of males in the ith sample household, we have

$$\overline{x} = \sum x_i/n = 52/50 = 2.6 \qquad SSx_i = 24.8 \qquad SPy_ix_i = 37.8.$$

The estimated total number of males is, from equation (2.43),

$$x_0^* = N\overline{x} = 600 \times 2.6 = 1560$$

and the estimated universe proportion of males is, from equation (2.48),

$$r = x_0^*/y_0^* = 1560/3060 = 0.5098$$

as in the inappropriate procedure.

The estimated variance of r is, from equation (2.49),

$$s_r^2 = (SSx_i + r^2 SSy_i - 2r\,SPy_ix_i)/n(n-1)\overline{y}^2 = 0.0005391082$$

so that $s_r = 0.02322$, considerably different from the value of s_p obtained by the inappropriate procedure.

2.14 Method of random groups for estimating variances and covariances

A short-cut method of estimating universe variances and covariances is to group the total of n sample units (selected with replacement) into k groups of m units each ($n = km$), and to compute estimates of the variances and covariances from the group means. The results follow from the fundamental theorems of section 2.7, but will be illustrated for a simple random sample (with replacement).

Let y_{jl} be the value of the study variable for the lth unit ($l = 1, 2, \ldots, m$) in the jth group ($j = 1, 2, \ldots, k$) and the mean for the jth group

$$\overline{y}_j = \frac{1}{m}\sum_{l=1}^{m} y_{jl} \tag{2.80}$$

The mean of all the n sample units is, as before,

$$\bar{y} = \frac{1}{n}\sum_{i=1}^{n} y_i = \frac{1}{km}\sum_{j=1}^{k}\sum_{l=1}^{m} y_{jl} = \frac{1}{k}\sum_{j=1}^{k}\bar{y}_j \qquad (2.81)$$

i.e. the mean of the group means.

The overall mean $\bar{y}$ is an unbiased estimator of the universe mean $\overline{Y}$ with variance σ^2/n, so also the group means $\bar{y}_j$ are each an unbiased estimator $\overline{Y}$ but with variance σ^2/m.

An unbiased estimator of $\sigma^2_{\bar{y}}$ is

$$s'^2_{\bar{y}} = \sum^{k}(\bar{y}_j - \bar{y})^2/k(k-1) \qquad (2.82)$$

and an unbiased estimator of the universe variance σ^2 is

$$s'^2 = ns'^2_{\bar{y}} = m\sum^{k}(\bar{y}_j - \bar{y})^2/k(k-1) \qquad (2.83)$$

For two study variables an unbiased estimator of the covariance of $\bar{y}$ and $\bar{x}$ is

$$s'_{\bar{y}\bar{x}} = \sum^{k}(\bar{y}_j - \bar{y})(\bar{x}_j - \bar{x})/k(k-1) \qquad (2.84)$$

Note: The variance and covariance estimators $s'^2_{\bar{y}}$ and $s'_{\bar{y}\bar{x}}$ are based on $(k-1)$ degrees of freedom, and are, therefore, less efficient than the variance and covariance estimators, obtained respectively from equations (2.44) and (2.47) based on $(n-1)$ degrees of freedom.

Example 2.6

The simple random sample of 20 households selected (with replacement) from the universe of 600 households in Example 2.2 is arranged in 4 random groups of 5 households each: the random groups were formed in order of selection of the sample households. The results are given in Table 2.13. Obtain estimates of the variance of the sample mean. How would the result differ if there were 2 groups of 10 households each?

Here $n = 20$, $k = 4$, $m = 5$. An unbiased estimate of the variance of $\bar{y}$ is, from equation (2.82), $s'^2_{\bar{y}} = 0.43$ so that the estimated standard error is $s'_{\bar{y}} = 0.656$.

If the 20 households were formed into 2 groups of 10 households each (taking again the households in the order of selection), then the two group means are

$$\bar{y}_1 = (24 + 29)/10 = 5.3 \qquad \bar{y}_2 = (17 + 32)/10 = 4.9$$

Table 2.13: Random groups of sample households: data of Example 2.2

Random group no.	Total number of persons $\sum_{l=1}^{4} y_{jl}$	Average size	$\bar{y}_j - \bar{y}$	$(\bar{y}_j - \bar{y})^2$
1	24	4.8	−0.3	0.09
2	29	5.8	+0.7	0.47
3	17	3.4	−1.7	2.89
4	32	6.4	+1.3	1.69
All groups	102	5.1	0	5.16

The estimated standard error is, from equation (2.82),

$$s'_y = \frac{1}{2}|\bar{y}_1 - \bar{y}_2| = 0.5$$

Further reading

Chaudhuri and Stenger, chapter 1; Cochran, chapters 2 and 3; Deming (1950), chapter 4; Hansen *et al.* (1953), Vols. I and II, chapter 4; Hedayat and Sinha, chapters 1 and 2; Kish (1965), chapter 2; Murthy (1967), chapter 2; Pathak (1988); Sukhatme *et al.* (1984), chapter 2; Yates, sections 6.1-6.4, 6.9, 7.1-7.4 and 7.8.

Exercises

1. A simple random sample of 10 agricultural plots drawn without replacement from 100 plots in a village gave the following areas (in acres) under a certain crop:

   2.4, 3.2, 2.9, 4.6, 1.9, 2.8, 3.1, 1.8, 3.6, 2.8

   Compute unbiased estimates of the mean and total area under the crop for the village, along with their standard errors and coefficients of variation.

2. From a list of 75,308 (= N) farms in a province, a simple random sample of 2072 (= n) farms was drawn. Data for the number of cattle for the sample were:

   $$\sum y_i = 25,883; \qquad \sum y_i^2 = 599,488.$$

Ignoring the finite sampling correction, estimate from the sample (a) the total number of cattle in the province and (b) the average number of cattle per farm, along with their standard errors and the 95% confidence limits. (United Nations (1972), *Manual*, Example 1).

3. From village number 8 in zone 1 (Appendix IV) draw a simple random sample of 5 households without replacement. Obtain estimates of the same characteristics (other than income) for the village as a whole as in Example 2.1, and compare the results.
4. From the list of 30 villages (Appendix IV) draw an srs of 4 villages with replacement and obtain estimates of the same universe characteristics as in Example 2.3. Compare your results with those in Example 2.3.
5. In a simple random sample of 50 households drawn with replacement from a total of 250 households in a village, only 8 were found to possess transistor radios. These households had respectively 3, 5, 3, 4, 7, 4, 4, and 5 members. Estimate the total number of households that possess transistor radios and the total number of persons in these households, along with their standard errors (Chakravarti, Laha, and Roy, Example 3.1, modified to "with replacement" scheme).
6. A survey was organized to determine the incidence of HIV (human immunodeficiency virus) seroconversion among first-time blood donors at a blood bank in a large city. During a particular month, 180 first-time donors gave blood there, of whom 175 were tested seronegative: a simple random sample of 60 persons was selected without replacement out of the seronegative first time donors and were asked to give blood again after a year, when 9 were found to be seropositive. Estimate the incidence of HIV seroconversion and its standard error (Levy and Lemeshaw, Exercise 7.10, adapted).
7. Show that in a simple random sample, unbiased estimators of the universe mean and universe total of a study variable have the same coefficient of variation, and that the CV of the estimated number of units possessing a certain attribute is the same as that of the estimated proportion of such units in the universe.
8. Table 2.14 gives the number of persons belonging to 43 *kraals* which form a random sample of 325 *kraals* in the Mondora Reserve in Southern Rhodesia (now Zimbabwe) and also the numbers of persons absent from these *kraals*.

    (a) Estimate (i) the total number of persons belonging to the reserve, (ii) the number absent from the reserve, and (iii) the proportion of persons absent, and their standard errors.

    (b) What would be the estimated standard error of the proportion of persons absent had the sample been a sample of individuals selected directly from all the persons belonging to the reserve? (Yates, Examples 6.9b and 7.8b, modified to "with replacement" sampling for computation of variance; note the differences in notations in Yates' book and this book).

Table 2.14: Total number of persons including absentees), y, and number of absentees, x; data from srs of 43 *kraals*

y	x	y	x	y	x	y	x
95	18	89	7	75	12	159	36
79	14	57	9	69	16	54	26
30	6	132	26	63	9	69	27
45	3	47	7	83	14	61	2
28	5	43	17	124	25	164	69
142	15	116	24	31	3	132	41
125	18	65	16	96	45	82	10
81	9	103	18	42	25	33	8
43	12	52	16	85	35	86	22
53	4	67	27	91	28	51	19
148	31	64	12	73	13		
					Total	3427	799

Source: Yates, Examples 6.9b and 7.8b (adapted).

The following corrected sums of squares and products are given:

$$SSy_i = 55,199.0; \; SSx_i = 7218.5; \; SPy_ix_i = 13,286.5.$$

9. Table 2.15 shows the number of persons ($\overline{y}_i$), the weekly family income (x_i), and the weekly expenditure on food (w_i) in a simple random sample of 33 families selected from 660 families. Estimate from the sample (a) the total number of persons, (b) the total weekly family income, (c) the total weekly expenditure on food, (d) the average number of persons per family, (e) the average weekly expenditure on food per family, (f) the average weekly expenditure on food per person, (g) the average weekly income per person, and (h) the proportion of income that is spent on food, along with their standard errors (adapted from Cochran, Example, pp. 33-34).

   The following raw sums of squares and products are given: $\sum y_i^2 = 533$; $\sum x_i^2 = 177,524$; $\sum w_i^2 = 28,224$; $\sum y_i x_i = 8889$; $\sum y_i w_i = 3595.3$; $\sum x_i w_i = 66,678.0$.

Table 2.15: Size, weekly income, and food cost of the srs of 33 families selected from 660 families

Family number i	Size y_i	Income x_i	Food cost w_i	Family number i	Size y_i	Income x_i	Food cost w_i
1	2	\$62	\$14.3	18	4	\$83	\$36.0
2	3	62	20.8	19	2	85	20.6
3	3	87	22.7	20	4	73	27.7
4	5	65	30.5	21	2	66	25.9
5	4	58	41.2	22	5	58	23.3
6	7	92	28.2	23	3	77	39.8
7	2	88	24.2	24	4	69	16.8
8	4	79	30.0	25	7	65	37.8
9	2	83	24.2	26	3	77	34.8
10	5	62	44.4	27	3	69	28.7
11	3	63	13.4	28	6	95	63.0
12	6	62	19.8	29	2	77	19.5
13	4	60	29.4	30	2	69	21.6
14	4	75	27.1	31	6	69	18.2
15	2	90	22.2	32	4	67	20.1
16	5	75	37.7	33	2	63	20.7
17	3	69	22.6				
				Total	123	\$2394	\$907.2

Source: Cochran, Example pp. 33-34.

CHAPTER 3

Ratio and Regression Estimators

3.1 Introduction

If ancillary information is available for each of the units of the universe, it can, under suitable conditions, be used in several ways to improve the efficiency of the estimators of the study variable. One way, to be described in this chapter, is to use the ancillary information, ***after sample selection***, in providing what are known as ***ratio estimators*** and ***regression estimators***. The other, described in Chapter 5, is to use the ancillary information ***in sample selection*** and also in estimation. The two could be used in combination for two ancillary variables.

3.2 Ratio method of estimation

From a simple random sample (with replacement) of n units out of the universe of N units, an unbiased estimator of the universe total Y has been seen to be

$$y_0^* = \frac{1}{n}\sum^n y_i^* = \frac{N}{n}\sum^n y_i = N\overline{y} \tag{3.1}$$

If, however, the values of the ancillary variable z are available for all the units of the universe, then

$$Z = \sum^N z_i \tag{3.2}$$

the universe total for the ancillary variable is also known, and this information can be utilized to improve the efficiency of the estimators relating to the study variable.

An unbiased estimator of the universe total Z is first obtained *from the sample*, namely,

$$z_0^* = \frac{1}{n}\sum^n z_i^* = \frac{N}{n}\sum^n z_i = N\bar{z} \text{ in srs} \tag{3.3}$$

As an estimator of the universe ratio

$$R_{Y/Z} = Y/Z \tag{3.4}$$

the ratio of the estimators of the universe totals of the study and the ancillary variables is taken,

$$r_{y/z} = y_0^*/z_0^* = \sum^n y_i \Big/ \sum^n z_i = \bar{y}/\bar{z} \text{ in srs} \tag{3.5}$$

As $Y = Z \cdot Y/Z = Z \cdot R_{Y/Z}$, the ratio estimator of Y is defined as

$$y_R^* = Z \cdot r_{y/z} = Z \cdot y_0^*/z_0^* = Z \cdot \bar{y}/\bar{z} \text{ in srs} \tag{3.6}$$

(the subscript 'R' in y_R^* standing for 'ratio').

The sampling variance of $r_{y/z}$ is, from equation (2.22),

$$\sigma^2_{r_{y/z}} = (\sigma^2_{y_0^*} + R^2_{Y/Z}\sigma^2_{z_0^*} - 2R_{Y/Z}\sigma_{y_0^* z_0^*})/Z^2$$

A variance estimator of $r_{y/z}$ is, from equation (2.49),

$$\begin{aligned} s^2_{r_{y/z}} &= (s^2_{y_0^*} + r^2_{y/z}s^2_{z_0^*} - 2r_{y/z}s_{y_0^* z_0^*})/Z^2 && (3.7)\\ &= N^2(SSy_i + r^2_{y/z}SSz_i - 2rSPy_iz_i)/n(n-1)Z^2 \\ & \qquad \text{in srswr} && (3.8)\end{aligned}$$

as the value of Z is known.

The variance of y_R^* is

$$\sigma^2_{y_R^*} = Z^2\sigma^2_{r_{y/z}} \tag{3.9}$$

an estimator of which is

$$\begin{aligned} s^2_{y_R^*} &= Z^2 s^2_{r_{y/z}} \\ &= (s^2_{y_0^*} + r^2_{y/z}s^2_{z_0^*} - 2r_{y/z}s_{y_0^* z_0^*}) && (3.10)\\ &= N^2(SSy_i + r^2_{y/z}SSz_i - 2r_{y/z}SPy_iz_i)/n(n-1) \\ & \qquad \text{in srswr} && (3.11)\end{aligned}$$

The following points should be noted:

1. Estimators of the type y_R^*, equation (3.6), are known as *ratio estimators*, obtained by the ratio method of estimation. Such ratio estimators should be distinguished from the *estimators of ratios* (of the type of equations (2.33), (2.48), or (3.5)), where both the numerator and the denominator are estimated from the sample. The two issues are related, but different.

2. The ratio estimator y_R^* is subject to the same observations as those for the estimation of a ratio (note 3, section 2.7). The ratio estimator y_R^* is biased generally: it becomes unbiased if the linear regression of y_0^* and z_0^* passes through the origin $(0,0)$. An estimator of the bias (this estimator is also generally biased) is

$$(r_{y/z}s^2_{z_0^*} - s_{y_0^* z_0^*})/Z \tag{3.12}$$

3. The variance estimator of y_R^* is valid for large sample generally, say when $n \geq 30$.

4. *Correction for bias.* If the sample is drawn in the form of k independent sub-samples, each of the same size m (so that $n = mk$) and giving unbiased estimators y_j^* and z_j^* $(j = 1, 2, \ldots, m)$ of the universe totals Y and Z respectively, two estimators of the universe ratio Y/Z are

$$r = y_0^*/z_0^* \tag{3.13}$$

and

$$r' = \sum^m r_j/m \tag{3.14}$$

where $y_0^* = \sum^m y_j^*/m$; $z_0^* = \sum^m z_j^*/m$; $r_j = y_j^*/z_j^*$.
Noting that in large samples the bias of the estimator r' is m times that of the estimator r, an approximately unbiased estimator of the bias of r is

$$(r' - r)/(m - 1)$$

and an almost unbiased estimator of the universe ratio (unbiased up to the second order of approximation) is

$$r'' = (mr - r')/(m - 1) \tag{3.15}$$

Thus an almost unbiased ratio estimator (Quenouille-Murthy-Nanjamma estimator) of Y is

$$Z \cdot r'' \tag{3.16}$$

Again, with the selected sample in the form of m independent and interpenetrating sub-samples, a ratio estimator of Y, corrected for bias is (vide T.J. Rao) is

$$y_R^{**} = Y_R^* + (y_0^* - r'z_0^*)/(m - 1) \tag{3.17}$$

where

$$y_R^* = Z \cdot r_{y/x} \tag{3.18}$$

For a simple random sample of n units out of the universe of N units, the ratio estimator of the universe total Y, corrected for bias, is

$$\begin{aligned} y_R^{**} &= y_R^* + (y_0^* - \bar{r}z_0^*)/(n-1) \\ &= y_R^* + N(\bar{y} - \overline{rz})/(n-1) \end{aligned} \tag{3.19}$$

where

$$\bar{r} = \sum^n r_i/n = \sum^n (y_i/z_i)/n \tag{3.20}$$

There are other methods of debiasing the ratio estimators.

5. In obtaining the ratio of two universe totals Y and X, the use of ratio estimators of the two totals is equivalent of using the unbiased estimators of the totals, for

$$r_{y/x} = y_R^*/x_R^* = Zr_{y/z}/Zr_{x/z} = y_0^*/x_0^* \tag{3.21}$$

We shall, however see later in Parts II-IV that in stratified and multistage sampling, ratio estimators may be applied at different levels of aggregation.

6. In a large srs, the ratio estimator would be more efficient than the simple unbiased estimator if the correlation coefficient between the study variable and the ancillary variable is greater than 1/2(CV of the ancillary variable/CV of the study variable). For example, if the ancillary variable represents the value of the study variable at some time past (such as the population in the previous census while estimating the current population), the two CVs could be taken as about equal and the ratio estimator would be superior if the correlation coefficient exceeds 1/2. On the other hand, suppose that the CV of the ancillary variable is double that of the study variable: the ratio estimator would then be less precise, for the correlation coefficient cannot exceed 1.

7. With a negative correlation between the study and the ancillary variables, the ratio method of estimation will be extremely inefficient: in that case, another method, the product method of estimation, may be used (see, Cochran, section 6.21; Murthy (1967), section 10.8; Singh and Chaudhary, section 6.10; and Sukhatme *et al.* (1984), section 5.11).

Example 3.1

For the same data as for Example 2.3, where a simple random sample of 4 villages was taken from 30 villages, use the information given in Table 3.1 on the

Table 3.1: Previous census population of the srs of 4 villages in Examples 2.3

Village serial no.	19	24	18	15
Previous census population z_i	122	97	102	98

population data of a census, conducted five years previously, to obtain ratio estimates of the total numbers of households and persons in the 30 villages. The total population of the 30 villages in the previous census was 2815.

Here $N = 30$; $n = 4$; $Z = 2815$. The reader should verify the following: $\sum^n z_i = 419$; $\bar{z} = 104.75$; $\sum^n z_i^2 = 44,301$; $\sum^n y_i z_i = 47,597$; $\sum^n h_i z_i = 9102$; and the corrected sums of squares and products: $SSz_i = 410.75$; $SPy_i z_i = 354.75$; $SPh_i z_i = 93.50$. The other required computations have already been made for Example 2.3.

From equation (3.5), $r_{y/z} = \sum^n y_i / \sum^n z_i = 451/419 = 1.076372$, so that the ratio estimate of the current total population, using the previous census population, is, from equation (3.6),

$$y_R^* = Z \cdot r_{y/z} = 2815 \times 1.076372 = 3029.99 \text{ or } 3030$$

The estimated variance of y_R^* is, from equation (3.11), $30^2 \times 3.079132$, so that the estimated standard error of Y_R^* is $30 \times 1.755 = 52.650$, and the estimated CV of Y_R^* is 52.65/3030 or 1.74 per cent.

Similarly, the ratio estimate of the total number of households is $h_R^* = Z \cdot r_{h/z} = 2815 \times 0.205251 = 577.78$ or 578, with estimated standard error of 24.345, and estimated CV of 4.21 per cent.

Note the following points:

1. The estimated average household size, using the ratio estimates, is the ratio estimate of the current total population divided by the ratio estimate of the total number of households, i.e. $3030 \div 578 = 5.24$, the same figures as obtained by using simple unbiased estimates of totals in Example 2.3 (note 5 of section 3.2).

2. The ratio estimates of the total numbers of persons and households have smaller standard errors than the unbiased estimates (in Example 2.3) and the ratio estimates are closer to the universe values.

3. *Correction for bias.* To obtain ratio estimates of totals corrected for bias, we need the additional computations shown in Table 3.2.

   The ratio estimate of the total current population, corrected for bias, is, from equation (3.19),

$$y_R^{**} = y_R^* + N(\bar{y} - \bar{r}_{y/z}\bar{z})/(n-1) = 3048.71 \text{ or } 3049.$$

Table 3.2: Computation of the ratio estimated, corrected for bias: data of Example 3.1

	Village serial no.				
	19	24	18	15	Average
Ratio h_i/z_i	0.204918	0.185567	0.225467	0.204082	0.2050085
Ratio y_i/z_i	1.040984	1.082474	1.039110	1.071429	1.0584992

Similarly, the ratio estimate of the total number of households, corrected for bias, is

$$h_R^{**} = h_R^* + N(\overline{h} - \overline{r}_{h/z}\overline{z})/(n-1) = 578.034 \text{ or } 578.$$

Here the correction is marginal.

3.3 Regression method of estimation

Another method of using the ancillary information at the estimating stage which is more general than the ratio method is the regression method of estimation.

The estimated linear regression coefficient β^* for the regression of y_0^* on z_0^* is

$$\begin{aligned} \beta^* &= SPy_i^*z_i^*/SSz_i^* & (3.22) \\ &= SPy_iz_i/SSz_i \text{ in srswr} & (3.23) \end{aligned}$$

The regressions estimator of the universe total Y is defined as

$$y_{Reg}^* = y_0^* + \beta^*(Z - z_0^*) \quad (3.24)$$

where Z is the known universe total of the ancillary variable.

The variance of y_{Reg}^* in large samples is

$$\sigma_{y_{Reg}^*}^2 = \frac{N^2}{n}\sigma^2(1-\rho^2) \quad (3.25)$$

where ρ is the correlation coefficient between the variables y and z.

An estimator of the variance of y_{Reg}^*, valid in large samples, is

$$s_{y_{Reg}^*}^2 = \frac{\sum^n[(y_i^* - y_0^*) - \beta^*(z_i^* - z_0^*)]^2}{n(n-2)} \quad (3.26)$$

$$= \frac{N^2[SSy_i - (SPy_iz_i)^2/SSz_i]}{n(n-2)} \quad \text{in srswr} \qquad (3.27)$$

Notes

1. The regression method of estimation is more general than the ratio method: the two become equivalent only when the (linear) regression of y on z passes through the origin (0,0). The ratio method is, however, simpler to compute and is preferred when the regression line is expected to pass through the origin, i.e. when y is expected to be proportional to z.
2. Although the regression method is more general than the ratio method, it is not much used in practice in large-scale surveys for two reasons: first, the computation of the estimates is more complex, and secondly, the gain in efficiency is not very marked in many cases as the regression line passes either through the origin or very close to it.
3. When the regression of y on z is perfectly linear, i.e. when the correlation coefficient $\rho = 1$, the variance of the regression estimator becomes zero; and when y and z are uncorrelated, the variance is the same as that of the unbiased estimator.
4. In some textbooks, in the formula for the estimated variance of the regression estimator, the divisor $n(n-1)$ is used instead of $n(n-2)$. Although in large samples the difference in these two estimators is negligible, the later divisor has been suggested here, as in the regression method two estimators - the y-intercept on z and the (linear) regression coefficient of y on z - are to be computed from the sample, leading to the loss of two degrees of freedom; it is also standard in regression theory and is known to give an unbiased estimator of the error variance in the universe regression equation if the universe is infinite and the regression is linear (Cochran, section 7.3).
5. In practice, the estimated variance of the regression estimator, namely, $s^2_{y^*_{Reg}}$ can be greater than that of the ratio estimator, namely, $s^2_{y^*_{Reg}}$; in fact, the greatest value the ratio $s^2_{y^*_{Reg}}/s^2_{y^*_{Reg}}$ can take is $(n-1)/(n-2)$, when the y-intercept on z happens to equal zero (that is, the regression line of y on z passes through the origin), so that $\beta^* = \sum^n y_i / \sum^n z_i$, and the numerators of the variance formulae (3.11) and (3.27) are identical.
6. If the regression of y on z is non-linear, the regression estimator is subject to a bias of the order $1/n$, so that the ratio of the bias to the standard error is small for large sample. The bias is equal to $-$covariance (β^*, z_0^*).

Example 3.2

For the same data as for Example 3.1, obtain the regression estimates of the total numbers of households and persons, using the data on the previous census population, along with their standard errors. Here $N = 30$; $n = 4$; $y_0^* = 3382.5$; $h_0^* = 645$; $z_0^* = 30 \times 104.75 = 3142.5$; $Z = 2815$; $SSy_i = 324.75$; $SSh_i = 29$; $SSz_i = 410.75$; $SPy_iz_i = 354.75$; $SPh_iz_i = 93.5$.

The estimated (linear) regressions coefficient of y on z is, from equation (3.23),

$$\beta^* = SPy_iz_i/SSz_i = 0.863664$$

The regression estimate of the current total population, is from equation (3.24),

$$\begin{aligned} y^*_{Reg} &= y^*_0 + \beta^*(Z - z^*_0) \\ &= 3382.5 + 0.863664(2815 - 3142.5) \\ &= 3099.65 \end{aligned}$$

or 3100, and the estimated variance of y^*_{Reg} is, from equation (3.27), $30^2 \times 2.295774$, so that the estimated standard error of y^*_{Reg} is $30 \times 1.515 = 45.45$, with the estimated CV of 1.47 per cent, somewhat less than that for the unbiased estimate (4.61 per cent).

Similarly, the regression estimate of the total number of households, using the previous census population, is 570 with an estimated standard error 29.46 (and an estimated CV of 5.16 per cent). Note that in this case the estimated CV though less than that for the unbiased estimate (7.00 per cent), is larger than that for the ratio estimate (4.21 per cent).

Further reading

Cochran, chapters 6 and 7; Deming (1950), chapter 4; Foreman, chapter 4; Hansen *et al.* (1953), vol. I, section 4C and 11.2, and vol. II, sections 4.11-4.19; Hedayat and Sinha, chapter 6; Kish (1965), chapter 2; Murthy (1967), chapters 10 and 11; Rao, P.S.R.S. (1988); Singh and Chaudhary, chapters 6 and 7; Sukhatme *et al.* (1984), chapters 5 and 6; Thompson, S.K., chapters 7 and 8; Yates, sections 6.8, 6.9, 6.12, 7.8 and 7.12.

Exercises

1. A simple random sample of 2055 ($= n$) farms, drawn from the universe of 75,308 ($= N$) farms to obtain the total number of cattle, gave the following data:

   Sample total number of cattle,
   $$\sum^n y_i = 25,751$$
   Sample total area of the 2055 farms,
   $$\sum^n z_i = 62,989 \text{ acres}$$
   Actual total area of the 75,308 farms,
   $$Z = 2,353,365 \text{ acres}$$

   The corrected sums of squares and products were

   $$\sum^n y_i^2 = 596,737; \quad \sum^n z_i^2 = 2,937,851; \quad \sum^n y_iz_i = 1,146,391.$$

   Obtain the ratio estimate of the total number of cattle, along with the standard error, and compare the results of the unbiased estimate (United Nations *Manual*, Example 7.i).

2. For the same sample as in exercise 1, additional information obtained from a census was available on the number of cattle (w) five years ago in the 75,308 farms.

   Sample total number of cattle in the 2055 farms five years ago, $\sum^n w_i = 23,642$; $\sum^n w_i^2 = 504,150$; $\sum^n y_i w_i = 499,172$.

   Actual total number of cattle in the 75,308 farms five years ago, $W = 882,610$.

   Obtain the ratio estimate of the total number of cattle and its standard error (United Nations *Manual*, Example 7.ii).

3. For the same data as for exercise 1, obtain the regression estimate of the total number of cattle and its standard error (United Nations *Manual*, Example 9.i).

4. For the same data as for exercise 3.3, obtain the regression estimate of the total number of cattle and its standard error (United Nations *Manual*, Example 9.ii).

5. For the same sample as for exercise 4, Chapter 2, use the previous census population figures, given in Appendix IV, to obtain ratio estimates of the universe characteristics, with the standard errors.

6. Using the same data as for exercise 5, obtain regression estimates of the universe characteristics, with the standard errors.

CHAPTER 4

Systematic Sampling

4.1 Introduction

One operationally convenient method of selecting a one in k sample from a list of units is to select first a random number between 1 and k and then select the unit with this serial number and every kth unit afterwards: thus, to take a 5 per cent sample of households during a population census, one would first choose a random number between 1 and 20 (here the sampling fraction is 0.05, so $k = 1/0.05 = 20$); if the random number (also called the *random start*) is 12, then households numbered 12, 32, 52, 72, 92, 112, and so on will constitute a 5 per cent systematic sample. This procedure is known as systematic sampling and ensures that each universe unit has the same chance of being included in the sample: the constant k is known as the *sampling interval* and is generally taken as the integer nearest to N/n, the inverse of the sampling fraction. This method has several advantages. First, it is operationally convenient, especially when, as in multi-stage sample designs, information on the lower stage units, especially the ultimate and the penultimate stages (such as households or families or farms) is not available at the central office and the enumerators have to list these units and to draw samples from them. Second, N, the total number of universe units, need not be known beforehand and a systematic sample may be selected along with the listing of the universe units or with a census, if the sampling fraction is fixed beforehand. Third, a systematic sample is spread out more evenly over the universe, so that it is likely to produce a sample that is more representative and more efficient than a simple random sample.

There are two main disadvantages of systematic sampling. First, variance estimators cannot be obtained from a single systematic sample; and second, a bad arrangement of the units may produce a very inefficient sample.

Systematic sampling often suggests itself when there is a sequence of units occurring naturally in space (trees in a forest) or time (landing of fishing crafts on the coast).

In this chapter, we deal with the selection and estimating procedure of systematic samples.

4.2 Linear and circular systematic sampling

To illustrate, let us consider the universe of six households, a, b, c, d, e and f, with respective sizes of 8, 6, 3, 5, 4, and 4. The *linear systematic sample* of 2 households from this universe consists of one of the following: ad, be, and cf, with the respective average household sizes of 6.5, 5.0, and 3.5. The average of the sample averages is 5 (the universe average), showing that the sample mean is an unbiased estimator of the universe mean, and so also for the totals.

Linear systematic sampling has one limitation when the sampling interval is not an integer. In the above example, if the universe had consisted of 5 households a, b, c, d, and e, the sampling interval 5/2 is not an integer, and we can take it either as 2 or 3. Taking it as 2, the possible samples are ace, and bd; and taking it as 3, the possible samples are ad, be, and c. Thus the sample size does not remain fixed in linear systematic sampling when the sampling interval is not an integer. Although unbiased estimators of totals and averages could still be obtained in such a case by using modified formulae, a more satisfactory and simpler procedure is that of *circular systematic sampling*, by selecting first a random number between 1 and N and then selecting this and every kth unit (where k is the integer nearest to N/n) in a cyclical manner until n sample units are selected. In the above example of sampling 2 households systematically from 5 households, the possible circular systematic samples with 2 as the sampling interval are ac, bd, ce, da, and eb; with 3 as the sampling interval, the possible circular systematic samples are ad, be, ca, db, and ec (Figure 4.1). The sample size thus remains constant. The usual estimators of totals and averages would also be unbiased. The circular systematic sample reduces to linear systematic sampling when N/n is an integer: it is thus more general, and is to be preferred.

Note that in the first example, unlike the selection of 2 units by simple random sampling without replacement, the combinations ab, ac, bc, bd, bf, cd, ce, de, df, and ef do not occur in systematic sampling.

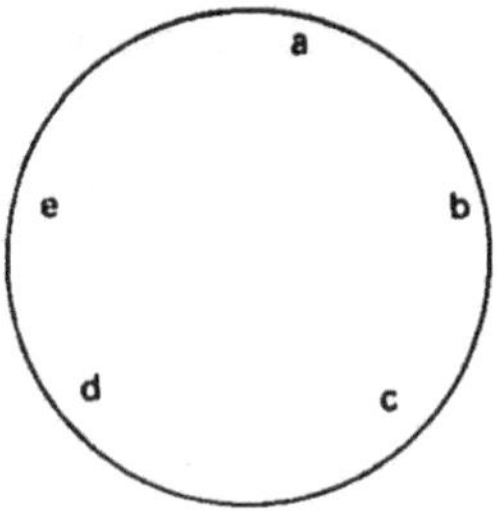

Figure 4.1: Diagrammatic representation of circular systematic sampling.

4.3 Sample mean and variance

4.3.1 Mean and variance estimators

From a universe comprising $N = nk$ units, it is proposed to select a systematic sample of 1 in k. The universe units are shown schematically in Table 4.1, with k columns and n rows, the columns 1 to k showing the random starts (or clusters), each column containing n units. A random number is chosen between 1 and k, and the set of n units in the selected column constitutes one systematic sample. Or in other words, every one of the k columns representing the k possible systematic samples has an equal chance ($= 1/k$) of being selected.

Let y_{rj} denote the value of the unit with the serial number $r + (j-1)k$ in the universe ($r = 1, 2, \ldots, k$; $j = 1, 2, \ldots, n$). If r is the chosen random number, then the sample mean is

$$\overline{y}_r = \frac{1}{n} \sum_{j=1}^{n} y_{rj} \tag{4.1}$$

and the universe mean is

$$\overline{Y} = \frac{1}{nk} \sum_{r=1}^{k} \sum_{j=1}^{n} y_{rj} = \frac{1}{k} \sum^{k} \overline{y}_r \tag{4.2}$$

i.e. the mean of the sample means.

It can be shown that the sample mean $\overline{y}_r$ is an unbiased estimator of the universe mean $\overline{Y}$.

Table 4.1: Serial numbering of the universe units showing the systematic samples of n units from N $(= nk)$ units

Row number	Random start (cluster number) 1	2	...	r	...	k
1	1	2	...	r	...	k
2	$1+k$	$2+k$	...	$r+k$	...	$2k$
.	.	.	.	.	.	.
.	.	.	.	.	.	.
.	.	.	.	.	.	.
j	$1+(j-1)k$	$2+(j-1)k$	...	$r+(j-1)k$	...	jk
.	.	.	.	.	.	.
.	.	.	.	.	.	.
.	.	.	.	.	.	.
n	$1+(n-1)k$	$2+(n-1)k$	...	$r+(n-1)k$	...	nk

Noting that

$$\begin{aligned}\sigma^2 &= \frac{1}{nk}\sum_{r=1}^{k}\sum_{j=1}^{n}(y_{rj}-\overline{Y})^2\\ &= \frac{1}{k}\sum_{r=1}^{k}(\overline{y}_r-\overline{Y})^2+\frac{1}{nk}\sum_{r=1}^{k}\sum_{j=1}^{n}(y_{rj}-\overline{y}_r)^2 \qquad (4.3)\\ &= \sigma_b^2+\sigma_w^2 \qquad (4.4)\end{aligned}$$

where σ_b^2 is the *between-sample variance* and σ_w^2 is the *within-sample variance*, defined respectively by the two terms on the right-hand side of equation (4.3), then the sampling variance of the sample mean is

$$\sigma_b^2 = \sigma^2 - \sigma_w^2 \qquad (4.5)$$

The sampling variance of the sample mean can also be expressed in terms of the intraclass correlation coefficient between pairs of sample units in a column

$$\rho_c = \frac{\sum_{r=1}^{k}\sum_{j'\neq j=1}^{n}(y_{rj}-\overline{Y})(y_{rj'}-\overline{Y})}{kn(n-1)\sigma^2} \qquad (4.6)$$

so that

$$\sigma_b^2 = \frac{\sigma^2}{n}\,[1+(n-1)\rho_c] \qquad (4.7)$$

Since $\sigma^2 \geq \sigma_b^2$, we note that ρ_c must lie between $-1/(n-1)$ and 1.

4.3.2 Comparison of systematic and simple random sampling

As the sampling variance of the sample mean in srs is σ^2/n in sampling with replacement and $[(N-n)/(N-1)]\sigma^2/n$ in sampling without replacement, the relative efficiency of systematic sampling compared to srs with replacement is

$$\frac{1}{1+(n-1)\rho_c} \tag{4.8}$$

and compared to srs without replacement is

$$\frac{N-n}{N-1} \cdot \frac{1}{1+(n-1)\rho_c} \tag{4.9}$$

4.3.3 Discussion of results

Expression (4.5) shows that the sampling variance of the sample mean in systematic sampling will be reduced if the within-sample variance is increased: this condition will be satisfied when the units within each systematic sample are as heterogeneous as possible with respect to study variable. This is also seen from Expression (4.7), for σ_b^2 will be zero when ρ_c takes the minimum value of $-1/(n-1)$, i.e. when there obtains the highest degree of negative correlation between pairs of units within each systematic sample. When $\rho_c = -1/(N-1)$, srswor and systematic sampling will bet equally efficient; systematic sampling will be more or less efficient than srswor according as ρ_c is less than or greater than $-1/(N-1)$. But when $\rho_c = 1$, systematic sampling will be most inefficient, for then one unit will provide as much information as n units.

In practice, it is difficult to know what value ρ_c will take in units occurring in a natural sequence. When the units can be re-arranged before sampling such that units homogeneous in respect of the study variable (or an ancillary variable that has a high degree of association with the study variable) are put together this will ensure heterogeneity within each systematic sample. Such a re-arrangement need not be physical in the sense that units that were neighboring before the re-arrangement need not remain so afterwards.

Thus, in the socio-economic inquiry of rural households in the Indian National Sample Survey (1964-65), in a sample village the households were classified according to size, namely, (I) 1-4 persons, and (II) 5 and above, and according to the means of livelihood, namely, (1) agricultural labor, (2) self-employed in agriculture, and (3) others. They were then re-arranged according to the following combinations of the household size and means of livelihood: I-3; I-2; I-1; II-1; II-2; and II-3. This ensured that small and large households with different means of livelihood were proportionately represented in a systematic sample of households.

4.3.4 Periodicity of units

If a universe has periodic variations, then a systematic sample whose interval coincides with the length of the period will be extremely inefficient, but if the interval is made to be an odd multiple of half the period, a very efficient systematic sample will result. If, for example, sales in markets are high on Saturdays and low on Tuesdays, a systematic sample of one in six working days that includes only Saturdays or Tuesdays will be very inefficient. In such cases, instead of selecting a sample random sample, it would be better to take advantage of the knowledge by appropriately choosing the interval in systematic selection so as to break step with periodicity by including both the high and the low phases.

4.4 Estimation of variance

In general, it is not possible to estimate from one sample the variance of the estimators obtained from the systematic sample. (This is because the procedure does not ensure the inclusion of each of the ${}^{N}C_2$ pairs of units of the universe at least in one of the samples, which is a necessary condition for the existence of an unbiased estimator of a variance.) If, however, two or more systematic samples are taken, each with a separate random start, the combined estimator and its variance are obtained from the fundamental theorems of section 2.7.

In practice, moreover, the variance estimators for the simple random sampling (without replacement) may be used: these estimators will be unbiased only if the arrangement of the units is at random.

Example 4.1

(a) From village number 8 in state I (Appendix IV), select a circular systematic sample of 5 households and for the 24 households in the village obtain estimates of the total number of persons, total monthly income and average daily income per person; also obtain estimates of standard errors of these estimates, assuming the data came from a simple random sample.

(b) Draw with a fresh random start a second circular systematic sample of 5 households from the village, and obtain estimates with standard errors for the same characteristics.

(c) From the two circular systematic samples, obtain combined estimates of the characteristics along with their standard errors.

Table 4.2: Size and daily income of the circular systematic sample of 5 households from 24 households in village no. 8 and state I (Appendix IV)

Household serial no.	5	10	14	19	24
Number of persons (y_i)	4	5	4	1	4
Daily income (\$) (x_i)	40	52	39	20	42

(a) The random start (i.e. the random number between 1 and 24) chosen was 14; the sampling interval is 24/5 or 5 to the nearest integer, so that the five sample households are those with serial numbers 14, $(14+5)=19$, $(19+5)=24$, 5 and $(5+5)=10$. The data for these five households are given in Table 4.2.

Here $N=24$ and $n=5$ so that $f=n/N=0.2083$. Hence we have $\sum^n y_i = 18$; $\bar{y}=\sum^n y_i/n = 18/5 = 3.6$; $\sum^n x_i = \$193$; $\bar{x}=\sum^n x_i/n = \$193/5 = \$38.6$; $\sum^n y_i^2 = 74$; $SSy_i = 9.2$; $\sum^n x_i^2 = 7989$: $SSx_i = 539.2$; $\sum^n y_i x_i = 764$; $SPy_ix_i = 69.2$.

The estimated total number of persons in the village is given by

$$y_0^* = N\bar{y} = 24 \times 3.6 = 86.4 \text{ or } 86$$

The estimated variance of y_0^* is, from equation (2.51),

$$\begin{aligned} s^2_{y_0^*} &= (1-f)N^2 SSy_i/n(n-1) = 0.79167 \times 24^2 \times 9.2/20 \\ &= 0.79167 \times 24^2 \times 0.46 = 24^2 \times 0.364168 \end{aligned}$$

Therefore, an estimated standard error of y_0^* is

$$s_{y_0^*} = 24 \times 0.603 = 14.472;$$

and an estimated CV of y_0^* is

$$s_{y_0^*}/y_0^* = 16.75 \text{ per cent.}$$

The estimated total daily income of all the 24 households in the village is

$$x_0^* = N\bar{x} = 24 \times \$38.6 = \$926.40 \text{ or } \$926$$

and estimated variance of x_0^* is

$$\begin{aligned} s^2_{x_0^*} &= (1-f)N^2 SSx_i/n(n-1) = 0.79167 \times 24^2 \times 26.95 \\ &= 24^2 \times 21.335506 \end{aligned}$$

Therefore, an estimated standard error of x_0^* is

$$s_{x_0^*} = 24 \times \$4.619 = \$110.86$$

Table 4.3: Size and daily income of the second circular systematic sample of 5 households selected from 24 households in village no. 8 in state I (Appendix IV)

Household serial no.	3	8	13	18	23
Number of persons (y_i)	3	6	6	5	4
Daily income ($) ($x_i$)	32	62	65	53	47

and an estimated CV of x_0^* is 11.97 per cent.

The estimated average daily income per person is given by the estimated total monthly income divided by the estimated total number of persons, i.e.

$$r = x_0^*/y_0^* = \$38.6/3.6 = \$10.72$$

An estimated variance of r is, from equation (2.53),

$$s_r^2 = \frac{(1-f)(SSx_i + r^2 SSy_i - 2rSPx_iy_i)}{n(n-1)\bar{y}^2} = 0.153267$$

Therefore, an estimated standard error of r is s_r = \$0.392 and an estimated CV of r is 3.66 per cent.

(b) The second circular systematic sample is drawn by selecting another random start, 51/24, remainder 3; so the selected households are numbered 3, 8, 13, 18, and 23. The data for these households are given in Table 4.3.

The reader should verify the following results: y_0^* = 115.2 persons; $s_{y_0^*}$ = 12.45 persons; x_0^* = \$1243.20; $s_{x_0^*}$ = \$125.86; r = \$10.79; s_r = \$0.20.

(c) From the fundamental theorem of section 2.7, the combined estimate of total number of persons is $\frac{1}{2}(86.4 + 115.2) = 100.8$ or 101 persons, with an estimated standard error of, from equation (2.8), $\frac{1}{2}|86.4 - 115.2| = 14.4$ and an estimated CV of 14.29 per cent.

The combined estimate of the total daily income is

$$\frac{1}{2}(\$926.4 + \$1243.2) = \$1084.8$$

with an estimated error

$$\frac{1}{2}|\$926.4 - \$1243.2| = \$158.4$$

and an estimated CV of 14.60 per cent.

The combined estimate of average daily income per person is

$$\$1084.8/100.8 = \$10.76$$

an estimated variance of which is

$$\frac{[(926.4-1243.2)^2+(10.76)^2(86.4-115.2)^2 \\ -10.76(926.4-1243.2)(86.4-115.2)]}{[4\times(100.8)^2]}=0.1176$$

so that an estimated standard error of \$0.333 and the estimated CV is 3.09 per cent.

Further reading

Bellhouse (1988b); Cochran, chapter 8; Hansen *et al.* (1953), vol. I, section 11.8; Kish (1965), chapter 4; Murthy (1967), chapter 5; Sukhatme *et al.* (1984), chapter 10; Yates, sections 3.6, 6.20, and 7.18.

Exercise

1. Out of 24 villages in an area, two linear systematic samples of 4 villages each were selected. The total area under wheat in each of these sample villages is given in Table 4.4. Estimate the total area under what in the area (Murthy (1967), Problem 7.3, adapted).

Table 4.4: Area under wheat of the two linear systematic samples of 4 villages each from 24 villages in an area

Linear systematic sample	Sample village			
	1	2	3	4
1	427	326	481	445
2	335	412	503	348

Source: Murthy (1967), Problem 7.3 (adapted).

CHAPTER 5

Varying Probability Sampling: Sampling with Probability Proportional to Size

5.1 Introduction

In simple random and systematic sampling, the only information required on the universe units prior to sampling is the serial listing. For systematic sampling, although it was indicated that a re-arrangement of the universe units according to the values of an ancillary variable might provide more efficient estimators, only the relative magnitudes rather than the actual values of the ancillary variable are required.

It would seem reasonable to suppose that if the values of an ancillary variable related to the study variable were known for all the N units, the information could be utilized in selecting the sample so as to provide estimators with greater efficiency than those from simple random or systematic sampling. In contrast to the sampling procedures considered so far that assign the same probability of selection to all the units of the universe, another procedure will now be outlined that utilizes the values of the ancillary variable such that unequal probabilities of selection are assigned to the universe units; the estimating procedure adopted ensures unbiased estimators of totals and averages with much greater efficiency, under favorable conditions, than those from simple random or systematic sampling.

In this chapter we shall deal with the most common type of sampling with varying (or unequal) probabilities, namely, ***sampling with probability proportional to "size"*** (pps), the "size" being the value of the ancillary variable. We shall often refer to a sample so selected as a pps sample, and, unless otherwise specified shall deal with pps sampling with replacement.

Table 5.1: Selection of one household with probability proportional to size from a hypothetical universe of six households in Table 2.1

Household no. i	Household identification	Household size y_i	Probability of selection $\pi_i = y_i / \sum_i^N y_i$
1	a	8	8/30
2	b	6	6/30
3	c	3	3/30
4	d	5	5/30
5	e	4	4/30
6	f	4	4/30
Total	-	30	1

5.2 Selection of one unit with probability proportional to size

Consider the universe of 6 households with respective household sizes given in Table 5.1. Let us draw one household from this universe with probability proportional to household size in order to estimate the total size (total number of persons): if, of course, the sizes of the households were known, we would not need to draw a sample to estimate the total size, but this will illustrate the logic of pps sampling. If households are selected with pps, the probability of selection of any household is its size divided by the total size of the universe (30); these probabilities are shown in the last column of Table 5.1. Note that the total of the probabilities is 1, as it should be.

In simple random sampling, the probability of selection of a unit at any draw is $1/N$, where N is the total number of universe units, and the unbiased estimator of the universe total Y for the study variable is obtained from any draw by multiplying the value of the unit drawn (y_i) by the total number of universe units (N), or in other words, on dividing y_i by the probability of selection ($= 1/N$, a constant in srs) of the unit i. Similarly, in varying probability sampling, the unbiased estimator of Y is obtained from the ith draw on dividing the value of the ith unit (y_i) by the probability of selection π_i (which varies from unit to unit in the universe); thus the unbiased estimator of the universe total Y in varying probability sampling is

$$y_i^* = y_i/\pi_i \tag{5.1}$$

If the values y_i of all the universe units were known before sampling and sampling is carried out with probability proportional to y_i, i.e.

$$\pi_i = y_i/\sum^N y_i \tag{5.2}$$

then the unbiased estimator y_i^* given in estimating equation (5.1) becomes

$$y_i^* = y_i/\pi_i = \sum^N y_i = Y$$

the universe total. In our example, if one draw of pps sampling gives household c, the probability of selection of which is $\frac{3}{30}$, then the unbiased estimator of the total universe size from this sampled household is $3 \div \frac{3}{30} = 30$, the actual universe total.

If, instead of drawing a unit with probability proportional to its actual value, we had drawn it with probability proportional to an ancillary variable, whose size (z_i) is related to the unit value (y_i) by the exact relation

$$z_i = \beta y_i$$

where β is a positive constant, the probability of selection

$$ppz_i = z_i/\sum^N z_i = \beta y_i/\beta\sum^N y_i = y_i/\sum^N y_i = ppy_i$$

remains the same, and would give the same results as probability proportional to the value of the study variable, i.e. there would be no sampling error.

The foregoing gives the clue to the determining factor for selection with pps. We cannot, of course, know the actual "sizes" of the study variable, but if we can find an ancillary variable, the values of which are known to be roughly proportional to the values of the study variable, then we may select the units with probability proportional to the values of the ancillary variable in order to obtain estimators with greater efficiency than those obtained from srs. the ancillary variable chosen should be such that its values are known prior to sampling and the two are linearly related with the regression line passing through the origin (0,0). If there is a perfect positive correlation between the study and the ancillary variables but the regression line does not pass through the origin (0,0), sampling with pps of the ancillary variable will not necessarily be more efficient than srs.

Some examples of study and ancillary variables are as follows:

Study variable	*Ancillary variable*
Current population	Previous census population
Current number of births	Geographical area (less suitable)
Current total income	Previous census population
Area under a crop	Previous census population
Factory production	Total geographical area or cultivated area
	Number of workers

5.3 Estimating procedures

If we draw a sample of n units ($n \geq 2$) with replacement out of the universe of N units, with the initial probability of selection of the ith unit ($i = 1, 2, \ldots, N$)

$$\pi_i = z_i / \sum^N z_i = z_i / Z \tag{5.2}$$

where z_i is the value ("size") of an ancillary variable, the ith selected sample unit ($i = 1, 2, \ldots, n$) having the value of the study variable y_i will provide an unbiased estimator of the universe total Y by the estimating equation

$$y_i^* = \frac{y_i}{\pi_i} = Z\,\frac{y_i}{z_i} \tag{5.1}$$

and where $Z = \sum^N z_i$.

By the fundamental theorem of section 2.7, a combined unbiased estimator of Y is

$$y_0^* = \frac{1}{n}\sum^n y_i^* = \frac{Z}{n}\sum^n \frac{y_i}{z_i} \tag{5.3}$$

The sampling variance of y_0^* is

$$\sigma^2_{y_0^*} = \frac{1}{n}\sum^N \left(\frac{Y_i}{\pi_i} - Y\right)^2 \pi_i = \frac{1}{N}\left(\sum^N \frac{Y_i^2}{\pi_i} - Y^2\right) \tag{5.4}$$

an unbiased estimator of which is

$$s^2_{y_0^*} = \sum^n (y_i^* - y_0^*)^2 / n(n-1) \tag{5.5}$$

An unbiased estimator of the universe mean $\overline{Y} = Y/N$ is

$$\overline{Y_0^*} = y_0^*/N \tag{5.6}$$

with an unbiased variance estimator

$$s^2_{\bar{y}_0^*} = s^2_{y_0^*}/N^2 \tag{5.7}$$

From the same sample, estimators of the total and the mean of another study variable may be computed from the estimating formulae of the above types. An unbiased estimator of the covariance of the estimators y_0^* and x_0^* of the two universe totals Y and X is

$$s_{y_0^* x_0^*} = \sum^n (y_i^* - y_0^*)(x_i^* - x_0^*)/n(n-1) \tag{5.8}$$

A consistent, but generally biased, estimator of the universe ratio $R = Y/X$ of two universe totals Y and X is the ratio of the sample estimators

$$r = y_0^*/x_0^* \tag{5.9}$$

and an estimator of the variance of r is

$$\begin{aligned} s_r^2 &= (s^2_{y_0^*} + r^2 s^2_{x_0^*} - 2r s_{y_0^* x_0^*})/x_0^{*2} \\ &= (SSy_i^* + r^2 SSx_i^* - 2rSPy_i^* x_i^*)/n(n-1)x_0^{*2} \end{aligned} \tag{5.10}$$

Notes

1. Theoretical proofs are given in Appendix II, section A2.3.7.
2. Estimator y_0^* (equation 5.3) was introduced by Hansen and Hurwitz (1943) and often referred to as "the **Hansen-Hurwitz estimator.**"
3. pps sampling can be made without replacement, but the estimating formulae are rather complicated. Two relatively simple estimators are, however, mentioned for the case of two sample units ($n = 2$).

   (a) *Ordered (Raj) estimator.* Suppose the two units selected (in order) with pps and without replacement have the respective values y_1 and y_2 and the probabilities of selection π_1 and π_2. Then an unbiased ordered estimator of the universe total Y is

   $$\frac{1}{2}\left[(1+\pi_1)\,\frac{y_1}{\pi_1} + (1-\pi_1)\,\frac{y_2}{\pi_2}\right]$$

   with an unbiased variance estimator

   $$\frac{1}{4}(1-\pi_1)\left(\frac{y_1}{\pi_1} - \frac{y_2}{\pi_2}\right)^2$$

   (b) *Unordered (Murthy) estimator.* If y_1 and y_2 are the values of the two sample units selected with pps and without replacement with probabilities of section π_1 and π_2 respectively, then an unbiased unordered estimator of Y is

   $$\frac{1}{2-\pi_1-\pi_2}\left[(1-\pi_2)\,\frac{y_1}{\pi_1} + (1-\pi_1)\,\frac{y_2}{\pi_2}\right]$$

with an unbiased variance estimator

$$\frac{(1-\pi_1)(1-\pi_2)(1-\pi_1-\pi_2)}{(2-\pi_1-\pi_2)^2}\left[\frac{y_1}{\pi_1}-\frac{y_2}{\pi_2}\right]^2$$

With the advent of electronic computers and the use of sample designs with two sample first-stage units in each stratum (after the strata have been formed in a desirable manner, to be detailed later), the unordered estimator is being employed increasingly. For detailed methods of ppswor, see the references under "*Further reading*" at the end of the chapter and consult a sampling statistician.

The unordered estimator is more efficient than the ordered, and both are more efficient than the unbiased pps "with replacement" estimator.

4. Note that the selection of the sample units with probability proportional to "size" is equivalent to the selection with probability proportional to the ratios of the sizes to their (a) total or (b) average. For in the latter cases, the selection probabilities are

$$\text{(a) } \pi_i' = \frac{z_i/Z}{\sum^N (z_i/Z)} = \frac{z_i}{Z} = \pi_i \qquad \text{(b) } \pi_i'' = \frac{z_i/\bar{Z}}{\sum^N (z_i/\bar{Z})} = \frac{z_i}{Z} = \pi_i$$

Thus in pps sampling it is not necessary to know the "sizes" if the ratios of these sizes to their total or average are known.

5. pps sampling can be made with a suitable function of the value of ancillary variable. For example, in surveys on fruit count, the selection of branches with probability proportional to the fourth power of the firth may be more efficient than ppg^3, ppg^2, or ppg ("g" indicating girth) or srs (Murthy (1967), Section 6.6); in a stratified two-stage design, selection with probability proportional to the square root of the number of ssu's will be reasonably close to the optimum and more efficient than probability proportional to the number of ssu's or srs if the costs vary substantially with both the number of fsu's and the number of ssu's per fsu (Hansen *et al.* (1953), vol. I, section 8.14, vol. II, section 8.3).

5.4 Procedures for the selection of sample units with pps

5.4.1 Selection from a list

For selection of sample units with pps (and replacement) from a list, two methods are generally used.

(a) *Cumulative method.* This method entails cumulation of the sizes of the units in the universe.

$$z_1;\ z_1+z_2;\ z_1+z_2+z_3;\ \ldots,;\ \sum^i z_i;\ \ldots;\ \sum^N z_i = Z$$

Random numbers are then drawn between 1 and Z. If the random number is greater than $\sum^{i-1} z_i$, but less than or equal to $\sum^{i} z_i$, then the ith unit is selected. The procedure is continued till the required number n of sample units have been drawn.

(b) *Selection of a pair of random numbers (Lahiri's method).* The other method does not need cumulation of the sizes of the universe units. A pair of random numbers is chosen, the first between 1 and N, and the other between 1 and z_{max}, where z_{max} is the maximum value of z_i, obtained on inspection. If for any pair of random numbers chosen, the first number is i and the second number is $\leq z_i$, then the ith unit is selected; otherwise it is rejected, i.e. no selection is made, but a fresh pair of random numbers chosen. This is continued until the required total number of sample units n have been drawn. To minimize the number of rejections, a very large z_{max} may be split up into more than one part, and the original unit selected whenever one of the split units is drawn. Both these methods of procedure will be illustrated with examples.

5.4.2 Selection from a map

A third procedure is available for selection of geographical area units from a map with probability proportional to area (ppa). Figure 5.1 of the map of 16 fields in a village gives an example. A pair of random numbers is chosen, the first between 1 and the length of the village (in our example, 12, in a certain unit) and the second between 1 and the breadth of the village (say 9). The selected pair of random numbers fixes a points on the map, and the field on which it falls is selected; if the point falls outside the village area, it is, of course, rejected. For example, let the pair of random numbers be 06 and 86 (remainder after division of 86 by 9 is 5); this point (6,5) is plotted on the map and is seen to fall in field number 6, which is selected. The procedure ensures selection with probability proportional to area, and does not require the values of the areas for selection, if a map is available; for estimation of the universe values, however, the areas of the selected fields and also the total area should be known.

5.4.3 pps systematic sampling

In this method of selection, the cumulative sizes $\sum^{i} z_i$ are first obtained ($i = 1, 2, \ldots, N$ for the universe). If n is the sample size, the sampling interval I is the integer nearest to Z/n. If r is the number chosen at random from 1 to I, the units corresponding to the numbers $(r + jI)$,

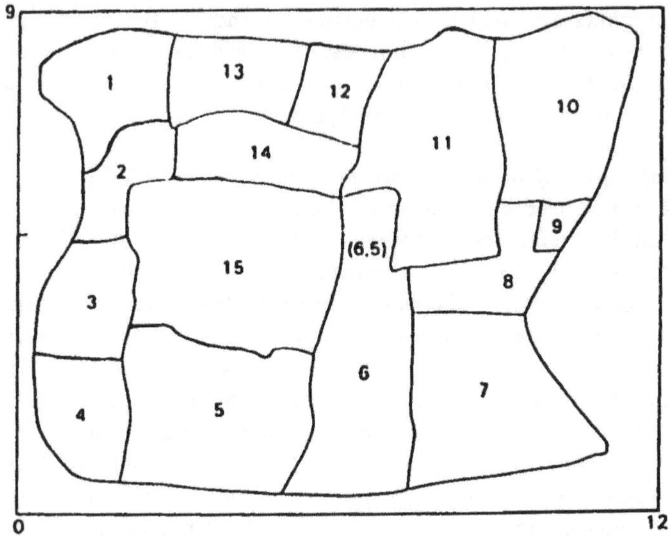

Figure 5.1: Selection of a farm with probability proportional to area.

$j = 0, 1, 2, \ldots, (n-1)$ are selected. The ith universe unit will be selected if

$$\sum^{i-1} z_i < r + jI \leq \sum^{i} z_i$$

for some value of j between 0 and $(n-1)$. If Z/n is not an integer, a pps circular systematic sample can be obtained by selecting a random start from 1 to Z and then proceeding cyclically with the integer nearest to Z/n as the interval.

The estimator for the universe total Y has the same form as that for ppswr sampling, and is unbiased, but as in simple systematic sampling (Chapter 4), a single pps systematic sample cannot provide an unbiased variance estimator. The variance can be estimated unbiasedly by selecting two or more pps systematic sample, each with a separate random start. This method will provide more efficient estimators than simple pps sampling when the units are arranged in ascending or descending order of y_i/z_i; in practice, the y_i values will not be known and those for a previous period or of a related variable will have to be used.

Example 5.1

From village number 8 in state I (Appendix IV), given the sizes of the 24 households in the village, select 5 households with probability proportional to size (and

Table 5.2: Cumulative household sizes and selection of households with probability proportional to size: village no. 8 in state I (Appendix IV)

Serial no.	Size	Cumulative size	Serial no.	Size	Cumulative size
1	5	5	13	6	63
2	3	8	14	4	67
3	3	11	15	5	72
4	7	18	16	6	78
5	4	22	17	3	81
6	4	26	18	5	86
7	6	32	19	1	87
8	6	38	20	3	99
9	4	42	21	5	95
10	5	47	22	6	101
11	3	50	23	4	105
12	7	57	24	4	109

with replacement) and from the data on total monthly income and food cost of these 5 households, obtain for the whole village estimates of the total daily income, the total daily food cost, and the proportion of income spent on food with their standard errors.

In this example, we shall illustrate the selection of the pps sample by cumulating the sizes. The cumulative sizes of the 24 households are shown in Table 5.2. As the total size of the 24 households is 109, we choose random numbers between 1 and 981 (the highest three digit number which is a multiple of 109); if the three-digit random number is greater than 109, it is divided by 109, and the remainder taken. The first three-digit random number is 213, which leaves a remainder of 104 when divided by 109; from Table 5.2, this random number is seen to be greater than the cumulative size up to the 22nd household (101), but less than the cumulative size up to the 23rd household (105): therefore the 23rd household is selected.

The procedure of selection of five sample households with pps is shown in Table 5.3.

The data on the selected five households and the required computations are shown in Table 5.4.

Here $N = 24$; $n = 5$. The probability of selection of a household is

$$\pi_i = y_i / \sum^{N} y_i = y_i/109$$

Table 5.3: pps sample of 5 households: data of Table 5.2

Random number		
Three-digit	Remainder after division by 109	Serial number of selected household
213	104	23
290	72	15
953	81	17
908	36	8
464	28	7

where y_i is the known size of the ith household. The corrected sums of squares and products are:

$$SSx_i^* = 109^2 \times 2.451704$$
$$SSw_i^* = 109^2 \times 3.7$$
$$SPx_i^*w_i^* = 109^2 \times 1.91822$$

An unbiased estimate of the total daily income in the 24 households is, from estimating equation (5.3),

$$x_0^* = \frac{1}{n}\sum x_i^* = \frac{1}{5} \times 109 \times 55.5167$$
$$= 109 \times 11.1033 = \$1210.26$$

An unbiased estimate of the variance of x_0^* is, from equation (5.5),

$$s_{x_0^*}^2 = SSx_i^*/n(n-1) = 109^2 \times 2.451704/20$$
$$= 109^2 \times 0.1225852$$

Therefore the estimated standard error of x_0^* is

$$s_{x_0^*} = 109 \times 0.34012 = \$38.16$$

and the estimated CV of x_0^* is 3.15 per cent.

Similarly, the estimated total daily food cost is

$$w_0^* = \sum^n w_i^*/n = 109 \times 5.4 = \$588.6$$

with estimated variance

$$s_{w_0^*}^2 = 109^2 \times 0.185$$

Table 5.4: Daily total income and food cost of 5 sample households selected with probability proportional to size in village no. 8, state I (Appendix IV) and computation of estimates

Household serial no.	Household sample no. i	Size y_i	Selection probability $\pi_i = y_i/\sum_1^N y_i$	Total daily: Income x_i	Total daily: Food cost w_i	$x_i^* = x_i/\pi_i$	$w_i^* = w_i/\pi_i$	x_i^{*2}	w_i^{*2}	$x_i^* w_i^*$
						109x	109x	109^2 x	109^2 x	109^2 x
7	1	6	6/109	$61	$27	10.1667	4.5	103.3611	20.25	37.7500
8	2	6	6/109	62	30	10.3333	5.0	106.7778	25.00	51.6667
15	3	5	5/109	58	25	11.6000	5.0	134.5600	25.00	58.0000
17	4	3	3/109	35	21	11.6667	7.0	136.1111	49.00	81.6667
23	5	4	4/109	47	22	11.7500	5.5	138.0625	30.25	64.6250
Total						109x 55.5167	109x 27.0	109^2 x 648.8725	109^2 x 149.50	109^2 x 301.7084
Mean						109x 11.1033 (x_0^*)	109x 5.4 (w_0^*)			

i.e. with estimated standard error

$$s_{w_0^*} = 109 \times 0.43012 = \$46.88$$

and estimated *CV* of 7.96 per cent.

The estimated proportion of income spent on food is, from equation (5.9),

$$\begin{aligned} r &= w_0^*/x_0^* = \$588.6/\$1210.26 \\ &= 0.4863 \text{ or } 48.63 \text{ per cent.} \end{aligned}$$

The estimated variance, s_r^2, of r is, from equation (5.10),

$$\frac{109^2(3.7 + 0.4863^2 \times 2.451704 - 2 \times 0.4863 \times 1.91822)}{109^2 \times 11.1033^2 \times 20}$$
$$= 0.120704/123.283271 = 0.0009791$$

so that the estimated standard error of r is $s_r = 0.03129$ and the estimated *CV* of r is 6.43 per cent.

Example 5.2

For the thirty villages listed in Appendix IV, the population data obtained from a census conducted fives years previously are available and given in Table 5.5. Select four samples villages with probability proportional to their previous census population, and on the basis of the current population and number of households in these four sample villages, obtain for the thirty villages estimates of the current total population, number of households, and average household size, along with their standard errors.

We shall follow the second procedure for selecting the pps sample which does not require cumulation of the sizes. As there are 30 villages and the maximum previous census population in any village is 122, we take five-digit random numbers, the first two digits referring to the village serial number, and the last three digits to the village census population. The first five-digit number is 06 733; the last three digits, divided by 122, leaves a remainder of 1, which is less than the census population of village serial number 6, namely, 65, so this village is selected. The second five-digit random number is 65 511; the first two-digits, on division by 30, leaves a remainder of 5, and the last three digits on division by 122, leaves a remainder of 23; which is less than the census population of the village serial number 5, namely 92, so this village too is selected. The third five-digit random number of 01 932, the last three digits, on division by 122, leaves a remainder of 78, which is greater than the census population of village serial number 1, namely, 69, so this random number is rejected. We continue in this manner until four villages have been selected, as shown in Table 5.6.

Note: We shall illustrate with this example the procedure of selection of a pps systematic sample. As the interval $I = Z/n = 2815/4 = 703.75$ is not an integer, we first select a random start between 1 and Z (i.e. 2815). Let this be 1938;

Table 5.5: Previous census population of the 30 villages: data of Appendix IV

State no.	Village serial no.	Continuous serial no.	Previous census population	State no.	Village serial no.	Continuous serial no.	Previous census population
I	1	1	69	II	6	16	84
	2	2	81		7	17	85
	3	3	110		8	18	102
	4	4	80		9	19	122
	5	5	92		10	20	102
	6	6	65		11	21	86
	7	7	72	III	1	22	78
	8	8	108		2	23	112
	9	9	106		3	24	97
	10	10	80		4	25	117
II	1	11	72		5	26	106
	2	12	102		6	27	115
	3	13	73		7	28	110
	4	14	84		8	29	104
	5	15	98		9	30	103
						Total	2815

Table 5.6: pps sample of 4 villages: data of Table 5.5

Random number for* Village	Random number for* Census population	Actual census population	Accept/reject
06	733 (R 1)	65	Accept
65 (R 5)	511 (R 23)	92	Accept
01	932 (R 78)	69	Reject
71 (R 11)	508 (R 20)	80	Accept
48 (R 18)	222 (R 100)	102	Accept

* R indicates remainder after division by 30 and 12 respectively.

then our random numbers are 1938; 1938 + 704 = 2642; (1938 + 2 × 704 =) 3346 − 2815 = 531; and 531 + 704 = 1235. From Table 5.5 the reader may verify that the cumulative previous census population is 1873 up to village no. 11 in state II and 1951 up to village no. 1 in state III; the random number 1938 therefore corresponds to the latter village, which is selected. Similarly the other three sample villages would be village no. 8 in state III, village no. 7 in state I, and village no. 5 in state II.

The data on the four sample villages and the required computations are shown in Table 5.7: the computational procedures are somewhat different from those in Example 5.1, and are generally to be preferred.

Here $N = 30$, $n = 4$, $Z = \sum_1^N z_i = 2815$. The corrected sums of squares and products are $SSy_i^* = 88945.57$; $SSh_i^* = 16714.07$; and $SPy_i^*h_i^* = 37524.44$.

An unbiased estimate of the present total population in the thirty villages is, from equation (5.3),

$$y_0^* = \frac{1}{n}\sum y_i^* = 3239.33 \text{ or } 3239$$

an unbiased estimate of variance of which is, from equation (5.5),

$$s_{y_0^*}^2 = SSy_i^*/n(n-1) = 7412.1308$$

so that the estimated standard error of y_0^* is

$$s_{y_0^*} = 86.09$$

and the estimated CV is 2.66 per cent.

Similarly, an unbiased estimate of the total number of households in the thirty villages is

$$h_0^* = \frac{1}{n}\sum h_i^* = 672.95 \text{ or } 673$$

an unbiased estimate of the variance of which is $s_{h_0^*}^2 = 1392.8392$, so that the estimated standard error of h_0^* is $s_{h_0^*} = 37.32$ and the estimated CV is 5.55 per cent.

The estimated average household size is, from equation (5.9),

$$\begin{aligned} r = y_0^*/h_0^* &= 3239.33/672.95 \\ &= 4.8136 \text{ or } 4.81 \end{aligned}$$

As an unbiased estimated of the covariance of y_0^* and h_0^* is

$$s_{y_0^*h_0^*} = SPy_i^*h_i^*/n(n-1) = 3127.0367,$$

the estimated variance of r is, from equation (5.10),

$$s_r^2 = (s_{y_0^*}^2 + r^2 s_{h_0^*}^2 - 2r s_{y_0^*h_0^*})/h_0^{*2} = 0.0211556$$

so that the estimates standard error of r is $s_r = 0.1454$ and the estimated CV 3.02 per cent.

Note that although this particular sample has not provided units with relatively large sizes, the CV's are much smaller than those for a simple random sample (Example 2.3).

Table 5.7: Present population and number of households of 4 sample villages selected with probability proportional to previous census population and computation of estimates for 30 villages

Village serial no.	Sample village no. i	Size (previous census population) z_i	Reciprocal of probability $1/\pi_i = Z/z_i$	Present population y_i	Present number of households h_i	$y_i^* = y_i/\pi_i$	$h_i^* = h_i/\pi_i$	y_i^{*2}	h_i^{*2}	$y_i^* h_i^*$
(1)	(2)	(3)	(4)	(5)	(6)	(7)	(8)	(9)	(10)	(11)
5	1	92	30.5978	112	24	3426.95	734.35	11 743 986.30	539 269.92	2 516 580.73
6	2	65	43.3077	77	17	3334.69	736.23	11 120 157.40	542 034.61	2 455 098.82
11	3	72	39.0972	78	15	3049.58	586.46	9 299 938.18	343 935.33	1 788 456.69
18	4	102	27.5980	114	23	3146.17	634.75	9 898 385.67	402 907.56	1 997 031.41
					Total	12 957.39	2691.79	42 062 467.55	1 828 147.42	8 757 167.65
					Mean	3239.33 (y_0^*)	672.95 (h_0^*)			

5.5 Special cases of crop surveys

5.5.1 Introduction

In a survey designed to estimate the total area under any particular crop and its total yield, sampling of fields (or farms or plots) with ***probability proportional to total (geographical) area*** (ppa) introduces simplifications in the estimating procedures in addition to possible improvements in the efficiency of the estimators. If a map of fields (or farms or plots) is available, selection with ppa may be made by the procedure described in section 5.4.2.

5.5.2 Area surveys of crops

Suppose n fields are selected with ppa

$$\pi_i = a_i/A$$

where

$$A = \sum^N a_i$$

and where a_i is the area of the ith field ($i = 1, 2, \ldots, N$ for the universe, and $i = 1, 2, \ldots, n$ for the sample). Let y_i denote the area under a particular crop for the ith sample field, then an unbiased estimator of the total area under the crop in the universe is, from estimating equation (5.3),

$$\begin{aligned} y_0^* &= \sum^n y_i^*/n = \sum^n y_i/n\pi_i \\ &= A\sum^n a_i p_i/na_i = A\sum^n p_i/n = A\bar{p} \end{aligned} \qquad (5.11)$$

where $p_i = y_i/a_i$ is the proportion of the area of the ith field under the particular crop, which varies from 0 to 1.

An unbiased estimator of the variance of y_0^* is, from equation (5.5),

$$s_{y_0^*}^2 = A^2 s_{\bar{p}}^2 = A^2 SSp_i/n(n-1) \qquad (5.12)$$

where SSp_i is the corrected sum of squares of p_i.

An unbiased estimator of the overall proportion under the crop is given by the estimated total area under the crop divided by the total geographical area, i.e.

$$A\bar{p}/A = \bar{p} = \sum^n p_i/n \qquad (5.13)$$

Thus a simple (unweighted) average of the sample proportions under the crop gives an unbiased estimator of the universe proportion.

An unbiased estimator of the variance of $\overline{p}$ is, from equation (5.12),

$$s_{\overline{p}}^2 = SSp_i/n(n-1) \tag{5.14}$$

Note: If the crop is such that it either occupies the whole of a field or no part of it, or if the fields are small enough for the this assumption to hold, such that the proportion of the total area under the crop (p_i) is either 1 (whole) or 0 (none), the estimator for the variance of the total area under the crop in equation (5.12) reduces to

$$s_{y_0^*} = A^2\overline{p}(1-\overline{p})/(n-1) \tag{5.15}$$

For let r of the n sample units have $p_i = 1$, and the rest $(n-r)$ have $p_i = 0$. Then $\sum^n p_i = \sum^r 1 = r$, also $\sum^n p_i^2 = \sum^r 1 = r$, so that $\overline{p} = \sum^n p_i/n = r/n$, and $SSp_i = \sum^n p_i^2 - n\overline{p}^2 = r - n\overline{p}^2 = n\overline{p} - n\overline{p}(1-\overline{p})$. Substituting this value of SSp_i in equation (5.12), we obtain equation (5.15).

Example 5.3

Ten farms selected with probability proportional to total area from the universe of 100 farms gave the following proportion of area under a crop:

(p_i): 0.20; 0.25; 0.10; 0.30; 0.15; 0.25; 0.20; 0.25; 0.10; 0.20

Estimate the total area under the crop for the 100 farms and its CV; the total geographical area is 16 124 acres.

Here $N = 100$; $n = 10$; $A = 16\ 124$; $\sum^n p_i = 2.00$; $\sum^n p_i^2 = 0.4175$; and $SSp_i = 0.0175$.

The estimated proportion of area under the crop is, from equation (5.13),

$$\overline{p} = \frac{1}{n}\sum p_i = 2.00/10 = 0.20$$

with estimated variance (from equation (5.14))

$$s_{\overline{p}}^2 = 0.0174/90 = 0.00019444$$

so that $s_{\overline{p}} = 0.01394$.

An unbiased estimate of the total area under the crop is (equation (5.11)) $y_0^* = A \cdot \overline{p} = 16\ 124 \times 0.20 = 3224.8$ or 3225, with an estimated standard error of y_0^* (equation (5.12)) of

$$s_{y_0^*} = As_{\overline{p}} = 16\ 124 \times 0.01394 = 224.77$$

and the estimated CV is 6.97 per cent.

As the t-value corresponding to probability 0.05 and degrees of freedom $(n - 1 =)$ 9 is 2.262, the 95 per cent probability limits to the total area under the crop are $y_0^* \pm 2.262 \times s_{y_0^*}$ or 3225 ± 508 or 2717 and 3733 acres.

5.5.3 Yield surveys of crops

Similar considerations apply for estimating the average yield of a crop per unit of area. If $r_i = x_i/a_i$ is the yell per unit are in the ith sample field (obtained on harvesting the crop and measuring the yield x_i), then an unbiased estimator of the average yield per unit area is (from an equation of the type (5.13))

$$\bar{r} = \frac{1}{n} \sum^{n} r_i \tag{5.16}$$

i.e. the simple (unweighted) average of the yields per unit area in the different fields. Also,

$$s_{\bar{r}}^2 = SSr_i/n(n-1) \tag{5.17}$$

An unbiased estimator of the total yield X is (from an equation of the type 5.11)

$$s_0^* = A\bar{r} \tag{5.18}$$

and

$$s_{x_0^*}^2 = A^2 s_{\bar{r}}^2 \tag{5.19}$$

A generally biased but consistent estimator of the average yield per unit of crop area is

$$r' = x_0^*/y_0^* = \bar{r}/\bar{p} \tag{5.20}$$

an estimator of the variance of which is

$$s_{r'}^2 = (s_{x_0^*}^2 + r'^2 s_{y_0^*}^2 - 2r' s_{y_0^* x_0^*})/y_0^{*2} \tag{5.21}$$

where $s_{y_0^* x_0^*}$ is an unbiased estimator of the covariance of y_0^* and x_0^*, given by

$$s_{y_0^* x_0^*} = A^2 SP p_i r_i/n(n-1) \tag{5.22}$$

Note: In the far less common situation when the areas under a particular crop are known for all the fields in the universe, sample fields can be selected with probability proportional to crop area, and an unbiased estimator of the average yield per unit of corp area is given by an equation of the type (5.16), namely

$$\bar{r}'' = \sum r_i''/n \tag{5.23}$$

and an unbiased estimator of the total yield of the crop by $Y\bar{r}''$, where $r_i'' = x_i/y_i$. Variance estimators of these two estimators are given by estimating equations of the types (5.17) and (5.19), respectively.

5.6 Ratio method of estimation

As with simple random sampling, so also with pps sampling, the ratio method of estimation may be used to improve the efficiency of estimators. The principle is the sample and will be illustrated with examples.

Note, first, however, that if sampling is with pps, and the ratio method is used with the help of the sizes themselves, the ratio estimator of the universe total becomes the same as the unbiased estimator from the pps sampling. For if y is the study variable and z the size variable, then with the usual notations,

$$y_0^* = \frac{1}{n}\sum^n y_i^* = \frac{Z}{n}\sum^n \frac{y_i}{z_i} \tag{5.3}$$

$$z_0^* = \frac{1}{n}\sum^n z_i^* = \frac{Z}{n}\sum^n \frac{z_i}{z_i} = Z \tag{5.24}$$

The ratio estimator of the total Y, using the size variable z, is therefore, from an estimating equation of the type (3.6),

$$y_R^* = Zy_0^*/z_0^* = Zy_0^*/Z = y_0^*$$

i.e. the unbiased estimator from the pps sample.

A corollary of the above is that the simple (i.e. unweighted) mean of the ratios y_i/z_i becomes the unbiased estimator of the universe ratio Y/Z, for the estimator of Y/Z is

$$\frac{y_0^*}{Z} = \frac{Z\sum^n (y_i/z_i)/n}{Z} = \frac{\sum^n r}{n} \tag{5.25}$$

where $r_i = y_i/z_i$. This result has special application in agricultural crop surveys (see section 5.5).

Although it is known that for sampling with pps to be more efficient than simple random sampling, the size variable should have a high, positive correlation with the study variable, and the linear regression line of the study variable on the size should pass through the origin, in a multi-subject inquiry the size variable chosen (as a necessary compromise owing to the conflicting desiderata of a number of variables) may be such that the above conditions are not fulfilled in respect of a particular study variable; in other situations, the required information on the desired size may not be available at the time of the sample selection (see exercise 3 at the end of this chapter). In these cases, the ratio method of estimation may be used in order to improve the efficiency of the estimators obtained from the pps sampling.

Thus, if w is the ancillary variable used for the ratio estimation and z is the size variable used for pps sampling, then with the usual notations

$$w_0^* = \frac{1}{n} \sum^n w_i^* = \frac{Z}{n} \sum^n \frac{w_i}{z_i} \tag{5.26}$$

The ratio estimator of the universe total Y is then, using an estimating equation of the type (3.6),

$$y_R^* = W(y_0^*/w_0^*) = Wr \tag{5.27}$$

where

$$W = \sum^N w_i \,; \qquad r = y_0^*/w_0^* \tag{5.28}$$

The variance estimator of y_R^* is, from equation (3.10),

$$s_{y_R^*}^2 = W^2 s_r^2 = (s_{y_0^*}^2 + r^2 s_{w_0^*}^2 - 2r s_{y_0^* w_0^*}) \tag{5.29}$$

Note: For the ratio method of estimation to be efficient, the selection probability should be appropriate for both y and w.

Example 5.4

A sample of 4 villages, drawn with probability proportional to area from the list of 30 villages given in Appendix IV, gave the data on the present population (Table 5.8). Obtain an unbiased estimate of the present total population in the 30 villages. Given the previous census population of the 30 villages in Table 5.6, obtain the ratio estimate of the present total population and compare the two estimates. The total area of the 30 villages is 270.0 km^2 and the total previous census population, 2815.

Here $N = 30$; $n = 4$; $Z = \sum^N z_i = 270.0$ km^2; $W = \sum^N w_i = 2815$.

An unbiased estimate of the present total population of the 30 villages is

$$y_0^* = \frac{1}{n} \sum y_i^* = 3506.14 \text{ or } 3506$$

An unbiased estimate of the variance of y_0^* is

$$s_{y_0^*}^2 = SSy_i^*/n(n-1) = 443\,565.2459$$

so that the estimated standard error of y_0^* is $s_{y_0^*} = 666.01$ and the estimated CV is 19.00 per cent.

These are the estimates we would obtain if no other information were available. If, however, the data on the previous census were available, we could use

Table 5.8: Present and previous census population of 4 villages selected with probability proportional to area and computation of estimates for 30 villages

Village serial no.	Sample village no. i	Area (km^2) z_i	Reciprocal of probability $1/\pi_i = Z/z_i$	Present population y_i	$y_i^* = y_i/\pi_i$	y_i^{*2}	Previous census population w_i	$w_i^* = w_i/\pi_i$	w_i^{*2}	$y_i^* w_i^*$
(1)	(2)	(3)	(4)	(5)	(6)	(7)	(8)	(9)	(10)	(11)
7	1	4.5	60,0000	88	5 280.00	27 878 400.00	72	4320.00	18 662 400.00	22 809 600.00
13	2	5.8	46.5517	80	3 724.14	13 869 218.74	73	3398.27	11 548 238.99	12 655 633.24
22	3	10.0	27.0000	83	2 241.00	5 022 081.00	78	2106.00	4 435 236.00	4 719 546.00
30	4	10.2	26.4706	105	2 779.41	7 725 119.95	103	2726.47	7 433 638.66	7 577 977.98
				Total	14 024.55	54 494 818.69		12 550.74	42 079 513.65	47 762 757.22
				Mean	3 506.14 (y_0^*)			3137.68 (w_0^*)		

these to improve our estimates. We first obtain the unbiased estimate of the total previous census population, from equation (5.26),

$$w_0^* = \frac{1}{n}\sum w_i^* = 3137.68 \text{ or } 3138$$

An unbiased estimate of the variances of w_0^* is

$$s^2_{w_0^*} = SSw_i^*/n(n-1) = 224,942.3142$$

Also,

$$s_{y_0^* w_0^*} = SPy_i^* w_i^*/n(n-1) = 313,175.4731$$

As the ratio of the two unbiased estimates y_0^* and w_0^* is

$$r = y_0^*/w_0^* = 3506.14/3137.68 = 1.11743$$

the ratio estimate of the present total population, using the previous census population is, from equation (5.27),

$$\begin{aligned} y_R^* &= Wy_0^*/w_0^* = Wr \\ &= 2825 \times 1.11743 = 3145.57 \text{ or } 3146 \end{aligned}$$

From equation (5.29), the estimated variance of y_R^* is 24,514.8162, so that the estimated standard error is 156.57 and the estimated CV is 4.98 per cent.

Note the tremendous improvement in the estimates by the ratio method of estimation as compared to the unbiased estimate.

Further reading

Brewer and Hanif; Chaudhury and Stenger, chapter 2; Godambe and Thompson; Hedayat and Sinha, chapters 1-3 and 5; Kish (1965), chapter 7; Murthy (1967), chapter 6 and section 15.5c; Singh and Chaudhary, chapter 5; Sukhatme *et al.* (1984), chapter 3; Yates, sections 3.9, 6.16, 7.15, 8.9 and 8.20.

Exercises

1. A sample of 10 villages was drawn with probability proportional to the 1951 Census population in a *tehsil* (a *tehsil* is an administrative sub-division) in India. The data on the 1951 Census population, and the sample data on cultivated area, are given in Table 5.9. Estimate the total cultivated area in the *tehsil* and its CV, given the total population in the *tehsil* in 1951 as 415,149 (Murthy, Problem 6.2).
2. From 35 farms with a total geographical area of 5759, a sample of 5 farms drawn with probability proportional to the total area of the farms gave the data shown in Table 5.10. Estimate the average and the total area under the crop with respective standard errors (Sampford, pp. 124-125).

Table 5.9: Cultivated area of 10 sample villages, selected with probability proportional to 1951 Census population in a *tehsil* in India

Village serial no.	1	2	3	4	5	6	7	8	9	10
1951 census population	551	865	2535	3523	8368	7357	5131	4654	1146	1165
Cultivated area (acres)	4824	924	1948	3013	7678	5506	4051	4060	809	1013

Source: Murthy, Problem 6.2.

Table 5.10: Area under a crop and the proportion to total area of 5 sample farms, selected with probability proportional to the total area

Farm No.	3	18	28	34	35
Total area	52	110	300	410	430
Area under a crop	10	24	59	72	103
Proportion of area under the crop	0.1923	0.2182	0.1967	0.1756	0.2385

Source: Sampford, pp. 124-125.

3. Table 5.11 shows for 1937 the area under wheat in 34 villages in Lucknow sub-division (India) selected out of 170 villages with probability proportional to the cultivated area as recorded in 1931.

   (a) Given that the total cultivated area in 1931 in the 170 villages was 78,019 acres, obtain an unbiased estimate of the total area under wheat and its standard error.

   (b) After the sample selection and enumeration, data on the area under wheat in 1936 became available. Using this information, and given that the total area under wheat in 1936 was 21,288 acres, obtain the ratio estimate of the area under wheat in 1937 and its standard error (Sukhatme and Sukhatme, 1970b, Example 4.4).

Table 5.11: Total cultivated area in 1931 and area under wheat in 1936 and 1937 for a sample 34 villages selected with probability proportional to cultivated area in Lucknow sub-division, India

Village serial no.	Total cultivated area, 1931 (acres)	Area under wheat in 1936 (acres)	Area under wheat in 1937 (acres)	Village serial no.	Total cultivated area, 1931 (acres)	Area under wheat in 1936 (acres)	Area under wheat in 1937 (acres)
1	401	75	52	18	186	45	27
2	634	163	149	19	1767	564	515
3	1194	326	289	20	604	238	249
4	1770	442	381	21	701	92	85
5	1060	254	278	22	524	247	221
6	827	125	111	23	571	134	133
7	1737	559	634	24	962	131	144
8	1060	254	278	25	407	129	103
9	360	101	112	26	715	192	179
10	946	359	355	27	845	663	330
11	470	109	99	28	1016	236	219
12	1625	481	498	29	184	73	62
13	827	125	111	30	282	62	79
14	96	5	6	31	194	71	69
15	1304	427	399	32	439	137	100
16	377	78	7	33	854	196	141
17	259	78	105	34	824	255	265

Source: Sukhatme and Sukhatme (1970b), Example 4.4.

CHAPTER 6

Choice of Sampling Units: Cluster Sampling

6.1 Introduction

In the cases of single-stage sampling so far considered, we have illustrated sampling procedures such as simple random sampling, systematic sampling, and varying probability sampling, with different sampling units such as villages, farms, households, and persons. When the recording units occur in clusters or groups, there may be some advantage in selecting a sample of clusters and completely surveying all the recording units within the selected clusters. This is known as *cluster sampling*. Thus, in a household inquiry, a sample of villages may be selected from the total list of villages and all the households in these sample villages surveyed; or in a demographic inquiry, where the elementary, recording unit is the individual for sex, age etc., all the members of the households may be surveyed in a sample of households; and so on. There may thus be a hierarchy of recording (and sampling) units.

The cluster may refer to naturally occurring groups such as individuals in a household, or households or farms in a village, or all the sheep in a flock; clusters may also be formed artificially by grouping together units that are neighboring or can be surveyed together conveniently. When the total geographical area of the universe is subdivided into smaller areas and a sample of areas is taken, the sampling plan is known as *area sampling*.

No new principles are involved in cluster sampling for the estimation of universe totals, averages, ratios etc., or for their variance estimators: the estimation procedure will of course depend on whether the sample of clusters is selected with equal or varying probabilities. However, the expression for the universe variances of the estimators and the sample estimators of variances may be formulated differently in order to facilitate the choice of the sampling unit – the cluster or its elements – and its size.

In this chapter, we shall consider the criteria for the choice of the sampling units, and for simplicity, only clusters of equal size (i.e. each with an equal number of elementary units) selected by simple random sampling.

We shall see that when, as it generally happens, the elementary units within a cluster tend to be similar in respect of the study variable, then cluster sampling will be less efficient than a direct (unrestricted) simple random sample of the elementary units, given the same total number of the units in the sample. However, cluster sampling reduces the costs and labor of travel, identification, contact, data collection etc., and may sometimes also reduce non-sampling errors and biases in data (section 25.7). Apart from the latter consideration, the question therefore arises of balancing the general increase in sampling error against the decrease in costs. This leads to the problem of determining the optimal size of cluster, i.e. the number of elementary units it should contain. This problem is considered in the next chapter.

Notes

1. Cluster sampling, as defined, may be used with single- as well as multi-stage designs, and unstratified as well as stratified designs. A single-stage cluster sampling may be considered as the case where all the second-stage units (i.e. the elementary units) in the selected first-stage units (i.e. clusters) are surveyed. Cluster sampling and two-stage sampling, to be considered in Part III, have many common considerations. The term "cluster sampling" is sometimes used to denote any multi-stage design; the difference in definitions is therefore well worth bearing in mind. The similarity in formulation of the theory in section 6.2 with that of systematic sampling in section 4.3 may also be noted.

2. Cluster sampling in general refers to geographical units; and a sample of households is not generally described as a cluster sample of household members.

6.2 Simple random sampling

6.2.1 Universe parameters

Let the universe consist of N mutually exclusive and exhaustive clusters, and in each cluster let there be the same number of M_0 elementary units, so that the total number of units in the universe is NM_0 (note that until now we had denoted the total number of universe units by N).

Let Y_{ij} denote the value of the study variable for the jth unit ($j = 1, 2, \ldots, M_0$) in the ith cluster ($i = 1, 2, \ldots, N$). The total value of the study variable in the ith cluster is

$$Y_i = \sum_{j=1}^{M_0} Y_{ij} \tag{6.1}$$

and the cluster mean

$$\overline{Y}_i = Y_i/M_0 \tag{6.2}$$

The universe total of the study variable is

$$Y \sum_{i=1}^{N} Y_i = \sum_{i=1}^{N} \sum_{j=1}^{M_0} Y_{ij} \tag{6.3}$$

and the overall universe mean per unit

$$\overline{Y} = Y/NM_0 = \sum^{N} \overline{Y}_i/N \tag{6.4}$$

is also the mean of the cluster means.

The universe variance

$$\sigma^2 = \frac{\sum_{i=1}^{N} \sum_{j=1}^{M_0} (Y_{ij} - \overline{Y})^2}{NM_0} = \sigma_w^2 + \sigma_b^2 \tag{6.5}$$

where

$$\sigma_w^2 = \frac{\sum_{i=1}^{N} \sum_{j=1}^{M_0} (Y_{ij} - \overline{Y}_i)^2}{NM_0} \tag{6.6}$$

is the *within-cluster variance*, and

$$\sigma_b^2 = \frac{\sum_{i=1}^{N} (\overline{Y}_i - \overline{Y})^2}{N} \tag{6.7}$$

is the *between-cluster variance*.

The between-cluster variance can also be expressed in terms of the intraclass correlation coefficient ρ_c between pairs of units within the clusters (or "intra-cluster" correlation coefficient), which is defined by

$$\rho_c = \frac{\sum_{i=1}^{N} \sum_{j \neq j=1}^{M_0} (Y_{ij} - \overline{Y})(Y_{ij'} - \overline{Y})}{NM_0(M_0 - 1)\sigma^2} \tag{6.8}$$

and it can be shown that

$$\sigma_b^2 = \sigma^2[1 + (M_0 - 1)\rho_c]/M_0 \tag{6.9}$$

Proofs are given in Appendix II, section A2.3.8.

6.2.2 Sample estimators

If n clusters are selected by srs with replacement and all the elementary units in the sample clusters surveyed, the total number of units in the sample is nM_0. If y_{ij} denotes the value of the study variable for the jth unit ($j = 1, 2, \ldots, M_0$) in the ith sample cluster ($i = 1, 2, \ldots, n$), the total value of the study variable in the ith sample cluster is

$$y_i = \sum_{j=1}^{M_0} Y_{ij} \tag{6.10}$$

and the mean of the ith cluster is

$$\overline{y}_i = y_i / M_0 = \overline{Y}_i \tag{6.11}$$

The overall mean per unit in the sample is also the mean of the sample cluster means

$$\overline{y}_c = \frac{\sum_{i=1}^{n} \sum_{j=1}^{M_0} Y_{ij}}{nM_0} = \frac{\sum^n \overline{y}_i}{n} \tag{6.12}$$

and it is an unbiased estimator of the universe mean $\overline{Y}$.

The sampling variance of $\overline{y}_c$ is

$$\sigma^2_{\overline{y}_c} = \sigma^2_b / n \tag{6.13}$$

An unbiased estimator of σ^2_b is the sample estimator

$$s^2_b = \sum_{i=1}^{n} (\overline{y}_i - \overline{y}_c)^2 / (n-1) \tag{6.14}$$

An unbiased estimator of $\sigma^2_{\overline{y}_c}$ is therefore

$$s^2_{\overline{y}_c} = s^2_b / n \tag{6.15}$$

From equation (6.9), the sampling variance of the sample estimator $\overline{y}_c$ in terms of the intraclass correlation coefficient ρ_c is

$$\sigma^2_{\overline{y}_c} = \sigma^2[1 + (M_0 - 1)\rho_c]/nM_0 \tag{6.16}$$

$$= \sigma^2_{\overline{y}_{srs}}[1 + (M_0 - 1)\rho_c] \tag{6.17}$$

where

$$\sigma^2_{\overline{y}_{srs}} = \sigma^2 / nM_0 \tag{6.18}$$

is the sampling variance of the mean of nM_0 units selected directly by srs (with replacement). Equation (6.17) is the fundamental formula in cluster sampling.

An unbiased estimator of the within-cluster variance σ_w^2 is the sample estimator

$$s_w^2 = \sum_{i=1}^{n} \sum_{j=1}^{M_0} (y_{ij} - \overline{y}_i)^2 / n(M_0 - 1) \tag{6.19}$$

From equations (6.5) and (6.9),

$$\rho_c = 1 - M_0\sigma_w^2/(M_0 - 1)\sigma^2 \tag{6.20}$$

A sample estimator of ρ_c is

$$\hat{\rho}_c = 1 - M_0 s_w^2/(M_0 - 1)(s_b^2 + s_w^2) \tag{6.21}$$

6.2.3 Design effect

From equation (6.17), the ratio of the sampling variance of the overall mean per unit in a cluster sample to that in an srswr (of the same total number of units nM_0) is

$$\sigma_{\overline{y}_c}^2/\sigma_{\overline{y}_{srs}}^2 = 1 + (M_0 - 1)\rho_c$$

This ratio is termed "**design effect**" or "**deff**" for short (Kish, 1965).

6.3 Discussion of the results

The following points emerge from the preceding formulae:

1. As $\sigma^2 \geq \sigma_w^2$ (from equation (6.5)), from equation (6.20) one sees that ρ_c lies between $-1/(M_0 - 1)$ (when $\sigma_w^2 = 0$) and 1 (with $\sigma_b^2 = 0$).

2. From equation (6.17), it can be seen that

   (i) If $M_0 = 1$, i.e. if each cluster consists of one elementary unit, then there is no clustering, and the variance formula for cluster sampling becomes the same as that for srs.

   (ii) If $\rho_c = 0$ (i.e. when the characteristic is distributed randomly on the ground, and there is no intra-cluster correlation), then the variance formula for cluster sampling becomes the same as that for srs; a sample of one cluster of size M_0 will then provide as much information on the study variable as an srs of M_0 selected directly.

(iii) If $\rho_0 = 1$ (the maximum value), i.e. if all the units in the cluster have the same value for the study variable such that the greatest degree of homogeneity obtains in a cluster, then $\sigma^2_{\bar{y}_c} = M_0 \sigma^2_{\bar{y}_{r,r'}}$; i.e. the variance for a cluster sampling will be M_0 times that for an srs. Cluster sampling will then be extremely inefficient.

(iv) The factor $(M_0 - 1)\rho_c$ of the "design effect", gives a measure of the relative change in the sampling variance due to sampling clusters instead of sampling the elementary units directly; for example, if clusters ($M_0 =$) 100 persons each are formed and $\rho_c = 0.01$, then the design effect is 1.99 or 2 approximately, so that the variance of cluster sampling will bet twice that of an srs of individuals. Thus, a relatively small value of the intracluster correlation coefficient, multiplied by the size of the cluster, could lead to a substantial increase in variance.

(v) ρ_c will be *negative* for the sex and age composition of members of a household (*see* Note 4, and exercises 3 and 4 at the end of this chapter); cluster sampling of households will then be more efficient than an srs of persons, and the cost-efficiency of cluster sampling greater still.

(vi) In general, however, ρ_c is positive, and decreases with the cluster size M_0, but the factor $(M_0 - 1)\rho_c$ increases with increasing cluster size, so that cluster sampling becomes less efficient than srs, and increasingly inefficient as the cluster size increases.

3. The comparison of the efficiencies of estimators from a direct (unrestricted) srs of the elementary units and an srs cluster sample of the units will be misleading for two reasons. First, unrestricted simple random samples are not generally realized in practice; and second, even when a cluster sample is taken (either single- or multi-stage), special procedures of sampling (such as stratification, pps selection of clusters) and of estimation (such as the ratio method of estimation) are often used to increase the efficiency of the estimators.

6.4 Notes

1. Similar considerations apply to unequal size clusters, selected with equal or varying probabilities, with $\overline{M}$, the average size of a cluster, replacing M_0 in equation (6.17) and others.

2. The above formulation also holds for the estimation of proportion of the units in the universe possessing a certain attribute. Defining, as in section 2.13, our study variable so that it takes the value 1, if the unit

has the attribute, the value zero otherwise, and if in the ith sample cluster, M_i' units possess the attribute, then the cluster proportion of units possessing the attribute is $M_i'/M_0 = P_i = p_i$; P_i and p_i denote respectively the universe proportion and the sample estimator in the ith sample cluster.

An unbiased estimator of the universe proportion P is

$$\overline{p}_c = \sum^{n} p_i/n \tag{6.22}$$

the mean of the sample cluster proportions. The unbiased estimator of the variance of $\overline{p}_c$ is

$$s^2_{\overline{p}_c} = s_b^2/n = \sum^{n}(p_i - \overline{p}_c)^2/n(n-1) \tag{6.23}$$

3. For proportions, $P(1 - P)/n$ would be the sampling variance if the units were selected directly from an srs. But design-based estimates of variance should be computed in every case. If, for example, an srs of clusters is taken and the ratio method of estimation used for the birth rate, the three variance estimators, namely from srs of clusters with and without ratio method of estimation, and the binomial variance, could be compared, which would show the loss of efficiency due to clustering on the one hand, and the gain due to using the ratio method of estimation.

4. When the study variable is a zero-one or yes-no characteristic, and a design-based estimate of its variance has been obtained, the value of ρ_c can be computed from equation (6.17). Thus, in Example 2.5, we have seen that the design-based estimate of the variance proportion of males is $s_r^2 = 0.0005391082$, whereas the estimated variance of the proportion had the sample persons been selected directly is $s_p^2 = 0.00206545$. From equation (6.17) and using the sample estimate of variances, we have

$$1 + (M_0 - 1)\rho_c = s_r^2/s_p^2 = 0.5391$$

taking M_0, the average number of persons per household from the sample, namely, 5.1, we have

$$\rho_c = (0.5391 - 1)/4.1 = -0.11$$

In this case, the efficiency of sampling households to that of sampling persons directly is $s_p^2/s_r^2 = 2.61$ or 261 per cent (*cf.* exercise 3a at the end of this chapter).

As another example, suppose that a simple random sample (with replacement) of 50 areal units (clusters) each of 300 persons estimated the birth rate at 0.045 per person with variance 0.00000395081, computed by the method of section 2.9.1. Had the $50 \times 300 = 15,000$ sample persons been selected directly from the universe of persons, then under some simplifying assumptions, from equation (2.74), the same birth rate would have an estimated variance 0.00000283689. Using equation (6.17), the intraclass correlation coefficient is then estimated at 0.00013.

5. In the different geographical strata with clusters of about 300 persons in Cameroon (1960-5), the value of the intraclass correlation coefficient for the birth rate ranged from −0.0013 to 0.0054, with a central value of 0.0013, and for the death rate from −0.0005 to 0.0101, with a central value of 0.0025 (Scott, 1967 and 1968). For the proportion of adult males employed, the coefficient was 0.1 for census enumeration areas (with an average of 1000 persons each) in the Ghana Census of 1960 (Scott, 1967): this indicates that in this case, cluster sampling of enumeration areas would be very inefficient and a sub-sample of households in selected enumeration areas should be taken. If for practical reasons (see section 26.13), a multi-subject survey covering both a demographic and a labor force inquiry is conducted, a possible solution would be to enumerate completely the selected geographical areas for population, births, and deaths, and to select a sub-sample of households for an "in-depth" inquiry into the size and characteristics of the labor force.

6. *Variance function.* The intra-cluster correlation coefficient, and therefore the variance of the cluster means, are not explicit functions of the size of the cluster; that is to say, the functions cannot be derived mathematically but depend on how the characteristic is in fact distributed on the ground. This makes it difficult to estimate the sampling variance for a sample of clusters of any one size, given the variance of an equivalent sample of cluster of any one size, given the variance of an equivalent sample of cluster of any other particular size. However, in area sampling for crops in India, the U.K., and the U.S.A., the within-cluster variance has been seen generally to follow the form aM_0^g where a and g are constants to be evaluated from the data (and g is a positive number less than unity). From this and using the sample estimator of the total variance, an expression for the between-cluster variance can be obtained in terms of M_0, if there are at least two values of the within-cluster variance for estimating a and g. Regarding, however, the total universe as a single cluster of

NM_0 units, the total variance becomes the same as the within-cluster variance in the finite universe and is equal to $a(NM_0)^g$, i.e. a and g can be estimated from a survey in which only one value of M_0 is used: this formulation may not hold for clusters with large M_0s.

Further reading

Cochran, chapter 9; Deming (1950), chapter 3C; Hansen *et al.* (1953), vol. I, chapter 6A-D, vol. II, chapter 6; Hedayat and Sinha, chapter 7; Kish (1965), sections 5.1, 5.2, 5.4, 6.1 and 6.2; Murthy (1967), chapter 8; Sukhatme *et al.* (1984), chapter 7; Yates, section 2.9.

Exercises

1. From the data and results of exercise 7, Chapter 2, estimate the value of intra-class correlation coefficient for the proportion of persons absent.
2. A bed of white pine seedlings contained six rows, each 434 ft long. Data for four types of sampling units into which the bed could be divided are shown in Table 6.1 along with estimates of cost (in terms of length of a row that could be covered in 15 minutes). Obtain the optimum sampling units after comparing the relative cost-efficiencies of the different sampling units (Cochran, pp. 235-236).

   (*Hint:* First compute the relative cost of measuring one unit, in this example, in terms of time required to count one unit: C_u = relative size of unit/length of a row (ft) that can be covered in 15 minutes; then compute the relative net precision of each unit which is defined as being inversely proportional to the variance obtained for fixed cost, namely $M_u^2/(C_u S_u^2)$ is the universe variance per unit. For details, *see* Cochran).

Table 6.1: Data for four types of sampling unit

Type of unit u	Relative size of unit M_u	Total number of units N_u	Universe variance per unit S_u^2	Length of a row (ft) that can be covered in 15 minutes
1-ft row	1	2604	2.537	44
2-ft row	2	1302	6.746	62
1-ft bed	6	434	23.094	78
2-ft bed	12	217	68.558	108

Source: Cochran, pp. 235-236.

3. (a) Consider households all of size 4, consisting of a couple and two children, and assume that the sex of a child is binomially distributed with the proportion of male children being one-half. Show by computing the between- and within-cluster variances or otherwise that the intraclass correlation coefficient between sexes of different members of the household is $-\frac{1}{6}$, and that the efficiency of sampling households (clusters of persons) to that of sampling persons directly for estimating the sex ratio is 200% (Sukhatme and Sukhatme (1970b), section 6.3).

   (b) Compute separately the values of the intraclass correlation coefficient between sexes of different members of households with 1 child, 3 children, and 4 children respectively in addition to a couple. Generalize the results for a household with M_0 members (a couple and the rest of their children) to show that the intraclass correlation coefficient between the sexes of different members of the households is $-1/{}^{M_0}C_2$.

   (*Hint:* Generalize the following statement for a household of size 4; 'The correlation between the sexes of husband and wife is -1, but for every other of the remaining five pairs is zero, since the sex of the husband or wife will not determine the sex of their children, nor will the sex of one child determine the sex of another. The average value of the correlation between sexes of different members in a household of 4 consisting of a husband, a wife and two children is therefore $-\frac{1}{6}$, Sukhatme and Sukhatme (1970b), section 7, also Deming (1950), pp. 209-11. (*Note:* As sampling of individuals is rarely done in the field for demographic inquiries, this and the following exercise have relevance mainly to sampling after the survey.)

4. Consider households all of size 5, consisting of a couple and three children (none adult). Show that the intraclass correlation between the age compositions of household members (taking only two broad age groups of adults and children) is $-\frac{1}{5}$, so that the efficiency of sampling households to that of sampling persons directly to obtain the age composition of the population is 500%. Generalize this result to households with M_0 members (a couple and the rest children, none of whom are adult) to show that the intraclass correlation coefficient is

$$[1 + {}^{(M_0-2)}C_2 - 2(M_0 - 2)]/{}^{M_0}C_2.$$

CHAPTER 7

Size of Sample: Cost and Error

7.1 Introduction

Two important questions on the designing of any sample inquiry are the total cost of the survey and the precision of the main estimates. Both these are related to the size of the sample, given the variability of the data, the type of sampling and the method of estimation. Obviously, the larger the sample, the smaller will be the sampling error, i.e. the greater the precision of the estimates, but the higher will also be the cost. The survey should be so designed as to provide estimates with minimum sampling errors (i.e. with maximum precision) when the total cost is fixed, and to result in the minimum total cost when the precision is preassigned: a sample size fulfilling these conditions is called the *optimal sample size*. We shall see later in Chapter 25 that other considerations such as the existence of non-sampling errors and biases in data have also to be taken into account: these generally increase with sample size beyond a certain point.

The size of a sample will be determined by the objective of the inquiry, and the permissible margin of error in the estimates. For example, during the depression of the Thirties in the U.S.A., when it was not known whether the unemployed numbered five million or fifteen million, the first sample did not necessarily have to be large to be useful; at present, however, larger samples with more sensitive measures are required to estimate if the unemployment rate (the number unemployed as a percentage of the total number in labor force) increases from say 3 to $3\frac{1}{4}$ per cent (Kish, 1971).

For simplicity, we shall in this chapter consider simple random sampling. In general, under conditions in which pps sampling and the ratio method of estimation are applied, these require smaller samples than an srs with the same efficiency.

7.2 Simple random sampling

7.2.1 Sampling for proportions

We have seen in section 2.13 that the sampling variance of a sample proportion p in an srs of n units is

$$\sigma_p^2 = P(1-P)/n \tag{7.1}$$

which depends only on the universe proportion P and the sample size n. The coefficient of variation of p is

$$e = \frac{\sigma_p}{P} = \sqrt{\frac{1-P}{nP}} \tag{7.2}$$

from which

$$n = \frac{P(1-P)}{\sigma_P^2} = \frac{(1-P)}{Pe^2} \tag{7.3}$$

This determines the sample size n required for any given coefficient of variation e.

As the CV per unit is

$$CV = \frac{\sqrt{[P(1-P)]}}{P} = \sqrt{\frac{1-P}{P}} \tag{7.4}$$

the required sample size n can also be expressed as

$$\begin{aligned} n &= \left[\frac{CV \text{ of one unit in universe}}{\text{desired } CV \text{ of sample estimator}}\right]^2 \\ &= \left[\frac{CV \text{ per unit}}{e}\right]^2 \end{aligned} \tag{7.5}$$

The sample size can also be determined such that the universe proportion P would lie within a given margin of error d on both sides of the sample estimator with a certain probability $(1-\alpha)$, which is equivalent to saying that the acceptable risk that the universe proportion P will lie outside the limits $p \pm d$ is α. For this, we use the assumption (which holds when n is large and the proportion P is not too small) that the sample proportion is normally distributed with mean P and standard deviation σ_p. Then the $(1-\alpha)$ per cent probability limits of the universe proportion P are

$$p \pm t_\alpha \sigma_p \tag{7.6}$$

where t_α is the value of the standard normal deviate that cuts off a total area α from both the tails taken together in the normal curve. Some illustrative values of t_α are given in Appendix III, Table 2; for example, for $\alpha = 0.05$ (1 in 20), $1 - \alpha = 0.95$ and $t = 1.96$ or approximately 2.

Setting the permissible margin of error

$$d = t_\alpha \sigma_p = t_\alpha \sqrt{[P(1-P)/n]} \tag{7.7}$$

and solving for n, we get

$$n = t_\alpha^2 P(1-P)/d^2 \tag{7.8}$$

In Appendix III, Table 3 have been tabulated the values of CV per unit in the universe for some values of the universe proportion P, and in Appendix III, Table 4, the required sample sizes (n) corresponding to given values of the universe CV per unit and the desired CV of the sample estimator. Given a value of P, the universe CV per unit can be obtained from equation (7.4) or read off Appendix III, Table 3; and the required sample size n can then be obtained from equation (7.5) or from Appendix III, Table 4, given the desired CV of the sample estimator e.

The margin of error d can be expressed as

$$d = t_\alpha \sigma_p = t_\alpha P e \tag{7.9}$$

Notes

1. As the CV of the estimated number of units possessing an attribute is the same as that of the estimated proportion of such units in the universe (exercise 6, Chapter 2), the same formulae for n will hold for the estimated number with the attribute as for the estimated proportion.
2. As the universe proportion P will not generally be known, an advance estimate p may be taken and used in the preceding equations.
3. As the value of $p(1-p)$ increases as p approaches $1/2$, a safer estimate of n is obtained on taking as an advance estimate that value of p which is nearer to $1/2$; for $p = 1/2$, the margin of error takes the maximum value (from equation 7.7)

   $$d = t_\alpha \sigma_p = 1.96 \times \sqrt{[(1/2 \times 1/2)/n]} = 0.98/\sqrt{n} = 1/\sqrt{n}$$

   approximately, for $\alpha = 0.05$ (i.e., 95% confidence limits).

   This is a good back-of-the-envelope formula to remember. For example, for a sample of size 1,296 persons, $d = 1/\sqrt{1,296} = 1/36$, i.e., about 3 per cent.

   For very small p, the advanced estimate should not be too rough; in this case, the Poisson approximation can be used, so that equation (7.8) is simplified to

   $$n = t_\alpha^2 P/d^2 \tag{7.10}$$

and equation (7.3) to

$$n = 1/Pe^2 \tag{7.11}$$

(The Poisson distribution can be regarded as the limiting distribution of a binomial when P becomes indefinitely small and n increases sufficiently to keep nP finite (say m), but not necessarily large; both the mean and the universe variance per unit are then equal to m).

4. When sampling is without replacement, and the sampling fraction n/N is not negligible, a more satisfactory estimate of the sample size is

$$n' = \frac{n}{1 + n/N} \tag{7.12}$$

where n is obtained from the previous equations on the basis of assumption of sampling with replacement.

5. In practice, estimates of proportion and ratios such as birth and death rates are seldom obtained from a simple random sample of individuals.

Example 7.1

A survey is to be made of the prevalence of the common diseases in a large population. For any disease that affects at least 1 per cent of the individuals in the population, it is desired to estimate the total number of cases with CV of not more than 20 per cent. What size of a simple random sample is needed, assuming that the presence of the disease can be recognized without mistake? (Cochran, exercise 4.3).

As the CV of a sample proportion is the same as the CV of the number of persons possessing the attribute, for $P = 0.01$, universe CV per unit $= 9.95$ (from equation (7.4)). As the desired CV is $e = 0.20$, the required sample size is, from equation (7.5), $n = 2475$. Or, from Appendix III, Table 3, $CV = 995\%$ for $P = 0.01$; and for CV 1000% (nearest to 995%) and $e = 20\%$, $n = 2500$.

Example 7.2

The (crude) birth rate is to be estimated in a country from a sample survey with 2.5 per cent CV. What is the sample size required if a rough estimate of the birth rate places it at 40 per 1000 persons?

Here $P = 0.04$, and the universe CV per unit is 4.9 (from equation (7.4)). As the desired CV is $e = 0.025$, $n = 38,400$ persons (from equation (7.5)).

7.2.2 Sampling of continuous data

If σ^2 is the universe variance per unit for the study variable, and $\overline{Y}$ the universe mean, the universe coefficient of variation per unit in the universe is

$$CV = \sigma/\overline{Y} \tag{7.13}$$

The sampling variance of the sample estimator $\overline{y}$, obtained from an srs of n units (with replacement), is

$$\sigma_{\overline{y}}^2 = \sigma^2/n \tag{7.14}$$

The CV of the sample estimator $\overline{y}$ is therefore

$$e = \sigma_{\overline{y}}/\overline{Y} = \sigma/\overline{Y}\sqrt{n} = (\text{universe } CV \text{ per unit})/\sqrt{n} \tag{7.15}$$

From equation (7.15), the sample size n required to obtain the sample mean with a given CV e is

$$\begin{aligned} n &= (\text{universe } CV \text{ per unit}/e)^2 \\ &= (\text{universe } CV \text{ per unit/desired } CV \text{ of sample mean})^2 \end{aligned} \tag{7.16}$$

In Appendix III, Table 4, have been given the required sample sizes (n) corresponding to some values of the universe CV per unit and the desired CV of the sample mean (e).

The sample size may also be determined such that the acceptable risk that the universe mean $\overline{Y}$ will lie outside the limits $\overline{Y} \pm d$ is α. For this, we assume that the sample mean $\overline{y}$ is normally distributed with mean $\overline{Y}$ and standard deviation $\sigma_{\overline{y}} = \sigma/\sqrt{n}$. The $(1 - \alpha)$ per cent confidence limits of the universe mean $\overline{Y}$ are

$$\overline{y} \pm t_\alpha \sigma_{\overline{y}} \tag{7.17}$$

where t_α is the value of the standard normal deviate that cuts off a total area α from both the tails taken together in the normal curve. Some illustrative values of t_α are given in Appendix III, Table 2.

Setting the permissible margin of error

$$d = t_\alpha \sigma_{\overline{y}} = t_\alpha \sigma/\sqrt{n} \tag{7.18}$$

and solving for n, we get

$$n = (t_\alpha \sigma/d)^2 = \sigma^2/V \tag{7.19}$$

where V is the desired variance of the sample mean. This determines the sample size n, given the values of σ, d and α.

Notes

1. If the universe total Y is to be estimated with margin of error D, the sample size required is

$$n = (Nt\sigma/D)^2 = (N\sigma)^2/V' \tag{7.20}$$

where V' is the desired variance of the sample total. Note that the unbiased estimators of the universe mean and total have the same CV (exercise 2, chapter 6), and equations (7.19) and (7.20) give the same value of n.

2. The above formulation assumes a knowledge of universe CV or σ. The universe CV remains remarkably stable over time and space and for characteristics of the same nature as the study variable; the CV for a previous study in the same or a different area for a related characteristic may therefore be taken. An advance estimate of σ may be taken from a pilot or some other study.

3. When sampling is without replacement, and the sampling fraction n/N is not negligible, a more satisfactory estimator of the sample size is

$$n' = \frac{n}{1 + n/N} \tag{7.21}$$

This applied both to equations (7.19) and (7.20).

4. If a sample of size n gives a CV e for a sample estimator (mean or total), then to obtain the sample estimator with CV e'', the required sample size is

$$n'' = n(e/e'')^2 \tag{7.22}$$

This follows from the relation

$$(CV \text{ per unit})^2 = ne^2 = n''e''^2 \tag{7.23}$$

which in turn follows from equation (7.15).

Example 7.3

The coefficients of variation per unit (an area of 1 mile square) obtained in a farm survey in Iowa, U.S.A. are given in Table 7.1. A survey is planned to estimate acreage items with a CV of 2.5 per cent, and the number of workers (excluding the unemployed) with a CV of 5 per cent. With simple random sampling, how many units are needed? How well would this sample be expected to estimate the number unemployed? (Cochran, exercise 4.5).

The maximum CV of the items (other than the number unemployed) is that of the number of hired workers, for which the CV is 1.10, the desired CV of which is 0.05. From equation (7.16),

$$\begin{aligned} n &= (CV \text{ per unit/desired } CV \text{ of sample estimator})^2 \\ &= (1.10/0.05)^2 = 22^2 = 484 \end{aligned}$$

Table 7.1: Coefficients of variation per unit (area of 1 mile square) in a farm survey in Iowa, U.S.A.

	Acres in farms	Acres in corn	Acres in oats	No. of family workers	No. of hired workers	No. of un-employed
Estimated CV per unit (%)	38	39	44	100	110	317

Source: Cochran (1977), exercise 4.5

The acreage items are required with CV of 0.025. Taking the acreage item with the maximum CV, namely, acres in oats, the required sample size is, from equation (7.16),

$$n = (0.44/0.025)^2 = 17.6^2 = 310$$

so that with a sample size of 484, required for the number of workers, the desired CV of the acreage items will also be attained.

For the number unemployed, we get, from equation (7.15), the CV of the sample estimator from a sample of size 484,

$$e = CV \text{ per unit}/\sqrt{n} = 3.17/22 = 0.144 \text{ or } 14.4\%$$

Example 7.4

In a village of 625 households, a simple random sample of 50 households were surveyed in order to estimate the average monthly household expenditure on toilet items. The estimate came out at \$0.88 with a standard error of \$0.10. Using this information, determine the sample size required to estimate the same characteristic in a neighboring village such that the permissible margin of error at 95 per cent probability level is 10 per cent of the true value (adapted from Murthy, problem 4.5).

We assume that the CV per unit is the same in both villages. From the first village, the CV of the sample estimator based on 50 households is $e = s_{\overline{y}}/\overline{y} =$ \$0.10/\$0.88 $= 0.1136$. The permissible margin of error of the estimator in the second village is

$d' = 10$ per cent of the true value at 95 per cent probability level, or

$$0.1\overline{Y}' = d' = t_{95\%}\sigma'_{\overline{y}} = 2\sigma'_{\overline{y}}$$

from equation (7.18); i.e. the desired CV for the sample estimator in the second village is

$$e' = \sigma'_{\bar{y}}/\bar{Y} = 0.1/2 = 0.05$$

From equation (7.22), the required sample size in the second village is

$$n' = n(e/e')^2 = 50 \times (0.1136/0.05)^2 = 50 \times 2.27^2 = 258$$

7.2.3 Sample sizes for sub-divisions of the universe

If estimates are required not only for the universe as a whole, but for sub-divisions such as geographical area, or sex and age groups of the population, obviously the sample size, obtained to estimate the overall universe value with a given precision, must be enlarged if estimators for the sub-divisions are required with the same precision as that of the overall universe estimators.

As a rough rule, if estimators with variance V are required for each of the k universe sub-divisions, the sample size should be

$$n'' = kn \tag{7.24}$$

where n is the required sample size for the overall universe estimate with the same variance V. The assumptions are that the per unit CVs of the sub-divisions are about equal, that the sub-divisions are approximately equal in size, and that the overall sample is large.

If the proportion of the universe units in any sub-division p_i, is known or could be estimated, the required sample size to estimate the average from the sub-division with variance V is

$$n_i = \sigma_i^2/p_iV \tag{7.25}$$

where σ_i^2 is the variance per unit in the ith sub-division and np_i is large. We have to take the maximum value of the right-hand side of equation (7.25) in order that it holds for all sub-divisions,

$$n = \text{maximum}\ (\sigma_i^2/p_iV) = (\sigma^2/V)\ \text{maximum}\ (1/p_i) \tag{7.26}$$

as the σ_i^2s will, on an average, be slightly smaller than the universe variance σ^2. If the sub-divisions are approximately equal in size, the p_i may be taken to be equal to $1/K$, and this relation is used in equation (7.25).

Example 7.5

In Example 7.1, what size of sample is required if total cases are wanted separately for males and females, with the same precision? (Cochran, exercise 4.3).

Here $k = 2$, and assuming that males and females are about equal in number in the population, we have, from equation (7.24), $n'' = 2 \times 2475 = 4590$.

7.3 Estimation of variance

An estimate of the universe variance or CV may be made from pilot studies or from data of related characteristics. The relations between the range (h) and the variance of mathematical distributions can also be utilized for estimating the variance. Deming (1960) has given the following rules (Table 7.2) for estimating the variance from the range if the shape of the distribution is known or could be guessed.

Table 7.2: Rules for estimating the variance from the range (h) depending on the shape of the distribution

Type	Mean	Variance	Standard deviation	Coefficient of variation
Binomial	ph	$p(1-p)/h^2$	$h/\sqrt{[p(1-p)]}$	$\sqrt{[(1-p)/p]}$
Rectangular	$1/2\,h$	$h^2/12$	$0.29\,h$	0.58
Right-triangle (I)	$1/3\,h$	$h^2/18$	$0.24\,h$	0.71
Right-triangle (II)	$2/3\,h$	$h^2/18$	$0.24\,h$	0.35
Symmetrical triangle	$1/2\,h$	$h^2/24$	$0.20\,h$	0.40
Normal	$1/2\,h$	$h^2/36$	$h/6$	1/3
		(Set $h = 6\sigma$)		

Source: Deming (1960), p. 260.

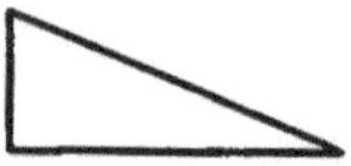

Right-triangle (I)

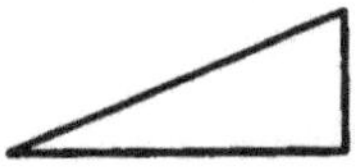

Right-triangle (II)

Note: The mathematical relations are not of much use if h is large or cannot be estimated closely. If h is large, the universe can be stratified (Part II) when, within a stratum, the shape of the distribution becomes simpler, closer to a rectangle, if stratification is effective and the mathematical relation between h and σ can be used for each stratum.

Example 7.6

The four-year colleges in the U.S.A. were divided into classes of four different sizes according to their 1952-3 enrollments. The standard deviations within each class are given in Table 7.3. If you know the class boundaries but not the value of σ, how well can you guess the σ values by using simple mathematical figures? No college has less than 200 students and the largest has about 50,000 students (Cochran, exercise 4.8).

Assuming a rectangular distribution within each class, the range (h) and the estimates of standard deviation ($= 0.29\,h$) are given in Table 7.3. If for Size class 1, we assume the right triangle distribution (II), the estimate s.d. $= 0.24\,h = 192$; and if for Size class 4, we assume the right triangle distribution (I), estimated s.d. $= 0.24\,h = 9600$.

Table 7.3: Actual and estimated standard deviations within each size class of four-year colleges in the U.S.A., 1952-3

	Size class of colleges			
	1	2	3	4
Number of students	up to 1000	1000-3000	3000-10,000	10,000+
Range (h)	800	2000	7000	40,000
Standard deviation				
(a) actual	236	625	2000	10,023
(b) estimate ($= 0.29\,h$)	232	580	2030	11,600
(c) estimated	192	580	2030	9600
	($0.24\,h$)	($0.29\,h$)	($0.29\,h$)	($0.24\,h$)

Source: Cochran, exercise 4.8.

7.4 Cost considerations

We have seen how to obtain the sample size required to provide estimators with a given precision in simple random sampling. To determine the implied total cost of the survey, we take the simplest type of cost function

$$c = c_0 + nc_1 \tag{7.27}$$

where c_0 is the overhead cost, and c_1 the cost of surveying one sample unit.

On the other hand, if the total cost C is fixed, the sample size is determined from the above relation, namely

$$n = (C - c_0)/c_1 \tag{7.28}$$

and the only thing left is to estimate the expected CV of the sample estimators from a sample of this size if prior information is available on the variability of data.

7.5 Balance of cost and error

With increase in sample size, the cost of the survey increases but the sampling error, and so the loss involved in basing any decision on the sample estimators, decreases; it is necessary to express this loss in monetary terms in order to find a balance between cost and error.

Taking an srs (with replacement) of n units, where the sample mean $\overline{y}$ is used to estimate the universe mean $\overline{Y}$, the error in the estimator $\overline{y}$ is

$$b = \overline{y} - \overline{Y}$$

If the loss due to this error b in the estimator $\overline{y}$ is taken as proportional to b^2, the expected loss for a given sample size is

$$L = lE(b^2) = l\sigma^2/n = l\sigma_{\overline{y}}^2 \tag{7.29}$$

where l is the constant of proportionality. With a total survey cost C, given by equation (7.27), the reasonable procedure would be to obtain the value of n that minimizes the total survey cost *plus* the loss involved with a sample of that size, namely, to minimize

$$C + L = c_0 + nc_1 + l\sigma^2/n \tag{7.30}$$

The minimizing value of n is obtained on partially differentiating equation (7.30) with respect to n, and equating the result to zero, when

$$n = \sigma\sqrt{\frac{l}{c_1}} \tag{7.31}$$

7.6 Cluster sampling

7.6.1 Sample size in cluster sampling

As cluster sampling is generally less efficient than an srs of the same size (Chapter 6), a larger total number of elementary units have to be included in a cluster sample in order to attain the same degree of precision as that of an unrestricted srs of the elementary units.

From equation (6.17) and section 6.2, it is seen that the relative change in variance due to sampling clusters instead of the elementary units directly is $(M_0 - 1)\rho_c \sim M_0\rho_c$, when M_0 (the cluster size) is large.

The required sample size due to sampling clusters is approximately

$$n_c = n(1 + M_0\rho_c) \tag{7.32}$$

where n is the sample size required for the given precision had the ultimate units been sampled directly.

Example 7.7

In Example 7.2, what would be the sample size required for estimating the birth rate with 2.5 per cent *CV* if clusters of 300 persons each are taken and the intraclass correlation coefficient is estimated at 0.001?

From equation (7.32), $n_c = 38,400\ (1 + 300 \times 0.001) = 49,920$, an increase of 30 per cent.

7.6.2 Optimal cluster size

The formulation in section 7.6.1 does not take into account the question of cost, for a cluster sample is less costly to enumerate than a direct sample of the elementary units. The optimal cluster size is so determined that the sampling variance is minimized for a given total cost or the total cost is minimized for a given precision.

A simple cost function with cluster sampling is

$$C = c_0 + nc_1 + nM_0c_2 \tag{7.33}$$

where c_0 is the overhead cost, c_1 the cost of travel, identification, contact etc., per cluster, and c_2 the cost of enumerating one ultimate unit. Generally, c_2 will be considerably smaller than c_1.

If the total cost is fixed at C, the value of n (the number of sample clusters) is (from equation 7.33)

$$n = (C - c_0)/(c_1 + M_0c_2) \tag{7.34}$$

The universe variance of the sample cluster mean $\overline{y}_c$ is, from equation (6.17),

$$\begin{aligned} \sigma^2_{\overline{y}_c} &= \sigma^2_b/n = \sigma^2[1+(M_0-1)\rho_c]/nM_0 \\ &= \sigma^2_b(c_1+M_0c_2)/(C-c_0) \text{ from equation (7.34)} \\ &= c_1\sigma^2(1+M_0c_2/c_1)[1+(M_0-1)\rho_c]/M_0(C-c_0) \qquad (7.35) \end{aligned}$$

If previous information is available from empirical or pilot studies, the values of the equation (7.35) can be computed and plotted for different values of M_0, to show the particular value of M_0 that minimizes the expression.

Putting

$$C-c_0 = C' = nc_1 + nM_0c_2 \qquad (7.36)$$

the optimal cluster size is

$$M_0 = \sqrt{\left[\frac{c_1}{c_2}\,\frac{1-\rho_c}{\rho_c}\right]} \qquad (7.37)$$

In practice, instead of dealing with the costs c_1 and c_2, we might consider the man-days required to be spent on the different operations. If the enumerators work singly, c_1 may be 2 to 3 man-days, and if 30 to 50 persons can be enumerated per day in a demographic inquiry, c_2 will be from 1/30 to 1/50 man-day; the ratio c_1/c_2 will then range from 60 to 150. The optimal size of cluster for some typical values of the ratio c_1/c_2 and of ρ_c are given in Table 7.4.

Table 7.4: Optimum size of cluster (number of persons) in typical surveys on birth and death rates

	Intraclass correlation coefficient ρ_c				
c_1/c_2	0.001	0.002	0.003	0.004	0.005
50	224	158	129	112	100
75	274	194	158	137	122
100	316	224	183	158	141
125	373	264	215	187	158
150	387	274	223	194	173

Taking a typical value of the ratio c_1/c_2 as 100, and the value of the intraclass correlation coefficient at 0.001 for the birth rate, the optimal cluster size is 316 persons; for the death rate, taking the intraclass correlation coefficient at 0.003, the optimal cluster size is 183 persons. The value of n (number of sample clusters) is determined from equation (7.34), given the values of C, c_0, c_1, c_2 and the optimum value of M_0.

Note that optimal cluster size is generally very broad (except for very small values of the intraclass correlation coefficient), so that substantial deviations from the optimal cluster size would not affect the cost very much.

Example 7.8

For a survey on birth rate, given that the total cost, neglecting overheads, is fixed at $20,000, and the enumerator cost per month is $300, what is the optimal size of sample if it is decided to select a cluster of persons, assuming that an enumerator has to spend, on average, two days in contacting the clusters and in other preliminary work; that he can enumerate an average of 40 persons a day; and that the intraclass correlation is estimated at 0.001?

Here $C = \$20,000$; 1 man-day = \$300/30 = \$10; $c_1 = 2.5$ man-days = \$25; $c_2 = 1/40$ man-day = \$0.25, so that $c_1/c_2 = 100$; $\rho_c = 0.001$. From Table 7.4, the optimal size of cluster is 316 persons. Taking M_0 at 300 persons, the total cost (neglecting overheads) is, from equation (7.33)

$$\$20,000 = \$25\,n + \$0.25 \times 300\,n = \$25\,n + \$75\,n = \$100\,n,$$

or $n = 200$ sample clusters. The total sample size is $nM_0 = 200 \times 300 = 60,000$ persons.

7.7 General remarks

Estimates of sample size required to obtain measures with a given precision will often be found to be quite large, when derived on the basis of unrestricted simple random sampling. But the "paean for large samples must be interrupted with caveats" (Kish, 1971). First, an unrestricted simple random sample is rarely used in practice; and, as noted earlier, special procedures (such as stratification, pps sampling etc.) and methods of estimation (such as ratio and regression methods) that require smaller sample sizes than srs, are used to provide estimators with the same or higher efficiency. Second, small samples have proved useful, not only as pilot studies to full-scale surveys, but also in providing interim estimates as in the earlier "rounds" of the Indian National Sample Survey where fairly accurate estimates of birth and death rates were obtained from only 3,000 to 8,000 sample households after special analytical techniques were applied (section 25.5.5): later, in 1958-9, the inquiry on population, births, and deaths was made to cover 2,600 sample villages with 234,000 households and 1.2 million persons. Third, a country with inadequate resources can start from a small sample and with increasing resources build up a fully adequate sample; the Current Population Survey of the U.S.A., for example, started in 1943 with 68 primary areas which were enlarged to the present 449. Fourth, it is possible to combine smaller monthly or quarterly estimates into yearly estimates, and the yearly estimates into estimates covering longer periods, to provide estimates with acceptable precision. And finally, in the interest of true accuracy, it may sometimes be better to conduct a smaller sample with adequate control than try to canvass a much larger sample but with poor quality data (see also Chapter 25).

Further reading

Cochran, chapter 4 and section 9.6; Deming (1950), chapter 14; Hansen *et al.* (1953), vol. I, section 4.11 and chapter 6D, and vol. II, section 4.9; Hedayat and Sinha, chapter 5; Levy and Lemeshaw, section 14.3; Kish (1965), section 2.6; Murthy (1967), sections 4.6 and 8.3; Sukhatme *et al.* (1984), section 1.9, 1.10, and 2.9-2.11; Yates, section 4.32.

Exercises

1. In newspapers in the U.S., items such as the following often appear:

   > "A poll taken recently of 1,296 persons nation-wide gave President Bill Clinton a 52/38 approval/disapproval rating; the margin of error was 3 per cent."

   Confirm the margin of error.

   (*Hint:* Use equation (7.7). See also Note 3 to section 7.2)

2. Material for the construction of 5000 wells was issued in a district in India in 1944. The list of cultivators to whom the material was issued was available along with the proposed location of each well. A large part of the material was, however, reported to have been misused, having been diverted to other purposes. It is proposed to estimate the proportion of wells not actually constructed, by taking a simple random sample of wells with the permissible margin of error of 10 per cent, and the degree of assurance desired 95 per cent. Determine the size of the sample required to estimate the proportion of wells not constructed for different values of the universe proportion ranging from 0.5 to 0.9 (Sukhatme *et al.* (1984), example 2.5, modified to with replacement plan in our answer.)

3. An anthropologist wishes to know the percentage of people with blood-group "O" in an island of 3200 persons with a margin of 5 per cent and 95 per cent degree of assurance. What is the required sample size? (Cochran, pp. 72-73.)

4. From the estimates obtained from the srs of 20 households in Example 2.2, compute the sample size required to provide an estimate of the total

number of persons in the universe within 10 per cent, apart from a chance of 1 in 20.

5. **For the 30 villages listed in Appendix IV, if no village is assumed to have less than 60 or more than 150 persons, estimate the s.d.**

CHAPTER 8

Self-weighting Designs

8.1 Introduction

To obtain estimates of totals, the sample observations have to be multiplied by factors, variously called weights, multipliers, or weighting, raising or inflation factors, which depend on the particular sample design and the method of estimation adopted. A sample design becomes self-weighting with respect to the particular linear estimator of a total (or an average) when the multipliers of the sample units are all equal: with only one multiplier, the tabulation becomes simpler, speedier and more economical, as the unbiased estimator of the universe total is obtained on multiplying the sample total by the constant, overall multiplier, and the estimator of the ratio of the universe totals of two variables is obtained on taking the ratio of the corresponding sample totals; the estimation of the variances, covariances, etc., is also simplified.

The problem of how to make a design self-weighting becomes particularly important in stratified multi-stage designs, but the principles are introduced in this chapter relating to unstratified single-stage samples.

8.2 Simple random sampling

For a simple random sample, the design is self-weighting with respect to the unbiased estimators of universe totals, averages, and ratios. For with the notations used in Chapter 2, an unbiased estimator of the universe total Y, obtained from a simple random sample of n units out of the universe total of N units, is

$$y_0^* = N \sum^{n} y_i / n \tag{8.1}$$

where y_i $(i = 1, 2, \ldots, n)$ is the value of the study variable for the ith sample unit. The weighting factor for the ith sample unit it, therefore,

$$w_i = N/n \tag{8.2}$$

which, when multiplied by y_i and added for the n sample units, provide the unbiased estimator y_0^* of the universe total Y. The multiplier is constant for all the sample units.

Similarly, the unbiased estimator of the universe mean $\overline{Y}$ is given by the sample mean

$$\overline{y} = \sum^n y_i/n \tag{8.3}$$

where the multiplier is $1/n$, constant for all the sample units.

In the estimation of the ratio of two universe totals, the sample estimator is

$$r = \sum^n y_i \Big/ \sum^n x_i \tag{8.4}$$

Here, because the design is self-weighting with respect to the two totals, the multiplier does not enter in the ratio; it does not also enter into the estimating equation for the variance of the ratio.

8.3 Varying probability sampling

In sampling with varying probability, the single-stage sample cannot in general be made self-weighting.

The unbiased estimator of the universe total Y obtained from a pps sample of n units is

$$y_0^* = \sum^n y_i/n\pi_i \tag{8.5}$$

where π_i is the probability of selection of the ith unit.

The multiplier of the ith sample unit for estimating the universe total Y is

$$w_i = 1/n\pi_i \tag{8.6}$$

which will not be the same for all the sample units, and the design will not, therefore, be self-weighting.

However, if $\pi_i = z_i/Z$, where z_i is the 'size' of the ith universe unit and $Z = \sum^n z_i$ is the total 'size' of the universe, then

$$y_0^* = \sum^n y_i/n\pi_i = Z\sum^n \frac{(y_i/z_i)}{n} = Z\sum^n r_i/n \tag{8.7}$$

where $r_i = y_i/z_i$.

If the ratios $r_i = y_i/z$ could be observed and recorded easily in the field, so that the r_is could be considered as the values of a newly defined study variable, then the design will be self-weighting with respect to the unbiased estimators, for the multipliers for the r_i values are $w_i' = Z/n$ which are the same for all the sample units. This has practical uses in crop surveys, as we have seen in section 5.5.

Notes

1. A pps sample design can be made self-weighting at the tabulation stage by selecting a sub-sample of the sample units with probability proportional to the multipliers, but the sampling variance will be increased (see section 13.5).
2. We shall see later in Chapters 18 and 23 how in a multi-stage design the sample can be made self-weighting even with pps sampling at some stages.

8.4 Rounding off of multipliers

When the original design is not self-weighting, one of the following procedures may be adopted to reduce the number of multipliers and thus to achieve at least partial self-weighting at the tabulation stage.

1. The multipliers w_is may be replaced by their simple average.
2. The multipliers may be rounded off to some convenient numbers as the nearest multiples of ten, hundred etc.
3. The multipliers may be rounded off to a small number of weights by a *random process* which would retain the unbiased character of the estimators (Note 3(c) to section 12.3.3).
4. A sub-sample of n' ultimate units may be selected from the original sample with probability proportional to their multipliers (Note 1 to section 8.3).

The first two procedures will lead to biased estimates with possible decrease in variance. However, the bias in the first procedure will be negligible if in the sample the covariance between the sample values and the multipliers is small. The third and the fourth procedures give unbiased estimators but with some increase in variance. In the third procedure, determination of optimum weights is very complicated if the number of such weights exceed three; on the other hand, the increase in variance will be large for a small number of weights.

PART II

STRATIFIED SINGLE-STAGE SAMPLING

CHAPTER 9

Stratified Sampling: Introduction

9.1 Introduction

If the universe is sub-divided into a number of sub-universes, called *strata*, and sampling is carried out independently in each stratum, the sampling plan is known as *stratified sampling*. Stratified sampling can be used in single- as well as multi-stage designs.

In this part, we shall illustrate the use of stratification in single-stage designs. The general theorems relating to stratified sampling will be dealt with in the present chapter; subsequent chapters in Part II will deal with stratified simple-random and varying-probability sampling; the size of the sample and allocation to different strata; formation of strata; and self-weighting designs.

9.2 Reasons for stratification

Stratified sampling is adopted in a number of situations:

1. when estimates are required for each sub-division of the universe separately, such as for geographical sub-divisions or for households in different social and economic groups in a household survey;

2. when estimates of universe characteristics are required with increased efficiency per unit of cost;

3. when a greater weightage is required to be given to some units that occur infrequently in the universe, such as households with very high incomes; and when the universe has a large variance, i.e. the units vary greatly in the values;

4. when different sampling procedures are to be adopted for different sub-universes, in which case the field work is easier to organize in

the different strata, formed according to the nature of the available ancillary information required for sample selection. For example, the population in many African countries can be classified according to their modes of living into urban, rural-sedentary, and rural-nomad; and these can be considered to constitute different strata; the sampling units – at least for the first-stage in a multi-stage design – could be town-blocks for the urban stratum, villages for the rural-sedentary stratum, and tribal hierarchies for the rural-nomad stratum. In sampling human populations, people in institutions such as boarding houses, hospitals, and jails may be considered separately from those living in households. Sampling might be simple random in one stratum, with probability proportional to size in another, and so on. The geographical strata may be further subdivided, each sub-stratum, for example, being allotted to a separate supervisor.

Notes

1. That stratification may lead to a gain in efficiency per unit of cost may be seen if we consider a universe composed of strata that are, with respect to the study variable, internally homogeneous, but heterogeneous with respect to each other: in this case, a very small sample from each strata would provide estimates with relatively small sampling variances.

   Taking an extreme case, consider the universe of six households with respective sizes 4, 4, 4, 5, 5, and 5. The reader will verify that the variance per unit in this universe is $\sigma^2 = 0.25$. If we were to draw a simple random sample of 4 households with replacement from this universe, an unbiased estimator of the total size is $N\bar{y}$ (where $N = 6$ and $\bar{y}$ is the sample mean, based on $n = 4$ sample units), with the sampling variance $N^2\sigma^2/n = 2.25$. Suppose, however, that the universe is subdivided into two strata, the first with the three households each of size 4, and the second with the other three households each of size 5; and an srs with replacement of 2 units is to be drawn from each stratum. Then, as we shall see later, an unbiased estimator of the total size is $(N_1\bar{y}_1 + N_2\bar{y}_2)$, where $N_1 = 3$, $N_2 = 3$, being respectively the total number of units in the two strata, and $\bar{y}_1$ and $\bar{y}_2$ the respective stratum means based on $n_1 = 2$ and $n_2 = 2$ sample units; the sampling variance of this unbiased estimator of the total is $N_1^2\sigma_1^2/n_1 + N_2^2\sigma_2^2/n_2 = 0$, as the variance per unit in the two strata $\sigma_1^2 = \sigma_2^2 = 0$.

2. Stratification may be carried out at different stages of sampling. The most common type of stratification is by administrative and geographical sub-division, such as by provinces, prefectures, counties, districts, and rural/urban categories. In household surveys, stratification is often carried out before the sample households are selected, by listing the households from which the sample is to be drawn, and recording such of other characteristics that may be readily obtained, e.g. size, and social and economic

classes; the households are then stratified on the basis of the characteristics recorded and then sampled (with a different sampling fraction) in each stratum.

3. Strata may be formed of units that are not geographically contiguous. Thus in a rural socio-economic survey, the villages may be stratified according to their population as given by the most recent census; or in a crop-yield survey, the fields may be classified according as they are irrigated or not; in the demographic inquiry in Mysore (a former State in India), conducted by the United Nations and the Government of India in 1951-2, three strata were formed in the rural areas; rural hills area with large-scale anti-malarial operations, rural hills area without large-scale anti-malarial operations, and rural plains (tank-irrigated area).
4. The data from a sample design that is not stratified, or is stratified according to some variable other than that desired, may, under some circumstances, be treated as if coming from a sample stratified according to the desirable stratification variable. This technique known as *stratification after sampling* or the *technique of post-stratification*, is explained in Chapter 10 in connection with stratified simple random sampling.

9.3 Fundamental theorems in stratified sampling

In the following the fundamental theorems given in section 2.7 are extended to the case of stratified sample designs.

1. The universe is sub-divided into L mutually exclusive and exhaustive strata. In the hth stratum ($h = 1, 2, \ldots, L$), there are n_h (≥ 2) independent and unbiased estimators t_{hi} ($i = 1, 2, \ldots, n_h$) of the universe parameter T_h; a combined unbiased estimator of T_h is, from estimating equation (2.28), the arithmetic mean

$$\bar{t}_h = \sum_{i=1}^{n_h} t_{hi}/n_h \tag{9.1}$$

An unbiased estimator of the variance of $\bar{t}_h$ is, from equation (2.29),

$$s^2_{\bar{t}_h} = \sum_{i=1}^{n_h}(t_{hi} - \bar{t}_h)^2/n_h(n_h - 1) = SSt_{hi}/n_h(n_h - 1) \tag{9.2}$$

where

$$\begin{aligned} SSt_{hi} &= \sum_{i=1}^{n_h}(t_{hi} - \bar{t}_h)^2 \\ &= \sum_{i=1}^{n_h} t^2_{hi} - \left(\sum_{i=1}^{n_h} t_{hi}\right)^2 \Big/ n_h \end{aligned} \tag{9.3}$$

is the corrected sum of squares of t_{hi}.

An unbiased estimator of the universe parameter for all the strata combined

$$T = \sum_{h=1}^{L} T_h \tag{9.4}$$

is obtained on summing up the stratum estimators $\bar{t}_h$, namely,

$$t = \sum_{h=1}^{L} \bar{t}_h \tag{9.5}$$

and an unbiased estimator of the variance of t is obtained on summing the L unbiased estimators of the stratum variances $s^2_{\bar{t}_h}$ (sampling in one stratum is independent of sampling in another, so that the stratum estimators $\bar{t}_h$ are mutually independent)

$$s^2_t = \sum_{h=1}^{L} s^2_{\bar{t}_h} \tag{9.6}$$

For another study variable similarly defined, an unbiased estimator of the universe parameter

$$U = \sum_{h=1}^{L} U_h \tag{9.7}$$

is the sum of the stratum estimators $\bar{u}_h$ of U_h,

$$u = \sum_{h=1}^{L} \bar{u}_h \tag{9.8}$$

and an unbiased variance estimator of u is the sum of the L unbiased estimators of the stratum variances $s^2_{\bar{u}_h}$, i.e.

$$s^2_u = \sum_{h=1}^{L} s^2_{\bar{u}_h} \tag{9.9}$$

2. An unbiased estimator of the covariance of $\bar{t}_h$ and $\bar{u}_h$ for the hth stratum is, from equation (2.32),

$$\begin{aligned} s_{\bar{t}_h \bar{u}_h} &= \sum_{i=1}^{n_h} (t_{hi} - \bar{t}_h)(u_{hi} - \bar{u}_h)/n_h(n_h - 1) \\ &= SPt_{hi}u_{hi}/n_h(n_h - 1) \end{aligned} \tag{9.10}$$

where

$$SPt_{hi}u_{hi} = \sum_{i=1}^{n_h}(t_{hi} - \bar{t}_h)(u_{hi} - \bar{u}_h)$$
$$= \sum_{i=1}^{n_h} t_{hi}u_{hi} - \left(\sum_{i=1}^{n_h} t_{hi}\right)\left(\sum_{i=1}^{n_h} u_{hi}\right) \Big/ n_h \qquad (9.11)$$

is the corrected sum of products of t_{hi} and u_{hi}.

For all the strata combined, an unbiased estimator of the covariance between t and u is the sum of the stratum estimators $s_{\bar{t}_h \bar{t}_h}$, i.e.

$$s_{tu} = \sum_{h=1}^{L} s_{\bar{t}_h \bar{u}_h} \qquad (9.12)$$

A consistent but generally biased estimator of the ratio of two universe parameters in the hth stratum $R_h = T_h/U_h$ is, from equation (2.33), the ratio of the sample estimators $\bar{t}_h$ and $\bar{u}_h$ in the stratum

$$r_h = \bar{t}_h/\bar{u}_h \qquad (9.13)$$

with estimated variance (from equation (2.34))

$$s^2_{r_h} = (s^2_{\bar{t}_h} + r^2_h s^2_{\bar{u}_h} - 2r_h s_{\bar{t}_h \bar{u}_h})/\bar{u}^2_h \qquad (9.14)$$

For all the strata combined, a consistent but generally biased estimator of the ratio of two universe parameters $R = T/U$ is the ratio of the respective sample estimators t and u,

$$r = t/u \qquad (9.15)$$

with estimated variance

$$s^2_r = (s^2_t + r^2 s^2_u - 2rs_{tu})/u^2 \qquad (9.16)$$

A generally biased but consistent estimator of the correlation coefficient ρ_h between the two study variables at the stratum level is, from equation (2.35),

$$\hat{\rho}_h = s_{\bar{t}_h \bar{u}_h} / s_{\bar{t}} s_{\bar{h}} \qquad (9.17)$$

and an estimator of the correlation coefficient ρ of the two study variables at the overall level is

$$\hat{\rho} = s_{tu}/s_t s_u \qquad (9.18)$$

Notes

1. The question of formation of strata, and the allocation of the total number of sample units into the different strata will be taken up later in Chapter 12.
2. Estimates are often required for higher levels of aggregation than strata, in addition to the overall universe estimators. For example, for the Indian National Sample Survey on Population, Births and Deaths (1958-9), the strata consisted of *tehsils* or groups of *tehsils* (a *tehsil* being an administrative unit comprising a number of villages and some small towns); a sample of villages was selected from each stratum and the population and the births and deaths during the preceding twelve months recorded. The country comprises a number of states, each state comprising a number of *tehsils*, and estimates were required not only for the country as a whole, but also for the states separately. For estimates for the whole country, estimating equations of the types (9.5), (9.6), (9.12), (9.15), and (9.16) were used; for the estimates for the states, the same types of equations were used excepting that for any state, the values of only those strata that fell within the state were considered (see Example 10.3).
3. As noted in section 9.2, the selection probabilities might be different in different strata.
4. For other notes, e.g. relating to estimators of ratios, see the notes to section 2.7.

9.4 Setting of confidence limits

We have seen in section 2.8 that for an unstratified single-stage sample, the $(100-\alpha)$ per cent confidence limits of a universe parameter T are set with the sample estimator $\bar{t}$ and its estimated standard error $s_{\bar{t}}$, thus:

$$\bar{t} \pm t'_{\alpha,n-1} s_{\bar{t}} \tag{2.40}$$

where $t'_{\alpha,n-1}$ is the 100α percentage point of the t-distribution for $(n-1)$ degrees of freedom. It was also noted that for large n, the percentage points of the normal distribution could be used for those of the t-distribution.

The above holds for each stratum separately. Thus the 100α per cent confidence limits of the universe parameter T_h for the hth stratum are

$$\bar{t}_h \pm t'_{\alpha,n_h-1} s_{\bar{t}_h} \tag{9.19}$$

the sample stratum estimator $\bar{t}_h$ being defined by equation (9.1) and its estimated standard error $s_{\bar{t}_h}$ by equation (9.2).

For the confidence limits of the overall universe parameter T, the normal distribution may be used if the estimators of variances of the stratum estimators $\bar{t}_h$, namely $s^2_{\bar{t}_h}$, are based on not too few degrees of freedom and it is assumed that the overall universe estimator t, defined by equation (9.5), is normally distributed. Then

$$t \pm t'_\alpha s_t \tag{9.20}$$

are the $(100 - \alpha)$ per cent confidence limits of the universe parameter T where s_t is defined by equation (9.5) and t'_α is the normal percentage point for probability α.

If, however, the variance estimators in the different strata $s^2_{\bar{t}_h}$ are based on small numbers of degrees of freedom, the normal distribution cannot be used. The t-distribution can be used by computing the effective number of degrees of freedom.

$$n'' = \frac{\left(\sum^L s^2_{\bar{t}_h}\right)^2}{\sum^L [s^4_{\bar{t}_h}/(n_h - 1)]} \tag{9.21}$$

which will lie between (the smallest of the) $n_h - 1$ and $\sum^L (n_h - 1)$. The $(100 - \alpha)$ per cent confidence limits of T are

$$t \pm t'_{\alpha,n''} s_t \tag{9.22}$$

where $t'_{\alpha,n''}$ is the 100α percentage point of the t-distribution of n'' degrees of freedom.

9.5 Selection of two units from each stratum

The computation of estimates is considerably simplified when, in each stratum, two units are selected with replacement out of the total number of units. Noting that $n_h = 2$ in this case,

$$\bar{t}_h = \frac{1}{2}(t_{h1} + t_{h2}) \tag{9.23}$$

where t_{h1} and t_{h2} are the two unbiased estimators of T_h from the two selected sample units.

An unbiased estimator of the variance of t_h is

$$s^2_{\bar{t}_h} = \frac{1}{4}(t_{h1} - t_{h2})^2 \tag{9.24}$$

so that

$$s_{\bar{t}_h} = \frac{1}{2}|t_{h1} - t_{h2}| \tag{9.25}$$

An unbiased estimator of the covariance of $\bar{t}_h$ and $\bar{u}_h$ is

$$s_{\bar{t}_h \bar{u}_h} = \frac{1}{4}(t_{h1} - t_{h2})(u_{h1} - u_{h2}) \tag{9.26}$$

Also,

$$r_h = t_h/u_h = (t_{h1} + t_{h2})/(u_{h1} + u_{h2}) \tag{9.27}$$

with a variance estimator

$$s^2_{r_h} = \frac{(t_{h1} - t_{h2})^2 + r_h^2(u_{h1} - u_{h2})^2 - 2r_h(t_{h1} - t_{h2})(u_{h1} - u_{h2})}{(u_{h1} + u_{h2})^2} \tag{9.28}$$

The overall estimators (for all the strata taken together) are:

$$t = \sum_{h=1}^{L} \bar{t}_h = \frac{1}{2}\sum_{1}^{L}(t_{h1} + t_{h2}) \tag{9.29}$$

$$s_t^2 = \sum_{h=1}^{L} s^2_{\bar{t}_h} = \frac{1}{4}\sum_{h=1}^{L}(t_{h1} - t_{h2})^2 \tag{9.30}$$

$$s_{tu} = \sum_{h=1}^{L} s_{\bar{t}_h \bar{u}_h} = \frac{1}{4}\sum_{h=1}^{l}(t_{h1} - t_{h2})(u_{h1} - u_{h2}) \tag{9.31}$$

Also,

$$r = t/u = \sum_{h=1}^{L}(t_{h1} + t_{h2}) \Big/ \sum_{h=1}^{L}(u_{h1} + u_{h2}) \tag{9.32}$$

with a variance estimator

$$s_r^2 = \frac{\sum_{h=1}^{L}[(t_{h1} - t_{h2})^2 + r^2(u_{h1} - u_{h2})^2 - 2r(t_{h1} - t_{h2})(u_{h1} - u_{h2})]}{\sum_{h=1}^{L}(u_{h1} + u_{h2})^2} \tag{9.33}$$

Thus the sums and differences of the two estimates obtained from the two (first-stage) sample units in the different strata supply the required data for computing the overall estimators and their variances, not only for the totals but also for the ratios. This procedure has been followed in the Indian National Sample Survey since its inception in 1950, and some standard errors of ratios and ratio estimates were published in 1956.

Note: Another unbiased estimator of the variance of t is

$$\frac{1}{4}\left(\sum_{h=1}^{L} t_{h1} - \sum_{h=1}^{L} t_{h2}\right)^2 \tag{9.34}$$

and the corresponding estimated standard error

$$\frac{1}{2}\left|\sum_{h=1}^{L} t_{h1} - \sum_{h=1}^{L} t_{h_2}\right| \tag{9.35}$$

These are simpler to compute, but being based on only one degree of freedom will be less efficient than the estimators defined earlier which are based on an effective number of degrees of freedom, given by equation (9.21), that will lie between 1 and $(L-1)$. Another unbiased estimator of the covariance of t and u can be defined similarly, and so also the estimated variance of r.

Further reading

Cochran, chapter 5; Hansen *et al.* (1953), vol. I, chapter 5A; Hedayat and Sinha, chapter 9; Murthy (1967), sections 7.1-7.3; Sukhatme *et al.* (1984), chapter 4; Yates, sections 3.3-3.5.

CHAPTER 10

Stratified Simple Random Sampling

10.1 Introduction

In this chapter we will consider the estimating methods for totals, means, ratios of study variables and their variances when a simple random sample has been drawn in each stratum. The methods of estimation of proportion of units in a category, the use of ratio estimators and stratification after sampling will also be considered.

10.2 Estimation of totals, means, ratios, and their variances

10.2.1 Structure of a stratified srs

Consider the total universe to be divided on certain criteria into L strata, and let the hth stratum ($h = 1, 2, \ldots, L$) contain N_h total number of units. The values of two study variables for the ith unit ($i = 1, 2, \ldots, N_h$) in the hth stratum will be denoted respectively by Y_{hi} and X_{hi}. The total number of units in the universe is

$$N = \sum_{h=1}^{L} N_h \tag{10.1}$$

In each stratum, a simple random sample is taken with replacement (or without replacement, but with a small sampling fraction, 10 per cent or less). Let the number of sample units in the hth stratum be n_h and the values of two study variables for the ith selected sample unit ($i = 1, 2, \ldots, n_h$) in the hth stratum be denoted respectively be y_{hi} and x_{hi}. The total number of sample units is

$$n = \sum_{h=1}^{L} n_h \tag{10.2}$$

10.2.2 Estimators of totals, means, and ratio of two totals

An unbiased estimator of the stratum total Y_h, obtained from the ith sample unit in the hth stratum is, from equation (2.42),

$$y^*_{hi} = N_h y_{hi} \tag{10.3}$$

and the combined unbiased estimator for the stratum total Y_h, obtained from the n_h sample units in the hth stratum, is, from equation (2.43) or (9.1),

$$y^*_{h0} = \sum_{i=1}^{n_h} y^*_{hi}/n_h = N_h \bar{y}_h \tag{10.4}$$

where $\bar{y}_h = \sum^{n_h} y_{hi}/n_h$ is the mean of the y_{hi} values in the hth stratum.

Universe totals, means, and ratios of two totals and their sample estimators are defined in Table 10.1. These follow from the results of section 2.9 for any stratum and of section 9.3 for all the strata combined. The sample estimators of totals and means are unbiased (theoretical proofs are given in Appendix II, section A2.3.9); those for ratios are consistent, but generally biased.

Table 10.1: Some universe parameters and their sample estimators for a stratified simple random sample

	For the hth stratum		For all strata combined		
	Universe parameter (a)	Sample estimator (b)	Universe parameter (c)	Sample estimator (d)	
For a study variable:					
Total	Y_h	y^*_{h0}	Y	y	
	$= \sum_{i=1}^{N_h} Y_{hi}$	$= \sum_{i=1}^{n_h} y^*_{hi}/n_h$	$= \sum_{h=1}^{L} Y_h$	$= \sum_{h=1}^{L} y^*_{h0}$	(10.5)
	$= N_h \bar{Y}_h$	$= N_h \bar{y}_h$			
Mean	$\bar{Y} = Y_h/N_h$	$\bar{y}_h$	$\bar{Y} = Y/N$	y/N	(10.6)
For two study variables:					
Ratio of totals	$R_h = Y_h/X_h$	$r_h = y^*_{h0}/x^*_{H0}$	$R = Y/X$ $= \bar{Y}_h/\bar{X}_h$	$r = y/x$	(10.7)

10.2.3 Sampling variances and their estimators for a stratum

In the hth stratum, the variance per unit is

$$\sigma_h^2 = \sum_{i=1}^{N_h} (Y_{hi} - \overline{Y}_h)^2 / N_h \tag{10.8}$$

and the variance of the sample mean $\overline{y}_h$ in a simple random sample (with replacement) of n_h units is, from equation (2.17),

$$\sigma_{\overline{y}_h}^2 = \sigma_h^2 / n_h \tag{10.9}$$

so that the variance of the sample estimator y_{h0}^* of the stratum total Y_h is

$$N_h^2 \sigma_{\overline{y}_h}^2 = N_h^2 \sigma_h^2 / n_h \tag{10.10}$$

An unbiased estimator of the variance of the estimator y_{h0}^* is, from equation (2.44) or (9.2),

$$\begin{aligned} s_{y_{h0}^*}^2 &= \sum_{i=1}^{n_h} (y_{hi}^* - y_{h0}^*)^2 / n_h(n_h - 1) \\ &= SSy_{hi}^* / n_h(n_h - 1) \\ &= N_h^2 \sum_{i=1}^{n_h} (y_{hi} - \overline{y}_h)^2 / n_h(n_h - 1) \\ &= N_h^2 SSy_{hi} / n_h(n_h - 1) \end{aligned} \tag{10.11}$$

An unbiased estimator of the variance of the sample mean $\overline{y}_h$ is, from equation (2.46),

$$s_{\overline{y}_h}^2 = s_{y_{h0}^*}^2 / N_h^2 \tag{10.12}$$

An unbiased estimator of the covariance of y_{h0}^* and x_{h0}^* (x_{h0}^* being defined similarly for another study variable) is from equation (2.47) or (9.10)

$$\begin{aligned} s_{y_{h0}^* x_{h0}^*} &= \sum_{i=1}^{n_h} (y_{hi}^* - y_{h0}^*)(x_{hi}^* - x_{h0}^*) / n_h(n_h - 1) \\ &= SPy_{hi}^* x_{hi}^* / n_h(n_h - 1) \\ &= N_h^2 \sum_{i=1}^{n_h} (y_{hi} - \overline{y}_h)(x_{hi} - \overline{x}_h) / n_h(n_h - 1) \\ &= N_h^2 SPy_{hi} x_{hi} / n_h(n_h - 1) \end{aligned} \tag{10.13}$$

An estimator of the variance of r_h is, from equation (9.14) or (2.49),

$$\begin{aligned} s_{r_h}^2 &= (s_{y_{h0}^*}^2 + r_h^2 s_{x_{h0}^*}^2 - 2r_h s_{y_{h0}^* x_{h0}^*})/x_{h0}^{*2} \\ &= (SSy_{hi} + r_h^2 SSx_{hi} - 2r_h SPy_{hi}x_{hi})/n_h(n_h-1)\bar{x}_h^2 \quad (10.14) \end{aligned}$$

10.2.4 Sampling variances and their estimators for all strata combined

The variance of the estimator y of the universe total Y is

$$\sigma_y^2 = \sum^L \sigma_{y_{h0}^*}^2 = \sum^L N_h^2 \sigma_h^2 / n_h \quad (10.15)$$

which is estimated unbiasedly by the sum of the unbiased variance estimators of y_{h0}^* from equation (9.6), namely, by

$$s_y^2 = \sum^L s_{y_{h0}^*}^2 \quad (10.16)$$

The sampling variance of the estimator of the universe mean is

$$\sigma_y^2 / N^2 \quad (10.17)$$

which is estimated unbiasedly by

$$s_y^2 / N^2 \quad (10.18)$$

Theoretical proofs of the above are given in Appendix II, section A2.3.9.

An unbiased estimator of the covariance of the estimators y and x is, from equation (9.12),

$$s_{yx} = \sum^L s_{y_{h0}^* x_{h0}^*} \quad (10.19)$$

A variance estimator of r is from equation (9.16)

$$s_r^2 = (s_y^2 + r^2 s_x^2 - 2r s_{yx})/x^2 \quad (10.20)$$

Notes

1. *Sampling without replacement.* In srs without replacement, the sample estimators of a universe total, mean, and ratio of two universe totals remain the same as for srs with replacement. The sampling variance of $y = \sum^L y_{h0}^*$ is, however, of the form

$$\sigma_y^2 = \sum^L \frac{N_h^2 \sigma_h^2}{n_h} \frac{(N_h - n_h)}{(N_h - 1)} = \sum^L \frac{N_h^2 S_h^2}{n_h} (1 - f_h) \tag{10.21}$$

where

$$S_h^2 = \frac{N_h}{N_h - 1} \sigma_h^2 \tag{10.22}$$

and

$$f_h = n_h / N_h \tag{10.23}$$

σ_y^2 is estimated unbiasedly by

$$s_y^2 = \sum^L \frac{N_h^2 s_h^2}{n_y} (1 - f_h) \tag{10.24}$$

where

$$s_h^2 = \sum_{i=1}^{n_h} (y_{hi} - \overline{y}_h)^2 / (n_h - 1) \tag{10.25}$$

The sampling variance of the universe mean and its unbiased estimator are obtained on dividing the respective expressions for the universe total by N^2.

Theoretical proofs are given in Appendix II, section A2.3.9.

An unbiased covariance estimator s_{yx} is similarly defined and so also the variance estimator of the ratio $r = y/x$.

2. The estimator of the overall universe mean $\overline{Y}$ is, from equation (10.6d)

$$\overline{y} = y/N = \sum_{h=1}^{L} y_{h0}^{*}/N = \sum_{h=1}^{L} n_h \overline{y}_h / N \tag{10.26}$$

i.e. the weighted average of the sample stratum means ($\overline{y}_h$), the weights being N_h/N, the proportion of the total number of universe units contained in each stratum. This will be the same as the simple unweighted mean of the values of the study variables (which is the same as the simple unweighted mean of the sample stratum means),

$$\sum_{h=1}^{L} \sum_{i=1}^{n_h} y_{hi}/n = \sum_{h=1}^{L} n_h \overline{y}_h / n \tag{10.27}$$

only when

$$N_h/N = n_h/n \quad \text{or} \quad n_h/N_h = n/N \tag{10.28}$$

i.e. when the sampling fraction is the same in all the strata, in which case

$$n_h = (n/N)N_h \tag{10.29}$$

Thus the allocation of the number of sample units to the different strata is in proportion to the total number of units in each stratum: this is known an *proportional allocation*. It leads to a self-weighting design for a stratified simple random sample.

3. For other notes relating for example to estimators of ratios, see the notes to section 2.7.

Example 10.1

From village 8 in state I (Appendix IV), assume that for a family budget inquiry, a preliminary listing of all the 24 households has been made along with information on size. Divide these 24 households into two strata - one with households of size 1-5 persons and the other with size 6 persons and above. Select a simple random sample (with replacement) of 3 households from the first stratum and of 2 households from the second stratum, and on the basis of the data on income and food cost of these sample households, estimate for all the 24 households the total income, total food cost, income per household and per person, food cost per household and per person and the proportion of income spent on food.

The details of the selected households in the two strata with the required information are given in Table 10.2. The required computations are shown in a summary form in Table 10.3 and the final results in Table 10.4.

Table 10.2: Size, total daily income, and food cost in the srs of households in village no. 8, in state I (Apppendix IV)

Household serial no.	Household size	Total daily income ($)	Daily food cost ($)
Stratum I (households of size 1-5); total number of households = 17			
2	3	33	21
9	4	39	24
11	3	35	22
Stratum II (households of size 6 and above); total number of households = 7			
16	6	62	31
22	6	61	29

Table 10.3: Computation of the required estimates for a stratified srs: data of Table 10.2

	Stratum I ($h = 1$)	Stratum II ($h = 2$)	Combined
(1) Number of units: total (N_h)	17	7	24 (N)
(2) Number of units: sample (n_h)	3	2	5 (n)
For total daily income			
(3) Total income in sample households $\left(\sum_1^{n_h} x_{hi}\right)$	\$107	\$123	
(4) Mean income per household $\left(\bar{x}_h = \sum_1^{n_h} x_{hi}/n_h\right)$	\$35.67	\$61.50	\$43.20†‡(x/n)
(5) Estimated total income ($x^*_{h0} = N_h\bar{x}_h$)	\$606.34	\$430.5	\$1036.84 (x)
(6) $\sum_1^{n_h} x^2_{hi}$	3835	7565	
(7) $\left(\sum_1^{n_h} x_{hi}\right)^2/n_h$	3816.3333	7564.5	
(8) $SSx_{hi} =$ row (6) − row (7)	18.6667	0.5	
(9) $SSx_{hi}/n_h(n_h - 1)$	3.1111	0.25	
(10) $s^2_{x^*_{h0}} = N^2_h \times$ row (9)	899.1111	12.25	911.36 (s^2_x)
(11) $s_{x^*_{h0}} =$ sq. root of row (10)	\$29.99§	\$3.5§	\$30.19† (x_s)
(12) CV of $x^*_{h0} =$ row (11)/row (5)	4.9%§	0.81%§	2.91%†
For Food Cost			
(13) $\left(\sum^{n_h} w_{hi}\right)$	\$67	\$60	
(14) Mean food cost per household $\bar{w}_h = \sum_1^{n_h} w_{hi}/n_h$	\$22.33	\$30	\$24.57¶ (w/N)
(15) Estimated total food cost ($w^*_{h0} = N_h\bar{w}_h$)	\$379.67	\$210	\$589.67 (w)
(16) $\sum_1^{n_h} w^2_{hi}$	1501	1802	

Table 10.3 continued

	Stratum I ($h = 1$)	Stratum II ($h = 2$)	Combined
(17) $\left(\sum_1^{n_h} w_{hi}\right)^2 / n_h$	1496.3333	1800	
(18) SSw_{hi} = row (16) − row (17)	4.3333	2	
(19) $SSw_{hi}/n_h(n_h - 1)$	0.777778	1	
(20) $s^2_{w^*_{h0}} = N_h^2 \times$ row (19)	224.7777788	49	273.777778 (s^2_w)
(21) $s_{w^*_{h0}}$ = sq. root of row (20)	15.00§	7§	16.55† (s_w)
(22) CV of $w^*_{h)}$ = row (21)/row (15)	3.9%§	3.33%§	2.81%†
(23) Proportion of income spent on food $r_h = w^*_{h0}/x^*_{h0}$ = row(15)/row (5)	0.626166§	0.487800§	0.568718 (r)†
(24) r_h^2	0.3092661§	0.2379488§	0.3234402 (r^2)†
(25) $2r_h$	1.243332§	0.975600§	1.137436 ($2r$)†
(26) $\sum_1^{n_h} x_{hi}w_{hi}$	2399	3691	
(27) $\left(\sum_1^{n_h} x_{hi}\right)\left(\sum w_{hi}\right) \Big/ n_h$	2389.666667	3690	
(28) $SPx_{hi}w_{hi}$ = row (26) - row (27)	9.333333	1	
(29) $SPx_{hi}w_{hi}/n_h(n-1)$	1.555556	0.5	
(30) $s_{x^*_{h0}w^*_{h0}} = N_h^2 \times$ row (29)	449.55556	24.5	474.055556 (s_{xw})
(31) $r_h^2 \times s^2_{x^*_{h0}}$	349.993432§	0.291480§	294.770278 ($r^2 \times x_x^2$)†
(32) $2r_h s_{x^*_{h0}} w^*_{h0}$	558.946009§	23.90220§	539.209802 ($2rs_{xw}$)†

Table 10.3 continued

	Stratum I ($h = 1$)	Stratum II ($h = 2$)	Combined
(33) row (20) + row (31) − row (32)	15.825201§	25.389286§	29.338254†
(34) x_{h0}^{*2}	367 648.1956§	185 330.25§	1 075 037.1856 (x^2)†
(35) $s_{r_h}^2$ = row (33)/x_{h0}^{*2}	0.0000430444§	0.000136995§	0.0000272904 (s_r^2)†
(36) s_{r_h} = sq. root of row (34)	0.00656§	0.0117§	0.00522 (s_r)†
(37) CV of r_h = row (36)/row (23)	1.05%§	2.40%§	0.91%†
(4) Average income per household $\overline{x}_h = \sum_1^{n_h} x_{hi}/n_h$	\$35.67§	\$61.50§	\$43.20 (x/N)†
(9) $s_{\overline{x}_h}^2 = s_{x_{h0}^*}^2/N_h^2$	3.1111§	0.25§	1.582222 (s_x^2/N^2)†
(38) $s_{\overline{x}_h}$	\$1.76§	\$0.50§	\$1.26 (s_x/N)†
(14) Average food cost per household ($\overline{w}_h$)	\$22.33§	\$30§	\$24.57 (w/N)†
(19) $s_{\overline{w}_h}^2 = s_{w_{h0}^*}^2/N_h^2$	0.777778§	1§	0.475309 (s_w^2/N^2)†
(39) $s_{\overline{w}_h}$	\$0.88	\$1	\$0.69 (s_w/N)†
(40) Total number of persons	65	44	109
(41) Average income per head = row (3)/row (41)	\$9.33§	\$9.78§	\$9.51†
(42) Estimated S.E. of average income per head = row (11)/row (40)	\$0.46§	\$0.08§	\$0.28†

Table 10.3 continued

	Stratum I ($h = 1$)	Stratum II ($h = 2$)	Combined
(43) Average food cost per head = row (15)/row (40)	\$5.84§	\$4.47§	\$5.41†
(44) Estimated S.E. of average food cost per head = row (21)/row (40)	\$0.23§	\$0.16§	\$0.15†

† Not additive, i.e. the combined estimate is not obtained by adding the stratum estimates.
‡ Obtained by dividing the value of x by N (= 24).
§ Computations not necessary if separate estimates are not required for each stratum.
¶ Obtained by dividing the value of w by N (= 24).

Table 10.4: Estimates and standard errors of income and food costs, computed from the data of Tables 10.2 and 10.3: Stratified srs of households

	Item	Estimate	Standard error	*CV* (%)
1.	Daily income			
	(i) Total	\$1037	\$30.19	2.91
	(ii) Per household	\$43.50	\$1.26	2.91
	(iii) Per person	\$9.51	\$0.28	2.91
2.	Daily Food cost			
	(i) Total	\$590	\$16.55	2.81
	(ii) Per household	\$24.57	\$0.69	2.81
	(iii) Per person	\$5.41	\$0.15	2.81
3.	Proportion of income spent on food	0.5687	0.00522	0.92

Example 10.2

From each of the three states (Appendix IV), select a simple random sample (with replacement) of two villages each, and on the basis of the data on the number of households and of persons in these sample villages, estimate for three states separately, and also for the three states combined, the total numbers of households and of persons, and the average household size with standard errors.

Here the states are the strata. The information for the selected sample villages is given in Table 10.5. As two sample villages are selected from each state, the simplified procedures of computation of section 9.5 will be used, where the sums and the differences of the values of the study variables for the two sample units will provide the required estimates. The required computations are shown in Tables 10.5-10.7 and the final results in Table 10.8.

Example 10.3

In the Indian National Sample Survey (1958-9), the inquiry into population, births and deaths in the rural areas was based on a single-stage stratified design. Rural India was divided into 218 strata (composed of *tehsils* or groups of *tehsils*, a *tehsil* being a administrative unit consisting of villages and a few towns), each stratum containing approximately equal populations as in the Census of 1951. A total period of survey of twelve months was divided into six sub-rounds each of two months' duration. A total of 2616 sample villages were covered, with 436 samples villages in each sub-round. In each stratum, two enumerators collected the information independently, each surveying a sub-sample of six villages, i.e. one village each in a sub-round. The allocation of the total sample of 2616 villages to the different States of India was made on the basis of various factors, including the total 1951 Census population, and the allocations were rounded to multiples of 12. In each stratum, the sample villages were selected systematically with a random start after arranging the tehsils in a particular order so as to increase the efficiency of the estimators. In the selected villages, all the households were surveyed in regard to demographic characteristics, including births and deaths during the 365 days preceding the enumeration. For any sub-round, the estimating equations will take the forms in Example 10.2 as there were two sample villages in each stratum in a sub-round. (For further details, see Murthy (1967), Chapter 15, and Som, De, Das, Pillai, Mukherjee, and Sarma, 1961).

Some estimates and their standard errors are given in Table 10.9 for the first sub-round (July-August 1958). The estimated rate of growth of population was 2 per cent per annum in 1958-9. The registration system for births and deaths is still defective in India, and this was the first national survey which showed that population in India was growing at a rate much faster than the annual 1.3 per cent rate recorded during the 1941-1951 census decade, which was used in the first two five-year plans covering the period 1951-61. This was later confirmed by the Census of 1961 and the third five-year plan took account of this accelerated rate of growth of population.

Table 10.5: Population and number of households in each of the two sample villages in the three states (Appendix IV) and computation of stratum estimates: stratified srs of villages

	Number of villages		Village serial no.	Sample village no.	Number of households		Population	
State h	Total N_h	Sample n_h		i	In sample village x_{hi}	Stratum estimate $x^*_{hi} = N_h x_{hi}$	In sample village y_{hi}	Stratum estimate $y^*_{hi} = N_h y_h$
(1)	(2)	(3)	(4)	(5)	(6)	(7)	(8)	(9)
I	10	2	2	1	18	180	82	820
			10	2	22	220	88	880
					Total	400		1700
					Mean (= stratum estimate of total)	200 (x^*_{h0})		850 (y^*_{h0})
					Difference	−40 (d_{xh})		−60 (d_{yh})
					$\frac{1}{2}$ |Difference| (= estimated standard error)	20 (s^*_{h0})		30 ($s_{y^*_{h0}}$)
II	11	2	5	1	20	220	105	1155
			11	2	18	198	94	1034
					Total	448		2189
					Mean (= stratum estimate of total)	209 (x^*_{h0})		1094.5 (y^*_{h0})
					Difference	22 (d_{xh})		121 (d_{yh})
					$\frac{1}{2}$ |Difference| (= estimated standard error)	11 (s^*_{h0})		60.5 ($s_{y^*_{h0}}$)
III	9	2	2	1	21	189	121	1089
			3	2	18	162	105	945
					Total	351		2034
					Mean (= stratum estimate of total)	175.5 (x^*_{h0})		1017 (y^*_{h0})
					Difference	27 (d_{xh})		144 (d_{yh})
					$\frac{1}{2}$ |Difference| (= estimated standard error)	13.5 (s^*_{h0})		72 ($s_{y^*_{h0}}$)

Table 10.6: Computation of estimated total population and number of households and their standard errors: data of Table 10.5

State	Number of households			Population		
	Estimate	Standard error	Variance	Estimate	Standard error	Variance
(h)	x^*_{h0}	$s_{x^*_{h0}}$	$s^2_{x^*_{h0}}$	y^*_{h0}	$s_{y^*_{h0}}$	$s^2_{y^*_{h0}}$
I	200.0	20.0	400.00	850.0	30.0	900.00
II	209.0	11.0	121.00	1094.5	60.5	3660.25
III	175.5	13.5	182.25	1017.0	72.0	5184.00
All zones combined	584.5 (x)	26.52† (s_x)	703.25 (s^2_x)	2961.5 (y)	98.71† (s_y)	9744.25 (s^2_y)

† Not additive.

Table 10.7: Computation of the estimated average household size and sampling variance: data of Tables 10.5 and 10.6

State (h)	Average household size $r_h = y^*_{h0}/x^*_{h0}$	$s_{y^*_{h0}x^*_{h0}} = \frac{1}{4} d_{yh} d_{xh}$	$r^2_h s^2_{x^*_{h0}}$	$2r_h s_{y^*_{h0}x^*_{h0}}$	$s^2_{y^*_{h0}} + r^2 s^2_{x^*_{h0}} - 2r_h s_{y^*_{h0}x^*_{h0}}$	$s^2_{r_h} = \frac{\text{col. (6)}}{x^{*2}_{h0}}$
(1)	(2)	(3)	(4)	(5)	(6)	(7)
I	4.2500	600.0	7225.0000	5100.0000	3025.0000	0.07562500
II	5.2368	665.5	3318.3161	6970.1808	8.3853	0.00015621
III	5.7949	972.0	6120.1190	11 265.2865	38.8334	0.00126081
All zones combined	5.0667 (r)	2237.5 (s_{yx})	18 053.4120 $(r^2 s^2_x)$	22 673.4825 $(2r s_{yx})$	5124.1795 $(s^2_y + r^2 s^2_x - 2r s_{yx})$	0.01499876 (s^2_r)

Table 10.8: Estimates and standard errors, computed from the data of Table 10.5: stratified srs of villages

	State I	State II	State III	All states combined
1. *Total population*				
(a) Estimate	850.0	1094.5	1017.0	2961.5
(b) Standard error	30.0	60.5	72.0	98.7
(c) Coefficient of variation (%)	3.52	5.53	7.08	3.33
2. *Total number of households*				
(a) Estimate	200.0	209.0	175.5	584.5
(b) Standard error	20.0	11.0	13.5	26.5
(c) Coefficient of variation (%)	10.00	5.26	7.69	4.54
3. *Average household size*				
(a) Estimate	4.25	5.24	5.79	5.07
(b) Standard error	0.2750	0.0125	0.0355	0.1225
(c) Coefficient of variation (%)	6.47	0.24	0.61	2.42

Table 10.9: Estimate and standard error of some demographic parameters: Indian National Sample Survey, July-August 1958 (404 sample villages with 36,365 households)

Parameter	Estimate	Standard error	*CV* (%)
Birth rate (per 1000 persons)	38.61	0.74	1.92
Death rate (per 1000 persons)	18.97	0.73	3.85
Growth rate (per cent)	1.96	0.08	4.08
Percentage of population employed	45.55	0.71	1.56
Household size	5.06	0.04	0.79

Source: Som *et al.*

Note that such a large-scale sample survey (with 2616 sample villages, about 234,000 sample households and 1.2 million sample persons) used the simple design described in this chapter, the data of which the reader should by now be able to analyze by himself.

Example 10.4

For crop surveys starting from 1937 in the Bengal province of pre-independence India and in the West Bengal State of India since 1948, the design has been stratified single-stage. The whole area is cadastrally surveyed (i.e. surveyed for tax purposes) so that village maps are available showing each plot. A number of sample grids were selected, the size varying in different years. (A "grid" is a square mesh on a plane formed by two sets of lines perpendicular to each other, each line of each set being at a constant interval from the adjacent lines.)

With the help of the village maps, the field enumerators identified the individual plots wholly or partially included in each grid, indicating against each plot the fraction of it that was sown with a particular crop. From these data and the total area of each plot (which was known), the proportion of the total area of the grid occupied by the crop was calculated. Let this proportion be denoted by p_{hi} $(h = 1, 2, \ldots, L;\ i = 1, 2, \ldots, n_h)$.

If A_h is the total area of the hth stratum, an unbiased estimator of the area under the crop in the stratum, obtained from the hith sample grid, is

$$y^{\bullet}_{hi} = A_h p_{hi}$$

and the combined unbiased estimator, obtained from all the n_h sample grids in the stratum is, from equation (10.5b),

$$y^{\bullet}_{h0} = \sum_{i=1}^{n_h} y^{\bullet}_{hi}/n_h = A_h \sum_{i=1}^{n_h} p_{hi}/n_h = A_h \bar{p}_h$$

where $\bar{p}_h$ is the simple (unweighted) average of the p_{hi} values in the hth stratum.

An unbiased variance estimator of $y^{\bullet}_{h0}$ is given by equation (10.11).

For the whole universe, an unbiased estimator of the total area under the crop is

$$y = \sum^{L} y^{\bullet}_{h0}$$

an unbiased variance estimator of which is given by equation (10.16).

In surveys for estimating crop areas, the sampling fraction was generally 1 in 140 to 250, but in estimating crop yields it was necessarily much smaller, about 1 in 6 million - for details, see Mahalanobis, 1944, 1946a, and 1968. (For the crop surveys recently being conducted in India by the National Sample Survey and the Indian Council of Agricultural Research, the design is stratified multi-stage).

Some results for the crop survey in the State of West Bengal, India (1962-4) are given in Table 10.10. The total geographical area of West Bengal is about 20

Table 10.10: Estimated area under Aus Paddy (1962-3); Aman paddy, and Jute (1963-4): West Bengal

Crop	Number of effective grids of 2.25 acres	Estimated area in thousand acres	CV (%)
Aus paddy	52,439	1352	0.94
Aman paddy	54,303	9668	0.32
Jute	52,439	1074	0.94

million acres (1 acre = 0.404 hectare), and square grids of size 2.25 acres each were the sampling units in each stratum.

10.3 Estimation of proportion of units

For the estimation of the proportion of the universe units that fall into a certain class or possess a certain attribute, the extension from the case of unstratified simple random sampling (dealt with in section 2.13) to stratified simple random sampling is straightforward.

Let N'_h be the number of units possessing the attribute, and N_h the total number of units in the hth stratum; a simple random sample of n_h units from the N_h units show that n'_h of the sample of n_h units possess the attribute. We have already seen in section 2.13 that the sample proportion

$$p_h = r_h/n_h \tag{10.30}$$

is an unbiased estimator of the universe proportion

$$P_h = N'_h/N_h \tag{10.31}$$

The sampling variance of the estimator p_h is

$$P_h(1 - P_h)/n_h \tag{10.32}$$

and an unbiased estimator of this is

$$p_h(1 - p_h)/(n_h - 1) \tag{10.33}$$

The overall (i.e. for all the strata combined) universe proportion is

$$P = N'/N = \sum N'_h \Big/ \sum N_h = \sum N_h P_h \Big/ \sum N_h \tag{10.34}$$

and an unbiased sample estimator of P is

$$p = \sum N_h p_h / N \tag{10.35}$$

The sampling variance of p is

$$\frac{1}{N^2} \sum \frac{N_h^2 P_h(1 - P_h)}{n_h} \tag{10.36}$$

an unbiased estimator of which is

$$\begin{aligned} s_p^2 &= \sum N_h^2 s_{p_h}^2 / N^2 \\ &= \frac{1}{N^2} \sum \frac{N_h^2 p_h(1 - p_h)}{(n_h - 1)} \end{aligned} \tag{10.37}$$

Note: p is a weighted average of the p_hs, the weights being N_h/N. If a constant sampling fraction is taken in each stratum, i.e. $n_h/N_h = n/N$, then

$$p = \sum r_h / n = r/n \tag{10.38}$$

where $r = \sum r_h$ is total number of sample units possessing the attribute.

10.4 Ratio method of estimation: Combined and separate ratio estimators

As for unstratified simple random sampling (Chapter 3), so also for stratified simple random sampling, the use of ancillary information may increase the efficiency of estimators: we shall illustrate the most common use of ancillary information, namely, the ratio method of estimation.

Following the methods of section 3.2, and using the same types of notations as for the study variables, the ratio estimator of Y_h, the stratum total of the study variable in the hth stratum, using the ancillary information, is, from equation (3.6),

$$\begin{aligned} y_{hR}^* &= Z_h r_{1h} = Z_h(y_{h0}^*/z_{h0}^*) \\ &= Z_h(\overline{y}_h/\overline{z}_h) \end{aligned} \tag{10.39}$$

where

$$r_{1h} = y_{h0}^*/z_{h0}^* \tag{10.40}$$

A consistent but generally biased estimator of the variance of y^*_{hR} is, from equation (3.11),

$$s^2_{y^*_{hR}} = Z_h^2 s^2_{r_{1h}} = s^2_{y^*_{h0}} + r^2_{1h} s^2_{z^*h0} - 2r_{1h} s_{y^*_{h0} z^*_{h0}} \quad (10.41)$$

$$= N_h^2(SSy_{hi} + r^2_{1h} SSz_{hi} - 2r_{1h} SPy_{hi}z_{hi})/n_h(n_h - 1) \text{ in srswr} \quad (10.42)$$

For the ratio estimator of the total of the study variable $Y = \sum^L Y_h$, two types of ratio estimators could be used:

1. *Separate ratio estimator.* Here, the ratio method of estimation is first applied at the stratum levels to obtain ratio estimators of the stratum totals y^*_{hR} (equation (10.39)); these stratum ratio estimators are then summed up to provide the ratio estimator of the overall total Y; thus

$$y_{RS} = \sum^L y^*_{hr} = \sum^L Z_h \bar{y}_h / \bar{z}_h \quad (10.43)$$

(the additional subscript 'S' in Y_{RS} standing for 'separate' ratio estimator).

The variance estimator of Y_{RS} is the sum of the estimated variances of the stratum ratio estimators; thus

$$s^2_{y_{RS}} = \sum^L s^2_{y^*_{hR}} \quad (10.44)$$

$s^2_{y^*_{hR}}$ being given by equation (10.42).

2. *Combined ratio estimator.* Here the ratio method of estimation is applied to the estimators of the overall totals; thus

$$y_{RC} = Zr = Zy/z \quad (10.45)$$

(the additional subscript 'C' in y_{RC} standing for 'combined' ratio estimator); y and z are the unbiased estimators of the universe totals Y and Z respectively, defined by equations of the type (10.5(d)), i.e.

$$z = \sum^L z^*_{h0} = \sum^L N_h \bar{z}_h \quad (10.46)$$

and

$$r = y/z \quad (10.47)$$

The variance estimator of y_{RC} is (from equation (3.10))

$$s^2_{y_{RC}} = Z^2 s^2_r = (s^2_y + r^2 s^2_z - 2r s_{yz}) \tag{10.48}$$

where s_{yz} is the unbiased estimator of the covariance of y and z, given by an equation of the type (10.19), i.e.

$$\begin{aligned} s_{yz} &= \sum^L s_{y^*_{h0} z^*_{h0}} \\ &= \sum^L N_h^2 SP y_{hi} z_{hi} / n_h(n_h - 1) \end{aligned} \tag{10.49}$$

Notes:

1. The combined ratio estimator Y_{RC} does not require a knowledge of the stratum totals Z_h of the ancillary variable, but only of the overall total Z.
2. The separate ratio estimator will be more efficient than the combined ratio estimator if the sample in each stratum is not too small and the universe stratum ratio $R_{1h} = Y_h/Z_h$ vary considerably from stratum to stratum. If the sample in each stratum is small, the separate ratio estimator will be subject to a large bias.
3. If stratification is made with respect to an ancillary variable, then the ratio method of estimation using the same ancillary variable is not likely to further improve the efficiency of estimators.
4. The ratio method of estimation can be applied at intermediate levels of aggregation between the individual strata and the overall universe; in the example given in note 2 of section 9.3, the ratio method could be used not only at the stratum and country levels, but also at the level of the states. However, the ratio estimate obtained for the country or for a similar higher level of aggregation by applying the ratio method of estimation at the lower levels is likely to contain a bias that may be relatively large in relation to the standard error.

Example 10.5

For the same data as for Example 10.2, given the additional information on the previous census population in Table 10.11, obtain ratio estimates of the present total population and the number of households for the three states separately and also for all the states combined.

The required computations are given in Tables 10.11-10.13 and the final estimates in Table 10.14. First, unbiased stratum estimates of the previous census population are computed in Table 10.11. For the ratio estimates of the present total population for the three states separately, we use equation (10.39); for all the strata combined, the separate ratio estimate is obtained using equation (10.43), and the combined ratio estimate by using equation (10.45), and the respective variance estimates by equation (10.44) and (10.48). The required computations are given in Table 10.12. Similar computations for the ratio estimates of the total number of households are given in Table 10.13.

Compared with the results of Example 10.2, the ratio estimates are seen to be more efficient, especially for total population. The combined ratio estimate is more efficient than the separate ratio estimate for estimating the population.

Example 10.6

For the Demographic Sample Survey in Chad (1964), the universe of population (excluding nomads) was divided into nine strata - one for the urban areas and eight for rural areas, where the stratification criterion was chosen on the basis of the most dominant ethnic group. In each rural stratum, the sub-universe was classified according to the administrative division (*Préfectures, Sous-préfectures* and *Cantons*) and the size of the village. The rural strata were composed of a number of *Sous-préfectures*. Each rural stratum was divided into three sub-strata according to the size of the village: (a) with up to 100 persons, (b) with 200 to 499 persons, and (c) with 500 or more persons, according to the previous administrative census. Primary units were constructed in sub-stratum (a) by grouping small villages, in (b) by considering the villages as the primary units, and in (c) by grouping the localities of a village so that they contain an average of 300 persons: a systematic sample of 1 in 20 primary units was chosen for the survey, and the population, and births and deaths during the preceding 365 days recorded. The relatively large sampling fraction was chosen in order to provide reliable estimates at the stratum level.

Note: Sampling in rural sub-stratum (c) was stratified two-stage, to be considered in Part IV.

The sample comprised 101 thousand persons in the rural areas and 11 thousand persons in the urban: the overall sampling fraction was 5 per cent. Using the previous administrative census data, the ratio estimates of population were 2365±87.1 thousand in the rural section and 2530±91.1 thousand in the country. On the basis of these results, the 95 per cent limits were placed at 2268 and 2444 thousand for the rural population and 2439 and 2629 thousand for the total population of the country (excluding nomads). (For further details, *see* the report by Behmoiras).

10.5 Ratio of ratio estimators of totals of two study variables

Ratio estimators of totals of two study variables may be used in estimating their ratio. However, it becomes the same as the ratio of the corresponding unbiased estimators of totals if computed at the stratum level, or for all the strata combined if the combined ratio estimator is used (see section 3.2, note 5).

Table 10.11: Previous census population in each of the two sample villages, selected in Example 10.2 in each of the three states and computation of stratum estimates: Stratified srs of villages

State	Number of villages		Village Serial no.	Sample village no.	Previous census population	
	Total	Sample			In sample village	Stratum estimate
h	N_h	n_h		i	z_{hi}	$z^*_{hi} = N_h z_{hi}$
(1)	(2)	(3)	(4)	(5)	(6)	(7)
I	10	2	2	1	81	810
			10	2	80	800
					Total	1610
					Mean (= stratum estimate of total)	805 (z^*_{h0})
					Difference	10 (d_{zh})
					$\frac{1}{2}$ |Difference| (= estimated standard error)	5 ($s_{z^*_{h0}}$)
II	11	2	5	1	98	1078
			11	2	86	946
					Total	2024
					Mean (= stratum estimate of total)	1012 (z^*_{h0})
					Difference	132 (d_{zh})
					$\frac{1}{2}$ |Difference| (= estimated standard error)	66 ($s_{z^*_{h0}}$)
III	9	2	2	1	112	1008
			3	2	97	873
					Total	1881
					Mean (= stratum estimate of total)	940.5 (z^*_{h0})
					Difference	135 (d_{zh})
					$\frac{1}{2}$ |Difference| (= estimated standard error)	67.5 ($s_{z^*_{h0}}$)

Table 10.12: Computation of ratio estimates of population and their variances, using the previous census population: data of Tables 10.5-10.7 and 10.11

State (h)	Z_h	$r_{1h} = y^*_{h0}/z^*_{h0}$	$y^*_{hR} = Z_h r_{1h}$	$s_{y^*_{h0}z^*_{h0}} = \frac{1}{4} d_{yh} d_{zh}$	$s^2_{y^*_{hR}} = s^2_{y^*_{h0}} + r^2_{1h} s^2_{z^*_{h0}} - 2r_{1h} s_{y^*_{h0}z^*_{h0}}$	$s_{y^*_{hR}}$
(1)	(2)	(3)	(4)	(5)	(6)	(7)
I	863	1.0559	911.24	−150	1244.6431	35.28
II	1010	1.0815	1092.34	3993	116.8691	10.81
III	942	1.0813	1018.62	4860	0.5988	0.77
Total	2815		3022.20 (y_{RS})		1362.1110 ($s^2_{y_{RS}}$)	36.91 ($s_{y_{RS}}$)
Combined	2815 (Z)	1.0734 (r_1)	3023.25 ($y_{RC} = Z r_1$)	8703 (s_{yz})	725.5130 ($s^2_{y_{RC}} = s^2_y + r^2_1 s^2_z - 2r_1 s_{yz}$)	26.98 ($s_{y_{RC}}$)

Table 10.13: Computation of ratio estimates of number of households and their variances, using the previous census population: data of Tables 10.5-10.7 and 10.11-10.12

State (h)	Z_h	$r_{2h} = x^*_{h0}/z^*_{h0}$	$x^*_{hR} = Z_h r_{2h}$	$s_{x^*_{h0}z^*_{h0}} = \frac{1}{4} d_{xh} d_{zh}$	$s^2_{x^*_{hR}} = s^2_{x^*_{h0}} + r^2_{2h} s^2_{z^*_{h0}} - 2r_{2h} s_{x^*_{h0}z^*_{h0}}$	$s_{x^*_{hR}}$
(1)	(2)	(3)	(4)	(5)	(6)	(7)
I	863	0.24845	214.41	−100.00	426.3882	20.65
II	1010	0.20652	208.59	726.00	6.9157	2.63
III	942	0.18660	175.78	911.25	0.8128	0.90
Total	2815		598.78 (x_{RS})		437.1167 ($s^2_{x_{RS}}$)	20.84 ($s_{x_{RS}}$)
Combined	2815 (Z)	0.21179 (r_2)	596.18 ($x_{RC} = Z r_2$)	1537.25 (s_{xz})	452.9819 ($s^2_{x_{RC}} = s^2_x + r^2_2 s^2_z - 2r_2 s_{xz}$)	21.28 ($s_{x_{RC}}$)

Table 10.14: Ratio estimates of population and number of households, using the previous census population, computed from the data of Tables 10.5 and 10.11: stratified srs of villages

State	Ratio estimates of population			Ratio estimates of number of households		
	Estimated number	Standard error	CV (%)	Estimated number	Standard error	CV (%)
I	911.2	35.28	3.89	214.4	20.65	9.63
II	1092.3	10.81	0.99	208.6	2.64	1.26
III	1018.6	0.77	0.076	175.8	0.90	0.51
All states:						
Separate ratio estimate	3022.2	36.91	1.22	598.8	20.84	3.48
Combined ratio estimate	3023.2	26.98	0.89	596.2	21.28	3.57

If separate ratio estimators are used for the totals of two study variables, namely

$$y_{RS} = \sum^{L} y^*_{hR} = \sum^{L} Z_h r_{1h} = \sum^{L} Z_h \bar{y}_h / \bar{z}_h \qquad (10.43)$$

$$x_{RS} = \sum^{L} x^*_{hR} = \sum^{L} Z_h r_{2h} = \sum^{L} Z_h \bar{x}_h / \bar{z}_h \qquad (10.50)$$

as respective estimators of the universe totals Y and X, then an estimator of the universe ratio

$$R = Y/X \qquad (10.7.c)$$

is

$$\begin{aligned} r' &= y_{RS}/x_{RS} = \sum^{L} y^*_{hR} \Big/ \sum^{L} x^*_{hR} \\ &= \sum^{L} Z_h r_{1h} \Big/ \sum^{L} Z_h r_{2h} = \sum^{L} Z_h \bar{y}_h \Big/ \sum^{L} Z_h \bar{x}_h \qquad (10.51) \end{aligned}$$

A variance estimator of r' is

$$s^2_{r'} = (s^2_{y_{RS}} + r'^2 s^2_{x_{RS}} - 2r' s_{y_{RS}x_{RS}})/x^2_{RS} \qquad (10.52)$$

where s^2_{yRS} and s^2_{xRS} are given by estimating equations of the type (10.44), and the estimated covariance is

$$s_{yRSxRS} = \sum^{L} s_{y^*_{hR}x^*_{hR}} \tag{10.53}$$

where

$$s_{y^*_{hR}x^*_{hR}} = \sum^{n_h} \frac{(y^*_{hi} - r_{1h}z^*_{hi})(x^*_{hi} - r_{2h}z^*_{hi})}{n_h(n_h - 1)} \tag{10.54}$$

For an example, see exercise 3 of this chapter.

10.6 Gain due to stratification

From the data of a stratified sample, it is possible to examine the gain, if any, due to stratification leading to a reduction in the variance of a sample estimator, as compared with an unstratified simple random sample. We will consider in the next section the question of estimating from the data of an unstratified srs, the gain had stratification been adopted.

An unbiased estimator of the variance of estimated total of the study variable y, obtained from a stratified simple random sample, is

$$s^2_y = \sum^{L} s^2_{y^*_{h0}} = \sum^{L} N^2_h \frac{\sum^{n_h} y^2_{hi} - \left(\sum^{n_h} y_{hi}\right)^2 / n_h}{n_h(n_h - 1)} \tag{10.16}$$

and an unbiased estimator of the variance of the estimator of the total, had the data come from an unstratified simple random sample, is

$$\frac{\left(N \sum^{L} N_h \sum^{n_h} y^2_{hi}/n_h - y^2 + s^2_y\right)}{n} \tag{10.55}$$

Example 10.7

For the data of Example 10.2, estimate the gain in the efficiency of the sample estimate of the total population due to stratification, as compared to that for an unstratified srs.

The required computations are given in Table 10.15. The value of the expression (10.55) is

$$(30 \times 297,072.5 - 2,961.5^2 + 9,744.25)/6 = 25,239.5$$

The efficiency of the estimate from the stratified sample as compared with that from the unstratified is thus

$$25,239.5/9,744.25 = 2.59 \text{ or } 259\%$$

Table 10.15: Computation of the gain due to stratification: Data of Example 10.2

Stratum	$\sum_{i=1}^{n_h} y_{hi}^2$	$\sum_{i=1}^{n_h} y_{hi}^2/n_h$	$N_h \sum_{i=1}^{n_h} y_{hi}^2/n_h$
I	14 468	7234.0	72 340.0
II	19 861	9930.5	109 235.5
III	25 666	12 833.0	115 497.0
Total			297 072.5 $\left(\sum_1^L N_h \sum^{n_h} y_{hi}^2/n_h\right)$

10.7 Stratification after sampling

For a stratified simple random sample, information on the total number and the listing of the units in each stratum for the "stratification variable" is required prior to sample selection. Such information on the desirable variables suitable for stratification, such as sex, age, tribe etc. in a demographic survey, is often not available prior to sampling; it is collected for the selected sample units only in the course of the actual enumeration.

In such a situation, under suitable conditions, the sample data from an unstratified simple random sample may be classified into a number of strata (say L) and analyzed in the same manner as a stratified simple random sample. Two cases may arise:

1. The total number of units in stratum h, N_h is known. The sample data (obtained from an unstratified simple random sample) are classified into L strata (***post-strata***); let the number of sample units falling in the hth post-stratum and the mean of the values of the study variable there be denoted respectively by n_h and $\bar{y}_h$. Then for the estimator of the universe total Y, the same estimating equation (10.5(d)) for a stratified simple random sample is used, namely

$$y \sum^{L} N_h \bar{y}_h \tag{10.56}$$

   and the variance estimator of y is given by equation (10.16). For an example, see exercise 4 of this chapter.

2. The total number of units in the post-stratum h, namely N_h, is not known, but the proportion is assumed to be known; e.g. if age is the

desirable stratification variable, the age structure of the population can be assumed to be effectively the same as in the last population inquiry. The estimator y of equation (10.56) can be written in the form

$$\begin{aligned} y &= \sum^{L} N_h \bar{y}_h = N \sum^{L} (N_h/N) \bar{y}_h \\ &= N \sum^{L} P'_h \bar{y}_h \end{aligned} \qquad (10.57)$$

where

$$P'_h = (N_h/N) \qquad (10.58)$$

are the "weights", i.e. the proportions of the total number of units in the different post-strata, the values of P'_h being known or taken from previous inquiries. This is a biased estimator.

A variance estimator of y is

$$s_y^2 = N^2 \sum^{L} P_h'^2 s_{\bar{y}_h}^2 \qquad (10.59)$$

$s_{\bar{y}_h}^2$, the variance estimator of the sample mean in the hth post-stratum, being given, as usual, by

$$s_{\bar{y}_h}^2 = SSy_{hi}/n_h(n_h - 1) \qquad (10.60)$$

The estimator of the universe mean is obtained, as usual, on dividing N into the estimator of the universe total, and so also for its estimated standard error.

Notes

1. If the sample is reasonably large, say over twenty in each stratum, and the errors in the weights negligible, the method of post-stratification can give results almost as efficient as *proportional stratified sampling* (see Chapter 12). This is also obvious from the fact that for large samples, the sample units are likely to be distributed in proportion to N_h.
2. The method can also be used when the sample is already stratified according to some variable other than that desired for the study in hand.

Further reading

Cochran, sections 5.1-5.4, 5.6, 5.10, 5A.8, 6.10-6.12; Hansen *et al.* (1953), vol. I, chapters 5A, 5B, and 5D, and vol. II, sections 5.1-5.3, 5.5, 5.6, 5.12, and 5.13; Hedayat and Sinha, chapter 9; Murthy (1967), sections 7.1-7.3, 7.5-7.7, 7.14, and 10.6; Sukhatme *et al.* (1984), sections 4.1-4.3, 4.5, and 4.6; Yates, sections 3.3-3.5, 6.5-6.7, 6.10, 6.11, 6.13, 6.14, 7.6, 7.6.1, 7.6.2, 7.9-7.10, 7.13, 8.3-8.5, 8.5.1, and 8.5.2.

Exercises

1. In a sample survey designed to estimate the total number of cattle, the universe of 2,072 farms was stratified into 5 strata on the basis of the total acreage of the farms. In the hth stratum ($h = 1, 2, \ldots, 5$), a simple random sample (with replacement) of farms (of size n_h) was taken in proportion to the total number of farms in the stratum (N_h), the total sample size being $n = \sum n_h = 500$; for each stratum, the total number of cattle in the sample farms ($\sum_{i=1}^{n_h} y_{hi}$) and the raw sum of squares ($\sum_{i=1}^{n_h} y_{hi}^2$) are given in Table 10.16; y_{hi} denotes the number of cattle in the ith sample farm $i = 1, 2, \ldots, n_h$). Estimate the total and the average number of cattle per farm, along with their standard errors (United Nations *Manual*, Process 4, Example 4b, modified to "with replacement" sampling).

2. For a socio-economic inquiry, the 23,671 villages in an area, including the uninhabited ones, were grouped into 4 strata on the basis of their altitude above sea-level and population density, and from each stratum a simple random sample of 10 villages was taken with replacement. The data on the number of households in each of the sample villages are given in Table 10.17.

   (a) Estimate the total number of households with its CV;

   (b) Examine whether there has been any gain due to stratification as compared to unstratified simple random sample with replacement (Murthy (1967), Problem 7.1).

3. For the data of Example 10.5, estimate the average household size from the separate ratio estimates of numbers of persons and households, and its standard error.

4. A simple random sample of 125 farms out of a total of 2496 farms in Hertfordshire, 1939, gave the data on (a) total acreage and (b) acreage of wheat, classified by districts after selection (Table 10.18). Estimate the total area of wheat and the number of farms growing wheat (i) directly from the sample, and (ii) by stratification by size, given the total number of farms in the specific size-groups in col. 2 in Table 10.2 to the answer (Yates, Examples 6.6 and 7.6.b, modified to "with replacement" sampling).

5. In the Micro-Census Sample Survey of the Federal Republic of Germany, conducted on a continuing basis from October 1957, the sample design for the annual survey has been stratified single-stage since October 1962. In each stratum, formed according to the size of the communities, a one per cent systematic sample of area units (enumeration districts) is selected, and all dwellings, households, families, and persons in the sample areas are enumerated.

   (a) The Federal Statistical Office uses the following estimators: the unbiased estimator of Y is $y = 100\sum_h n_h\bar{y}_h$, and the unbiased variance estimator of y is $0.99 \times 10^4 \sum_h n_h SSy_{hi}/(n_h - 1)$. Verify these, noting that the finite multiplier is not ignored for the variance estimator (Herberger, 1971).

   (b) Also verify that

$$s_r^2 = 0.99\sum_h \left[\frac{n_h}{n_h - 1}\,(SSy_{hi} + r^2 SSx_{hi} - 2r\,SPy_{hi}x_{hi}\right] \Big/ \left(\sum_h n_h\bar{x}_h\right)^2$$

(*Hint*: (a) Use respectively equations (10.5.d) and (10.24), noting that $f_h = n_h/N_h = 0.01$.)

6. To estimate the number of AIDS (acquired immunodeficiency syndrome) admissions in 24 counties that are not metropolitan areas in a large state in the U.S., the 24 counties were grouped into 6 strata, with 4 counties each, and a sample of two counties was selected out of the total number of counties in each stratum and in each selected county, one hospital was taken, sampling being simple random without replacement. Table 10.19 gives the data on the number of hospital beds and the number of AIDS admissions in each sample hospital.

   (a) Estimate the total number and proportion of AIDS admissions for the 34 county area;

   (b) How would you estimate the number of AIDS admissions by utilizing the data on the total number of hospital beds for all the hospitals in each stratum (given also in the bottom half of Table 10.19) (Levy and Lemeshow, Exercise 12.2, adapted)?

Table 10.16: Total number of cattle in the srs of farms and the raw sum of square in different strata

Stratum (acres)	No. of farms in stratum (N_h)	No. of farms sampled (n_h)	$\sum_{i=1}^{n_h} y_{hi}$	$\sum_{i=1}^{n_h} y_{hi}^2$
(1)	(2)	(3)	(4)	(5)
0-15	635	153	619	5579
16-30	570	138	1423	24 253
31-50	475	115	1758	34 082
51-75	303	73	1691	51 419
76-100	89	21	603	18 305
All strata	2072 (N)	500	6094	133 658

Source: United Nations *Manual*, Process 4, Example 4b (adapted).

Table 10.17: Number of households in the srs of 10 villages in the different strata

Stratum	Total number of villages	Total number of households in sample villages									
		1	2	3	4	5	6	7	8	9	10
I	1411	43	84	98	0	10	44	0	124	13	0
II	4705	50	147	62	87	84	158	170	104	56	160
III	2558	228	262	110	232	139	178	334	0	63	220
IV	14 997	17	34	25	34	36	0	25	7	15	31

Source: Murthy (1967), Problem 7.1.

Table 10.18: Total acreage (a) and acreage in wheat (b) in the srs of farms in Hertfordshire, U.K., 1939

District 1	(a)	188	60	192	48	44	79	14	465	197	163	198
(15 farms)		78	6	35	168	(1935)						
	(b)	16	0	0	0	0	33	0	92	0	0	0
		0	0	0	0	(141)						
District 2	(a)	8	294	597	8	2	200	14	262	(1385)		
(8 farms)	(b)	0	29	107	0	0	65	0	58	(259)		
District 3	(a)	370	26	369	212	153	287	28	14	4	17	2
(40 farms)		3	7	6	335	4	1	4	180	120	40	28
		221	31	6	34	316	116	4	409	6	115	19
		274	3	144	3	482	156	302	(4851)			
	(b)	67	0	58	45	20	44	0	0	0	0	0
		0	0	0	82	0	0	0	0	11	0	0
		59	0	0	0	75	33	0	102	0	0	0
		6	0	0	0	62	28	71	(763)			
District 4	(a)	11	6	543	822	654	3	158	4	68	55	4
(24 farms)		2	192	4	491	224	280	90	3	3	6	4
		161	246	(4034)								
	(b)	0	0	80	265	112	0	50	0	27	12	0
		0	24	0	24	28	75	0	0	0	0	0
		80	60	(837)								
District 5	(a)	4	312	8	11	(335)						
(4 farms)	(b)	0	102	0	0	(102)						
District 6	(a)	8	87	6	44	4	614	192	10	24	2	9
(24 farms)	(b)	3	2	120	58	20	30	197	14	32	2	285
		138	126	(2027)								
	(b)	0	14	0	0	0	72	20	0	0	0	0
		0	0	24	0	0	0	6	3	6	0	29
		0	0	(174)								
District 7	(a)	128	4	46	181	17	24	10	36	12	89	(547)
(10 farms)	(b)	5	0	0	20	0	0	0	0	0	0	(25)
Grand	125 farms											
total:	(a)	15,114										
	(b)	2301										

Source: Yates, Examples 6.6 and 7.6.b (adapted).

Table 10.19: Number of beds and AIDS admissions in hospitals in an AIDS Admission Survey

Stratum	Sample County	Sample Hospital	
		Total Beds	AIDS
1	1	72	20
	2	87	49
2	1	99	38
	2	48	23
3	1	99	38
	2	131	78
4	1	42	7
	2	38	28
5	1	42	26
	2	34	9
6	1	30	18
	2	76	20

Stratum	Number of counties	Number of	
		Hospitals	Beds
1	4	11	588
2	4	9	421
3	4	8	776
4	4	6	196
5	4	4	175
6	4	8	375

Source: Levy and Lemeshow (1991), Exercise 12.2 (adapted).

CHAPTER 11

Stratified Varying Probability Sampling

11.1 Introduction

As for simple random sampling, so also for varying probability sampling, the extension from unstratified to stratified sampling is straightforward and follows directly from the fundamental theorems of Chapter 9. The use of different schemes of selection - srs, pps etc. - in the different strata, the special cases of crop surveys, the ratio method of estimation, and the gain due to stratification will also be considered in the chapter.

11.2 Estimation of totals, mean, ratios and their variances

11.2.1 Structure of a stratified varying probability sampling

Consider the total universe to be divided, on certain criteria, into L strata, and let the hth stratum ($h = 1, 2, \ldots, L$) contain a total of N_h units. The values of the two study variables for the ith unit ($i = 1, 2, \ldots, N_h$) in the hth stratum will be denoted, as in section 10.2, respectively by Y_{hi} and X_{hi}. The total number of units in the universe is, as before,

$$N = \sum_{h=1}^{L} N_h \tag{11.1}$$

In the hth stratum, a sample of n_h units is selected (with replacement) with varying probabilities, the probability of selection of the ith unit ($i = 1, 2, \ldots, N_h$) being π_{hi}. Let the values of the two study variables for the ith selected sample unit ($i = 1, 2, \ldots, n_h$) be denoted respectively by y_{hi} and

x_{hi}. The total number of sample units is

$$n = \sum_{h=1}^{L} n_h \tag{11.2}$$

11.2.2 Estimators of totals, means, and ratio of two totals

An unbiased estimator of the stratum total Y_h, obtained from the ith sample unit in the hth stratum is, from equation (5.2),

$$y^*_{hi} = y_{hi}/\pi_{hi} \tag{11.3}$$

and the combined unbiased estimator of Y_h, obtained from the n_h sample units in the hth stratum, is, from equation (5.3) or (9.1),

$$y^*_{h0} = \sum_{i=1}^{n_h} y^*_{hi}/n_h \tag{11.4}$$

Thus, if the sample units are selected with probability proportional to size z_{hi}, i.e. $\pi_{hi} = z_{hi}/Z_h$, where $Z_h = \sum_{i=1}^{N_h} z_{hi}$, then

$$y^*_{hi} = y_{hi}/\pi_{hi} = Z_h\, y_{hi}/z_{hi} \tag{11.5}$$

Some universe parameters and their sample estimator are defined in Table 11.1. These follow from the results of section 5.3 for any stratum and section 9.3 for all strata combined.

Table 11.1: Some universe parameters and their sample estimators for a stratified varying probability sample

	For the hth stratum		For all strata combined		
	Universe parameter	Sample estimator	Universe parameter	Sample estimator	
	(a)	(b)	(c)	(d)	
For a study variable: Total	$Y_h = \sum_{i=1}^{N_h} Y_{hi}$	$y^*_{h0} = \sum_{i=1}^{n_h} y^*_{hi}/n_h$	$Y = \sum_{h=1}^{L} Y_h$	$y = \sum_{h=1}^{L} y^*_{h0}$	(11.6)
Mean	$\overline{Y}_h = Y_h/N_h$	$\overline{y}^*_{h0} = y^*_{h0}/N_h$	$\overline{Y} = Y/N$	y/N	(11.7)
For two study variables: Ratio of totals	$R_h = Y_h/X_h$	$r_h = y^*_{h0}/x^*_{h0}$	$R = Y/X$	$r = y/x$	(11.8)

11.2.3 Sampling variances and their estimators for a stratum

The sampling variance of the sample estimator y^*_{h0} of the stratum total Y_h in ppswr is, from equation (5.4),

$$\sigma^2_{y^*_{h0}} = \frac{1}{n_h}\sum_{i=1}^{N_h}\left(\frac{Y_{hi}}{\pi_{hi}} - Y_h\right)^2 \pi_{hi} = \frac{1}{n_h}\left(\sum_{i=1}^{N_h}\frac{Y_{hi}^2}{\pi_{hi}} - Y_h^2\right) \tag{11.9}$$

an unbiased estimator of which is (from equation (5.5) or (9.2))

$$s^2_{y^*_{h0}} = \sum_{i=1}^{n_h}(y^*_{hi} - y^*_{h0})^2/n_h(n_h-1) = SSy^*_{hi}/n_h(n_h-1) \tag{11.10}$$

The sampling variance of the sample estimator of the hth stratum mean is

$$\sigma^2_{\bar{y}^*_{h0}} = \sigma^2_{y^*_{h0}}/N_h^2 \tag{11.11}$$

which is estimated unbiasedly from equation (5.7) by

$$s^2_{\bar{y}^*_{h0}} = s^2_{y^*_{h0}}/N_h^2 \tag{11.12}$$

An unbiased estimator of the covariance of y^*_{h0} and x^*_{h0} (x^*_{h0} being defined similarly for another study variable) is, from equation (5.8) or (9.10),

$$\begin{aligned} s_{y^*_{h0}x^*_{h0}} &= \sum_{i=1}^{n_h}(y^*_{hi} - y^*_{h0})(x^*_{hi} - x^*_{h0})/n_h(n_h-1) \\ &= SPy^*_{hi}x^*_{hi}/n_h(n_h-1) \end{aligned}$$

An estimator of the variance of r_h is, from equation (5.10) or (9.14),

$$\begin{aligned} s^2_{r_h} &= (s^2_{y^*_{h0}} + r_h^2 s^2_{x^*_{h0}} - 2r_h s_{x^*_{h0}y^*_{h0}})/x^{*2}_{h0} \\ &= (SSy^*_{hi} + r_h^2 SSy^*_{hi} - 2r_h SPy^*_{hi}x^*_{hi})/n_h(n_h-1)x^{*2}_{h0} \end{aligned} \tag{11.13}$$

11.2.4 Sampling variances and their estimators for all strata combined

The sampling variance of the estimator y of the universe total Y is

$$\sigma^2_y = \sum_{h=1}^{L}\sigma^2_{y^*_{h0}} = \sum_{h=1}^{L}\frac{1}{n_h}\sum_{i=1}^{N_h}\left(\frac{Y_{hi}}{\pi_{hi}} - Y_h\right)^2 \pi_{hi} \tag{11.14}$$

which is estimated by the sum of the unbiased variance estimators of y^*_{h0} from equation (9.6), by

$$s^2_y = \sum^{L} s^2_{y^*_{h0}} \tag{11.15}$$

The sampling variance of the estimator of the universe mean is

$$\sigma_y^2/N^2 \tag{11.16}$$

which is estimated unbiasedly by

$$s_y^2/N^2 \tag{11.17}$$

An unbiased estimator of the covariance of the estimators y and x is, from equation (9.12),

$$s_{yx} = \sum^{L} s_{y^*_{h0}x^*_{h0}} \tag{11.18}$$

A variance estimator of r is (from equation (9.16))

$$s_r^2 = (s_y^2 + r^2 s_x^2 - 2rs_{yx})/x^2 \tag{11.19}$$

For the notes relating to the estimation of ratios, see the notes to section 2.7.

Example 11.1

From each of the three states (Appendix IV), select with probability proportional to the previous census population and with replacement a sample of two villages each, and on the basis of the data on the current number of households and of persons in these sample villages, estimate for the three states separately and also for the three states combined, the total numbers of households and of persons and average household size, with standard errors.

Here the states are the strata. The sample villages are selected with probability proportional to the previous census population by adopting the procedure of cumulation of the previous census population in each state (section 5.4). The information on the selected sample villages is given in Table 11.2; the Z_h values (the total previous census population in the states) are 863, 1010, and 942 respectively. As two sample villages are selected from each state, the simplified procedure of section 9.5 for estimation will be followed, as was done in Example 10.2 relating to a stratified srs. The required computations are shown in Tables 11.2-11.4 and the final results in Table 11.5.

Example 11.2

For the Indian National Sample Survey (1957-8) in the rural section, the total number of 2522 *tehsils* were grouped into a number of geographical strata. In each stratum, two villages were selected from the total number of villages with probability proportional to the 1951 Census population; in the sample villages

Table 11.2: Population and number of households in each of the two sample villages selected with probability proportional to previous census population in each of the three states (Appendix IV): Stratified pps sample of villages

State	Sample village	Village serial no.	Size (previous census population)	Reciprocal of probability $1/\pi_{hi} = Z_h/z_{hi}$	Number of households: In sample	Number of households: Stratum estimate	Population: In sample	Population: Stratum estimate
h	i		z_{hi}		x^*_{hi}	$x^*_{hi} = x_{hi}/\pi_{hi}$	y_{hi}	$y^*_{hi} = y_{hi}/\pi_{hi}$
(1)	(2)	(3)	(4)	(5)	(6)	(7)	(8)	(9)
I	1	5	92	9.3804	24	225.1296	112	1050.6048
	2	7	72	11.9861	20	239.7220	88	1054.7768
					Total	464.8516		2105.3816
					Mean (= stratum estimate of total)	232.45 (x^*_{h0})		1052.6908 (y^*_{h0})
					Difference	−14.59 (d_{xh})		−4.1720 (d_{yh})
					$\frac{1}{2}$ |Difference| (= estimated standard error)	7.295 ($s_{x^*_{h0}}$)		2.0860 ($s_{y^*_{h0}}$)
II	1	3	73	13.8356	17	235.2052	80	1106.8480
	2	7	85	11.8824	19	225.7656	95	1128.8280
					Total	460.9708		2235.6760
					Mean (= stratum estimate of total)	230.49 (x^*_{h0})		1117.8380 (y^*_{h0})
					Difference	9.44 (d_{xh})		−21.9800 (d_{yh})
					$\frac{1}{2}$ |Difference| (= estimated standard error)	4.72 ($s_{x^*_{h0}}$)		10.9900 ($s_{y^*_{h0}}$)
III	1	2	112	8.4107	21	176.6247	121	1017.6947
	2	4	117	8.0513	21	169.0773	129	1038.6177
					Total	345.7020		2056.3124
					Mean (= stratum estimate of total)	172.85 (x^*_{h0})		1028.1562 (y^*_{h0})
					Difference	7.55 (d_{xh})		−20.9230 (d_{yh})
					$\frac{1}{2}$ |Difference| (= estimated standard error)	3.775 ($s_{x^*_{h0}}$)		10.4614 ($s_{y^*_{h0}}$)

Table 11.3: Computation of the estimated total population and number of households and their standard errors: data of Table 11.2

State	Number of households			Population		
	Estimate	Standard error	Variance	Estimate	Standard error	Variance
h	x^*_{h0}	$s_{x^*_{h0}}$	$s^2_{x^*_{h0}}$	y^*_{h0}	$s_{y^*_{h0}}$	$s^2_{y^*_{h0}}$
I	232.43	7.295	53.2170	1052.69	2.086	4.3514
II	230.49	4.720	22.2784	1117.84	10.990	120.7801
III	172.85	3.775	14.2506	1028.16	10.462	109.4534
All states combined	635.77 (x)	9.473* (s_x)	89.7460 (s_x^2)	3198.69 (x)	15.316* (s_x)	234.5849 (s_x^2)

* Not additive

Table 11.4: Computation of the estimated average household size and sampling variance: data of Tables 11.2 and 11.3

State h	Average household size $r_h = y^*_{h0}/x^*_{h0}$	$s_{y^*_{h0}x^*_{h0}} = \frac{1}{4} d_{yh} d_{xh}$	$s^2_{y^*_{h0}} + r_h^2 s^2_{x^*_{h0}} - 2r_h s_{y^*_{h0}x^*_{h0}}$	$s^2_{r_h} =$ col.(4)$/x^{*2}_{h0}$
(1)	(2)	(3)	(4)	(5)
I	4.5291	15.0861	959.3228	0.0177574
II	4.8499	−51.8728	1147.9573	0.0216084
III	5.9482	−39.4918	1083.4655	0.0362654
All states combined	5.0312 $(r = y/x)$	−76.2785 (s_{yx})	3273.8702 $(s_y^2 + r^2 s_x^2 - 2r s_{yx})$	0.00809956 (s_r^2)

Table 11.5: Estimates and standard errors computed from the data of Table 11.2: Stratified pps sample of villages

	Item	State I	State II	State III	All states combined
1.	*Total population*				
	(a) Estimate	1052.7	1117.8	1028.2	3198.7
	(b) Standard error	2.09	10.99	10.46	15.32
	(c) Coefficients of variation (%)	0.20	0.98	1.02	0.48
2.	*Total number of households*				
	(a) Estimate	232.4	230.5	172.8	635.8
	(b) Standard error	7.30	4.72	3.78	9.47
	(c) Coefficient of variation (%)	3.14	2.05	2.18	1.49
3.	*Average household size*				
	(a) Estimate	4.53	4.85	5.95	5.03
	(b) Standard error	0.1333	0.1470	0.1904	0.0900
	(c) Coefficient of variation (%)	2.95	3.04	3.20	1.79

thus selected, the existing households were listed for the purpose of sub-sampling of households for socio-economic inquiries, the sampling design for which was thus stratified two-stage (see section 1.9 and Chapter 14). However, at the time the list of households in the sample villages was constructed, information on the total number of births during the preceding 365 days in the premises occupied by the households was collected, along with the household size from all the households. The sample design for this inquiry on births was thus stratified single-stage pps. The estimating equations for the total population, births and the birth rate and their estimated variances will thus take the forms given in section 11.2.

Estimates of the birth rates and estimated standard errors for rural India as a whole as also for the five zones are given in Table 11.6. The zones comprised one or more states, each state containing several strata: for the zonal estimates, the same types of formulae as for the overall estimates for the country as a whole were used, excepting that only those strata constituting a zone were considered (for further details, see Som *et al.*, 1961). There were 924 villages, over 135,000 households, and over 680,000 persons in the sample.

By definition, births occurring in hospitals and other institutions were not recorded, nor, for households occupying the premises for less than a year, the births occurring in the previous residence. Even with the limited definition, it appears likely that a number of births were not reported because of the lack of emphasis and of probes and other associate items and of cross-checks on this item in the "listing schedule". The methods of obtaining adjusted estimates from demographic data are described briefly in section 25.6.

Table 11.6: Estimated birth rate (per 1000 persons): Indian National Sample Survey, rural sector, 1957-8

Zone	Estimate	Standard error	*CV* (%)
North	37.08	0.86	2.32
Central	42.36	0.89	2.10
East	32.96	1.25	3.79
South	28.07	0.75	2.67
West	32.81	0.52	2.80
India (rural)	35.16	0.45	1.28

Source: Som *et al.* (1961).

11.3 Different schemes of selection in different strata

As mentioned in section 9.2, it is possible to adopt different sampling schemes – srs, pps, or systematic – in different strata. The fundamental theorems of section 9.3 permit the estimation of the total of a study variable and its variance for all the strata combined by adding up the corresponding unbiased stratum estimators. The ratio of the totals of two study variables and its variance can also be estimated, as also can other related measures such as the correlation coefficient, from the results of section 9.3. The procedure will be illustrated with an example.

Example 11.3

The universe of 112 villages was divided into three strata with 51, 37, and 24 villages respectively. From the first stratum, a simple random sample of 6 villages was selected without replacement, from the second stratum a sample of 5 villages with probability proportional to the cultivated area (the total cultivated area in the stratum was 26,912 acres) and with replacement, and from the third stratum two linear systematic samples of 4 villages each. For each selected sample village, the total area under wheat was observed: this information is given in Table 11.7, along with the total areas of the sample villages of stratum II. Estimate the total area under wheat in each stratum separately and also for all strata combined, along with the standard errors (Murthy (1967), Problem 7.3, adapted).

Table 11.7: Area under wheat (y_{hi}) for all the sample villages and cultivated area (x_{hi}) for the sample villages of stratum II

Sample village	Stratum (h)				
	Stratum I	Stratum II		Stratum III	
i	y_{1i}	x_{2i}	y_{2i}	y_{3i}	$y_{3'i}$
1	75	729	247	427	335
2	101	617	238	326	412
3	5	870	359	481	503
4	78	305	129	445	348
5	78	569	223		
6	45				

Source: Murthy (1967), Problem 7.3 (adapted).

For stratum I, the methods of Example 2.1 are followed. The estimated area under wheat is 3247 acres with estimated variance 435,540 $(\text{acres})^2$.

For stratum II, we follow the methods of section 11.2. From equations (11.5b) and (11.10), the estimated area under wheat is 10,507 acres with estimated variance 153,506 $(\text{acres})^2$.

For stratum III, the methods of Example 4.1 are followed. The estimated area under wheat is 9831 acres with variance 59,049 $(\text{acres})^2$.

The estimated total area under wheat in the three strata separately and combined are shown in Table 11.8, along with their estimated variance, standard errors, and coefficients of variation.

11.4 Special cases of crop surveys

11.4.1 Introduction

As for unstratified pps sampling (section 5.5), so also for stratified sampling, the sampling of fields (or farms or plots) with ***probability proportional to total (geographical) area*** (ppa) simplifies the estimating procedures, in addition to possibly improving the efficiency of the estimators.

Table 11.8: Estimated wheat acreage, obtained from the data of Table 11.7

	Area under wheat	Stratum I	Stratum II	Stratum III	All strata combined
1.	Estimate (acres)	3247	10,507	9831	23,585
2.	Estimated variance (acre)^2	435,540	153,506	59,049	648,095
3.	Estimated standard error (acres) [square root of row (2)]	660	392	243	843
4.	Estimated CV = row (3)/row(1)	20.3(%)	3.7(%)	2.5(%)	3.4(%)

11.4.2 Area survey of crops

The results for any stratum follow from those in section 5.5.2. If in the hth stratum, n_h fields are selected with ppa

$$\pi_{hi} = a_{hi} \Big/ \sum_{i=1}^{N_h} a_{hi} = a_{hi}/A_h$$

where a_{hi} is the area of the ith field ($i = 1, 2, \ldots, N_h$ for the universe, and $i = 1, 2, \ldots, n_h$ for the sample) and $\sum_{i=1}^{N_h} a_{hi} = A_h$, the total area of all the N_h fields in the stratum, and if for the ith sample field, y_{hi} denotes the area under a particular crop, then an unbiased estimator of the average proportion under the crop in the stratum is, from equation (5.13),

$$\overline{p}_h = \sum_{i=1}^{n_h} p_{hi}/n_h \tag{11.20}$$

i.e. the simple (unweighted) average of the sample proportions $p_{hi} = y_{hi}/a_{hi}$ in the sample fields, with an unbiased variance estimator (from equation (5.14)) of

$$s^2_{\overline{p}_h} = SSp_{hi}/n_h(n_h - 1) \tag{11.21}$$

An unbiased estimator of the total area under the crop in the hth stratum is, from equation (5.11),

$$y^*_{h0} = A_h \overline{p}_h \tag{11.22}$$

with an unbiased variance estimator (from equation (5.12))

$$s^2_{y^*_{h0}} = A_h^2 s^2_{\bar{p}_h} \tag{11.23}$$

For all the strata combined, the unbiased estimator of the total area under the crop Y is, from equation (11.6.d),

$$y = \sum^L y^*_{h0} = \sum^L A_h \bar{p}_h \tag{11.24}$$

with an unbiased variance estimator (from equation (11.15))

$$s_y^2 = \sum^L s^2_{y^*_{h0}} = \sum^L A_h^2 s^2_{\bar{p}_h} \tag{11.25}$$

An estimator of the average proportion under the crop in the universe is

$$\bar{p} = y/A = \sum^L A_h \bar{p}_h / A \tag{11.26}$$

where $A = \sum^L A_h$ is the total area of all the strata, and an unbiased variance estimator of $\bar{p}$ is

$$s^2_{\bar{p}} = s_y^2 / A^2 \tag{11.27}$$

Note: If the crop is such that it either occupies the whole of a field or no part of it, or if the fields are small enough for this assumption to hold, i.e. the proportion of the total area under the crop p_{hi} is either 1 (whole) or 0 (none), then the estimator of the variance of the total area under the crop in equation (11.25) reduces to

$$s_y^2 = \sum^L A_h^2 p_h (1 - p_h)/(n_h - 1) \tag{11.28}$$

(see note to section 5.5.2).

11.4.3 Yield surveys of crops

Similar considerations apply for estimating the average yield of a crop per unit of area. The results for any stratum follow from those in section 5.5.3. If $r_{hi} = x_{hi}/y_{hi}$ is the yield per unit area in the ith sample field (obtained on harvesting the crop and measuring the yield x_{hi}) in the hth stratum, then an unbiased estimator of the average yield per unit area is

$$\bar{r}_h = \sum_{i=1}^{n_h} r_{hi}/n_h \tag{11.29}$$

with an unbiased variance estimator

$$s^2_{\bar{r}_h} = SSr_{hi}/n_h(n_h - 1) \tag{11.30}$$

An unbiased estimator of the total yield X_h in the hth stratum is

$$x^*_{h0} = A_h \bar{r}_h \tag{11.31}$$

with an unbiased variance estimator

$$s^2_{x^*_{h0}} = A^2_h s^2_{\bar{r}_h} \tag{11.32}$$

For all the strata combined, the unbiased estimator of the total yield is

$$x = \sum^L x^*_{h0} \tag{11.33}$$

with an unbiased variance estimator

$$s^2_x = \sum^L s^2_{x^*_{h0}} \tag{11.34}$$

An unbiased estimator of the average yield per unit area in the universe is

$$r = x/A \tag{11.35}$$

with an unbiased variance estimator

$$s^2_r = s^2_x/A^2 \tag{11.36}$$

A generally unbiased but consistent estimator of the average yield per unit of crop area in the hth stratum is

$$r'_h = x^*_{h0}/y^*_{h0} = \bar{r}_h/\bar{p}_h \tag{11.37}$$

with a variance estimator

$$s^2_{r'_h} = (s^2_{x^*_{h0}} + r'^2_h s^2_{y^*_{h0}} - 2r'_h s_{y^*_{h0}x^*_{h0}})/y^{*2}_{h0} \tag{11.38}$$

where

$$s_{y^*_{h0}x^*_{h0}} = A^2_h[SPp_{hi}r_{hi}/n_h(n_h - 1)] \tag{11.39}$$

For all the strata combined, a generally biased but consistent estimator of the average yield per unit of crop area is

$$r' = x/y \tag{11.40}$$

with a variance estimator

$$s^2_{r'} = (s^2_x + r'^2 s^2_y - 2r' s_{yx})/y^2 \tag{11.41}$$

where s_{yx} is an unbiased estimator of the covariance of y and x, given by equation (11.18).

11.5 Ratio method of estimation: Combined and separate ratio estimators

As for unstratified pps sampling (section 5.6), so also for stratified pps sampling the ratio method of estimation may be used to improve the efficiency of estimators; and as for stratified srs (section 10.4) for the whole universe, "combined" and "separate" ratio estimators may be used.

If w is the ancillary variable used for the ratio estimation, then an unbiased estimator of the universe total W_h in the hth stratum from a pps (the "size" variable being z) sample of n_h units is, from equation (11.6.b),

$$\begin{aligned} w_{h0}^{*} &= \sum_{i=1}^{n_h} w_{hi}^{*}/n_h = \sum_{i=1}^{n_h} w_{hi}/(\pi_{hi} n_h) \\ &= Z_h \sum_{i=1}^{n_h} w_{hi}/(z_{hi} n_h) \end{aligned} \tag{11.42}$$

where $Z_h = \sum_{i=1}^{n_h} z_{hi}$, the total of the values of the size variable in the hth stratum.

The ratio estimator of the total Y_h in the stratum is (from equation (5.27))

$$y_{hR}^{*} = W_h y_{h0}^{*}/w_{h0}^{*} = W_h r_{1h} \tag{11.43}$$

where

$$W_h = \sum_{i=1}^{N_h} w_{hi} \text{ and } r_{1h} = y_{h0}^{*}/w_{h0}^{*} \tag{11.44}$$

A consistent but generally biased estimator of the variance of y_{hR}^{*} is (from equation (5.29))

$$s_{y_{hR}^{*}}^{2} = W_h^2 s_{r_{1h}}^2 = s_{y_{h0}^{*}}^{2} + r_{1h}^2 s_{w_{h0}^{*}}^{2} - 2 r_{1h} s_{y_{h0}^{*} w_{h0}^{*}} \tag{11.45}$$

where

$$\begin{aligned} s_{y_{h0}^{*}}^{2} &= SSy_{hi}^{*}/n_h(n_h - 1) \\ s_{w_{h0}^{*}}^{2} &= SSw_{hi}^{*}/n_h(n_h - 1) \\ s_{y_{h0}^{*} x_{h0}^{*}} &= SPy_{hi}^{*} w_{hi}^{*}/n_h(n_h - 1) \end{aligned} \tag{11.46}$$

For the ratio estimator of the total of the study variable $Y = \sum^{L} Y_h$, two types of ratio estimators could be used.

1. *Separate ratio estimator.* This is obtained by summing up the stratum ratio estimators, namely

$$y_{RS} = \sum^{L} y^*_{hR} \qquad (11.47)$$

the variance estimator of y_{RS} being provided by summing the stratum variance estimators, namely

$$s^2_{y_{RS}} = \sum^{L} s^2_{y^*_{hR}} \qquad (11.48)$$

2. *Combined ratio estimator.* This is obtained by applying the ratio method of estimation at the overall level, namely,

$$y_{RC} = Wr = Wy/w \qquad (11.49)$$

where y and w are the unbiased estimators of the universe totals Y and W respectively, defined by equations of the type (11.6.d), namely,

$$w = \sum^{L} w^*_{h0}$$

w^*_{h0} being given by equation (11.6.b), or more specifically by equation (11.42) and

$$r = y/w \qquad (11.50)$$

The variance estimator of y_{RC} is (from equation (5.29))

$$s^2_{y_{RC}} = W^2 s^2_r = s^2_y + r^2 s^2_w - 2r s_{yw} \qquad (11.51)$$

where

$$\begin{aligned} s_{yw} &= \sum_{h=1}^{L} s_{y^*_{h0} w^*_{h0}} \\ &= \sum_{h=1}^{L} \left[\sum_{i=1}^{n_h} y^*_{hi} w^*_{hi} - \left(\sum_{i=1}^{n_h} y^*_{hi} \right) \left(\sum_{i=1}^{n_h} w^*_{hi} \right) \Big/ n_h \right] \end{aligned} \qquad (11.52)$$

Note: For the ratio method of estimation to be efficient, the probability of selection should be appropriate for both y and w. See also the notes in section 10.4.

11.6 Ratio of ratio estimators of totals of two study variables

For two or more study variables, the ratio method of estimation may be applied, using the same ancillary variable. However, in computing the ratio of the totals of two study variables, the ratio method is equivalent to using the unbiased estimators of the totals for the stratum levels, as also for all the strata combined if combined ratio estimators for totals are used.

If separate ratio estimators are used for the totals of the two study variables, namely

$$y_{RS} = \sum^{L} y^*_{hR} = \sum^{L} W_h y^*_{h0}/w^*_{h0} = \sum^{L} W_h r_{1h} \tag{11.47}$$

$$x_{RS} = \sum^{L} x^*_{hR} = \sum^{L} W_h x^*_{h0}/w^*_{h0} = \sum^{L} W_h r_{2h} \tag{11.53}$$

are respective estimators of the universe total Y and X, where x_{RS} is defined as for y_{RS} and

$$r_{2h} = x^*_{h0}/w^*_{h0} \tag{11.54}$$

An estimator of the universe ratio $R = Y/X$ is

$$r' = \frac{y_{RS}}{x_{RS}} = \frac{\sum^{L} y^*_{hR}}{\sum^{L} x^*_{hR}} = \frac{\sum^{L} W_h r_{1h}}{\sum^{L} W_h r_{2h}} \tag{11.55}$$

A variance estimator of r' is

$$s^2_{r'} = (s^2_{y_{RS}} + r'^2 s^2_{x_{RS}} - 2r' s_{y_{RS}x_{RS}})/x^2_{RS} \tag{11.56}$$

where $s^2_{y_{RS}}$ and $s^2_{x_{RS}}$ are given by estimating equations of the type (11.48), and the estimated covariance by

$$s_{y_{RS}x_{RS}} = \sum^{L} s_{y^*_{hR}x^*_{hR}} \tag{11.57}$$

where

$$s_{y^*_{hR}x^*_{hR}} = \frac{\sum^{n_h} [(y^*_{hi} - r_{1h}w^*_{hi})(x^*_{hi} - r_{2h}x^*_{hi})]}{n_h(n_h - 1)} \tag{11.58}$$

11.7 Gain due to stratification

As for stratified srs (section 10.6), so also for stratified pps samples it is possible to examine the gain, if any, due to stratification as compared to unstratified pps sampling.

We define

$$\pi_{h0} = \sum^{N_h} \pi_i = \sum_{i=1}^{N_h} z_{hi}/Z$$

as the sum of the probabilities of selection of the units of an unstratified pps sample, the summation being over the N_h units in the hth stratum.

The gain due to stratification in the variance estimator of the total of the study variable is

$$\sum^{L}(1/n\pi_{h0} - 1/n_h)s^2_{y^*_{h0}} + \frac{\left(\sum^{L} y^{*2}_{h0}/\pi_{h0} - y^2\right)}{n} - \frac{\sum^{L}(1/\pi_{h0} - 1)s^2_{y^*_{h0}}}{n} \qquad (11.59)$$

When $\pi_{h0} = N_h/N$, the expression (11.59) takes the form

$$\frac{N\sum^{L}(n_h s^2_{y^*_{h0}}/N_h)}{n} - \sum^{L} s^2_{y^*_{h0}} + \frac{N\sum^{L} N_h(\bar{y}^*_{h0} - \bar{y})^2}{n} - \frac{\sum^{L}(N/N_h - 1)s^2_{y^*_{h0}}}{n} \qquad (11.60)$$

Note: The condition under which stratification would lead to a gain will be discussed in Chapter 12.

Further reading

Cochran, sections 9A.13, 11.5, and 11.16; Hedayat and Sinha, chapter 9; Murthy (1967), sections 7.8, 7.8b, and 10.6; Sukhatme *et al.* (1984), sections 4.17-4.19, 5.13, and 6.8; Yates, sections 3.10, 6.17, 7.5, and 7.16.

Exercise

1. Table 11.9 gives the acreage of wheat (y_{hi}) in sample parishes, selected with probability proportional to size (the total acreage of crops and grass, x_{hi}) in four districts of Hertfordshire, U.K. Estimate the total acreage of wheat in these four districts taken together with its standard error, given the total sizes of the districts at the bottom of the Table (Yates, Examples 6.17 and 7.16, modified).

Table 11.9: Acreage of wheat (y_{hi}) in sample parishes selected with probability proportional to size (the total acreage of crops and grass, x_{hi}) in four districts of Hertfordshire, U.K.

District I (3 parishes)		District II (4 parishes)		District III (5 parishes)		District IV (2 parishes)	
y_{hi}	x_{hi}	y_{hi}	x_{hi}	y_{hi}	x_{hi}	y_{hi}	x_{hi}
766	3040	311	2370	558	2300	225	2520
701	3440	228	3330	775	4430	738	3740
503	2040	249	2290	495	2890		
		686	2930	565	2420		
				862	4160		
Total	43 591		57 263		73 946		34 437

Source: Yates, Example 6.17 and 7.16 (adapted).

CHAPTER 12

Size of Sample and Allocation to Different Strata

12.1 Introduction

In this chapter we will consider first the problems of the allocation of the total sample size into different strata and of the formation of strata, and second, the determination of the total size of the sample.

12.2 Principles of stratification

Equation (10.10) shows that the sampling variance of y (unbiasedly estimating the universe total Y) in a stratified srs depends on the within-stratum variances σ_h^2; and from equation (11.14), we see that in a stratified pps sample, the sampling variance of y depends on the within-stratum variability of Y_{hi}/z_{hi}. For efficient stratification, therefore, the strata should be so formed as to be internally homogeneous with respect to Y_{hi} in srs and with respect to Y_{hi}/z_{hi} in pps sampling, so that within-stratum variability of Y_{hi} (in srs) or of Y_{hi}/z_{hi} (in pps sampling) is minimized.

12.3 Allocation of sample size to different strata

12.3.1 Formulation of the problem

Given a total sample of size n, its allocation to the different strata would be based on the same principle as in Chapter 7, namely, that for a specified total cost of surveying the sample, the sampling variance should be the minimum or *vice versa*.

The sampling variance of the sample estimator y of the universe total

Y in srs and pps (both with replacement) takes the form

$$\sum_{h=1}^{L} N_h^2 V_h / n_h \tag{12.1}$$

where V_h is the variance per unit in the hth stratum, and n_h is the sample size in the hth stratum. Thus in stratified srs (with replacement),

$$V_h = \sigma_h^2 = \sum_{i=1}^{N_h} (Y_{hi} - \overline{Y}_h)^2 / N_h \tag{12.2}$$

(see also equation (10.8)) and in stratified srs (with replacement) for proportions

$$V_h = \sigma_h^2 = P_h(1 - P_h) \tag{12.3}$$

from equation (2.68).

In stratified pps (with replacement)

$$V_h = \sum_{i=1}^{N_h} (Y_{hi}/\pi_{hi} - Y_h)^2 \pi_{hi} / N_h^2 \tag{12.4}$$

where $\pi_{hi} = z_h/Z_h$ is the probability of selection of the ith universe unit ($i = 1, 2, \ldots, N_h$) in the hth stratum, z_{hi} is the "size" of the hith universe unit, and Z_h is the total "size" of the hth stratum from equation (5.4).

A simple cost function in stratified sampling is

$$C = c_0 + \sum_{1}^{L} n_h c_h \tag{12.5}$$

where c_0 is the overhead cost, and c_h the average cost of taking a sample unit in the hth stratum, which may vary from stratum to stratum, depending on field conditions.

12.3.2 Optimum allocation

It can be shown that with the general variance function (12.1) and the cost function (12.5), the optimum value of n_h that minimizes the variance for a given total cost or *vice versa* is proportional to $N_h\sqrt{(V_h/c_h)}$, and is given by

$$\frac{n_h}{n} = \frac{N_h\sqrt{(V_h/c_h)}}{\sum N_h\sqrt{(V_h/c_h)}}$$

When the total cost is fixed at C',

$$n_h = (C' - c_0)\,\frac{N_h\sqrt{(V_h/c_h)}}{\sum N_h\sqrt{(V_h/c_h)}} \tag{12.6}$$

For a fixed variance V', the optimum allocation is

$$n_h = \left[\sum N_h\sqrt{(V_h c_h)}\right]\left[N_h\sqrt{(V_h/c_h)}\right]\Big/ V' \tag{12.7}$$

These are the ***optimum allocations*** of the total sample size n into different strata and show that a large sample should be taken in a stratum which is large (N_h large), is more variable internally (V_h large), and sampling is inexpensive (c_h small).

Notes

1. Theoretical proofs are given in Appendix II, section A2.3.10.
2. The optimum n_h values would have been the same had we considered the estimation of the universe mean Y/N, rather than that of the total Y. In comparing expression (12.7) relating to the total with equivalent expressions for the mean in other textbooks, note that the variance for the total is N^2 times the variance for the mean.
3. For stratified srs without replacement, the optimum allocation is similarly obtained by taking the variance function (10.21) and the cost function (12.5). The optimum n_h is proportional to $N_h S_h/\sqrt{c_h}$ (where S_h is defined by equation (10.22)), and is given by the relation

$$\frac{n_h}{n} = \frac{N_h S_h/\sqrt{c_h}}{\sum N_h S_h/\sqrt{c_h}}$$

   For a fixed total cost C', the optimum n_h is obtained on substituting S_h for $\sqrt{V_h}$ in equation (12.6). For a fixed variance V', the optimum n_h is obtained on substituting in the above relation the value of n given in note (4) to section 12.6.2.
4. A sample survey usually includes more than one important item and what may be an optimum stratification for one item will not necessarily be for another item. Several solutions have been suggested in this common situation:

   (a) A few items may be selected for optimization and an average taken of the allocations taken of the optimum allocations for these items (Cochran, section 5A.3).

(b) Another compromise, suggested by Chatterjee (1967), would be to choose the n_k that minimizes the average of the proportional increases in variances, i.e. average of

$$\left[\sum_{h=1}^{L}(\hat{n}_h - n_h')^2/\hat{n}_h\right]\Big/ n$$

where $\hat{n}_h$ is the actual and n_h' the optimum sample size in stratum h. This amounts to choosing

$$n_h = n\sqrt{\left[\sum_j n_{jh}'^{\,2}\right]}\Big/\sum_h\sqrt{\left[\sum_j n_{jh}'^{\,2}\right]}$$

where n_{jh}' is the optimum sample size in stratum h for variable j (Cochran, section 5A.4).

(c) Other options have been suggested by Yates (1981); see also Cochran, section 5A.4.

Some other methods of allocation will now be considered.

12.3.3 Neyman allocation

If the cost per unit c_h is assumed to be the same ($= \bar{c}$) in all the strata, then the cost function (12.5) becomes

$$C = c_0 + n\bar{c} \tag{12.8}$$

(see also equation (7.27)) which determines the total sample size, given the total cost C', thus

$$n = (C' - c_0)/\bar{c} \tag{12.9}$$

(see also equation (7.28)). The optimum allocation for a fixed total cost is then (from equation (12.6))

$$n_h = \frac{nN_h\sqrt{V_h}}{\sum N_h\sqrt{V_h}} \tag{12.10}$$

This is known as the *Neyman allocation*.

In both the optimum and the Neyman allocations, the values of V_h will not be known, and estimates would have to be obtained from a previous or a pilot survey relating to the study variable or from a survey relating to an ancillary variable, which has a high positive correlation with it. Lacking any information on the universe standard deviation, the range of the values of the study variable (R_h) may be substituted for $\sqrt{V_h}$ in the Neyman allocation (also see section 7.3), i.e.

$$n_h = \frac{nN_h R_h}{\sum N_h R_h} \tag{12.11}$$

Notes

1. The Neyman allocation is sometimes called the optimum allocation. For theoretical proof, see Appendix II, section A2.3.10.
2. In the Neyman allocation, the stratum sampling fraction n_h/N_h is proportional to $\sqrt{(V_h)}$ (i.e., proportional to σ_h in srs).

12.3.4 Proportional (Bowley) allocation

In the absence of any information on V_h, if it can be assumed to be the same in all the strata, then the Neyman allocation takes the simple form

$$n_h = nN_h/N \tag{12.12}$$

i.e. n_h is proportional to N_h, or that the sample is allocated to the different strata in proportion to the number of universe units N_h.

Also, since $\sum n_h = n\sum N_h/N = n$, the design will be self-weighting stratified srs with proportional allocation (see section 13.2).

12.3.5 Allocation proportional to the stratum total

When the stratum coefficients of variation are assumed not to vary considerably, the Neyman allocation (12.10) becomes

$$n_h = nY_h/Y \tag{12.13}$$

i.e. n_h is proportional to the stratum total Y_h. The condition is likely to be met when stratification is adopted for administrative or operational conveniences. However, as the stratum totals Y_h will not be known, in practice the allocation will have to be made in proportion to the total of an ancillary variable.

For skew universes, such as industrial establishments, income of individuals, and agricultural farms, allocation proportional to the stratum total is more efficient than proportional allocation.

12.3.6 Allocation in srs

A stratified srs with Neyman or proportional allocation is more efficient than an unstratified srs, but the efficiency decreases with departure from these allocations. While moderate departures do not affect the efficiency seriously, an extremely wide departure may lead to a loss due to stratification.

12.3.7 Allocation in pps sampling

In stratified pps sampling, the sample units are selected within each stratum with probability proportional to the size of an ancillary variable ($\pi_{hi} = z_{hi}/Z_h$). The formulae for the different types of allocation can be derived from equations (12.4)-(12.13); the difficulty in applying the optimum and the Neyman allocations lies, as in stratified srs, in the non-availability of advanced estimates of the stratum variances. A simple and appropriate allocation is, however, that proportional to the stratum total sizes Z_h (see section 12.3.5 above), i.e.

$$n_h = nZ_h/Z \tag{12.14}$$

In such a case, stratified pps sampling is always more efficient than unstratified pps sampling. The wider the variation of the stratum proportions Y_h/Z_h, the greater will be the efficiency.

Note: If, further, n_h is a constant, then the design can be made self-weighting by selecting a fixed number of sample units with pps in each stratum, the "size" being the stratum total Y_h (see section 13.3).

12.3.8 Acreage surveys of crops

In section 11.4.2 was considered stratified sampling with probability proportional to total area in the different strata for crop area surveys: an unbiased estimator of the total area under a crop in such a plan is

$$y = \sum_{h=1}^{L} A_h \sum_{i=1}^{n_h} p_{hi}/n_h \tag{12.15}$$

where A_h is the total geographical area of the hth stratum, n_h the number of sample units (fields, farms, or plots) in the hth stratum, and p_{hi} is the proportion of the total area under the crop in the hith sample unit.

If the allocation of the sample units is made proportional to the total areas of the strata, i.e. if n_h is proportional to A_h or

$$n_h = nA_h/A \tag{12.16}$$

where $A = \sum^L A_h$ is the total area of the universe, then the design becomes self-weighting (section 13.3).

Notes

1. Similar considerations apply to yield surveys of crops if the sample units are selected with probability proportional to area (section 11.4.3).

2. A special case of the above is when the crop is such that it either occupies the whole of a plot ($p_{hi} = 1$) or not at all ($p_{hi} = 0$), when the sampling variance of the estimated total area under the crop is

$$\sum A_h^2 P_h(1 - P_h)/n_h \tag{12.17}$$

where P_h is the universe proportion of the area under the crop in the hth stratum.

With n_h proportional to A_h, the variance formula (12.17) becomes

$$A \sum A_h P_h(1 - P_h)/n \tag{12.18}$$

The optimum allocation then is

$$n_h = \frac{n A_h \sqrt{[P_h(1 - P_h)]}}{\sum A_h \sqrt{[P_h(1 - P_h)]}} \tag{12.19}$$

If the universe proportion P_h could be estimated from the data of an earlier survey, an approximation to the optimum allocation could be obtained.

3. *Fractional values of n_h*. The allocation (n_hs) will mostly turn out to be fractional. They may be rounded off to whole numbers in one of the following three ways:

   (a) The usual rule of rounding off of the decimal system may be followed.

   (b) The smaller n_hs may be rounded off to higher integers at the cost of larger n_hs.

   (c) n_h may be rounded off to the neighboring integers in a randomized manner such that its expected value remains the same. Thus let $n_h = k_h + d_h$, where k_h is an integer and d_h is a fraction; then n_h should be rounded off to k_h with probability $(1 - d_h)$ and to $(k_h + 1)$ with probability d_h. For example, suppose in one stratum $n_h = 5.3$ up to one decimal figure; then a one digit random number is chosen; if it falls between 1 and $(10 - 3 =)7$, then n_h is taken as 5, and if the random number is 8, 9 or 0, then n_h is taken as 6. This procedure may be necessary in self-weighting designs (Chapter 13), but otherwise procedure (b) may be more reasonable.

12.4 Formation of strata

12.4.1 Formulation of the problem

In large-scale surveys, the primary strata are usually compact administrative areas, within which ultimate strata are constituted. The problem is to determine the number of strata and to demarcate them. In stratified srs, the strata should be so formed as to maximize principally the variation between stratum means and, to a lesser extent, the stratum standard deviations; in stratified pps sampling, on the other hand, the units within each stratum should be homogeneous with respect to Y_{hi}/z_{hi}; in practice, the values of the study variable relating to an earlier survey or those of a related ancillary variable would have to be used. The variable on basis of which stratification is made is called the ***stratification variable.***

12.4.2 Demarcation of strata

There are several practical rules for the demarcation of the different strata.

(i) *Equalization of* $N_h\sqrt{V_h}$ *(Dalenius-Gurney rule).* The strata may be so formed as to equalize the values of $N_h\sqrt{V_h}$ (= $N_h\sigma_h$ in srswr) in the different strata, and then an equal allocation of the total sample to the different strata may be made. This requires, of course, a prior knowledge of $\sqrt{V_h}$ for different sets of stratification points.

(ii) *Equalization of* N_hR_h *(Ekman rule).* In srs, lacking information on σ_h, the values of N_hR_h may be equalized, where R_h is the range of the study variable in the hth stratum.

(iii) *Equalization of stratum totals* Y_h *(Mahalanobis-Hansen-Hurwitz-Madow rule).* The strata may be formed by equalizing the stratum totals Y_h (or the stratum totals of the stratification variable, in practice) with an equal allocation. This is a good rule when the stratum coefficients of variation are approximately the same (see also section 12.3.5) and would not change appreciably with changes in stratum boundaries; this situation is likely to be met when geographical stratification is made on administrative considerations. The rule is simple and also helps in achieving a self-weighting design, when a fixed number of sample units is selected in each stratum (see section 13.3).

For example, in the Indian National Sample Survey (1958-9), with the main emphasis on a demographic inquiry, the primary strata in the rural areas consisted of geographically compact areas formed by equalizing the 1951 Census populations.

Note: For estimates of ratios (such as birth rate, death rate etc., in a vital rate survey, or average income per person in a household budget inquiry), the stratum sizes may be made equal with respect to a measure of the size that is highly correlated with the denominator of the ratios (total population in our examples).

(iv) *Equalization of Cumulatives of* $\sqrt{[f(y)]}$ *(Dalenius-Hodges rule).* If the number of strata is large and stratification effective, the distribution of the study variable can be assumed to be rectangular within the different strata. A very good rule in this case is to form the strata by equalizing the cumulatives of $\sqrt{[f(y)]}$ where $f(y)$ is the frequency distribution of the study variable. This makes $N_h\sigma_h$ approximately constant in srs, so that the Neyman allocation gives a constant sample size $n_0 = n/L$ in all strata. As the optimum is generally flat with respect to variations in n_h, the use of the Dalenius-Hodges rule would be highly efficient.

12.4.3 Number of strata

Stratification with proportional allocation is always more efficient than unstratified sampling. But stratification may be carried up to a point where only one unit is selected from each stratum: estimates of sampling variances cannot be computed then without resort to approximate methods (see section 24.8). One consideration is therefore that the number of strata should at most be $\frac{1}{2}n$ if n is even and $\frac{1}{2}(n-1)$ if n is odd, so that at least two units are selected from each stratum. The other consideration is that although in general the efficiency increases with the number of strata, beyond a certain point the gain in efficiency will not be worth the efforts required to increase the number of strata.

12.5 Multiple stratification

If a universe is stratified according to two or more factors, it is said to be multiple stratified. In practice, this may be difficult to apply even when the required data are available, due to possible conflicts in the different stratification criteria. For example, for surveys on areas under different crops, the total geographical area may be a suitable stratification variable, but population may be more suitable for purposes of stratification in demographic or labor force surveys.

A compromise in such a case would be to stratify the universe first according to the most important stratification variable, and then within each stratum so formed to sub-stratify according to the next most important variable, and so on. This is known as *multiple stratification* or *deep stratification*.

12.6 Determination of the total sample size

12.6.1 Specified cost

For the optimum allocation, with the given total cost C', the total sample size is

$$n = \frac{(C' - c_0)\sum N_h\sqrt{(V_h/c_h)}}{\sum N_h\sqrt{(V_h c_h)}} \tag{12.20}$$

For proof, see Appendix II, section A2.3.10.

For the Neyman allocation, from equation (12.8) the total sample size is determined as

$$n = (C' - c_0)/\bar{c} \tag{12.9}$$

For the proportional allocation (equation (12.12)), the cost function (12.5) gives the total sample size as

$$n = (C' - c_0)\Big/\sum w_h c_h \tag{12.21}$$

where for the proportional allocation $w_h = n_h/n = N_h/N$. If cost function (12.8) is taken, then the total sample size is given by equation (12.9) with proportional allocation also.

Note: For stratified srs without replacement, the value of n is obtained on replacing $\sqrt{V_h}$ by S_h in equation (12.20) for optimum allocation.

12.6.2 Specified variance: continuous data

For continuous data, if V is the desired variance of a sample estimator y of the universe total Y in a stratified sample, then the sampling variance of the estimator y equation (10.15) could be written as

$$V = \sigma_y^2 = \sum(N_h^2\sigma_h^2/w_h)/n \tag{12.22}$$

where $n_h = nw_h$, where the w_h values have been chosen already.

From equation (12.22) the required sample size n for a specified V is

$$n = \sum(N_h^2\sigma_h^2/w_h)/V \tag{12.23}$$

For the Neyman allocation, i.e. when $w_h \propto N_h\sigma_h$, equation (12.23) becomes

$$n = \left(\sum N_h\sigma_h)\right)^2\Big/V \tag{12.24}$$

and for proportional allocation, i.e. when $w_h = N_h/N$, the required total sample size is

$$n = \sum N_h^2 \sigma_h^2 / V \tag{12.25}$$

Notes

1. As σ_h will not in general be known, it has to be estimated from a pilot inquiry or from other available information, such as the range.
2. Estimation of the universe mean Y/N requires the same sample size as that for the total Y.
3. If the margin of error d is specified (see section 7.2.2), then

$$V = (d/t)^2 \tag{12.26}$$

   where t is the normal deviate corresponding to the permissible probability that the error will exceed the specified margin.
4. For stratified srs without replacement, the value of n in Neyman allocation is

$$n = \frac{\left(\sum N_h S_h / c_h\right) \sum N_h S_h / \sqrt{c_h}}{V + \sum N_h S_h^2}$$

12.6.3 Specified variance: estimation of proportions

If V is the desired variance of the proportion P for the whole universe, then the required sample size n for the Neyman allocation is

$$n = \frac{\sum N_h \sqrt{(P_h Q_h)}}{N^2 V} \tag{12.27}$$

and for proportional allocation

$$n = \sum (N_h P_h Q_h) / NV \tag{12.28}$$

where P_h is the proportion in the hth stratum, and $Q_h = 1 - P_h$.

Notes

1. In practice, an estimate p_h of P_h will have to be used in formulae (12.27) and (12.28).
2. Estimation of the total number of universe units in a certain category, namely, NP, requires the same sample size as for that P.

12.7 Examples

Example 12.1

The data on the number of cattle obtained from a recent census are given in Table 12.1, in the 5 strata according to the total acreage of the farms, along with the present total number of farms in these strata. The problem is to estimate the present total number of cattle in the universe and its variance, by taking a sample of 500 farms.

(a) Determine the allocations of the sample in the different strata according to the following principles: (i) Neyman allocation, (ii) proportional allocation, and (iii) allocation proportional to the total number of cattle in the different strata;

(b) Also compute the expected variance for each of these (United Nations *Manual*, Process 4, Example 4, adapted).

(a) The required computations are shown in Table 12.2. Formula (12.10) is used for the Neyman allocation, and formula (12.12) for the proportional allocation. For allocation proportional to the stratum totals, the stratum total $Y'_h = N'_h \bar{y}'_h$ for the previous census is used, and the formula (12.13), modified as $n_h = nY'_h/Y'$, is applied ($Y' = \sum Y'_h$).

Table 12.1: Number of cattle, obtained from a previous census, and the present total number of farms in each stratum

Stratum (acres)	Previous census			Present total number of farms
	Total number of farms	Average number of cattle	Estimated s.d. per unit	
h	N'_h	$\bar{y}'_h$	s'_h	N_h
I: 0-15	625	3.91	4.5	635
II: 16-30	564	10.38	7.3	570
III: 31-50	476	14.72	9.6	475
IV: 51-75	304	21.99	12.2	303
V: 76-100	86	27.38	15.8	89
All strata combined	2055 (N')			2072 (N)

Source: United Nations *Manual*, Process 4, Example 4 (adapted).

Table 12.2: Computation of the allocation to different strata of the sample 500 farms: data of Table 12.1

Stratum	$N'_h s'_h$	$Y'_h = N'_h \bar{y}_h$	Allocations (n_h) Neyman*	Propor-tional†	Propor-tional to Y'_h ‡
I	2812.5	2444	84	153	50
II	4117.2	5854	125	138	120
III	4569.6	7007	138	115	144
IV	3708.8	6685	112	73	137
V	1358.8	2355	41	21	49
All strata combined	16,566.9 ($\sum N'_h s'_h$)	24,345 ($\sum Y'_h$)	500 (n)	500 (n)	500 (n)

* $n_h = \dfrac{500\, N'_h S'_h}{\sum N'_h s'_h}$ (12.10)

† $n_h = 500\, N_h/N$ (12.12)

‡ $n_h = 500\, Y'_h/\sum Y'_h$ (12.13)

(b) The expected variances are computed in Table 12.3. The relevant formula for any stratum is

$$s^2_{y^*_{ho}} = N_h^2 s'^2_h / n_h$$

where s'^2_h are the variances per unit obtained from the previous census, and N_h and n_h respectively the total and sample number in the present inquiry.

Note that in this case, allocation proportional to the stratum totals will be almost as efficient as the Neyman allocation. The results of an actual sample, using proportional allocation, is given in exercise 1, Chapter 10; the actual computed variance of the estimated total number of cattle was 575,597, as compared with the expected value of 640,997.

Example 12.2

An inquiry is to be designed to estimate the expected proportion of families with transistor ratios in two cities in Ethiopia. Rough estimates of the total number of families, the proportion with radios, and the cost of surveying family are given in Table 12.4. Treating the cities as strata, and assuming srs (with replacement) in each stratum, obtain the total optimum sample size and its allocation to the two cities, if the total cost (excluding overheads) is fixed at Eth. $20,000.

Table 12.3: Expected variances of total number of cattle according to different allocations: data of Tables 12.1 and 12.2

Stratum h	Estimated variance* in		
	Neyman allocation	Proportional allocation	Allocation proportional to Y'_h
I	97,206	53,368	163,306
II	138,511	125,463	144,283
III	150,678	180,814	144,400
IV	122,008	187,190	99,743
V	48,229	94,162	40,355
All strata combined	566,632	640,997	592,087

* $N_h^2 s'^2_h / n_h$

Table 12.4: Rough estimates of the total number of families, proportion with transistor ratios, and the cost of enumerating one family in two cities in Ethiopia

City h	Number of families N_h	Proportion with transistor radios P_h	Cost of surveying one family c_h
I	140,000	0.10	Eth $2.25
II	30,000	0.25	1.00

Table 12.5: Computation of the optimum allocations from the data of Table 12.4: Stratified srs of families

City	$V_h = P_h(1-P_h)$	$\sqrt{(V_h c_h)}$	$N_h\sqrt{(V_h c_h)}$	$\sqrt{(V_h/c_h)}$	$N_h\sqrt{(V_h/c_h)}$	$n_h = 20,000 \times N_h\sqrt{(V_h/c_h)}/24,525$
I	0.0900	0.1350	18,900	0.0600	6400	6850
II	0.1875	0.1875	5625	0.1875	5625	4587
Total			24,525		14,025 (n)	11,437

The required computations are shown in Table 12.5. With V_h given by equation (12.3), the optimum allocations are given by equation (12.6). There are 6850 families in the first city and 4587 in the second, with a total of 11,437 sample families. (As a check on the computations, it can be verified that the total cost $= 6850\times$ Eth. \$2.25 $+ 4587\times$ Eth. \$1.00 $=$ \$19,999 or Eth. \$20,000).

Further reading

Cochran, sections 5.5-5.12, and chapter 5A; Deming (1950), chapter 6, and (1960), chapter 20; Hansen *et al.* (1953), vols. I and II, chapter 5; Hedayat and Sinha, chaper 9; Kish (1965), chapter 3; Murthy (1967), chapter 7; Sukhatme *et al.* (1984), sections 4.4-4.8, and 4.11-4.14.

Exercises

1. In a survey, using a stratified srs with five strata, rough estimates of the universe units (N_h), the standard deviation (σ_h) and the cost of surveying one unit (c_h in a certain unit) in the different strata are given in Table 12.6. The total cost is fixed at 10,000 and the overhead cost at 500. Determine the optimum total sample size and its allocations to different strata (Chakravarti *et al.*, Illustrative Example 4.1).
2. A survey is designed to estimate the proportion of illiterate persons in three communities. Rough estimates of the total number of persons and the proportion illiterate are given in Table 12.7. Assuming a stratified srs, with the communities as the strata, how would you allocate a total sample of 2000 persons in the strata so as to estimate the overall proportion of illiterates?

Table 12.6: Rough estimates of the total number of units, standard deviation, and cost of surveying one unit in the different strata

Stratum h	N_h	σ_h	c_h	Optimum allocation
I	37,800	28.5	3.50	587
II	52,600	18.6	2.75	601
III	82,000	27.6	2.25	1537
IV	41,600	21.2	3.00	519
V	28,800	16.8	2.50	312

Source: Chakravarti *et al.*, Illustrative Example 4.1.

Table 12.7: Rough estimates of the total number of persons, and proportion illiterate in three communities

Community h	Total number of persons N_h	Proportion illiterate P_h
I	60,000	0.4
II	10,000	0.2
III	30,000	0.6

CHAPTER 13

Self-weighting Designs in Stratified Single-stage Sampling

13.1 Introduction

This chapter deals with the problems of making a stratified design self-weighting. We first consider stratified srs and then stratified varying probability sampling: the method of making any stratified design self-weighting at the tabulation stage will also be outlined.

13.2 Stratified simple random sampling

In stratified srs (as also in stratified circular systematic sampling with one sample), an unbiased estimator y of the universe total Y is

$$y = \sum_{h=1}^{L} N_h \sum_{i=1}^{n_h} y_{hi}/n_h \tag{13.1}$$

(see also (10.5.d)) where N_h and n_h are the total number of units and the number of sample units respectively in the hth stratum, and y_{hi} the value of the study variable for the ith selected sample unit in the hth stratum.

The weighting factor (or multiplier) for the hith sample unit is

$$w_{hi} = N_h/n_h \tag{13.2}$$

and for the design to be self-weighting with respect to y, this should be constant w_0. Then

$$w_0 = N_h/n_h = N/n \tag{13.3}$$

where N $(= \sum^L N_h)$ is the total number of units in the universe and n $(= \sum^L n_h)$ is the total number of units in the sample, i.e.

$$n_h = \frac{n}{N} N_h = \frac{N_h}{w_0} \tag{13.4}$$

This is the ***proportional allocation*** case (see section 10.2, note 2, and section 12.3.4).

Thus, a stratified srs will be self-weighting only with proportional allocation. Although the optimum and the Neyman allocations result in more efficient estimators, these require a prior knowledge of the variability in the different strata, which may not often be available. In such situations, there may be some advantage in the proportional allocation.

With a self-weighting stratified srs (i.e. a stratified srs with proportional allocations), some unbiased estimators defined in section 10.2 take the following form:

$$\begin{aligned} y^*_{hi} &= N_h y_{hi} = w_0 n_h y_{hi}\,; \\ y^*_{h0} &= N_h \overline{y}_h = w_0 \sum^{n_h} y_{hi}\,; \end{aligned} \tag{13.5}$$

$$y = \sum^L y^*_{h0} = w_0 \sum^L \sum^{n_h} y_{hi} = w_0 \sum_{\text{sample}} y_{hi}\,; \tag{13.6}$$

$$\begin{aligned} s^2_{y^*_{h0}} &= N^2_h \sum^{n_h} (y_{hi} - \overline{y}_h)^2 / n_h(n_h - 1) \\ &= w_0^2 n_h^2 \sum^{n_h} (y_{hi} - \overline{y}_h)^2 / n_h(n_h - 1)\,; \end{aligned} \tag{13.7}$$

and

$$s^2_y = \sum^L s^2_{y^*_{h0}} \tag{13.8}$$

The estimators of the ratio of the two universe totals (defined in estimating equations (10.7.b) and (10.7.d)) are

$$r_h = y^*_{h0}/x^*_{h0} = \sum^{n_h} y_{hi} \Big/ \sum^{n_h} x_{hi} \tag{13.9}$$

and

$$r = \sum^L \sum^{n_h} y_{hi} \Big/ \sum^L \sum^{n_h} x_{hi} = y/x \tag{13.10}$$

13.3 Stratified varying probability sampling

In stratified varying probability sampling, the unbiased estimator y of the universe total Y of a study variable is

$$y = \sum^{L} \frac{1}{n_h} \sum^{n_h} \frac{y_{hi}}{\pi_{hi}} = \sum^{L} \frac{Z_h}{n_h} \sum^{n_h} \frac{y_{hi}}{z_{hi}} \quad \text{in pps sampling} \tag{13.11}$$

(see also (11.6.d)) where π_{hi} is the (initial) probability of selection of the ith universe unit ($i = 1, 2, \ldots, N_h$) in the hth stratum.

In pps sampling $\pi_{hi} = z_{hi}/Z_h$, where z_{hi} is the value of the "size" variable of the hith universe unit, and $Z_h = \sum^{N_h} z_{hi}$ is the total "size" of the hth stratum.

The multiplier for the hith sample unit is

$$w_{hi} = 1/(n_h \pi_{hi}) = Z_h/(n_h z_{hi}) \tag{13.12}$$

If the ratio y_{hi}/z_{hi} can be observed and recorded easily in the field, then the design will be self-weighting when the ratios Z_h/n_h are the same in all the strata and equal to Z/n. This will be so if n_h is proportional to Z_h, i.e.

$$n_h = \frac{n}{Z} Z_h \tag{13.13}$$

A special case of this is when the strata are made equal with respect to the total "sizes" and an equal number of sample units is selected with pps from each stratum (sections 12.3.7 and 12.4).

The above rule is especially helpful in acreage and yield surveys of crops, when a sample of fields (or farms or plots) is selected with probability proportional to total (geographical) area in each stratum. If the number of sample units allocated to a stratum is made proportional to the total geographical area of the stratum, the design becomes self-weighting as y_{hi}/z_{hi} (the proportion of the area under the crop or the crop-yield per unit area) can be observed easily (sections 11.4 and 12.3.8).

13.4 Fractional values of n_h

The values of n_h required to make a stratified srs or pps sample self-weighting will not in general be integers. In such cases, one of the following procedures may be adopted.

1. n_h may be rounded off to the neighboring integers in a randomized manner such that its expected value remains the same (note 3 to section 12.3).

2. In stratified srs, the samples may be selected systematically with the same fractional interval N/n (see section 4.2), or the interval may be rounded off to N/n or $N/n+1$ in a randomized manner so that the expected value of n_h remains the same; similarly, in stratified pps sampling the samples may be selected with the same fractional interval Z/n, or the intervals may be rounded off in a randomized manner.

3. Samples may be selected linear systematically with the same interval I (integer nearest to N/n) in each stratum, when the estimator y will become $I\sum^{L}\sum^{n_h} y_{hi}$.

13.5 Self-weighting designs at the tabulation stage

A design that is not originally self-weighting can be made so at the tabulating stage by selecting a sub-sample with probability proportional to the multipliers: the pps sampling can be with replacement or systematic. The procedure will, however, lead to an increase in variance.

When a sub-sample of size n' ($< n$) is taken out of the original n ($= \sum^{L} n_h$) sample units in a stratified srs or pps sample with probability proportional to multipliers (w_{hi}) and with replacement, then

$$y_0'^{*} = \frac{1}{n'}\sum_{j=1}^{n'} y_j'^{*} \tag{13.14}$$

gives an unbiased estimator of Y, where

$$y_j'^{*} = \left(\sum^{L}\sum^{n_h} w_{hi}\right) y_j' \tag{13.15}$$

y_j' being the value of the study variable of the jth unit in the sub-sample ($j = 1, 2, \ldots, n'$).

An unbiased estimator of the variance of $y_0'^{*}$ is

$$\sum^{n'}(y_j'^{*} - y_0'^{*})^2/n'(n'-1) \tag{13.16}$$

Estimators of ratios and their variance estimators follow from the fundamental theorems of section 2.7.

Further reading

Murthy (1967), sections 12.2 and 12.5

PART III

MULTI-STAGE SAMPLING

CHAPTER 14

Multi-stage Sampling: Introduction

14.1 Introduction

In multi-stage sampling, the sample is selected in stages, the sampling units in each stage being sub-sampled from the (larger) units chosen at the previous stage, with appropriate methods of selection of the units – simple random sample (with or without replacement), systematic, probability proportional to size etc. – being adopted at each stage. In other words, the universe is divided into a number of first-stage (or primary sampling) units, which are sampled; then the selected first-stage units are sub-divided into a number of smaller second-stage (or secondary sampling) units, which are again sampled: the process is continued until the ultimate sampling units are reached.

This part of the book will deal with (unstratified) multi-stage sample designs. In this chapter, we shall consider the general theorems relating to multi-stage sampling; subsequent chapters will deal with two- and three-stage sampling with srs and pps sampling at different stages; the size and allocation of the sample; the estimation of the sampling variances at different stages; and self-weighting designs.

14.2 Reasons for multi-stage sampling

Multi-stage sampling is adopted in a number of situations:

1. Sampling frames may not be available for all the ultimate observational units in the universe, and it is extremely laborious and expensive to prepare such a complete frame. Here, multi-stage sampling is the only practical method. For example, in a rural household sample survey, conducted at intercensal periods, the households in rural areas could be reached after selecting a sample of villages (first-stage

units), after which a list of households within the selected villages only is prepared, and then selecting a sample of households (second-stage units) in the selected villages. In a crop survey, villages may be first selected (first-stage units); and next a list of fields prepared within the selected villages and a sample of the fields taken (second-stage units); and finally, a list of plots prepared within the selected fields and a sample of these plots taken (third-stage units). In this way, great savings are achieved as sampling frames need be constructed only for the selected sampling units and not for all the sampling units in the universe. Moreover in multi-stage sampling, ancillary information collected on the sampling units while listing the units at each stage could help in improving the efficiency of sample designs, either by stratification of the units, or by selecting the sample with probability proportional to size when the ancillary information is available before sample selection at that stage, or by using the ratio or regression method of estimation.

2. Even when suitable sampling frames for the ultimate units are available for the universe, a multi-stage sampling plan may be more convenient than a single-stage sample of the ultimate units, as the cost of surveying and supervising such a sample in large-scale surveys can be very high due to travel, identification, contact, etc. This point is closely related to the consideration of cluster sampling (Chapter 6). For instance, in a large-scale agricultural survey conducted in a developed country, although an up-to-date list of farms may be readily available from which a simple random sample of farms can be drawn, the cost of travelling and supervision of work on the widely scattered farms may be extremely high. Therefore, the procedure to be adopted would be to try to confine the sample of farms to certain area segments.

3. Multi-stage sampling can be a convenient means of reducing response errors and improving sampling efficiency by reducing the intra-class correlation coefficient observed in natural sampling units, such as households or villages. Thus, in opinion and marketing research, it becomes necessary to select only one individual in a sample household in order to avoid conditioned response, and also to spread the sample over a greater number of sample households because of the general homogeneity of responses of individuals in a household, even if the "true" responses of all the members of the household could be obtained.

Note: Multi-stage sampling is cheaper and operationally easier than srs but not more than single-stage cluster sampling; considering sampling variability, however, multi-stage sampling is generally less efficient than srs but more efficient than cluster sampling: this is, of course, based on the assumption that the total sample size is fixed. Some of the lost efficiency may, however, be regained by using ancillary information.

14.3 Structure of a multi-stage design

Figure 14.1 illustrates the structure of a two-stage design for a household survey in rural areas for which there is no current list of households. Of the total N villages, a sample of n villages is selected in an appropriate way; all the extant households (M_i in number; $i = 1, 2, \ldots, n$) in these n sample villages are listed by the enumerators by actual field visit; and in the ith sample village, a sample of households m_i in number) is finally selected for the survey and the required household characteristics recorded. This is a two-stage design, with villages as the first-stage and households as the second-stage units. The total number of households in all the villages is $\sum^N M_i$, which is generally not known; the sample number is $\sum^n m_i$.

Note: In such a plan, if all the N villages are included in the survey and a sample of households m_i ($i = 1, 2, \ldots, N$) selected in each village, the design becomes stratified single-stage (with total number of sample households, $\sum^N m_i$), and the methods of Part II of the book will apply. If in the sample of n villages, all the M_i households are included in the survey, the design becomes single-stage cluster sampling (with total number of sample households, $\sum^n M_i$), and the methods of Part I of the book will apply. Of course, if all the N villages are included in the survey and all the M_i households in all these villages surveyed, the survey is one of complete enumeration without any sampling (with total number of households, $\sum^N M_i$).

14.4 Fundamental theorems in multi-stage sampling

The fundamental theorems for the estimation of universe parameters in multi-stage sampling are analogous to those in section 2.7 (see note 1 to the section).

1. If n (≥ 2) units are selected out of the total N first-stage units with replacement (and with equal or varying probabilities), and t_i ($i = 1, 2, \ldots, n$) is an unbiased estimator of the universe parameter T, obtained from the ith selected first-stage unit, then a combined

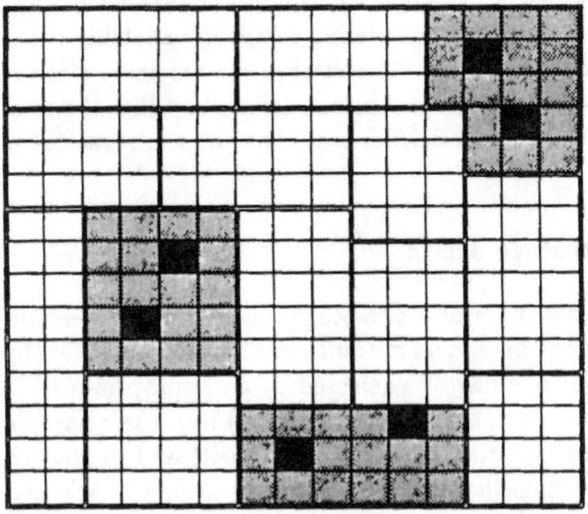

Figure 14.1: Schematic representation of a two-stage sample design. Villages (first-stage units) are bounded by double lines and households (the second-stage units) by thin lines; selected villages are indicated by grey shading and sample households (in the selected villages) by black shading.

unbiased estimator of T is the arithmetic mean

$$\bar{t} = \sum^{n} t_i/n \tag{14.1}$$

and an unbiased estimator of the variance of $\bar{t}$ is

$$s_{\bar{t}}^2 = \sum^{n} (t_i - \bar{t})^2/n(n-1) = SSt_i/n(n-1) \tag{14.2}$$

We shall show later how to compute the t_i values for a multi-stage design.

If u_i ($i = 1, 2, \ldots, n$) are the n independent unbiased estimators, obtained also from the n first-stage sample units, of another universe parameter U, then a combined unbiased estimator of U is similarly

$$\bar{u} = \sum^{n} u_i/n \tag{14.3}$$

and an unbiased estimator of the variance of $\bar{u}$ is

$$s_{\bar{u}}^2 = \sum^{n} (u_i - \bar{u})^2/n(n-1) = SSu_i/n(n-1) \tag{14.4}$$

2. An unbiased estimator of the covariance of $\bar{t}$ and $\bar{u}$ is the estimator

$$\begin{aligned} s_{\bar{t}\bar{u}} &= \sum^{n}(t_i - \bar{t})(u_i - \bar{u})/n(n-1) \\ &= SPt_u u_i/n(n-1) \end{aligned} \tag{14.5}$$

3. A consistent, but generally biased, estimator of the ratio of the universe parameters $R = T/U$ is the ratio of the sample estimators $\bar{t}$ and $\bar{u}$

$$r = \bar{t}/\bar{u} \tag{14.6}$$

and an estimator of the variance of r is

$$s_r^2 = (s_{\bar{t}}^2 + r^2 s_{\bar{u}}^2 - 2r s_{\bar{t}\bar{u}}/\bar{u}^2 \tag{14.7}$$

4. A consistent, but generally biased, estimator of the universe correlation coefficient ρ between the two study variables is the sample correlation coefficient

$$\hat{\rho}_{\bar{t}\bar{u}} = s_{\bar{t}\bar{u}}/s_{\bar{t}}s_{\bar{u}} \tag{14.8}$$

5. For unbiased estimators of universe totals, obtained from the first-stage units we change the notation and also deal with the totals of a study variable.

In a multi-stage sample design with u states ($u > 1$), at the tth stage ($t = 1, 2, \ldots, u$), let $n_{1\,2\,\ldots\,(t-1)}$ sample units be selected with replacement out of the total $N_{1\,2\,\ldots\,(t-1)}$ units and let the (initial) probability of selection of the $(1\,2\,\ldots\,t)$th unit be $\pi_{1\,2\,\ldots\,t}$. We define

$$f_t \equiv n_{1\,2\,\ldots\,(t-1)}\pi_{1\,2\,\ldots\,t} \tag{14.9}$$

An unbiased estimator of the universe total Y of the study variable is

$$y = \sum_{\text{sample}} \left(\prod_{1}^{u} \frac{1}{f_t}\right) y_{1\,2\,\ldots\,u} = \sum_{\text{sample}} w_{1\,2\,\ldots\,u} y_{1\,2\,\ldots\,u} \tag{14.10}$$

where $y_{1\,2\,\ldots\,u}$ is the value of the study variable in the $(1\,2\,\ldots\,u)$th ultimate-stage sample units, and the factors

$$\prod_{1}^{u} \frac{1}{f_t} = w_{1\,2\,\ldots\,u} \tag{14.11}$$

are the weighting-factors (or the multipliers) for the ($12\ldots u$)th ultimate-stage sample units; the sum of the products of these multipliers and the values of the study variable for all the ultimate-stage sample units provides the unbiased linear estimator y of the universe total Y in equation (14.10).

An unbiased estimator of the variance of the sample estimator y in the general estimating equation (14.10) is

$$s_y^2 = \sum^n \left[n \sum_{2,3,\ldots,u} w_{12\ldots u} y_{12\ldots u} - \sum_{1,2,\ldots,u} w_{12\ldots u} y_{12\ldots u} \right]^2 \Big/ n(n-1) \qquad (14.12)$$

where n (≥ 2) is the number of first-stage units selected with replacement, the summation outside the square brackets being over these first-stage units.

The formulae can be put in simpler forms. Using a slightly different notation, an unbiased estimator of the universe total Y, obtained from the ith first-stage unit ($i = 1, 2, \ldots, n$), is

$$y_i^* = n \sum_{2,3,\ldots,u} w_{12\ldots u} y_{12\ldots u} \qquad (14.13)$$

the summation being over the second- and subsequent-stage sampling units.

The combined unbiased estimator of the universe total Y from all the n sample first-stage units is, as we have seen from equation (14.1), the arithmetic mean

$$y_0^* = \sum^n y_i^* / n \qquad (14.14)$$

which is the same as the estimator y in equation (14.10), the summation in equation (14.14) being over the n sample first-stage units.

An unbiased estimator of the variance of y_0^* is, from equation (14.2),

$$s_{y_0^*}^2 = \sum^n (y_i^* - y_0^*)^2 / n(n-1) \qquad (14.15)$$

which is the same as equation (14.11).

Notes

1. Estimating equation (14.10) applies to single- as well as multi-stage designs, sample selection being done with equal probability with or without replacement, or with varying probability with replacement or systematic sampling with equal or varying probability.
2. For srs, the fs are the sampling fractions at different stages.
3. The above formulae, which will be illustrated in later chapters for two- and three-stage designs, are adequate in obtaining estimates of universe totals, means, ratios, etc. of study variables, and also estimates of their variances.
4. The variance of a universe estimator is built up of the variances at different stages of a multi-stage design. The equations for estimating variances given above do not show these components separately and these are not required for estimating universe totals, means, ratio etc., with their respective standard errors. However, the decomposition of the total variance into the component stage-variances is essential in improving the design of subsequent sample surveys (Chapter 17).
5. For the estimation of the variance of y by equation (14.12), it is not necessary that the number of sample units at stages other than the first be two or more; they must, however, be so if estimates of stage-variances are required.

14.5 Setting of confidence limits

The same results as for single-stage sampling given in section 2.8 apply also for multi-stage sampling. The $(100 - \alpha)$ per cent confidence limits of the universe parameter T are set with the sample estimator $\bar{t}$ and its estimated standard error $s_{\bar{t}}$, thus

$$\bar{t} \pm t'_{\alpha,n-1} s_{\bar{t}} \tag{14.16}$$

where $t'_{\alpha,n-1}$ is the $100\,\alpha$ percentage point of the t-distribution for $(n - 1)$ degrees of freedom. For large n, the percentage points of the normal distribution could be using in place of those for the t-distribution.

Further reading

Murthy, sections 9.1 and 9.2.

CHAPTER 15

Multi-stage Simple Random Sampling

15.1 Introduction

In this chapter will be presented the estimating methods for totals, mean, and ratios of study variables and their variances in a multi-stage sample with simple random sampling at each stage: the methods for two- and three-stage designs are illustrated in some detail and the general procedure for a multi-stage design indicated. The methods of estimation of proportion of units in a category and the use of ratio estimators will also be considered.

15.2 Two-stage srs: Estimation of totals, means, ratios and their variances

15.2.1 Universe totals and means

Let the number of first-stage units (fsu's) be N, and the number of second-stage units (ssu's) in the ith fsu ($i = 1, 2, \ldots, N$) be M_i; the total number of ssu's in the universe is $\sum^N M_i$ (Table 15.1). Denoting by Y_{ij} the value of the study variable of the jth ssu ($j = 1, 2, \ldots, M_i$) in the ith fsu, we define the following universe parameters:

Total of the values of the study variable in the ith fsu:

$$Y_i = \sum_{j=1}^{M_i} Y_{ij} \tag{15.1}$$

Mean for the ith fsu:

$$\overline{Y}_i = Y_i / M_i \tag{15.2}$$

Table 15.1: Sampling plan for a two-stage simple random sample with replacement

State (t)	Unit	No. in universe	No. in sample	Selection method	Selection probability	f_t
1	First-stage	N	n	srswr	Equal $= 1/N$	n/N
2	Second-stage	M_i	m_i	srswr	Equal $= 1/M_i$	m_i/M_i

Grand total:

$$Y = \sum^{N} \sum^{M_i} Y_{ij} = \sum^{N} Y_i = \sum^{N} M_i \overline{Y}_i \tag{15.3}$$

Mean per fsu:

$$\overline{Y} = Y/N = \sum^{N} M_i \overline{Y}_i / N \tag{15.4}$$

Mean per ssu:

$$\overline{\overline{Y}} = Y \Big/ \left(\sum^{N} M_i \right) = \sum^{N} M_i \overline{Y}_i \Big/ \left(\sum^{N} M_i \right) \tag{15.5}$$

15.2.2 Structure of a two-stage srs

Out of the N fsu's, n are selected; and out of the M_i ssu's in the ith selected fsu ($i = 1, 2, \ldots, n$), m_i are selected; sampling at both stages in srs with replacement (also see section 14.3 and Figure 14.1). The total number of ssu's in the sample, i.e. the sample size, is $\sum^n m_i$.

Let y_{ij} denote the value of the study variable in the jth selected ssu ($j = 1, 2, \ldots, m_i$) in the ith selected fsu ($i = 1, 2, \ldots, n$).

15.2.3 Estimation of the universe total Y and the variance of the sample estimator

The problem is first to estimate the universe total Y and the variance of the sample estimator. This will be shown in three ways:

(a) Estimating downwards from the fsu's;
(b) estimating upwards from the ssu's; and
(c) estimating from the general estimating equations given in section 14.4.

(a) ***Estimating downwards from the first-stage units.*** If all the M_i ssu's in each of the n sample fsu's are completely surveyed and y_i is the value of the study variable (i.e. the sum of the values for the M_i ssu's) in the ith sample fsu, then the estimating methods of Chapter 2 will apply. Thus, an unbiased estimator of the universe total Y from the ith selected fsu is, from equation (2.42),

$$y_i^* = Ny_i \tag{15.6}$$

The combined unbiased estimator of Y, obtained from all the n sample fsu's, is the simple arithmetic mean (from equation (2.43))

$$y_0^* = \frac{1}{n}\sum^n y_i^* = \frac{N}{n}\sum^n y_i \tag{15.7}$$

An unbiased estimator of the variance of y_0^* is, from equation (2.44),

$$s_{y_0^*}^2 = \sum^n (y_i^* - y_0^*)^2/n(n-1) \tag{15.8}$$

If, however, all the M_i ssu's in the ith sample fsu are not completely surveyed, but a sample of m_i ssu's is taken, then the value of the study variable in the sample fsu Y_i (or y_i above) has to be estimated on the basis of these sample m_i ssu's. Let the value of the study variable in the jth sample ssu ($j = 1, 2, \ldots, m_i$) in the ith sample fsu be y_{ij}. Then an unbiased estimator of the total Y_i (or y_i above) from the ijth sample ssu is, from equation (2.42),

$$y_{ij}^* = M_i y_{ij} \tag{15.9}$$

and the combined unbiased estimator of Y_i (from equation (2.43)) is the simple arithmetic mean

$$y_{i0}^* = \frac{1}{m_i}\sum_{j=1}^{m_i} y_{ij}^* = \frac{M_i}{m_i}\sum_{j=1}^{m_i} y_{ij} = M_i\overline{y}_i \tag{15.10}$$

where

$$\overline{y}_i = \sum_{j=1}^{m_i} y_{ij}/m_i \tag{15.11}$$

is the unbiased estimator of the universe mean $\overline{Y}_i$ in the ith fsu.

This value of y_{i0}^* is substituted for y_i in equation (15.6) to provide the unbiased estimator of Y, from the ith sample fsu, namely

$$y_i^* = N y_{i0}^* = N \frac{M_i}{m_i} \sum_{j=1}^{m_i} y_{ij} \tag{15.12}$$

and in equation (15.7) to provide the combined unbiased estimator of Y from all the n sample fsu's, namely

$$\begin{aligned} y_0^* &= \frac{1}{n} \sum^{n} y_i^* = \frac{N}{n} \sum^{n} y_{i0}^* \\ &= \frac{N}{n} \sum_{i=1}^{n} \frac{M_i}{m_i} \sum_{j=1}^{m_i} y_{ij} = \frac{N}{n} \sum^{n} M_i \overline{y}_i \end{aligned} \tag{15.13}$$

An unbiased estimator of the variance of y_0^* is, from equation (14.2),

$$s_{y_0^*}^2 = \sum^{n} (y_i^* - y_0^*)^2 / n(n-1) \tag{15.14}$$

(b) *Estimating upwards from the second-stage units.* If y_{ij} is the value of the study variable in the jth sample ssu of the ith sample fsu, the average value of the study variable in an srs of ssu's is obtained on summing up these y_{ij} values for the m_i sample ssu's and dividing by m_i, namely

$$\overline{y}_i = \frac{1}{m_i} \sum_{j=1}^{m_i} y_{ij} \tag{15.11}$$

To obtain the unbiased estimator of the total Y_i in the ith sample fsu in an srs, we multiply this average by the actual total number of ssu's, M_i, to give

$$y_{i0}^* = M_i \overline{y}_i \tag{15.10}$$

An unbiased estimator of the universe total Y obtained from the ith sample fsu is given in an srs on multiplication of y_{i0}^* by N, the total number of fsu's, namely

$$y_i^* = N y_{i0}^* \tag{15.12}$$

and the combined unbiased estimator of Y from all the n sample villages is the simple arithmetic mean

$$y_0^* = \frac{1}{n} \sum_{i=1}^{n} y_i^* = \frac{N}{n} \sum_{i=1}^{n} y_{i0}^*$$

$$= \frac{N}{n} \sum_{i=1}^{n} \frac{M_i}{m_i} \sum_{j=1}^{m_i} y_{ij} = \frac{N}{n} \sum^{n} M_i \bar{y}_i \tag{15.13}$$

an unbiased variance estimator of which is given by equation (15.14).

(c) *From the general estimating equations for a multi-stage design.* We use the results of section 14.4(5). Noting that in simple random sampling, the (initial) probability of selection of an fsu is $\pi_i = 1/N$, and that of an ssu is $\pi_{ij} = 1/M_i$ (see table 15.1), we find from equation (14.9),

$$\begin{array}{ll} \text{For the first-stage units:} & f_1 \equiv n\pi_i = n/N \\ \text{For the second-stage units:} & f_2 \equiv m_i \pi_{ij} = m_i/M_i \end{array} \tag{15.15}$$

These are shown in the last column of Table 15.1.

An unbiased estimator of the universe total Y is, from equation (14.10),

$$y = \sum_{\text{sample}} \frac{1}{f_1 \cdot f_2} y_{ij} = \sum_{i,j} w_{ij} y_{ij} \tag{15.16}$$

where y_{ij} is the value of the study variable in the jth sample ssu in the ith sample fsu, and

$$w_{ij} = \frac{1}{f_1 \cdot f_2} = \frac{N}{n} \frac{M_i}{m_i} \tag{15.17}$$

Putting this value of w_{ij} in equation (15.16), we get

$$y = \frac{N}{n} \sum_{i=1}^{n} \frac{M_i}{m_i} y_{ij} \tag{15.18}$$

which is the same as the estimating equation (15.13).

Also, from equation (14.13), the unbiased estimator of Y from the ith sample fsu is

$$y_i^* = n \sum_{i} w_{ij} y_{ij} = \frac{nNM_i}{nm_i} \sum_{j=1}^{m_i} y_{ij} = \frac{NM_i}{m_i} \sum_{j=1}^{m_i} y_{ij} \tag{15.19}$$

which is the same as the estimating equation (15.12).

An unbiased estimator of the variance of y (or y_0^*) is given by equation (14.15).

Notes

1. If all the first-stage units are known to have the same total number of ssu's, i.e. if $M_i = M_0$ say, then the combined unbiased estimator y_0^* in equation (15.13) becomes

$$y_0^* = \frac{NM_0}{n}\sum_{i=1}^{n}\frac{1}{m_i}\sum_{j=1}^{m_i} y_{ij} = \frac{NM_0}{n}\sum^{n}\overline{y}_i \tag{15.20}$$

2. If, in addition, the sample number of second-stage units in each sample fsu is fixed, $m_i = m_0$ say, then the design becomes self-weighting and

$$y_0^* = \frac{NM_0}{nm_0}\sum_{i=j}^{n}\sum_{j=1}^{m_i} y_{ij} \tag{15.21}$$

3. In some textbooks, the number of ssu's M_i is referred to as the "size" of the ith fsu. We shall, however, use the term "size" to mean the value of the ancillary variable of pps selection (Chapters 5, 16, and 21); a special case of the size is M_i (sections 16.3.2 and 21.3.2).

15.2.4 Sampling variance of the estimator y_0^*

The sampling variance of y_0^* in srswr at both stages is

$$\sigma_{y_0^*}^2 = \frac{N^2\sigma_b^2}{n} + \frac{N}{n}\sum^{N}\frac{M_i^2\sigma_{wi}^2}{m_i} \tag{15.22}$$

where

$$\sigma_b^2 = \sum^{N}(Y_i - \overline{Y})^2/N \tag{15.23}$$

is the variance between the fsu's, and

$$\sigma_{wi}^2 = \sum^{M_i}(Y_{ij} - \overline{Y}_i)^2/M_i \tag{15.24}$$

is the variance between the ssu's within the ith fsu. $\sigma_{y_0^*}^2$ is estimated unbiasedly by $s_{y_0^*}^2$, defined by equation (15.14).

Notes

1. If sampling is simple random without replacement at both stages, y_0^*, defined in equation (15.13), still remains the (combined) unbiased estimator of Y. The sampling variance and its unbiased estimator are given in section 17.2.4.

2. If sampling is simple random with replacement at the first stage and simple random without replacement at the second stage, the estimator y_0^*, defined in equation (15.13), remains unbiased for Y. Its sampling variance is given in section 17.2.4, note 4, and the unbiased variance estimator is $s_{y_0^*}^2$, defined in equation (15.14).

15.2.5 Estimation of the ratio of the totals of two study variables

For another study variable, the unbiased estimator x_0^* of the universe total X can be obtained in the same manner as in section 15.2.3. Thus, the unbiased estimator of X, obtained from the ith sample fsu is, as in equation (15.12),

$$x_i^* = N\frac{M_i}{m_i}\sum_{j=1}^{m_i} x_{ij} \tag{15.25}$$

where x_{ij} is the value of the other study variable in the jth sample ssu in the ith sample fsu; and the combined unbiased estimator of X from all the n sample fsu's is, as in equation (15.13),

$$x_0^* = \frac{1}{n}\sum^{n} x_i^* \tag{15.26}$$

with an unbiased variance estimator $s_{x_0^*}^2$ defined as for $s_{y_0^*}^2$ in equation (15.14).

A consistent but generally biased estimator of the ratio of two universe totals $R = Y/X$ is the ratio of the sample estimators y_0^* and x_0^* (from equation (14.6)),

$$r = y_0^*/x_0^* \tag{15.27}$$

and a variance estimator of r is, from equation (14.7),

$$s_r^2 = (s_{y_0^*}^2 + r^2 s_{x_0^*}^2 - 2r s_{y_0^* x_0^*})/x_0^{*2} \tag{15.28}$$

where

$$s_{y_0^* x_0^*} = \sum^{n}(y_i^* - y_0^*)(x_i^* - x_0^*)/n(n-1) \tag{15.29}$$

is the unbiased estimator of the covariance of y_0^* and x_0^* (from equation (14.5)).

Notes

1. If $M_i = M_0$, then from equations of the type (15.20), the estimator of the universe ratio $R = Y/X$ becomes

$$r = \frac{y_0^*}{x_0^*} = \frac{\left(\sum_{i=1}^{n}\sum_{j=1}^{m_i} y_{ij}/m_i\right)}{\left(\sum_{i=1}^{n}\sum_{j=1}^{m_i} x_{ij}/m_i\right)} = \frac{\sum_{i=1}^{n} \overline{y}_i}{\sum_{j=1}^{n} \overline{x}_i} \tag{15.30}$$

2. If in addition, the same number of sample units m_0 is selected in the sample fsu's, then the design becomes self-weighting and

$$r = y_0^*/x_0^* = \sum_{i=1}^{n}\sum_{j=1}^{m_0} y_{ij} \Bigg/ \sum_{i=1}^{n}\sum_{j=1}^{m_0} x_{ij} \tag{15.31}$$

i.e. the ratio of the sample totals.

15.2.6 Estimation of the means of a study variable

(a) *Estimation of the mean per first-stage unit.* In two-stage srs, the universe mean per first-stage unit $\overline{Y}$ (defined by equation (15.4)) is estimated by

$$\overline{y} = \frac{y_0^*}{N} = \frac{\sum^n M_i \overline{y}_i}{n} = \frac{\sum^n y_{i0}^*}{n} \tag{15.32}$$

i.e. the average of the unbiased estimators of the fsu totals Y_i.

Note the similarity of the structures of the universe value and its estimator in srs.

An unbiased estimator of the variance of $\overline{y}$ is

$$s_{y_0^*}^2/N^2 \tag{15.33}$$

where $s_{y_0^*}^2$ (defined by equation (15.14)) is the unbiased estimator of the variance of y_0^*.

(b) *Estimation of the mean per second-stage unit.* Two situations may arise: (i) the total number of ssu's M_i is known for all the N fsu's, the total being $\sum^N M_i$; (ii) the total number of ssu's is known only for the n sample fsu's, after these n first-stage units are enumerated for the list of ssu's.

(i) *Unbiased estimator.* When the total number of ssu's in the universe is known, namely, $\sum^N M_i$, then the unbiased estimator of the universe mean per second-stage unit $\overline{\overline{Y}}$ (defined by equation (15.5)) is obtained on dividing the unbiased estimator y_0^* by $\sum^N M_i$, namely

$$\overline{\overline{y}} = y_0^* \Big/ \sum^N M_i \tag{15.34}$$

and the unbiased estimator of the variance of this sample estimator $\overline{\overline{y}}$ is obtained on dividing $s^2_{y_0^*}$ (as given by equation (15.14)) by $\left(\sum^N M_i\right)^2$, i.e.

$$s^2_{\overline{\overline{y}}} = s^2_{y_0^*} \Big/ \left(\sum^N M_i\right)^2 \tag{15.35}$$

Notes

1. If all the fsu's are known to have the same number of ssu's, i.e. if $M_i = M_0$, then $\overline{\overline{y}}$ becomes (from equations (15.20) and (15.34)),

$$\overline{\overline{y}} = y_0^*/NM_0 = \sum_{i=1}^{n}\sum_{j=1}^{m_i} y_{ij}/nm_i = \sum^n \overline{y}_i/n \tag{15.36}$$

   i.e. the simple mean of means (*see* (iii) of this section).

2. If, in addition, a fixed number m_0 of ssu's is selected in the sample ssu's then from equations (15.21) and (15.34) the design becomes self-weighting and $\overline{\overline{y}}$ becomes

$$\overline{\overline{y}} = y_0^*/NM_0 = \sum_{i=1}^{n}\sum_{j=1}^{m_0} y_{ij}/nm_0 \tag{15.37}$$

3. In crop surveys, if the design is two-stage srswr with villages as first-stage units, and fields as the second-stage units, and the total geographical area is A, and y_{ij} denotes the area under a particular crop in the ijth sample field, then the proportion of the area under the crop in the universe is estimated unbiasedly by

$$y_0^*/A \tag{15.38}$$

   and its sampling variance by

$$s^2_{y_0^*}/A^2 \tag{15.39}$$

   Similarly, if x_{ij} denotes the yield of a crop in the ijth sample field, then the average yield per field is estimated unbiasedly by an equation of the type (15.34) and its variance by an equation of the type (15.35). Similar considerations apply when the total number of fields in the universe $\left(\sum^N M_i\right)$ is known.

(ii) *Ratio estimator.* When the total number of ssu's is known only for the n sample fsu's, the unbiased estimator of universe number of ssu's $\sum^N M_i$ is obtained from the sample. In two-stage srs, the average number of ssu's in the sample fsu's,

$$\sum^n M_i/n \tag{15.40}$$

is multiplied by the total number N of fsu's to provide the unbiased estimator, thus,

$$m_0^* = N\sum^n M_i/n \tag{15.41}$$

This can also be derived from the estimating equations of section 15.2.3 by putting $y_{ij} = 1$ for the selected sample units. Thus putting $y_{ij} = 1$ in equation (15.12), we get as the unbiased estimator of the total number of ssu's, obtained from the ith sample fsu,

$$m_i^* = NM_i m_i/m_i = NM_i \tag{15.42}$$

and the combined unbiased estimator (from equation (15.13))

$$m_0^* = \frac{1}{n}\sum^n m_i^* = \frac{N}{n}\sum^n M_i \tag{15.43}$$

An unbiased estimator of the variance of m_0^* is (from equation (15.14))

$$\begin{aligned} s^{2*}_{m_0} &= \sum^n (m_i^* - m_0^*)^2/n(n-1) \\ &= \frac{N^2\sum^n [M_i - (\sum^n M_i/n)]^2}{n(n-1)} \end{aligned} \tag{15.44}$$

A consistent but generally biased estimator of the universe mean, $\overline{\overline{Y}}$, is from equation (15.27), the ratio

$$r = y_0^*/m_0^* = \sum^n M_i\bar{y}_i \Big/ \sum^n M_i \tag{15.45}$$

A variance estimator of r is (from equation (15.28))

$$s_r^2 = (s^2_{y_0^*} + r^2 s^2_{m_0^*} - 2rs_{y_0^* m_0^*})/m_0^{*2} \tag{15.46}$$

where

$$s_{y_0^* m_0^*} = \sum^n (y_i^* - y_0^*)(m_i^* - m_0^*)/n(n-1) \tag{15.47}$$

is an unbiased estimator of the covariance of y_0^* and m_0^*.

Note: In crop surveys, even when the total geographical area or the total number of farms in the universe is known, in estimating the proportion of area under a particular crop, or the average yield per field, estimating equations of the ratio type, namely (15.27) and (15.45) respectively, may be used (see part (iv) of this section).

(iii) *Unweighted mean of means.* In estimating means, a third estimator – the simple (unweighted) mean of means – may be used. Denoting by $\overline{y}_i$ the mean $\sum_{j=1}^{m_i} y_{ij}/m_i$, this third estimator is

$$\sum^{n} \overline{y}_i/n = \sum_{i=1}^{n}\sum_{j=1}^{m_i} y_{ij}/nm_i \tag{15.48}$$

This is a biased estimator, unless all the M_is are the same (see note 1 to (i) of this section) or the design is self-weighting.

(iv) *Comparison of the estimators.* Of the three estimators, only the estimator given by equation (15.34) is unbiased. However, this estimator can be used only if the total number of ssu's in the universe is known, and even then its sampling variance is likely to be high. On the other hand, the ratio-type estimator, given by equation (15.45), although biased, is consistent. The third estimator, the simple mean of means, is neither unbiased nor consistent; however, if it is known that the number of ssu's (M_i) in the fsu's does not vary considerably and there is no appreciable correlation between M_i and the study variable, this estimator would be the most efficient. Crop surveys in India have shown the absence of any such correlation, but the specific situations should be considered for other studies.

15.3 Two-stage srs: Estimation for sub-universes

As for single-stage srs (section 2.12), the unbiased estimator of the total of the values of the study variable in the sub-universe Y' is obtained from the estimating equation (15.13) by defining

$$y'_{ij} = y_{ij} \text{ if the sample unit belongs to the sub-universe;}$$
$$y'_{ij} = 0 \text{ otherwise.}$$

An unbiased estimator of the variance of the estimator of the total for the sub-universe is similarly given by equation (15.14).

15.4 Two-stage srs: Estimation of proportion of units

Let M_i' be the total number of ssu's possessing a certain attribute in the ith fsu ($i = 1, 2, \ldots, N$); the total number of such ssu's in all the N fsu's is

$$\sum^{N} M_i' \tag{15.49}$$

The proportion of such ssu's with the attribute in the ith fsu in the universe is

$$P_i = M_i'/M_i \tag{15.50}$$

and the universe proportion is

$$P = \sum^{N} M_i' \Big/ \sum^{N} M_i = \sum^{N} M_i P_i \Big/ \sum^{N} M_i \tag{15.51}$$

In a two-stage srs (as for single-stage srs, see section 2.13), the estimated total number of ssu's in the universe with the attribute and its estimated variance could be obtained respectively from estimating equations (15.13) and (15.14) by putting $y_{ij} = 1$ if the sample unit has the attribute, and 0 otherwise.

Thus in an srs if m_i' of the m_i sample ssu's in the selected ith fsu has the attribute, then the sample proportion

$$p_i = m_i'/m_i \tag{15.52}$$

is an unbiased estimator of P_i.

The unbiased estimator of M_i' is $M_i p_i$; and the unbiased estimator of $\sum^N M_i'$, the universe number of ssu's with the attribute, from the ith sample fsu, is (from equation (15.12))

$$m_i'^{*} = N M_i p_i \tag{15.53}$$

The combined unbiased estimator of $\sum^N M_i'$ from all the n sample fsu's is (from equation (15.13)) the arithmetic mean

$$m_0'^{*} = \sum^{n} m_i'^{*}/n = N \sum^{n} M_i p_i/n \tag{15.54}$$

An unbiased estimator of the variance of $m_0'^{*}$ is (from equation (15.14))

$$s^2_{m_0'^{*}} = \sum^{n} (m_i'^{*} - m_0'^{*})^2/n(n-1) \tag{15.55}$$

If the M_i values are known for all the N fsu's, then the unbiased estimator of the universe proportion P is

$$m_0'^* \Big/ \left(\sum^N M_i\right) \tag{15.56}$$

and an unbiased estimator of the variance of this estimator is

$$s^2_{m_0'^*} \Big/ \left(\sum^N M_i\right)^2 \tag{15.57}$$

If $\sum^N M_i$ is not known, it is estimated by

$$m_0^* = N \sum^n M_i/n \tag{15.43}$$

an unbiased estimator of whose variance is given by equation (15.44).

A consistent but generally biased estimator of the universe proportion P is then

$$m_0'^*/m_0^* = \sum^n M_i p_i \Big/ \sum^n M_i \tag{15.58}$$

The estimated variance of this estimator is given by an equation of the type (15.46).

Notes

1. Even if $\sum^N M_i$ is known, the use of the estimator given by equation (15.58) may be preferred to that given in equation (15.56) as the former is likely to more efficient.
2. If all the M_i values are known for the universe and are the same, M_0, then $\sum^N M_i = NM_0$, and the estimator in equation (15.56) becomes

$$m_0'^*/NM_0 = \sum^n p_i/n = \bar{p} \tag{15.59}$$

i.e. the simple (unweighted) arithmetic mean of the p_i sample values.
3. If, in addition, a fixed number m_0 of ssu's is selected in the sample fsu's, then the design is self-weighting and the estimator (15.59) becomes

$$\bar{p} = \sum^n p_i/n = \sum^n m_i'/nm_0 \tag{15.60}$$

i.e. the sample number of units possessing the attribute divided by the total number of sample units.

An unbiased estimator of the variance of $\overline{p}$ (equation (15.59) or (15.60)) is

$$s_{\overline{p}}^2 = \sum^n (p_i - \overline{p})^2/n(n-1) \qquad (15.61)$$

In particular, the unbiased estimator of the variance of $\overline{p}$ in equation (15.60) is

$$s_{\overline{p}}^2 = \frac{SSm_i'/n(n-1)}{m_0^2} \qquad (15.62)$$

For an example, see exercise 1(b) of this chapter.

15.5 Two-stage srs: Ratio method of estimation

If information on an ancillary variable is available for all the second-stage units in the universe, then the universe total

$$Z = \sum_{i=1}^{N} \sum_{j=1}^{M_i} z_{ij} \qquad (15.63)$$

is also known, where z_{ij} is the value of the ancillary variable of the ijth universe unit. The unbiased estimator of Z in a two-stage srs is (from equation (15.13))

$$z_0^* = \frac{N}{n} \sum_{i=1}^{n} \frac{M_i}{m_i} \sum_{j=1}^{m_i} z_{ij} = \frac{N}{n} \sum^{n} z_i^* \qquad (15.64)$$

and its unbiased variance estimator is (from equation (15.14))

$$s_{z_0^*}^2 = \sum_{i=1}^{n} (z_i^* - z_0^*)^2/n(n-1) \qquad (15.65)$$

where z_{ij} is the value of the ancillary variable in the ijth sample unit and

$$z_i^* = NM_i \sum^{m_i} z_{ij}/m_i$$

The ratio estimator of the universe total Y, using the ancillary information, is (from equation (3.6))

$$y_R^* = Zy_0^*/z_0^* = Zr_1 \qquad (15.66)$$

where

$$r_1 = y_0^*/z_0^* \qquad (15.67)$$

A consistent but generally biased estimator of the variance of y_R^* is (from equations (3.10) and (3.11))

$$s_{y_R^*}^2 = (s_{y_0^*}^2 + r_1^2 s_{z_0^*}^2 - 2r_1 s_{y_0^* z_0^*})$$
$$= (SSy_i^* + r_1^2 SSz_i^* - 2r_1 SPy_i^* z_i^*)/n(n-1) \qquad (15.68)$$

Notes

1. If x is another study variable, a ratio estimator x_R^* for the universe total X can be similarly defined as y_R^*. However, in estimating the ratio $R = Y/X$, the use of ratio estimators of the totals Y and X leads to the same result as those of the unbiased estimators, for $y_R^*/x_R^* = y_0^*/x_0^*$ (see note 5 to section 3.2).
2. The ratio estimator is likely to be very efficient when (i) the M_i values vary considerably, (ii) the ancillary variable is highly correlated with the study variable, and (iii) n is large. The ratio estimator of the universe mean $\overline{Y}$, namely $\overline{Z}(y_0^*/z_0^*) = \overline{Z}r_1$ (where $\overline{Z}$, the universe mean of the ancillary variable is known) will then be more efficient than the other three estimators of the universe mean, considered in section 15.2.6.

Example 15.1

For the Rural Household Budget Survey in the Shoa province of Ethiopia, 1966-7, ten geographical strata were formed comprising a number of sub-divisions; in each stratum, three sub-divisions were selected with equal probability out of the total number of sub-divisions; in each selected sub-division, of the total number of households listed by the enumerators, twelve were selected with equal probability for the inquiry. Although the sample was not specifically designed to provide estimates of demographic parameters, the example shows the method of computation for estimating the average age of the household heads in one particular stratum and its standard error. Table 15.2 gives the required data for the stratum, which had a total of eighty sub-divisions.

The required computations are shown in Table 15.3, denoting by y_{ij} the age of the head of the ijth sample household ($i = 1, 2, 3$ for the sample sub-division; $j = 1, 2, \ldots, 12$ for the sample households); N denotes the total number of sub-divisions in the stratum $= 80$, and M_i and m_i $(= 12)$ respectively the total and sample number of households in the ith sample sub-division.

From equation (15.13), the combined unbiased estimate of the total of the ages of the households heads in the stratum is

$$y_0^* = \frac{N}{n}\sum_{i=1}^{n}\frac{M_i}{m_i}\sum_{j=1}^{m_i} y_{ij} = 80 \times 18,472.89 \text{ years}$$

(shown at the bottom of column 5), and from equation (15.43), the combined unbiased estimate of the total number of households in the stratum is

$$m_0^* = N\sum^{n} M_i/n = 80 \times 392.33 = 31,387 \text{ households}$$

Table 15.2: Total and sample number of households and total ages of the sample household heads in one stratum: Rural Household Budget Survey, Shoa Province, Ethiopia, 1966-7

Sample sub-division (i)	Total no. of households (M_i)	No. of households in sample (m_i)	Total ages (in years) of the sample household heads $\left(\sum_{j=1}^{m_i} y_{ij}\right)$
1	222	12	474
2	42	12	503
3	913	12	590

(shown at the bottom of column 7).

The estimated average age of household heads is (from equation (15.45))

$$r = y_0^*/m_0^* = 18,472.89/392.33 = 47.08 \text{ years}$$

An estimate of the variance of r is given by equation (15.46), namely,

$$s_r^2 = (s_{y_0^*}^2 + r^2 s_{m_0^*}^2 - 2r s_{y_0^* m_0^*})/m_0^{*2}$$

Since

$$\begin{aligned} SSy_i^* &= \sum^n y_i^{*2} - \left(\sum^n y_i^*\right)^2 \Big/ n \\ &= 80^2 \times [2,095,034,998 - 1,023,744,103] \\ &= 80^2 \times 1,071,290,895 \end{aligned}$$

so, from equation (15.14),

$$s_{y_0^*}^2 = SSy_i^*/n(n-1) = 80^2 \times 1,071,290,895/6$$

Similarly from equation (15.44)

$$\begin{aligned} s_{m_0^*}^2 &= SSy_i^*/n(n-1) \\ &= 80^2 \times [884,617 - 461,776]/6 \\ &= 80^2 \times 422,841/6, \end{aligned}$$

Table 15.3: Computation of the average age of household heads and its sampling variance: data of Table 15.2

Sub-division i (1)	Total no. of households M_i (2)	No. of households in sample m_i (3)	Total of ages of household heads $\sum_{j=1}^{m_i} y_{ij}$ (4)	$y_i^* = NM_i \sum^{m_i} y_{ij}/m_i$ (5)	y_i^{*2} (6)
					$80^2 \times$
1	222	12	474	80×8769.00	76 895.361
2	42	12	503	1760.50	3099.360
3	913	12	590	44 889.17	2 015 040.277
					$80^2 \times$
			Total	$80 \times 55,418.67$	2 095 035.998
			Mean	$80 \times 18\,472.89$ (y_0^*)	

Sub-division i (1)		For the number of households $m_i^* = NM_i$ (7)	m_i^{*2} (8)	For the sum of products $y_i^* m_i^*$ (9)
				$80^2 \times$
1		80×222	$80^2 \times 49\,284$	1 946 718
2		42	1764	73 941
3		913	833 569	40 983 809
				$80^2 \times$
	Total	80×1177	$80^2 \times 884\,617$	43 004 468
	Mean	80×392.33 (m_0^*)		

and from equation (15.47)

$$s_{y_0^* m_0^*} = \frac{\left[\sum^n y_i^* m_i^* - \left(\sum^n y_i^*\right)\left(\sum^n m_i^*\right) \Big/ n\right]}{n(n-1)}$$
$$= 80^2[43,004,468 - 21,742,590]/6$$
$$= 80^2 \times 21,262,878/6.$$

Noting that $r = 47.08$, $2r = 94.16$, and $r^2 = 2216.5264$, we get for s_r^2 (from equation (15.46))

$$\frac{80^2[(1,071,290,895 + 937,238,070 - 2,002,018,432)/6]}{80^2 \times (392.33)^2}$$
$$= (6,510,533/6)/153,925.44 = 7.0494 \text{ years}^2$$

so that the standard error of estimate of r is $s_r = 2.66$ years, and the estimated CV of r is $2.66/47.08 = 0.0565$ or 5.65 per cent.

Example 15.2

In each of the simple random sample of 4 villages selected in Example 2.3 from the list of 30 villages (Appendix IV), select a simple random sample (with replacement) of 4 households from the total number of households, and on the basis of collected data on the households, the total daily income, and the number of adults (over 18 years of age) of the sample households, estimate for the 30 villages the total number of persons, the total daily income, the per capita daily income, and the total number and proportion of adults, along with their standard errors.

The sample data are shown in Table 15.4 and the required computations in Table 15.5, denoting by y_{ij} the household size, by x_{ij} the daily income, and by y'_{ij} the number of adults, in the ijth sample household (i = 1, 2, 3, 4 for sample villages numbered 15, 18, 19, and 24; j = 1, 2, 3, 4 for the sample households in the selected villages); the respective means in the ith sample village are denoted by $\overline{y}_i$, $\overline{x}_i$, and $\overline{y}'_i$ (defined by an equation of the type 15.11), $N = 30$ is the total number of villages, and M_i and m_i (= 4) denote respectively the total and the sample numbers of households in the ith sample village.

From equation (15.13), the combined unbiased estimate of the total number of persons in the 30 villages is

$$y_0^* = \frac{N}{n}\sum_{i=1}^{n}\frac{M_i}{m_i}\sum_{j=1}^{m_i} y_{ij} = 30 \times 108.5 = 3255 \text{ persons;}$$

Table 15.4: Size, total daily income, number of adults (18 years or over) and the agreement by the selected adult to an increase in state taxes for education in sample households in each of the four sample villages of Example 2.3: two-stage simple random sample

Household serial no.	Household size (y_{ij})	Daily income (x_{ij})	Number of adults (y'_{ij})	Agreement by the selected adult to an increase of state taxes for education ($q'_{ij} = 1$ for Yes; $= 0$ for No)
Village serial no. 15; total number of households = 20				
1	8	\$92	4	0
2	4	41	2	1
5	6	55	3	0
14	3	36	2	1
Village serial no. 18; total number of households = 23				
13	4	39	2	0
15	5	58	3	1
16	6	61	3	0
21	7	63	4	0
Village serial no. 19; total number of households = 25				
5	5	58	2	0
11	5	50	3	0
23	3	33	2	0
24	5	47	2	1
Village serial no. 24; total number of households = 18				
3	4	48	2	0
7	6	70	4	0
14	6	49	2	1
15	4	45	2	0

Table 15.5: Computation of the estimated total number of persons, total and *per capita* daily income, number and proportion of adults for a two-stage srs of villages and households: data of Table 15.4

Sample village i	Number of households total M_i	sample m_i	$\sum_{j=1}^{m_i} y_{ij}$	For the number of persons $\bar{y}_i = \sum_{j=1}^{m_i} y_{ij}/m_i$	$y_i^* = NM_i\bar{y}_i$	y_i^{*2}
(1)	(2)	(3)	(4)	(5)	(6)	(7)
1	20	4	21	5.25	30 × 105.0	30^2 × 11 025.00
2	23	4	22	5.50	126.5	16 002.25
3	25	4	18	4.50	112.5	12 656.25
4	18	4	20	5.00	90.0	8 100.00
				Total	30 × 434.0	30^2 × 47 783.50
				Mean	30 × 108.5 (y_0^*)	

Sample village	$\sum_{j=1}^{m_i} x_{ij}$	$\bar{x}_i = \sum_{j=1}^{m_i} x_{ij}/m_i$	For the total daily income $x_i^* = NM_i\bar{x}_i$	x_i^{*2}
(1)	(8)	(9)	(10)	(11)
1	\$224	\$56.00	\$30 × 1120.00	30^2 × 1 254 400.0000
2	221	55.25	1270.75	1 614 805.5625
3	188	47.00	1175.00	1 380 625.0000
4	212	53.00	954.00	910 116.0000
		Total	\$30 × 4519.75	30^2 × 5 159 946.5625
		Mean	\$30 × 1129.9375 (x_0^*)	

Sample village	$\sum_{j=1}^{m_i} y'_{ij}$	$\bar{y}'_i = \sum_{j=1}^{m_i} y'_{ij}/m_i$	For the number of adults $y_i'^* = NM_i\bar{y}'_i$	$y_i'^{*2}$
(1)	(12)	(13)	(14)	(15)
1	11	2.75	30 × 55.00	30^2 × 3025.0000
2	12	3.00	69.00	8728.5000
3	9	2.25	56.25	3164.0625
4	10	2.50	45.00	2025.0000
		Total	30 × 225.25	30^2 × 12 975.0625
		Mean	30 × 56.3125 ($y_0'^*$)	

Table 15.5 (continued)

Sample village i	For the sum of products $y_i^* x_i^*$	$y_i^* y_i'^*$
(1)	(16)	(17)
1	$30^2 \times$ 117 600.000	$30^2 \times$ 5775.000
2	160 749.875	8728.500
3	132 187.500	6328.500
4	85 860.000	4050.000
Total	$30^2 \times$ 496 397.375	$30^2 \times$ 24 881.625

and from equation (15.14), an unbiased estimate of the variance of y_0^* is

$$\begin{aligned} s^2_{y_0^*} &= SSy_i^*/n(n-1) \\ &= 30^2 \times [47,783.5 - 47,089]/12 \\ &= 30^2 \times 694.5/12 \end{aligned}$$

so that the estimated standard error of y_0^* is

$$s_{y_0^*} = 30 \times 7.608 = 228 \text{ persons}$$

Similarly, the combined unbiased estimate of the total daily income is

$$x_0^* = \sum^n x_i^*/n = 30 \times \$1129.9375 = \$33,938$$

with estimated variance

$$\begin{aligned} s^2_{x_0^*} &= 30^2[5,159,946.5625 - 5,107,035.0156]/12 \\ &= 30^2 \times 52,911.5469/12 \end{aligned}$$

so that the estimated standard error of x_0^* is

$$s_{x_0^*} = 30 \times \$66.4025 = \$1992$$

The combined unbiased estimate of the total number of adults is

$$y_0'^* = \sum^n y_0'^*/n = 30 \times 56.3125 = 1689$$

with estimated variance of

$$\begin{aligned} s^2_{y_0'^*} &= 30^2 \times [12,975.0625 - 12,684.3906]/12 \\ &= 30^2 \times 290.6719/12 \end{aligned}$$

so that the estimated standard error of $y_0'^*$ is $s_{y_0'^*} = 148$.

The estimated daily income per capita is equal to the estimated total daily income divided by the estimated total number of persons or

$$r = \frac{x_0^*}{y_0^*} = \frac{(30 \times \$1129.9375)}{(30 \times 108.50)} = \$10.414$$

As

$$\begin{aligned} s_{y_0^* x_0^*} &= SPy_0^* x_0^*/n(n-1) \\ &= 30^2 \times [496,397.3750 - 490,392,8750]/12 \\ &= 30^2 \times 6004.5/12 \end{aligned}$$

the estimated variance of r is (from equation (15.28))

$$\begin{aligned} s_r^2 &= (s_{x_0^*}^2 + r^2 s_{y_0^*}^2 - 2r s_{y_0^* x_0^*})/y_0^{*2} \\ &= \frac{30^2}{12}\left[\frac{52,911.5469 + (10.414)^2 \times 694.5 - 2 \times 10.414 \times 6004.5}{(30 \times 108.50)^2}\right] \\ &= 3169.3182/(12 \times 11,772.25) = 0.02243495, \end{aligned}$$

so that the estimated standard error of r is $s_r = 0.1498$.

The estimated proportion of adults is equal to the estimated total number of adults divided by the estimated total number of persons, or

$$p = \frac{y_0'^*}{y_0^*} = \frac{(30 \times 56.3125)}{(30 \times 108.50)} = 0.5190$$

As

$$\begin{aligned} s_{y_0^* y_0'^*} &= 30^2 \times [24,881.625 - 24,439.625]/12 \\ &= 30^2 \times 442/12 \end{aligned}$$

the estimated variance of p is, from equation (15.28),

$$\begin{aligned} s_p^2 &= (s_{y_0'^*}^2 + p^2 s_{y_0^*}^2 - 2p s_{y_0^* y_0'^*})/y_0^{*2} \\ &= \frac{(30^2/12)[290.6719 + (0.519)^2 \times 694.5 - 2 \times 0.519 \times 442]}{(30 \times 108.50)^2} \\ &= 0.0013411 \end{aligned}$$

so that the estimated standard error of p is $s_p = 0.0116$.

15.6 Three-stage srs: Estimation of totals, means, ratios and their variances

15.6.1 Universe totals and means

Extending the notations of section 15.2.1, let the number of third-stage units (tsu's) in the jth ssu ($j = 1, 2, \ldots, M_i$) of the ith fsu ($i = 1, 2, \ldots, N$)

be Q_{ij}; the total number of tsu's in the universe is $\sum_{i=1}^{N}\sum_{j=1}^{M} Q_{ij}$. Denoting by Y_{ijk} the value of the study variable of the kth tsu ($k = 1, 2, \ldots, Q_{ij}$) in the jth ssu in the ith fsu, we define the following universe parameters:

Total of the values of the study variable in the jth ssu in the ith fsu:

$$Y_{ij} = \sum_{k=1}^{Q_{ij}} Y_{ijk} \tag{15.69}$$

Mean for the jth ssu in the ith fsu:

$$\overline{Y}_{ij} = Y_{ij}/Q_{ij} \tag{15.70}$$

Total for the ith fsu:

$$Y_i = \sum_{j=1}^{M_i} Y_{ij} \tag{15.71}$$

Mean for the ith fsu per ssu:

$$\overline{Y}_i = Y_i/M_i \tag{15.72}$$

Grand total:

$$\begin{aligned} Y &= \sum_{i=1}^{N}\sum_{j=1}^{M_i}\sum_{k=1}^{Q_{ij}} Y_{ijk} \\ &= \sum_{i=1}^{N}\sum_{j=1}^{M_i} Y_{ij} = \sum_{i=1}^{N} Y_i \\ &= \sum_{i=1}^{N}\sum_{j=1}^{M_i} Q_{ij}\overline{Y}_{ij} = \sum_{i=1}^{N} M_i\overline{Y}_i \end{aligned} \tag{15.73}$$

Mean per fsu:

$$\overline{Y} = Y/N = \sum^{N} M_i\overline{Y}_i/N \tag{15.74}$$

Mean per ssu:

$$\overline{\overline{Y}} = Y\bigg/\left(\sum^{N} M_i\right) = \sum^{N} M_i\overline{Y}_i\bigg/\left(\sum^{N} M_i\right) \tag{15.75}$$

Mean per tsu:

$$\overline{\overline{\overline{Y}}} = Y\bigg/\left(\sum^{N}\sum^{M_i} Q_{ij}\right) = \sum^{n}\sum^{M_i} Q_{ij}\overline{Y}_{ij}\bigg/\left(\sum^{n}\sum^{M_i} Q_{ij}\right) \tag{15.76}$$

Table 15.6: Sampling plan for a three-stage simple random sample with replacement

Stage (t)	Unit	No. in universe	No. in sample	Selection method	Selection probability	f_t
1	First-stage	N	n	srswr	Equal = $1/N$	n/N
2	Second-stage	M_i	m_i	srswr	Equal = $1/M_i$	m_i/M_i
3	Third-stage	Q_{ij}	q_{ij}	srswr	Equal = $1/Q_{ij}$	q_{ij}/Q_{ij}

15.6.2 Structure of a three-stage srs

Extending the structure of a two-stage srs given in section 15.2.2, the design will be three-stage srswr if in the ijth sample ssu ($i = 1, 2, \ldots, n$; $j = 1, 2, \ldots, m_i$), q_{ij} sample tsu's are selected as an srs (and with replacement) out of the total Q_{ij} tsu's, the total number of tsu's in the sample, i.e. the total sample size is $\sum_{i=1}^{n} \sum_{j=1}^{m_i} q_{ij}$.

An example of a three-stage design for a crop survey has been given in section 14.2. Extending the example of a two-stage design for a household survey given in the same section, the design will be three-stage srs if a sample of persons (tsu's) is selected in the sample households (ssu's) in the sample villages (fsu's), sampling at the three stages being simple random. Or, to take another example, in an urban household inquiry, where all the towns constitute the urban area, an srs of towns (fsu's) may be first selected, then a sample of blocks (ssu's) in the selected towns, and finally a sample of households (tsu's) in the selected town-blocks. The sampling plan of a three-stage srs is shown in summary form in Table 15.6.

Let y_{ijk} denote the value of the study variable in the kth selected tsu ($k = 1, 2, \ldots, q_{ij}$) in the ijth sample ssu.

15.6.3 Estimation of the universe total Y and the variance of the sample estimator

Extending the methods of section 15.2.3 for a two-stage srs, it can be seen that an unbiased estimator of the universe total Y, obtained form the ith sample fsu ($i = 1, 2, \ldots, n$) is

$$y_i^* = \frac{NM_i}{m_i} \sum_{j=1}^{m_i} \frac{Q_{ij}}{q_{ij}} \sum_{k=1}^{q_{ij}} y_{ijk} = \frac{NM_i}{m_i} \sum_{j=1}^{m_i} Q_{ij} \overline{y}_{ij} \qquad (15.77)$$

where

$$\overline{y}_{ij} = \frac{1}{q_{ij}} \sum_{k=1}^{q_{ij}} \overline{y}_{ijk} \tag{15.78}$$

is the simple arithmetic mean of the y_{ijk} values.

The combined unbiased estimator of Y from all the n fsu's is the simple arithmetic mean

$$y_0^* = \frac{1}{n} \sum^{n} y_i^* \tag{15.79}$$

an unbiased variance estimator of which is, from equation (14.15),

$$s_{y_0^*}^2 = \sum^{n} (y_i^* - y_0^*)^2 / n(n-1) \tag{15.80}$$

15.6.4 Estimation of the ratio of the totals of two study variables

The extension from the method in the two-stage srs is straightforward. The unbiased estimator of the universe total X of another study variable, obtained from the ith sample fsu, is (from equation (15.77))

$$x_i^* = \frac{NM_i}{m_i} \sum_{j1}^{m_i} \frac{Q_{ij}}{q_{ij}} \sum_{k=1}^{q_{ij}} x_{ijk} = \frac{NM_i}{m_i} \sum_{j=1}^{m_i} Q_{ij} \overline{x}_{ij} \tag{15.81}$$

where x_{ijk} is the value of the study variable in the ijkth sample third-stage unit, and

$$\overline{x}_{ij} = \sum_{k=1}^{q_{ij}} x_{ijk} / q_{ij} \tag{15.82}$$

The combined unbiased estimator of X, from all the n sample fsu's, is (from equation (15.79))

$$x_0^* = \frac{1}{n} \sum^{n} x_i^* \tag{15.83}$$

with an unbiased estimator $s_{x_0^*}^2$ defined as for $s_{y_0^*}^2$ in equation (15.80).

A consistent but generally biased estimator of the universe ratio $R = Y/X$ is the ratio of the sample estimators

$$r = y_0^* / x_0^* \tag{15.84}$$

with a variance estimator

$$s_r^2 = (s_{y_0^*}^2 + r^2 s_{x_0^*}^2 - 2r s_{y_0^* x_0^*})/x_0^{*2} \tag{15.85}$$

where

$$s_{y_0^* x_0^*} = \sum^n (y_i^* - y_0^*)(x_i^* - x_0^*)/n(n-1) \tag{15.86}$$

is the unbiased estimator of the covariance of y_0^* and x_0^*, from equation (14.5).

15.6.5 Estimation of the means of a study variable

For the estimation of the means of a study variable per fsu and ssu, see section 15.2.6(a) and (b): estimating equations of the same type, but relating to a three-stage srs, will apply. We consider here the estimating method for the mean per tsu $\overline{\overline{\overline{Y}}}$, as defined in equation (15.76).

The estimating procedure will depend on whether the total number of tsu's in the universe, namely $\sum_{i=1}^{N}\sum_{j=1}^{M_i} Q_{ij}$ is known or has to be estimated from the sample.

(a) *Unbiased estimator.* When the universe number of tsu's is known, then the universe mean per tsu $\overline{\overline{\overline{Y}}}$ is estimated unbiasedly by

$$\overline{\overline{\overline{y}}} = y_0^* \Bigg/ \left(\sum_{i=1}^{N}\sum_{j=1}^{M_i} Q_{ij}\right) \tag{15.87}$$

with an unbiased variance estimator

$$s_{y_0^*}^2 \Bigg/ \left(\sum_{i=1}^{N}\sum_{j=1}^{M_i} Q_{ij}\right)^2 \tag{15.88}$$

(b) *Ratio estimator.* When the universe number of third-stage units is not known, it is estimated from the sample by putting $y_{ijk} = 1$ in the estimating equations (15.77) and (15.79), thus from the ith sample fsu

$$q_i^* = \frac{NM_i}{m_i}\sum_{j=1}^{m_i} Q_{ij} \tag{15.89}$$

and from all the n sample fsu's

$$q_0^* = \sum^n q_i^*/n \tag{15.90}$$

An unbiased variance estimator of q_0^* is

$$s_{q_0^*}^2 = \sum^n (q_i^* - q_0^*)^2/n(n-1) \tag{15.91}$$

A consistent but generally biased estimator of the universe mean per third-stage unit is the ratio

$$r' = y_0^*/q_0^* \tag{15.92}$$

with a variance estimator

$$s_{r'}^2 = (s_{y_0^*}^2 + r'^2 s_{q_0^*}^2 - 2r's_{y_0^* q_0^*})/q_0^{*2} \tag{15.93}$$

where

$$s_{y_0^* q_0^*} = \sum^n (y_i^* - y_0^*)(q_i^* - q_0^*)/n(n-1) \tag{15.94}$$

is an unbiased estimator of the covariance of y_0^* and q_0^*.

(c) *Unweighted mean of means.* A third estimator of the universe mean per third-stage unit is the unweighted mean of means

$$\sum_{i=1}^n \sum_{j=1}^{m_i} \overline{y}_{ij} \Big/ \sum_{i=1}^n \sum_{j=1}^{m_i} q_{ij} \tag{15.95}$$

where

$$\overline{y}_{ij} = \sum_{k=1}^{q_{ij}} y_{ijk}/q_{ij} \tag{15.96}$$

This estimator is both biased and inconsistent, unless the design is self-weighting (see note 3).

Notes

1. *Comparison of the estimators.* Although biased, the ratio-type estimator is likely to have a smaller sampling variance than the unbiased estimator: the latter is, of course applicable only when the number of second- and third-stage units in the universe are known. The unweighted mean of means will be both biased and inconsistent, unless the design is self-weighting, but the observation made in section 15.2.6(b-iv) may also be noted.

2. In crop-surveys, when the total geographical area (A) or the total numbers of fields ($\sum^N M_i$) and plots ($\sum_{i=1}^N \sum_{j=1}^{M_i} Q_{ij}$) are known in the universe, then in a three-stage srs with villages as the first-stage, fields as the second-stage, and plots as the third-stage units, the proportion of area under a particular crop in the universe is estimated unbiasedly by

$$y_0^*/A \tag{15.97}$$

where y_0^* is the unbiased estimator of the area under the crop in the universe, given by equation (15.79), y_{ijk} denoting the area under the crop in the ijkth sample plot. An unbiased variance estimator of y_0^*/A is

$$s_{y_0^*}^2/A^2 \tag{15.98}$$

$s_{y_0^*}^2$, the unbiased variance estimator of y_0^*, being defined in equation (15.80). The average yield per unit area can be estimated similarly; and the average yield per plot is estimated unbiasedly by equation (15.87), and its unbiased variance estimator by equation (15.88), y_{ijk} denoting the yield of ijkth sample plot.

15.7 Three-stage srs: Estimation for sub-universes

As for two-stage srs, estimators for sub-universes are obtained from the general results for the universe by defining

$$\begin{aligned} y'_{ijk} &= y_{ijk} \text{ if the sample unit belongs to the sub-universe; and} \\ y'_{ijk} &= 0 \text{ otherwise.} \end{aligned}$$

15.8 Three-stage srs: Estimation of proportion of units

Let Q'_{ij} be the total number of tsu's in the ijth second-stage unit ($i = 1,2,\ldots,N$; $j = 1,2,\ldots,M_i$) possessing a certain attribute; the universe number of such tsu's is

$$\sum_{i=1}^{N}\sum_{j=1}^{M_i} Q'_{ij} \tag{15.99}$$

and the universe proportion of the tsu's possessing the attribute is

$$P = \sum_{i=1}^{N}\sum_{j=1}^{M_i} Q'_{ij} \Big/ \sum_{i=1}^{N}\sum_{j=1}^{M_i} Q_{ij} \tag{15.100}$$

As for two-stage srs, the estimated number of tsu's with the attribute could be obtained by putting $y_{ijk} = 1$, if the sample unit has the attribute, and 0 otherwise.

Thus if q'_{ij} of the q_{ij} sample third-stage units in the selected ijth ssu has the attribute, then the unbiased estimator of the total number of tsu's in the universe possessing the attribute $(\sum_{i=1}^{N}\sum_{j=1}^{M_i} Q'_{ij})$ from the ith sample fsu is

$$q_i'^* = \frac{NM_i}{m_i}\sum_{j=1}^{m_i}\frac{Q_{ij}}{q_{ij}}q'_{ij} = \frac{NM_i}{m_i}\sum_{j=1}^{m_i} Q_{ij}p_{ij} \tag{15.101}$$

and from all the n sample fsu's

$$q_0'^* = \sum^{n} q_i'^*/n \tag{15.102}$$

where

$$p_{ij} = q'_{ij}/q_{ij} \tag{15.103}$$

is an unbiased estimator of the universe proportion $P_{ij} = Q'_{ij}/Q_{ij}$.

An unbiased variance estimator of $q_0'^*$ is

$$s^2_{q_0'^*} = \sum^{n}(y_i'^* - y_0'^*)^2/n(n-1) \tag{15.104}$$

A consistent but generally biased estimator of the universe proportion P is (from the general estimating equation (14.6))

$$p = q_0'^*/q_0^* \tag{15.105}$$

where q_0^* is defined by equation (15.90).

A variance estimation of p is given by the general estimating equation (14.7).

Note: If all the Q_{ij} values for the universe are known, then an unbiased estimator of the universe proportion P is

$$q_0'^* \Big/ \sum_{i=1}^{N}\sum_{j=1}^{M_i} Q_{ij} \tag{15.106}$$

However, this estimator is likely to have a much larger sampling variance than the estimator defined in estimation equation (15.105).

15.9 Three-stage srs: Ratio method of estimation

As for two-stage srs (section 15.5), so also for three-stage srs the ratio method of estimation may be used to increase the efficiency of the estimators. The estimating procedures are the same as in section 15.5 for two-stage srs, except that for the unbiased estimators of the study and the ancillary variables and their estimated variances, formulae of the type (15.79) and (15.80) would be used.

Table 15.7: Computation of the estimated number and proportion of adults who agree to an increase in state taxes for education: data of Tables 15.4 and 15.5

Sample village i	Number of households		For the number of adults who agree to an increase in state taxes for education				
	Total M_i	Sample m_i	$\sum_{j=1}^{m_i} Q_{ij}q'_{ij}$	$\frac{\text{Col.(4)}}{m_i}$	$q_i'^* =$ $NM_i\times$ col.(5)	$q_i'^{*2}$	$q_i^* q_i'^*$
(1)	(2)	(3)	(4)	(5)	(6)	(7)	(8)
						$30^2\times$	$30^2\times$
1	20	4	4	1.00	30 × 20.00	400.0000	1100.000
2	23	4	3	0.75	17.15	297.5625	1190.250
3	25	4	2	0.50	12.50	156.2500	703.125
4	18	4	2	0.50	9.00	81.0000	405.000
						$30^2\times$	$30^2\times$
				Total	30 × 58.75	934.8125	3398.375
				Mean	30 × 14.6875 ($q_0'^*$)		

Example 15.3

In example 15.2 are given the number of adults in each of the sample households in the srs of 4 villages out of the total 30 villages. In each of these sample households, one adult member was further selected at random from the total number of adults in the household, and asked if he/she agrees to an increase in state taxes for education. This information is given also in Table 15.4. Estimate the total number of adults who agree to an increase in state taxes for education and the proportion they constitute of the total number of adults.

Extending the notation of example 15.1, let Q_{ij} ($= y'_{ij}$ in example 15.1) denote the total number of adults, and q_{ij} ($= 1$) the number selected for interview in the ijth sample household ($i = 1, 2, 3, 4$; $j = 1, 2, 3, 4$). As only one sample third-stage unit is selected, we can dispense with the subscript k in y_{ijk} by which we had denoted the value of the study variable for ijkth third-stage sample unit.

Following the method of section 15.8, we put $q'_{ij} = 1$, if the selected adult in the ijth sample household agrees to an increase in state taxes for education, and 0, otherwise. The required computations are shown in Table 15.7.

From estimating equation (15.102), the unbiased estimate of the total number of adults who agree to an increase in state taxes for education is

$$q_0'^* = \sum^n q_i'^*/n = 30 \times 14.6875 = 441,$$

where (from equation (15.101)

$$q_i'^* = \frac{NM_i}{m_i}\sum_{j=1}^{m_i} Q_{ij}q_{ij}'$$

An unbiased estimate of the variance of $q_0'^*$ is (from equation (15.104))

$$\begin{aligned} s^2_{q_0'^*} &= SSq_i'^*/n(n-1) \\ &= 30^2[934.8125 - 862.8906]/12 \\ &= 30^2 \times 71.9219/12 \end{aligned}$$

so that the estimated standard error of $q_0'^*$ is $30 \times 2.448 = 73$, and the estimated CV $73/441 = 16.55$ per cent.

The estimated proportion of adults who agree to an increase in state taxes for education is equal to the estimated number of such adults divided by the estimated total number of adults, or

$$\begin{aligned} p' &= q_0'^*/q_0^* = (30 \times 14.6875)/(30 \times 56.3125) \\ &= 0.2608 \text{ or } 26.08 \text{ per cent} \end{aligned}$$

As

$$s_{q_0^* q_0'^*} = SPq_i^* q_i^*/n(n-1) = 30^2 \times 90.0156/12$$

the estimated variance of p' is

$$\begin{aligned} s^2_{p'} &= (s^2_{q_0^*} + p'^2 s^2_{q_0'^*} - 2p' s_{q_0^* q_0'^*})/q_0^{*2} \\ &= \frac{30^2}{12}\,\frac{71.9212 + (0.2608)^2 \times 290.6719 - 2 \times 0.2608 \times 90.0156}{(30 \times 56.3125)^2} \\ &= \frac{54.1636}{12 \times 3171.0977} = 0.0014234 \end{aligned}$$

so that the estimated standard error of p' is $s_{p'} = 0.0377$, with estimated CV of 14.46 per cent.

Further reading

Cochran, chapters 10 and 11; Deming (1950), chapter 5A; Hansen *et al.* (1953), vols. I and II, chapter 6, 8, 9, and 10; Hedayat and Sinha, sections 7.3 and 7.7; Kish (1965), section 5.3; Murthy (1967), sections 9.3 and 9.10; Singh and Chaudhury, sections 9.1, 9.2, 9.4, and 9.6; Sukhatme *et al.* (1984), chapter 8; Yates, sections 3.8, 6.18, 6.19, and 7.17.

Exercises

1. A hospital has received 1000 bottles of 100 tablets each. A simple random sample of 6 bottles is taken and from each sample bottle a sample of 20 tablets is taken at random. Given the data in Table 15.8, estimate (a) the average weight per tablet, (b) the proportion of sub-standard tablets, and (c) the ratio between the composition of two active substances A and B in the tablet, with their standard errors (Weber, Example 4.4).
2. Of 53 communes in an area, 14 were selected at random; from each of the selected communes, of the total number of farms, a sample of farms (1 in 4) was taken also at random. Table 15.9 gives the required information on the sample, including the total number of cattle in the sample farms in the selected communes.
   (a) Estimate the total number of cattle in the area;
   (b) Estimate the average number of cattle per farm, by (i) using the ratio of the unbiased estimators of the total number of cattle and the total number of farms; (ii) using the unweighted mean of means; and (iii) using the additional information that the average number of farms per commune in the universe is 39.09.

Table 15.8: Total weight, number of sub-standard tablets, and the percentage composition of two active substances in 20 sample tablets selected from 100 tablets in 6 out of 1000 bottles received by a hospital

Bottle number	Total weight (gr) of The sample 20 tablets	Number of sub-standard tablets	Composition in percentage for the sample of 20 tablets	
			Substance A	Substance B
1	21.8	1	5.42	31.2
2	19.3	0	5.01	30.3
3	22.8	1	5.43	30.5
4	23.1	3	5.43	32.4
5	22.9	1	5.20	31.4
6	19.7	1	5.56	32.9
Total	129.6	7	32.05	188.7
Mean per tablet	1.08	0.058	0.267	1.572
Raw sum of squares	2813.68	13	171.3999	5939.91
Raw sum of products			1008.658	

Source: Weber, Example 4.4.

Table 15.9: Number of cattle in sample farms

Sample commune i	Number of farms: Total M_i	Number of farms: Sample m_i	Number of cattle: Total in sample $\sum_{j=1}^{m_i}$	Number of cattle: Average per farm $\bar{y}_i = \sum_{j=1}^{m_i} y_{ij}/m_i$	$M_i\bar{y}_i$
1	46	11	88	8.0000	368.0000
2	39	10	114	11.4000	444.6000
3	25	6	96	16.0000	400.0000
4	23	5	82	16.4000	377.2000
5	32	8	83	10.3750	332.0000
6	31	8	207	25.8750	802.1250
7	60	15	208	13.8667	832.0020
8	28	7	73	10.4286	292.0008
9	59	14	195	13.9286	821.7874
10	24	6	73	12.1667	292.0008
11	84	21	191	9.0952	763.9968
12	30	7	79	11.2857	338.5710
13	64	16	226	14.1250	904.0000
14	66	17	166	9.7647	644.4702
Total	611	151	1881	182.7112	7612.7540

Source: United Nations *Manual*, Process 20-22 (adapted).

Estimate also the standard errors of these estimates, assuming sampling with replacement. (United Nations *Manual*, Processes 20, 21, and 22, modified to "with replacement" sampling.)

3. To estimate the total number of beetles in a square plot of land, 250 ft by 250 ft, the plot was divided into 625 square cells, 10 ft by 10 ft. A random sample of 10 cells was taken, and for each cell two independent estimates of the number of beetles were obtained by actual survey of two randomly located circles within the cell. The data are given in Table 15.10. (Chakravarti *et al.*, Vol. II, Example 3.4).

For checking and facility of computation, the following are given:

$\sum_{i=1}^{n}(M_i\bar{y}_i)^2 = 4,857,967.5243$; $m_0^* = 2313$; $\sum_{i=1}^{n} M_i^2 = 31,625$;

$s^2_{m_0^*} = 53^2 \times 4959.2143/182$; $\sum_{i=1}^{n}(M_i\overline{y}_i)M_i = 376,746.97$;

$s_{y_0^* m_0^*} = 53^2 \times 44,504.6347/182$; $\sum_{i=1}^{n} \overline{y}_i^2 = 2644.3481$.

Table 15.10: Estimated number of beetles in cells 10 ft by 10 ft obtained from two randomly located circles within each cell

Cell	1	2	3	4	5	6	7	8	9	10
Estimate 1	61	38	25	0	71	95	32	50	10	0
Estimate 2	29	81	32	45	46	69	26	39	24	5
Total	90	119	57	45	117	164	58	89	34	5

Source: Chakravarti *et al.*, Example 3.4.

Note: The figures in Table 15.8 show $y^*_{ij} = M_0 y_{ij}$.

CHAPTER 16

Multi-stage Varying Probability Sampling

16.1 Introduction

As for single-stage, so also for multi-stage sampling, sampling with varying probabilities – more specifically with probability proportional to size – would, under favorable conditions, lead to an improvement in the efficiency of estimators if ancillary information is available for such selection. Although pps sampling can be resorted to in all stages of sampling, in general it is restricted to the stages that can be covered in offices; the selection of sample units at other stages, generally the ultimate or the penultimate stages or both which can readily be sampled either systematically or by simple random sampling, is left to the enumerators in the field.

The general estimating equations for a multi-stage pps design are given in section 14.4. In this chapter we will consider first the specific cases of two-stage and three-stage designs with pps either at all stages or only at the first stage, and then the special cases of crop surveys.

16.2 General estimating equations: Two- and three-stage pps designs

16.2.1 Two-stage pps design

The structure of a two-stage pps is given in Table 16.1. In the first stage, a sample of n units is selected with pps (and replacement) from N units, the initial probability of selection of the ith first-stage unit ($i = 1, 2, \ldots, N$) being

$$\pi_i = z_i/Z \tag{16.1}$$

where z_i is the value ("size") of an ancillary variable, $Z = \sum^N z_i$ being known. In each of the n first-stage sample units (fsu's) thus selected, a

Table 16.1: Sampling plan for a two-stage pps sample design with replacement

Stage (t)	Unit	No. in universe	No. in sample	Selection method	Selection probability	f_t
1	First-stage	N	n	ppswr	$\pi_i = z_i/Z$	$n\pi_i$
2	Second-stage	M_i	m_i	ppswr	$\pi_{ij} = w_{ij}/W_i$	$m_i\pi_{ij}$

sample of m_i second-stage units (ssu's) is selected with pps (and replacement) from the M_i units, the (initial) probability of selection of the jth ssu ($j = 1, 2, \ldots, M_i$) being

$$\pi_{ij} = w_{ij}/W_i \tag{16.2}$$

where w_{ij} is the value of another ancillary variable and $W_i = \sum_{j=1}^{M_i} w_{ij}$ is known.

The universe totals and means are defined as for a two-stage srs in section 15.2.1.

If y_{ij} is the value of the study variable in the jth selected ssu ($j = 1, 2, \ldots, m_i$) in the ith selected fsu ($i = 1, 2, \ldots, n$), then from the general estimating equation (14.13), or an extension from single-stage pps to two-stage pps in the same manner as was done for srs in section 15.2.3(a), it will be seen that an unbiased estimator of the universe total Y of the study variable, obtained from the ith sample fsu, is

$$y_i^* = n\sum_j \frac{1}{f_1 f_2} y_{ij} = \frac{1}{\pi_i m_i}\sum_{j=1}^{m_i} \frac{y_{ij}}{\pi_{ij}} = \frac{ZW_i}{z_i m_i}\sum_{j=1}^{m_i} \frac{y_{ij}}{w_{ij}} \tag{16.3}$$

and the combined unbiased estimator of the universe total Y from all the n sample fsu's is

$$y_0^* = \frac{1}{n}\sum^{n} y_i^* \tag{16.4}$$

An unbiased estimator of the variance of y_0^* is (from equation (14.15))

$$s_{y_0^*}^2 = \sum^{n} (y_i^* - y_0^*)^2/n(n-1) \tag{16.5}$$

Estimators of the means of study variables, and the ratio of two totals follow directly from the fundamental theorems of section 14.4.

Notes

1. *Sampling variance of the estimator y_0^*.* The sampling variance of y_0^* in ppswr at both stages is

$$\sigma_{y_0^*}^2 = \frac{1}{n}\left(\sum_{i=1}^{N}\frac{Y_i^2}{\pi_i} - Y^2\right) + \frac{1}{n}\sum_{i=1}^{N}\frac{1}{\pi_i m_i}\left(\sum_{j=1}^{M_i}\frac{Y_{ij}^2}{\pi_{ij}} - Y_i^2\right) \tag{16.6}$$

which is estimated unbiasedly by $s_{y_0^*}^2$, defined by equation (16.5).

2. If the same ancillary "size" variable is used for pps in both the stages, i.e. if $\pi_i = z_i/Z$, and also $\pi_{ij} = z_{ij}/Z_i$ (note that $z_i = Z_i$ in our notation), then estimating equation (16.3) simplifies to

$$y_i^* = \frac{Z}{m_i}\sum^{m_i}\frac{y_{ij}}{z_{ij}} \tag{16.7}$$

and estimating equation (16.4) becomes

$$y_0^* = \frac{Z}{n}\sum^{n}\frac{1}{m_i}\sum^{m_i}\frac{y_{ij}}{z_{ij}} \tag{16.8}$$

3. If, in the above case, a fixed number m_0 of ssu's is selected in each sample fsu, then estimating equation (16.7) becomes

$$y_i^* = \frac{Z}{m_0}\sum^{m_0}\frac{y_{ij}}{z_{ij}} \tag{16.9}$$

and estimating equation (16.8) becomes

$$y_i^* = \frac{Z}{nm_0}\sum^{n}\sum^{m_0}\frac{y_{ij}}{z_{ij}} \tag{16.10}$$

The ratio Y/Z is then estimated unbiasedly by

$$\frac{y_0^*}{Z} = \frac{1}{nm_0}\sum^{n}\sum^{m_0}\frac{y_{ij}}{z_{ij}} \tag{16.11}$$

i.e. by the simple (unweighted) mean of the ratios y_{ij}/z_{ij}. This result is particularly useful in crop surveys (see section 16.5).

16.2.2 Three-stage pps design

The sampling plan for a three-stage pps sample design is given in Table 16.2. The universe totals and means are defined as for a three-stage srs in section 15.6.1. If y_{ijk} is the value of the study variable in the kth selected third-stage unit (tsu) of the jth selected ssu of the ith selected fsu ($i = 1, 2, \ldots, n$; $j = 1, 2, \ldots, m_i$; $k = 1, 2, \ldots, q_{ij}$) in a three-stage pps sample, then from

Table 16.2: Sampling plan for a three-stage pps sample design with replacement

Stage (t)	Unit	No. in universe	No. in sample	Selection method	Selection probability	f_t
1	First-stage	N	n	ppswr	$\pi_i = z_i/Z$	$n\pi_i$
2	Second-stage	M_i	m_i	ppswr	$\pi_{ij} = w_{ij}/W_i$	$m_i\pi_{ij}$
3	Third-stage	Q_{ij}	q_{ij}	ppswr	$\pi_{ijk} = v_{ijk}/V_{ij}$	$q_{ij}\pi_{ijk}$

the general estimating equation (14.13) or an extension from the case of the two-stage pps design, it will be seen that an unbiased estimator of the universe total Y of the study variable, obtained from the ith selected fsu, is

$$y_i^* = n\sum_{j,k}\frac{1}{f_1f_2f_3}\,y_{ijk} = \frac{1}{\pi_i m_i}\sum_{j=1}^{m_i}\frac{1}{\pi_{ij}q_{ij}}\sum_{k=1}^{q_{ij}}\frac{y_{ijk}}{\pi_{ijk}}$$

$$= \frac{ZW_i}{z_i m_i}\sum^{m_i}\frac{V_{ij}}{w_{ij}q_{ij}}\sum^{q_{ij}}\frac{y_{ijk}}{v_{ijk}} \tag{16.12}$$

where $Z = \sum^N z_i$; $W_i = \sum^{M_i} W_i$, and $V_{ij} = \sum^{Q_{ij}} v_{ijk}$.

The combined unbiased estimator of the universe total Y from all the n sample fsu's is

$$y_0^* = \frac{1}{n}\sum^{n} y_i^* \tag{16.13}$$

with an unbiased variance estimator

$$s_{y_0^*}^2 = \sum^{n}(y_i^* - y_0^*)^2/n(n-1) \tag{16.14}$$

Estimators of the means of study variables and the ratio of two totals follow directly from the fundamental theorems of section 14.4.

Note: As for two-stage pps sampling, so also for three-stage sampling, great simplifications result if the same ancillary ("size") variable is used for selection in all three stages, i.e. if

$$\pi_i = z_i/Z \qquad \pi_{ij} = z_{ij}/Z_i \qquad \pi_{ijk} = z_{ijk}/Z_{ij} \tag{16.15}$$

(In our notation, $z_i = Z_i$ and $z_{ij} = Z_{ij}$). This is particularly useful in crop surveys, as will be seen in section 16.5.

Table 16.3: Sampling plan for a two-stage sample design with pps sampling at the first-stage and srs at the second-stage

Stage (t)	Unit	No. in universe	No. in sample	Selection method	Selection probability	f_t
1	First-stage	N	n	ppswr	$\pi_i = z_i/Z$	$n\pi_i$
2	Second-stage	M_i	m_i	srswr	Equal $= 1/M_i$	m_i/M_i

16.3 Two-stage design with pps and srs

16.3.1 General case

As mentioned in section 16.1, from operational and technical considerations in a two-stage design, the first-stage units may be selected with pps, and the second-stage with srs (or systematically). The sampling plan for such a design is given in Table 16.3.

An unbiased estimator of the universe total Y of a study variable, obtained from the ith selected fsu, is (from the general estimating equation 14.13 or from the combination of sections 16.2.1 and 15.2.3)

$$y_i^* = n\sum_j \frac{1}{f_1 f_2}\, y_{ij} = \frac{M_i}{\pi_i m_i}\sum_{j=1}^{m_i} y_{ij} = \frac{M_i}{\pi_i}\,\bar{y}_i \tag{16.16}$$

where

$$\bar{y}_i = \sum_{j=1}^{m_i} y_{ij}/m_i \tag{16.17}$$

is the unbiased estimator of the mean $\overline{Y}_i = Y_i/M_i$ in the ith sample fsu (as sampling is simple random), and $\pi_i = z_i/Z$ is the (initial) probability of selection of the ith fsu.

The combined unbiased estimator of Y from all the n sample fsu's is the arithmetic mean

$$y_0^* = \frac{1}{n}\sum^{n} y_i^* \tag{16.18}$$

with an unbiased estimator of the variance of y_0^*

$$s_{y_0^*}^2 = \sum^{n}(y_i^* - y_0^*)^2/n(n-1) \tag{16.19}$$

Estimators of the ratio of the totals of two study variables, their means etc. as also of their variances follow from the fundamental theorems in section 14.4, and will be illustrated by an example.

Example 16.1

In each of the four sample villages selected out of 30 villages with probability proportional to the previous census population in Example 5.2, select a simple random sample of 4 households each from the total number of households in the villages, and on the basis of the collected data on household size, the total daily income and the number of adults (over 18 years of age) of these sample households (given in Table 16.4), estimate for the 30 villages the total number of persons, the total and per capita daily income, and the proportion of adults, along with their standard errors.

The required computations are shown in Table 16.5, denoting by the y_{ij} the household size, by x_{ij} the daily income, and by y'_{ij} the number of adults in the jth sample household of the ith sample village ($i = 1, 2, 3, 4$ for sample villages numbered 5, 6, 11, and 18 respectively; $j = 1, 2, 3, 4$). The means in the ith sample villages are denoted by $\bar{y}_i$, $\bar{x}_i$, and $\bar{y}'_i$ respectively, defined by equations of the type (16.17).

$N = 30$ is the total number of villages, and M_i and m_i (= 4) denote respectively the total and the sample numbers of households in the ith sample village.

From equation (16.18), the combined unbiased estimate of the total number of persons in the 30 villages is $y_0^* = \sum^n y_i^*/n = 3,349$ (shown at the bottom of column 7 of Table 16.5). An unbiased variance estimate of y_0^* is (from equation (16.19))

$$s^2_{y_0^*} = (45,056,019.02 - 44,870,538.61)/12 = 15,456.70$$

Similarly, the combined unbiased estimate of the total daily income is $x_0^* = \sum^n x_i^*/n = \$33,548$ (shown at the bottom of column 11 of Table 16.5), and the combined unbiased estimate of the total number of adults is

$$y_0'^* = \sum^n y_i'^*/n = 1930$$

(shown at the bottom of column 15 of Table 16.5) with an unbiased variance estimate 11,916.32.

The estimated daily income per capita is, from equation (14.6), the estimated total daily income divided by the estimated total number of persons, or

$$r = x_0^*/y_0^* = \$10.02$$

As

$$s_{x_0^* y_0^*} = (450,668,357 - 449,446,581)/12 = 101,814.67$$

the estimated variance of r is (from equation (14.7))

$$s_r^2 = (s^2_{x_0^*} + r^2 s^2_{y_0^*} - 2r s_{y_0^* x_0^*})/y_0^{*2} = 0.00961286$$

Table 16.4: Size, total monthly income, number of adults (18 years or over) and agreement by the selected adult to an increase of state taxes for education in the srs of the sample of households in each of the four sample villages of Example 5.2 selected with probability proportional to previous census population

Household serial no.	Household size y_{ij}	Total daily income x_{ij}	Number of adults y'_{ij}	Agreement by the selected adult to an increase of state taxes for education (q'_{ij} = 1 for Yes; = 0 for No)
Village serial No. 5				
10	4	$37	3	0
14	5	53	3	0
20	4	40	2	1
23	5	48	3	0
Village serial No. 6				
4	6	52	3	0
5	5	42	2	1
15	3	35	2	0
17	6	60	4	0
Village serial No. 11				
2	4	45	2	1
4	6	69	3	0
10	5	46	3	0
13	6	57	3	0
Village serial No. 18				
3	5	60	3	0
12	5	43	3	1
18	5	52	3	0
20	6	65	4	0

Table 16.5: Computations of the estimated total number of persons, total and per capita monthly income, number and proportion of adults for a two-stage design: data of Table 16.4

Sample village i	Reciprocal of probability $1/\pi_i$	Number of households M_i	M_i/π_i †	For the number of persons $\sum_{j=1}^{m_i} y_{ij}$	$\bar{y}_i = \sum_{j=1}^{m_i} y_{ij}/m_i$	$y_i^* = (M_i/pi_i)\bar{y}_i$	y_i^{*2}
(1)	(2)	(3)	(4)	(5)	(6)	(7)	(8)
1	30.5978	24	734.35	18	4.50	3304.58	10 920 248.98
2	43.3077	17	736.23	20	5.00	3681.15	13 550 865.32
3	39.0972	15	586.46	21	5.25	3078.92	9 479 748.37
4	27.5980	23	634.75	21	5.25	3332.44	11 105 156.35
					Total	13 397.09	45 056 019.02
					Mean	3349.27 (y_0^*)	

Sample village i	For the total daily income $\sum_{j=1}^{m_i} x_{ij}$	$\bar{x}_i = \sum_{j=1}^{m_i} x_{ij}/m_i$	$x_i^* = (M_i/\pi_i)\bar{x}_i$	x_i^{*2}
(1)	(9)	(10)	(11)	(12)
1	\$178	\$44.50	\$32 678.58	1 067 890 244
2	189	47.25	34 786.87	1 210 127 368
3	217	54.25	31 815.46	1 012 224 768
4	220	55.00	34 911.25	1 218 793 631
		Total	\$134 192.16	4 509 036 011
		Mean	\$33 548.04 (x_0^*)	

Sample village i	For the number of adults $\sum_{j=1}^{m_i} y'_{ij}$	$\bar{y}_i = \sum_{j=1}^{m_i} y'_{ij}/m_i$	$y_i^* = (M_i/\pi_i)\bar{y}'_i$	$y_i'^{*2}$	For the sum of products $y_i^* x_i^*$	$y_i^* y_i'^*$
(1)	(13)	(14)	(15)	(16)	(17)	(18)
1	11	2.75	2019.19	4 077 128.26	108 315.768	6 672 574.89
2	11	2.75	2024.63	4 099 126.64	128 055.687	7 452 966.72
3	11	2.75	1612.76	2 600 994.82	97 957.256	4 965 559.02
4	13	3.25	2062.94	4 255 721.44	116 339.646	6 874 623.77
		Total	7719.32	15 032 971.16	450 668.357	25 965 724.40
		Mean	1929.88 ($y_0'^*$)			

† Also the estimated number of households (h_i^*) in Table 5.7.

Similarly, the estimated proportion of adults is the estimated total number of adults divided by the estimated total number of persons, or

$$p = y_0'^* / y_0^* = 0.5762$$

and the estimated variance of p is

$$s_p^2 = (s_{y_0'^*}^2 + p^2 s_{y_0^*}^2 - 2p s_{y_0^* y_0'^*}) / y_0^{*2} = 0.0005643645$$

as

$$s_{y_0^* y_0'^*} = (25\,965\,724.40 - 25\,854\,125.48)/12 = 9299.91\,.$$

16.3.2 First-stage units selected with probability proportional to the number of second-stage units

In a two-stage sample design, with pps sampling at the first-stage and srs at the second, the estimating procedure becomes simpler if for the pps sampling at the first stage, the "size" is the number of second-sage units, i.e. if the fsu's are selected with probability proportional to the number of ssu's contained in each fsu: this assumes, of course, that the information is available for all the fsu's in the universe.

In this case, the (initial) probability of selection of the ith fsu ($i = 1, 2, \ldots, N$) is

$$\pi_i = M_i / M$$

where $M = \sum^N M_i$ is the total number of second-stage units in the universe.

Estimating equation (16.16) then becomes

$$y_i^* = M(M_i/M_i)\bar{y}_i = M\bar{y}_i \tag{16.20}$$

and the combined unbiased estimator of Y from all the n sample fsu's, given by equation (16.18), becomes

$$y_0^* = \frac{1}{n}\sum^n y_i^* = \frac{M}{n}\sum^n \bar{y}_i = M\bar{y} \tag{16.21}$$

where

$$\bar{y} = \sum^n \bar{y}_i / n \tag{16.22}$$

is the simple (unweighted) mean of the $\bar{y}_i$ values.

An unbiased estimator of the mean $\bar{\bar{Y}} = Y/M$ is

$$M\bar{y}/M = \bar{y} = \sum^n \bar{y}_i / n \tag{16.23}$$

which is the simple (unweighted) mean of the $\bar{y}_i$ values, and an unbiased estimator of the variance of $\bar{y}$ is

$$s_{\bar{y}}^2 = \sum^{n}(\bar{y}_i - \bar{y})^2/n(n-1) \tag{16.24}$$

Note: If $m_i = m_0$, a constant, then $y_i^* = M\sum y_i/m_0$, i.e., the design will be self-weighting (see Example 16.2).

Example 16.2

In an area there are 315 schools with a total of 27,215 students. Eight schools were first selected with probability proportional to the number of students, and in each selected school, 50 students were selected at random. Table 16.6 gives for each school the number of students with trachoma and the number with multiple scars. Estimate (a) the proportion of student with trachoma, and (b) among those with trachoma, the proportion with multiple scars (Weber, Example 4.5).

Here $N = 315$ schools, and $M = 27,215$ students. As the sample of 8 schools is selected with probability proportional to the number of students (M_i) and in each selected school, a sample of students is taken at random, the method of section 16.3.2 will apply here. The estimating equations are further simplified as a fixed number $m_0 = 50$ of students are selected in the sample schools.

(a) Defining $y_{ij} = 1$, if the jth sample student in the ith sample school has trachoma, and $y_{ij} = 0$, otherwise ($i = 1, 2, \ldots, 8$; $j = 1, 2, \ldots, 50$), an unbiased estimate of the proportion of students with trachoma is (from equation (16.23))

$$\bar{y} = \sum^{n} \bar{y}_i/n = \sum^{n} y_i/nm_0 = 325/(8 \times 50) = 0.8125 \text{ or } 81.25 \text{ per cent.}$$

where $y_i = \sum_{j=1}^{m_0} y_{ij}$.

An unbiased estimate of the variance of $\bar{y}$ is (from equation (16.24))

$$\begin{aligned} s_{\bar{y}}^2 &= \sum^{n}(y_i - \bar{y})^2/n(n-1) = (SSy_i/m_0^2)/n(n-1) \\ &= (13,421 - 13,203.125)/(50^2 \times 8 \times 7) = 0.00155625, \end{aligned}$$

so that the estimated standard error of $\bar{y}$ is $s_{\bar{y}} = 0.03945$ with CV 4.86 per cent.

(b) Similarly, defining $y'_{ij} = 1$, if the ijth sample student has trachoma with multiple scars, and $y'_{ij} = 0$, otherwise, an unbiased estimate of the total number of students with multiple scars is (from equation (16.23))

$$\bar{y}'_i/n = \sum^{n} y'_i/nm_0 = 55/(8 \times 50) = 0.1375$$

where $y'_i = \sum_{j=1}^{m_0} y'_{ij}$.

Table 16.6: Number of students with trachoma and number with multiple scars among samples of 50 students each selected at random from each of the 8 schools selected with probability proportional to the total number of students

School no. i	Number of students with trachoma	
	Total number y_i	Number with multiple scars y'_i
1	40	3
2	31	0
3	47	16
4	41	8
5	43	8
6	36	5
7	39	2
8	48	13
Total	325	55
Raw sum of squares	13,421	591
Raw sum of products	2426	

Source: Weber, Example 4.5.

Table 16.7: Sampling plan for a three-stage sample design with pps sampling at the first-stage and srs at the second- and third-stages

Stage (t)	Unit	No. in universe	No. in sample	Selection method	Selection probability	f_t
1	First-stage	N	n	ppswr	$\pi_i = z_i/Z$	$n\pi_i$
2	Second-stage	M_i	m_i	srs	Equal $= 1/M_i$	m_i/M_i
3	Third-stage	Q_{ij}	q_{ij}	srs	Equal $= 1/Q_{ij}$	q_{ij}/Q_{ij}

An unbiased estimate of the variance of $\bar{y}'$ is (from equation (16.24))

$$\begin{aligned} s_{\bar{y}'}^2 &= \sum^n (y_i' - \bar{y}')^2/n(n-1) = (SSy_i'/m_0^2)/n(n-1) \\ &= (591 - 378.125)/(50^2 \times 8 \times 7) = 0.00152054 \end{aligned}$$

The estimated proportion of students with multiple scars among those with trachoma is $r' = \bar{y}'/y = 55/325 = 0.1692$ or 16.92 per cent.

An estimate of the variance of r' is, from the general estimating equation (14.7),

$$s_{r'}^2 = (s_{\bar{y}'}^2 + r'^2 s_{\bar{y}}^2 - 2r' s_{\bar{y}\bar{y}'})/\bar{y}^2$$

Since

$$\begin{aligned} s_{\bar{y}\bar{y}'} &= \sum^n (y_i - \bar{y})(y_i' - \bar{y})/n(n-1) = (SPy_i y_i'/m_0^2)/n(n-1) \\ &= (2426 - 2232.125)/(50^2 \times 8 \times 7) = 193.875/(50^2 \times 8 \times 7) \end{aligned}$$

therefore

$$\begin{aligned} s_{r'}^2 &= \frac{212.875 + (0.1692)^2 \times 217.875 - 2 \times 0.1692 \times 193.875/(50^2 \times 8 \times 7)}{[325/(8 \times 50)]^2} \\ &= \frac{212.975 + 6.237543 - 65.607300}{325^2 \times 7/8} = 0.0016609 \end{aligned}$$

so that the estimated standard error of r' is 0.04075, with CV of 24.08 per cent.

16.4 Three-stage design with pps, srs, and srs

In this section will be extended the method of section 16.3.1 to a three-stage sample design with pps sampling at the first stage and srs at the second and the third stages. The sampling plan is shown in Table 16.7.

An unbiased estimator of the universe total Y of a study variable, obtained from the ith selected fsu, is (from the general estimating equation 16.12)

$$y_i^* = n \sum_{j=1}^{m_i} \sum_{k=1}^{q_{ij}} \frac{1}{f_1 f_2 f_3} y_{ijk} = \frac{M_i}{\pi_i m_i} \sum_{j=1}^{m_i} \sum_{k=1}^{q_{ij}} y_{ijk} \frac{Q_{ij}}{q_{ij}} \qquad (16.25)$$

where y_{ijk} is the value of the study variable in the kth sample tsu in the jth sample ssu in the ith sample fsu ($i = 1, 2, \ldots, n$; $j = 1, 2, \ldots, m_i$; $k = 1, 2, \ldots, q_{ij}$) and π_i is the (initial) probability of selection of the ith first-stage unit in the universe ($i = 1, 2, \ldots, N$).

The combined unbiased estimator of Y from all the n sample fsu's is the arithmetic mean

$$y_0^* = \frac{1}{n} \sum^n y_i^* \qquad (16.26)$$

with an unbiased variance estimator

$$s^2_{y_0^*} = \sum^{n}(y_i^* - y_0^*)^2/n(n-1) \tag{16.27}$$

Estimators of the ratio of the totals of the values of two study variables, their means etc. as also their variances follow from the fundamental theorems in section 14.4 and will be illustrated with an example.

Example 16.3

In Example 16.1 are given the data on the number of adults in the srs of 4 households in each of the 4 sample villages, selected with probability proportional to the previous census population. In each of these sample households, one adult member was further selected at random from the total number of adults in the household, and asked if he/she agrees to an increase in state taxes for education. This information is given also in Table 16.4. Estimate the total number of adults who agree to an increase in state taxes for education and the proportion they constitute of the total number of adults.

Extending the notation of Example 16.1, let Q_{ij} ($= y'_{ij}$ in Example 16.1) denote the total number of adults, and q_{ij} ($= 1$) the number selected for interview in the ijth sample household ($i = 1, 2, 3, 4$; $j = 1, 2, 3, 4$). As only one sample third-stage unit is selected, we can dispense with the subscript k in y_{ijk} by which we had denoted the value of the study variable for the ijkth third-stage sample unit.

We define $q'_{ij} = 1$, if the selected adult in the ijth sample household agrees to an increase in state taxes for education, and 0 otherwise.

The computations are shown in Table 16.8.

From equation (16.26), the unbiased estimate of the total number of adults who agree to an increase in state taxes for education is $q_0'^* = \sum^n q_i'^*/n = 376$ (shown at the bottom of column 4 of Table 16.8), where

$$q_i'^* = \frac{M_i}{\pi_i m_i}\sum^{m_i} Q_{ij}q'_{ij}$$

from equations (16.25) and (16.26)

An unbiased variance estimate of $q_0'^*$ is (from equation (16.27))

$$s^2_{q_0'^*} = (582,950.43 - 565,947.77)/12 = 1,416.8883$$

so that the estimated standard error of $q_0'^*$ is 37.64, and the estimated CV is 10.00 per cent.

The estimated proportion of adults who agree to an increase in state taxes for education equals the estimated total number of such adults divided by the estimated total number of adults, or

Table 16.8: Computation of the estimated number and proportion of adults who agree to an increase in state taxes for education: data of Tables 16.4 and 16.5

Sample village i	$\sum_{j=1}^{m_i} Q_{ij}q'_{ij}$	$\sum_{j=1}^{m_i} Q_{ij}q'_{ij}/m_i$	$q_i'^* = (M_i/\pi_i)\times$ col.(3)	$q_i'^{*2}$	$q_i^* q_i'^{*}$†
(1)	(2)	(3)	(4)	(5)	(6)
1	2	0.50	367.18	134,821.15	714,406.18
2	2	0.50	368.12	135,512.33	745,306.80
3	2	0.50	293.23	85,983.83	472,909.61
4	3	0.75	476.06	226,633.12	982,083.22
Total	-	-	1,505.59	582,950.43	2,914,705.81
Mean	-	-	376.15	-	-
			($q_0'^*$)	-	-

† q_i^* ($= y_i'^*$ in Table 16.5) denotes the estimates total number of adults.

$$p' = q_0'^*/q_0^* = 376.15/1,929.88 = 0.1949$$

since

$$s_{q_0^* q_0'^*} = (2,914,705.81 - 2,903,602.92)/12 = 11,102.89$$

the estimated variance of p' is

$$s_{p'}^2 = (s_{q_0'^*}^2 + p'^2 s_{q_0^*}^2 - 2p' s_{q_0^* q_0'^*})/q_0^{*2} = 0.0004051344,$$

so that the estimated standard error of p' is 0.02013, with estimated CV of 10.33 per cent.

16.5 Special cases of crop surveys

16.5.1 Introduction

As observed in the notes to section 16.2.1 and 16.2.2, great simplifications result if the same "size" variable is used in all the stages of a multi-stage pps sample. This is particularly useful in crop surveys when ancillary information is available on each of the universe units.

16.5.2 Area surveys of crops

When the areas of all the fields in the universe are known, let the sample design for estimating the total area under a crop be pps two-stage with villages as the fsu's, selected with probability proportional to total (geographical) area, and fields as the ssu's, selected with probability proportional to their areas: the fields may be selected from a list of fields showing the areas or from a cadastral map of the fields, following the procedure of section 5.4. We further assume that the same number m_0 of fields is selected in each sample village.

If a_{ij} is the total area and y_{ij} is the area under the crop in the ijth sample field ($i = 1, 2, \ldots, n$; $j = 1, 2, \ldots, m_0$), then an unbiased estimator of the total area under the crop, from the ith sample village, is (from equation (16.9))

$$y_i^* = \frac{A}{m_0} \sum^{m_0} p_{ij} = A\bar{p}_i \tag{16.28}$$

and from all the n sample villages is (from equation (16.10))

$$y_0^* = \sum^{n} y_i^*/n = A \sum^{n} \bar{p}_i/n = A\bar{p} \tag{16.29}$$

where $p_{ij} = y_{ij}/a_{ij}$ is the proportion of the area under the crop in the ijth sample field and A the total area in the universe

$$\bar{p}_i = \sum^{m_0} p_{ij}/m_0 \tag{16.30}$$

is an unbiased estimator of the proportion of the area under the crop in the ith sample village; and

$$\bar{p} = \sum^{n} \bar{p}_i/n = \sum^{n} \sum^{m_0} p_{ij}/nm_0 \tag{16.31}$$

is the simple (unweighted) average of the sample p_{ij} values and an unbiased estimator of the proportion of the area under the crop in the universe.

An unbiased variance estimator of $\bar{p}$ is

$$s_{\bar{p}}^2 = \sum^{n} (\bar{p}_i - \bar{p})^2/n(n-1) \tag{16.32}$$

and an unbiased variance estimator of $y_0^* = A\bar{p}$ is

$$s_{y_0^*}^2 = A^2 s_{\bar{p}}^2 \tag{16.33}$$

16.5.3 Yield surveys of crops

Similar considerations apply for estimating the yield of a crop. If $r_{ij} = x_{ij}/a_{ij}$ is the yield per unit area in the ijth sample field, then an unbiased estimator of the total yield in the universe is

$$x_0^* = A\bar{r} \tag{16.34}$$

where

$$\bar{r} = \sum^{n}\sum^{m_0} r_{ij}/nm_0 \tag{16.35}$$

is the simple average of the sample r_{ij} values and an unbiased estimator of the average yield per unit area in the universe. Unbiased variance estimators of x_0^* and $\bar{r}$ are given respectively by estimating equations of the types (16.33) and (16.32).

A consistent but generally biased estimator of the yield rate per unit of crop area is the ratio

$$x_0^*/y_0^* = \sum^{n}\sum^{m_0} r_{ij} \Big/ \sum^{n}\sum^{m_0} p_{ij} \tag{16.36}$$

a variance estimator of which is given by the general estimating equation (14.7).

Notes

1. If one crop cut of prescribed size and shape is located in the selected fields at random and the yield rate obtained from it, then although the design becomes three-stage, the same estimating procedure as above will apply.
2. In the far less common situation when the crop areas of all the fields in the universe are known, the villages and fields may be selected with probability proportional to their respective crop areas, and the total crop yield in the universe will be estimated by

$$x_0^* = Y\bar{r}' \tag{16.37}$$

where

$$\bar{r}' = \sum^{n}\sum^{m_0} r'_{ij}/nm_0 \tag{16.38}$$

is an unbiased estimator of the average yield per unit of crop area; $r'_{ij} = x_{ij}/y_{ij}$ is the yield per unit of crop area in the ijth sample field; and Y is the total crop area in the universe. Unbiased variance estimators of x_0^* and $\bar{r}'$ are given respectively by equations of the types (16.33) and (16.32).

3. As the area of a crop is likely to have a high positive correlation with the total area, sampling with probability proportional to total area (ppa) is likely to be more efficient than srs. However, yield rates may not be highly correlated with the area of the field, and in a crop survey with ppa sampling, estimators of crop areas will generally be more efficient than an srs, but not necessarily so for estimators of crop yields. In that case, for estimating the yield rate, a biased estimator of the unweighted mean of means may be used, after examining from a pilot survey that the bias is not appreciably high; some Indian surveys have shown that it was safe to use such biased estimators there.

Further reading

Cochran, section 11.9; Hansen *et al.* (1953), vol. I, sections 8.6 and 8.14, vol. II, section 8.1; Hedayat and Sinha, sections 7.1-7.4 and 7.7; Murthy (1967), sections 9.4, 9.5, 9.8c, 9.9, and 9.10; Singh and Chaudhary, sections 9.8 and 9.10; Sukhatme *et al.* (1984), 9.1, 9.2, 9.5-9.7, 9.9, and 9.10; Yates, sections 6.19, 7.17, and 8.11.

Exercises

1. Of 53 communes in an area, 14 were selected with probability proportional to the area of the communes (and with replacement), and in each selected commune, farms were randomly selected using a sampling fraction of $\frac{1}{4}$. For each selected commune, the probability of its selection, the total and the sample number of farms, and the number of cattle in the sample farms are given in Table 16.9. Estimate the total number of cattle in the area with its standard error (United Nations *Manual* Process 24).
2. Of the 53 communes in the area, 14 were selected with probability proportional to the number of farms in the communes (and with replacement), and in each selected commune, farms were randomly selected using a sample fraction of $\frac{1}{4}$. For each selected commune the total number of farms (M_i), the number of sample farms (m_i), and the number of cattle in the sample farms are given in Table 16.10. Estimate the total number of cattle in the area with its standard error; the total number of farms in the area is $M = \sum^N M_i = 2072$ (United Nations *Manual* Process 23).
3. For estimating the total number of cultivators (Y), the sample of n villages is selected from the N villages in the universe with ppswr, the "size" being the current number of households (M_i), and from each sample village m_0 households are selected circular systematically. The number of cultivators in each of the nm_0 sample households is determined. Suggest an unbiased estimator of Y and obtain an unbiased variance estimator of it (Murthy, Problem 9.8).

   (*Hint:* See section 16.3.2).

Table 16.9: Number of cattle in sample farms selected at random from sample communes, selected with probability proportional to area

Sample commune i	Probability of selection π_i	Total number of farms M_i	No. of farm samples m_i	Number of cattle in sample farms $\sum_{j=1}^{m_i} y_{ij}$
1	0.0026	19	5	14
2	0.0098	23	5	82
3	0.0146	31	8	207
4	0.0167	40	10	124
5	0.0187	54	13	113
6	0.0187	54	13	113
7	0.0220	39	10	114
8	0.0249	55	14	242
9	0.0258	46	12	203
10	0.0298	83	20	256
11	0.0362	74	19	272
12	0.0370	70	17	131
13	0.0465	60	15	208
14	0.0465	60	15	208

Source: United Nations *Manual*, Process 24.

Table 16.10: Number of cattle in sample farms selected at random from sample communes, selected with probability proportional to the number of farms

Sample commune i	Total no. of farms M_i	No. of farm samples m_i	Number of cattle in sample farms $\sum_{j=1}^{m_i} y_{ij}$
1	13	3	30
2	15	3	58
3	19	5	14
4	28	7	73
5	39	10	162
6	41	11	88
7	46	12	102
8	46	12	102
9	48	12	203
10	51	13	134
11	59	14	195
12	74	19	272
13	83	20	242
14	83	20	242

Source: United Nations *Manual*, Process 23.

CHAPTER 17

Size of Sample and Allocation to Different Stages

17.1 Introduction

In a multi-stage design, in addition to the determination of the total sample size, there arises the problem of its allocation to the different stages: this requires a knowledge of the variability of the data and the unit costs at the different stages. These will be considered in this chapter, especially relating to a two-stage design; those for a three-stage design will also be indicated.

For a two-stage design, the problem and its treatment are similar to those for single-stage cluster sampling dealt with in Chapter 6 and section 7.6. As noted in section 6.1, single-stage cluster sampling may be considered to be a special case of two-stage sampling with all the second-stage units being surveyed in the selected first-stage units.

17.2 Two-stage design

17.2.1 Variance function

For simplicity, we shall consider a two-stage design with n (≥ 2) of fsu's being selected (with replacement) out of the N total fsu's with equal or varying probabilities, and a fixed number m_0 (≥ 2) of ssu's being selected (with replacement) out of the M_i total ssu's in the ith selected fsu with equal or varying probabilities.

We have seen in sections 15.2.3 (for a two-stage srs) and 16.2.1 (for a two-stage pps) that if y_i^* is an unbiased estimator from the ith sample fsu ($i = 1, 2, \ldots, n$) of the universe total Y, then the combined unbiased estimator of Y from all the n sample fsu's is

$$y_0^* = \frac{1}{n}\sum^{n} y_i^* \tag{17.1}$$

with an unbiased variance estimator

$$s_{y_0^*}^2 = \sum^{n}(y_i^* - y_0^*)^2/n(n-1) \tag{17.2}$$

It is also seen from equations (15.22) and (16.6) that the sampling variance of y_0^* in both srs and pps sampling (with replacement at both stages) is of the form

$$V = \frac{V_1}{n} + \frac{V_2}{nm_0} \tag{17.3}$$

The value of the study variable of the second-stage unit is considered to be the sum of two independent parts. One term, associated with the fsu's, has the same value for all the ssu's in an fsu, and varies from one fsu to another with variance

$$V_1 = N^2\sigma_b^2 = N^2 \frac{\sum^N (Y_i - \overline{Y})^2}{N} \quad \text{in srs} \tag{17.4}$$

$$= \sum^{N} \frac{Y_i^2}{\pi_i} - Y^2 \quad \text{in ppswr} \tag{17.5}$$

where

$$Y_i = \sum_{j=1}^{M_i} Y_{ij} \tag{17.6}$$

is the universe total for the ith fsu, and

$$\overline{Y} = Y/N = \sum^{N} Y_i/N \tag{17.7}$$

is the universe mean per fsu and σ_b^2 the variance between the fsu's.

The second term in equation (17.3), which serves to measure the differences between the ssu's, varies independently from one unit to another with variance

$$V_2 = \sum^{N} M_i^2 \sigma_{wi}^2$$

$$= N\sum^{N} M_i^2 \frac{\sum_{j=1}^{M_i}(Y_{ij} - \overline{Y}_i)^2}{M_i} \quad \text{in srs} \tag{17.8}$$

$$= \sum^{N} \frac{1}{\pi_i}\left(\sum_{j=1}^{M_i} \frac{Y_{ij}^2}{\pi_{ij}} - Y_i^2\right) \quad \text{in pps} \tag{17.9}$$

where σ_{wi}^2 is the variance between the ssu's within the ith fsu, and $\overline{Y}_i$ the mean for the ith fsu.

Thus, the sample as a whole consists of n independent values of the first term, and nm_0 independent values of the second term.

Note: The variance estimator of y_0^* in estimating equation (17.2) does not explicitly take account of the stage-variances; the estimation of the latter is shown in section 17.2.4.

17.2.2 Cost function

We consider a simple cost function

$$C = nc_1 + nm_0c_2 \tag{17.10}$$

where c_1 is the average cost per fsu and c_2 that per ssu. The overhead cost, not shown in the above cost function, may later be added to C. In general, the cost per fsu (c_1) will be greater than that per ssu (c_2).

17.2.3 Optimum size and allocation

With the above cost and variance functions, the optimum values of n and m_0 are those that minimize the variance V for a given total cost C or *vice versa*, and are given by

$$m_0 = \frac{\sqrt{(V_2c_1)}}{\sqrt{(V_1c_2)}} \tag{17.11}$$

and

$$n = \frac{C\sqrt{(V_1/c_1)}}{\sqrt{(V_1c_1)} + \sqrt{(V_2c_2)}} \tag{17.12}$$

if the cost is preassigned (from equation (17.10)); or

$$n = \frac{\sqrt{(V_1c_1)} + \sqrt{(V_2c_2)}}{V\sqrt{(c_1/V_1)}} \tag{17.13}$$

if the variance is preassigned (from equation (17.3)).

The optimum number of ssu's can also be expressed in terms of the intra-class correlation coefficient ρ between the second-stage units

$$m_0 = \sqrt{\left[\frac{c_1}{c_2}\,\frac{1-\rho}{\rho}\right]} \tag{17.14}$$

Note that this is of the same form as that of (7.37) for the optimal cluster size.

Equations (17.11) and (17.14) show that a larger sub-sample should be taken when sub-sampling is relatively inexpensive (c_1 large in relation to c_2) and the variability between the ssu's within the fsu's (V_2) is larger than that between the fsu's (V_1) i.e. the fsu's are internally heterogeneous.

These equations often lead to the same value of m_0 for a wide range of ratios c_1/c_2. Thus a choice of the sub-sampling fraction can often be made even when information on relative costs is not too definite.

17.2.4 Estimation of stage-variances

We have seen from the fundamental theorems of section 14.4 that if sampling is with replacement at least at the first stage, then the sampling variance is estimated properly from the estimators obtained from the first-stage units by estimating equation (14.2); the variability of the data at the different stages does not enter explicitly in this equation. In estimating the stage-variances, to avoid the consistent appearance of negative variances (as it would happen with the "with replacement" sampling when the variability between the fsu's is smaller than that between the ssu's), we assume that sampling is without replacement at both stages; for simplicity, we also assume that sampling is simple random, m_i ssu's being selected in the ith selected fsu's. The unbiased estimator y_0^* (estimating equation (17.1)) of the universe total Y remains the same, but its sampling variance is given by

$$V(y_0^*) = N^2\overline{M}^2(1-f_1)\,\frac{S_b^2}{n} + \frac{N}{n}\sum^{n} M_i^2(1-f_{2i})\,\frac{S_{wi}^2}{m_i} \qquad (17.15)$$

where $\overline{M}$ is the average number of ssu's per fsu; $f_1 = n/N$; $f_{2i} = m_i/M_i$; S_b^2 is the variance between the fsu's and S_{wi}^2 that between the ssu's within the ith fsu:

$$S_b^2 = \frac{1}{N-1}\sum^{N}\left(\frac{M_i}{\overline{M}}\,\overline{Y}_i - \overline{Y}\right)^2 \qquad (17.16)$$

$$S_{wi}^2 = \frac{1}{M_i-1}\sum_{j=1}^{M_i}(Y_{ij} - \overline{Y}_i)^2 \qquad (17.17)$$

Y_{ij} is the value of study variable of the ijth universe unit ($i = 1, 2, \ldots, N$; $j = 1, 2, \ldots, M_i$):

$$\overline{Y}_i = \sum_{j=1}^{M_i} Y_{ij}/M_i; \qquad \overline{Y} = Y/N\overline{M} = \sum^{N} M_i\overline{Y}_i/N\overline{M}$$

An unbiased estimator of S_{wi}^2 is

$$s_{wi}^2 = \frac{1}{m_i - 1} \sum_{j=1}^{m_i} (y_{ij} - \overline{y}_i)^2 \tag{17.18}$$

where y_{ij} is the value of the study variable of the ijth sample unit ($i = 1, 2, \ldots, n$; $j = 1, 2, \ldots, m_i$) and $\overline{y}_i = \sum_{j=1}^{m_i} y_{ij}/m_i$.

Denoting by

$$s_b^2 = \frac{1}{n-1} \sum^n \left(\frac{M_i}{\overline{M}}\, \overline{y}_i - \overline{\overline{y}} \right)^2 \tag{17.19}$$

where $\overline{\overline{y}} = y_0^*/N\overline{M}$, an unbiased variance estimator of S_b^2 is

$$s_b^2 - \frac{1}{n} \sum^n \left(\frac{M_i}{\overline{M}} \right)^2 (1 - f_{2i})\, \frac{s_{wi}^2}{m_i} \tag{17.20}$$

An unbiased estimator of $V(y_0^*)$ in (17.15) is

$$N^2 \overline{M}^2 (1 - f_1)\, \frac{s_b^2}{n} + \frac{N}{n} \sum^n M_i^2 (1 - f_{2i})\, \frac{s_{wi}^2}{m_i} \tag{17.21}$$

$$= \frac{(1 - f_1)}{n(n-1)} \sum^n (y_i^* - y_0^*)^2 + \frac{N}{n} \sum^N M_i^2 (1 - f_{2i})\, \frac{s_{wi}^2}{m_i} \tag{17.22}$$

where $y_i^* = N M_i \overline{y}_i$ is the unbiased estimator of Y from the ith sample fsu (estimating equation (15.12)).

Notes

1. For the estimator of the mean, the stage-variances are obtained by dividing the components in equations (17.15), (17.21), and (17.22) by $N^2\overline{M}^2$.
2. When $M_i = M_0$, a constant, and $m_i = m_0$ is fixed, the unbiased variance estimator (equation (17.21)) for the total becomes

$$N^2 M_0^2 (1 - f_1)\, \frac{s_b^2}{n} + \frac{N}{n}\, \frac{M_0^2}{m_0}\, (1 - f_2) s_w^2 \tag{17.23}$$

where $f_2 = m_0/M_0$; and

$$s_w^2 = \sum^n s_{wi}^2 \tag{17.24}$$

For the variance estimator of the mean, equation (17.23) divided by $N^2 M_0^2$ becomes

$$(1 - f_1)\, \frac{s_b^2}{n} + \frac{1}{N} (1 - f_2)\, \frac{s_w^2}{n m_0} \tag{17.25}$$

3. When $M_i = M_0$, and $m_i = m_0$, unbiased estimators of the stage-variances can be obtained readily from an analysis of variance of the sample observations, assuming sampling from infinite universes at both stages. In that case, the universe variance of y_0^* is of the form (17.3) and is estimated unbiasedly by

$$N^2 M_0^2 \text{ (mean square between fsu's)}/nm_0 = N^2 M_0^2 s_b^2/n \tag{17.26}$$

where

$$\text{mean square between fsu's } = \frac{1}{n(n-1)} \sum^{n} (\bar{y}_i - \bar{\bar{y}})^2;$$

$$\bar{y}_i = \sum_{j=1}^{m_0} y_{ij}/m_0; \text{ and } \bar{\bar{y}} = \sum^{n} \bar{y}_i/n$$

Unbiased estimators of V_1 and V_2 in equation (17.3) are then respectively

$$\hat{V}_1 = N^2 M_0^2 \text{ (mean square between fsu's} - \text{mean square within fsu's)}/m_0 \tag{17.27}$$

$$\hat{V}_2 = N^2 M_0^2 \text{ (mean square within fsu's)} \tag{17.28}$$

where

$$\text{mean square within fsu's } = \frac{1}{n(m_0-1)} \sum^{n} \sum^{m_0} (y_{ij} - \bar{y}_i)^2$$

For the mean, its sampling variance is also of the form (17.3), and is estimated unbiasedly by

$$\text{(mean square between fsu's)}/nm_0 \tag{17.29}$$

and V_1 and V_2 are estimated unbiasedly by

$$\hat{V}_1 = \text{(mean square between fsu's} - \text{mean square within fsu's)}/m_0 \tag{17.30}$$

$$\hat{V}_2 = \text{mean square within fsu's} \tag{17.31}$$

4. When sampling is simple random with replacement at the first stage and simple random without replacement at the second stage, the sampling variance of y_0^*, the (combined) unbiased estimator of Y, is

$$\sigma_{y_0^*}^2 = N^2 \overline{M}^2 \frac{S_b^2}{n} + \frac{N}{n} \sum^{N} M_i^2 (1 - f_{2i}) \frac{S_{wi}^2}{m_i} \tag{17.32}$$

S_b^2 and S_{wi}^2 being defined by equation (17.16) and (17.17). $\sigma_{y_0^*}^2$ is estimated unbiasedly by $s_{y_0^*}^2$, defined in equation (17.2).

Table 17.1: Analysis of variance

Source of variation	Degrees of freedom	Mean square
Between first-stage units	39	634
Between second-stage units within first-stage units	160	409

Source: Hendricks, pp. 190-191.

Example 17.1

Table 17.1 gives the analysis of variance of a sample of $n = 40$ from $N = 1000$ fsu's and $m_0 = 5$ sample ssu's in each selected fsu ($M_0 = 50$), sampling at both stages being simple random.

(a) Assuming that sampling has been from infinite universes at both the sages, estimate unbiasedly the variance of the mean and its components (Hendricks, pp. 190-191).

(b) Given the total cost (excluding overheads) of a survey, $C = \$6,000$, the cost of including an fsu in the sample, $c_1 = \$50$; and the cost of getting the data from an ssu within each selected fsu, $c_2 = \$10$, obtain the optimum sample numbers of fsu's and ssu's within each selected fsu, and the expected variance of the mean.

(a) From equation (17.29), the unbiased variance estimate of the mean is

$$(\text{mean squre between fsu's})/nm_0 = 634/200 = 3.1700$$

From equations (17.30) and (17.31),

$$\begin{aligned} \hat{V}_1 &= (\text{mean square between fsu's - mean square within fsu's})/m_0 \\ &= (634 - 409)/5 = 45, \text{ and} \end{aligned}$$

$$\hat{V}_2 = \text{mean square within fsu's } = 409,$$

so that the components of the variance of the mean are (from equation (17.3))

$$\hat{V}_1/n = 45/40 = 1.1250, \text{ and } \hat{V}_2/nm_0 = 409/200 = 2.0450,$$

totaling 3.1700, as before.

(b) From equation (17.11) the optimum number of ssu's per sample fsu is

$$m_0 = \frac{\sqrt{(V_2 c_1)}}{\sqrt{(V_1 c_2)}} = \frac{\sqrt{(409 \times 50)}}{\sqrt{(45 \times 10)}} = \sqrt{(45.44)} = 6.7$$

or, to the nearest whole number, 7.

From equation (17.10),

$$n = C/(c_1 + m_0 c_2) = 6000/(50 + 70 \times 1) = 50$$

so that the total sample size is $nm_0 = 50 \times 7 = 350$ ssu's.

The estimated (expected) variance of the mean with this sample is

$$\hat{V} = \frac{\hat{V}_1}{n} + \frac{\hat{V}_2}{nm_0} = \frac{45}{50} + \frac{409}{350} = 0.9 + 1.1686 = 2.0686$$

Example 17.2

For Example 15.2 relating to a two-stage srs, and assuming sampling without replacement, estimate the variance of the estimated total number of persons in the form of expression (17.22). Also obtain an unbiased estimate of S_b^2.

The required computations are shown in summary form in Table 17.2. Taking Expression (17.22), the first term is, from the data of Example 15.2.

$$\frac{(1 - f_1)}{n(n-1)} \sum^{n} (y_i^* - y_0^*)^2 = \frac{(1 - 4/30)}{4 \times 3} \times 625,050 = 45,142.67,$$

and the second term is

$$\frac{N}{n} \sum^{n} M_i^2 (1 - f_{2i}) \frac{s_{wi}^2}{m_i} = \frac{30}{4} \times 790.67 = 5930.02,$$

so that the estimated variance of the estimated total number of persons (y_0^*) is $45,142.67 + 5930.02 = 51,072.69$. The difference between the fsu's (villages) accounts for $45,142.67/51,072.69 = 88$ per cent of this variance and the variability within the fsu's between the ssu's (households) for 12 per cent.

The estimated standard error is 226.0, and the estimated CV is 6.94 per cent.

Note that the variance estimated on the basis of a "without replacement" scheme is, as known, less than that for a "with replacement" scheme in Example 15.2, namely,

$$s_{y_0^*}^2 = \sum^{n} (y_i^* - y_0^*)^2 / n(n-1) = 625,050/12 = 52,087.5$$

the estimated standard error was 228.2, and the estimated CV 7.01 per cent.

Table 17.2: Computation of the stage-variances of the estimated total number of persons in Example 15.2, assuming sampling without replacement

Sample village (fsu) i	SSy_{ij}	$s_{wi}^2 = \frac{SSy_{ij}}{(m_i-1)}$	$f_{2i} = m_i/M_i$	$(1-f_{2i}) \times s_{wi}^2/m_i$	M_i^2	$M_i^2(1-f_{2i}) \times s_{wi}^2/m_i$	$M_i/\overline{M}$†	$(M_i/\overline{M})^2$	$(M_i/\overline{M})^2 \times (1-f_{2i})s_{wi}^2/m_i$
(1)	(2)	(3)	(4)	(5)	(6)	(7)	(8)	(9)	(10)
1	14.75	4.916667	0.2000	0.983333	400	393.33	0.9302	0.865272	0.850851
2	5.00	1.666667	0.1739	0.344209	529	182.09	1.0698	1.144472	0.393938
3	3.00	1.000000	0.1600	0.210000	625	131.25	1.1628	1.352104	0.283942
4	4.00	1.333333	0.2222	0.259266	324	84.00	0.8372	0.700904	0.181721
					Total	790.67		1.710452	

† $\overline{M} = \frac{1}{4}(20 + 23 + 25 + 18) = 21.5$ households per village.

From the first terms of (17.21) and (17.22), we find

$$s_b^2 = \frac{\sum^n (y_i^* - y_0^*)^2}{N^2\overline{M}^2(n-1)} = \frac{625,050}{416,025 \times 3} = 0.500811$$

so that an unbiased estimate of S_b^2 is (from equation (17.20))

$$s_b^2 - \frac{1}{n}\sum^n (M_i^2/\overline{M})^2(1 - f_{2i})s_{wi}^2/m_i = 0.500811 - 0.427613 = 0.073198$$

17.3 Three-stage design

The above procedures can be extended to sample designs with three and higher stages. Those for a three-stage design are given below: the procedures for higher-stage designs follows from symmetry:

Assume that the sample at the three stages is drawn with varying probabilities and with replacement, the respective numbers being n, m_0 and q_0. Then the universe variance of the estimator y_0^* (estimating equation (16.13)) is

$$\begin{aligned} V &= \frac{1}{n}\left(\sum_{i=1}^{N}\frac{Y^2}{\pi_i} - Y^2\right) + \frac{1}{nm_0}\sum^{N}\frac{1}{\pi_i}\left(\sum_{j=1}^{M_i}\frac{Y_{ij}^2}{\pi_{ij}} - Y_i^2\right) \\ &+ \frac{1}{nm_0q_0}\sum^{N}\frac{1}{\pi_i}\sum_{j=1}^{M_i}\frac{1}{\pi_{ij}}\left(\sum_{k=1}^{Q_{ij}}\frac{Y_{ijk}^2}{\pi_{ijk}} - Y_{ij}^2\right) \end{aligned} \tag{17.33}$$

where $Y_{ij} = \sum^{Q_{ij}} Y_{ijk}$; $Y_i = \sum^{M_i} Y_{ij}$; and $Y = \sum^N Y_i$ is the universe total. This is of the form

$$V = \frac{V_1}{n} + \frac{V_2}{nm_0} + \frac{V_3}{nm_0q_0} \tag{17.34}$$

where V_1, V_2 and V_3 are respectively the variations (1) between the fsu's, (2) between the ssu's within the fsu's, and (3) between the third-stage units within the ssu's, for an estimator based on any of the nm_0q_0 possible third-stage sample units.

Assuming the simple cost function (neglecting the overhead costs)

$$C = nc_1 + nm_0c_2 + nm_0q_0c_3 \tag{17.35}$$

where c_1 is the average cost per fsu, c_2 that per ssu, and c_3 that per third-stage unit, the optimum values of m_0 and q_0 are obtained as

$$m_0 = \frac{\sqrt{(V_2 c_1)}}{\sqrt{(V_1 c_2)}} \qquad q_0 = \frac{\sqrt{(V_3 c_2)}}{\sqrt{(V_2 c_3)}} \tag{17.36}$$

If the cost is preassigned, the optimum value of n is

$$n = \frac{C\sqrt{(V_1/c_1)}}{\sqrt{(V_1 c_1)} + \sqrt{(V_2 c_2)} + \sqrt{(V_3 c_3)}} \tag{17.37}$$

If the variance is preassigned, the optimum value of n is

$$n = \frac{\sqrt{(V_1 c_1)} + \sqrt{(V_2 c_2)} + \sqrt{(V_3 c_3)}}{V\sqrt{(c_1/V_1)}} \tag{17.38}$$

Further reading

Cochran, sections 10.6, 10.7, 11.13, and 11.14; Deming (1950), chapter 5A; Hansen *et al.* (1953), Vols. I and II, chapter 6; Hedayat and Sinha, sections 7.5 and 7.6; Hendricks, chapter 7; Kish (1965), section 5.6; Murthy (1973), sections 9.3-9.10; Singh and Chaudhary, sections 9.3, 9.4, and 9.7; Sukhatme *et al.* (1984), sections 8.3, 8.7, 9.3, and 9.4; Yates, sections 7.17, 8.10.1, and 8.11-8.21.

Exercises

1. In a sample survey for the study of the economic conditions of the population, it is proposed to take a two-stage sample with villages as the first-stage units and m_0 households per village as second-stage units, sampling at both the stage units to be with equal probability and with replacement. The variance of an estimate is $V = \sigma_b^2/n + \sigma_w^2/nm_0$, where $\sigma_b^2 = 5.2$ and $\sigma_w^2 = 11.2$. The cost of the survey is given by $C = c_0 + nc_1 + nm_0c_2$, with $c_0 = Rs\,600$, $c_1 = Rs\,160$, and $c_2 = Rs\,4$. Determine the optimum values of n and m_0 for $C = Rs\,10.000$ (Chakravarti *et al.*, Illustrative Example 4.2).
2. For Example 15.2, obtain estimates of the stage-variances for the total daily income, assuming sampling without replacement.
3. To develop the sampling technique for the determination of the sugar percentage in field experiments on sugar beets, ten beets were chosen from each of 100 plots in a uniformity trial, the plots being the first-stage units. The sugar percentage was obtained separately for each beet. Table 17.3 gives the analysis of variance between plots and between beets within plots.

Table 17.3: Analysis of variance of sugar percentage of beets (on a single-beet basis)

Source of variation	Degrees of freedom	Mean square
Between plots (first-stage units)	80	2.9254
Between beets (second-stage units) within plots	900	2.1374

Source: Snedecor and Cochran, pp. 529-531.

(a) Estimate directly the estimated variance of the mean, and also separately its components.

(b) How reliable are the treatment means with 6 replications and 5 beets per plot? (Snedecor and Cochran, pp. 529-531).

CHAPTER 18

Self-weighting Designs in Multi-stage Sampling

18.1 Introduction

In this chapter we will discuss the problem of how to make a multi-stage design self-weighting. The general case will first be covered, followed by the specific cases of two-stage designs, and sampling for opinion surveys and marketing research along with the situation where it is desirable to select a fixed number of ultimate stage units in each of the selected penultimate stage units.

18.2 General case

Taking the multi-stage design of section 14.4, an unbiased estimator of the universe total Y of a study variable is

$$y = \sum_{\text{sample}} \left(\prod_{1}^{u} \frac{1}{f_t} \right) y_{12\ldots u} = \sum_{\text{sample}} w_{12\ldots u} y_{12\ldots u} \tag{18.1}$$

(see also equation (14.13)) where $y_{12\ldots u}$ is the value of the study variable in the $(12\ldots u)$th ultimate stage sample unit, the factor

$$\prod_{1}^{u} \frac{1}{f_t} \equiv w_{12\ldots u} \tag{18.2}$$

(see also (14.14)) is the multiplier (or the weighting-factor) for the $(12\ldots u)$th ultimate-stage unit; and $f_t \equiv n_{12\ldots(t-1)}\pi_{12\ldots t}$, there being $n_{12\ldots(t-1)}$ sample units each selected with probability $\pi_{12\ldots t}$ out of the total $N_{12\ldots(t-1)}$ units at the tth stage.

The design will be self-weighting with respect to y when the multiplier $w_{12...u}$ defined above is a pre-determined constant, w_0, for all the ultimate-stage sample units. It will be so if for the ultimate stage sampling, the factor f_u could be made proportional to the product of the reciprocals of the corresponding factors for the preceding stages, i.e.

$$f_u \propto \prod_1^{u-1} \frac{1}{f_t} \tag{18.3}$$

For example, in a two-stage srs, where $f_1 = n/N$ and $f_{2i} = m_i/M_i$, the multiplier for the sample ssu's is

$$w_{ij} = \frac{1}{f_1 \cdot f_{2i}} = \frac{N}{n} \cdot \frac{M_i}{m_i}$$

and this will be a constant w_0, when

$$m_i = \frac{NM_i}{w_0 n}$$

In this case, the sampling fraction at the second stage should be the same ($= N/w_0 n$) in all the n sample fsu's.

In general, a multi-stage srs will be self-weighting if, in each stage, the sampling fraction is a constant for selecting the next-stage sample units. That is $n_{12...t}/N_{12...t}$ should be a constant; e.g. the second-stage sampling fraction n_{12}/N_{12} should be a constant, the third-stage sampling fraction n_{123}/N_{123} another constant, and so on.

We shall further consider the requirement of having a fixed number, say n_0, of the ultimate stage sample units in each of the selected penultimate stage units. If these n_0 units are selected with equal probability, then

$$\pi_{12...(u-1)} \propto N_{12...(u-1)} \tag{18.4}$$

i.e. the penultimate stage units should be selected with probability proportional to the total number of ultimate stage units contained in each, and the sample number of penultimate stage units is given by

$$n_{12...(u-2)} = \frac{N_{12...(u-2)}}{w_0 n_0} \left(\prod_1^{u-2} \frac{1}{f_t} \right) \tag{18.5}$$

If the n_0 units are selected with probability proportional to "size", i.e.

$$\pi_{12...u} = \frac{z_{12...u}}{z_{12...(u-1)}} \tag{18.6}$$

where the numerator refers to the "size" of the particular units to be selected and the denominator to the total size of all the units, and the design will be self-weighting when:

1. The variate value is the ratio of the characteristic under study $y_{12...u}$ to the size in the ultimate stage units $z_{12...u}$; and

2. the penultimate stage units are selected with probability proportional to total "size", and the number of penultimate stage units is given by

$$\pi_{12...(u-2)} = \frac{1}{w_0 n_0} \left(\prod_1^{u-2} \frac{1}{f_t} \right) z_{12...u} \tag{18.7}$$

Notes

1. We have already seen in section 16.5 how two- and three-stage pps designs for crop surveys can be made self-weighting by using the same size variable, such as area, in all stages of sample selection.
2. For fractional values of the sample sizes at different stage, see section 13.4.
3. For making a multi-stage design self-weighting at the tabulating stage, the same principles as for a stratified sample, given in section 13.5, will apply, namely, selecting a sub-sample with probability proportional to the multipliers.

18.3 Two-stage sample design with pps and srs

Let us consider the sampling plan of section 16.3, with n fsu's selected with pps (π_i) out of a total N, and m_i ssu's selected with equal probability out of a total M_i units in each selected fsu. Here $f_1 = n\pi_i$ and $f_2 = m_i/M_i$. The multiplier for the ijth sample ssu ($i = 1, 2, \ldots, n$; $j = 1, 2, \ldots, m_i$) is

$$w_{ij} = \frac{1}{f_1 f_2} = \frac{M_i}{n\pi_i m_i} \tag{18.8}$$

and the design will be self-weighting with a constant multiplier w_0 when

$$m_i = M_i/(w_0 n \pi_i) \tag{18.9}$$

Suppose the above sample design refers to a household inquiry with villages as the fsu's and households the ssu's. The difficulty is that the ratios M_i/π_i will vary from village to village and so will the sample number of households m_i, given by equation (18.9) required for making the design self-weighting; i.e. the work load per enumerator in each of the sample villages will vary greatly: this is apart from the fact that the total number of households (M_i) in the ith village will not be known beforehand, unless the inquiry is conducted simultaneously with a census. The problem is to make the work load approximately equal, retaining a self-weighting design.

If a fixed number of m_0 of households is to be sampled in each village, then with a self-weighting design

$$w_0 = M_i/(n\pi_i m_0) \tag{18.10}$$

or

$$\pi_i = M_i/(w_0 n m_0) = M_i/M \tag{18.11}$$

where $M = \sum^N M_i$ is the total number of households in the universe, i.e. the villages are to be selected with probability proportional to the number of households. It can be seen that

$$\begin{aligned} w_0 &= M/(nm_0) \\ &= \text{total number of households/number of sample households} \\ &= 1/\text{sampling fraction} \end{aligned} \tag{18.12}$$

In practice, however, the number of households (M) in the universe is not known and the number of households existing at the time of enumeration in the ith sample village (M_i) will be known only after the listing of households when the village is reached and the households listed. In such a situation, the values which are known from the latest census may be used for selecting the sample villages i.e. $\pi_i = M_i'/M'$, where M_i' is the number of census households in the ith village and $M' = \sum^N M_i'$ the total number of census households in the universe. The overall constant multiplier can be fixed on the basis of the expected number of households (M'') in the universe and the total sample size (nm_0); the value of M'' can, for example, be obtained by projecting the number of total households in the censuses. By using these values it is seen that

$$\begin{aligned} w_0 &= \frac{\text{estimated total number of households}}{\text{number of sample households}} \\ &= M''/(nm_0) = M'M_i/nM_i'm_i \end{aligned} \tag{18.13}$$

so that

$$m_i = m_0 M' M_i / M'' M_i' = m_0 I_i / I \tag{18.14}$$

where

$I = M''/M' =$ estimated increase in the total number of households,

$I_i = M_i/M_i' =$ actual increase in the number of households in the ith village from the latest census.

This procedure is expected to restrict the variation of the number of sample households in the ith village within a short range. The operational steps will be:

1. Fix n and m_0 from cost and error considerations.

2. Find the overall sampling fraction which is the ratio of the number of households to be sampled to the expected total number of households in the universe. The reciprocal of this will be the constant multiplier w_0.

3. Select villages with probability proportional to the number of households as in the previous census.

4. Compute the estimated increase in the number of households in the universe from the census, $I = M''/M'$; and compute $Mm'_0 = m_0/I$.

5. Give to the enumerators the values m'_0 and M'_i (the number of households in the village at the time of the census) with instruction to select at random (or systematically) $m_i = m'_0 M''_i/M'_i$ households out of the M_i households listed in the ith village.

Note: The expected number of sample households $= nm_0 M/M''$. For fractional m_i, see section 13.4.

18.4 Opinion and marketing research

In opinion and market research, to obtain a valid, independent response it is important to interview only one individual in a selected household so that the response of other household members, which are likely to be conditioned, are not included in the survey. Even if the "true" response of all the household members could be obtained, a situation not usually realized in practice, the sample may be more profitably spread over a greater number of households, because of the homogeneity of responses in a household often observed in such surveys. It is easy to see that with the selection of one individual from a household, the sampling design can be made self-weighting if the households are selected with probability proportional to the respective sizes.

To illustrate, consider a three-stage sample design, as in Examples 15.3 and 16.2, for the type of opinion and marketing surveys usually conducted in cities and towns with small first-stage units - city or town-blocks. Suppose that n sample blocks are selected with probability π_i out of N blocks; in the ith sample block ($i = 1, 2, \ldots, n$), m_i sample households are selected with probability π_{ij} ($j = 1, 2, \ldots, m_i$) out of the total M_i households, and finally, one adult is selected with equal probability out of the total Q_{ij} adults in the ijth sample household: sampling is with replacement at the first two stages. Here $f_1 = n\pi_i$; $f_2 = m_i\pi_{ij}$; and $f_3 = 1/Q_{ij}$; so that the multiplier for the selected adult in the ijth sample household is

$$1/f_1 f_2 f_3 = Q_{ij}/(n\pi_i m_i \pi_{ij}) \tag{18.15}$$

If the sample households are selected with probability proportional to the number of adults, i.e. if $\pi_{ij} = Q_{ij}/Q_i$, where $Q_i = \sum^{M_i} Q_{ij}$ is the total number of adults in the ith block, the multiplier becomes $Q_i/n\pi_i m_i$. If this is to be a constant, w_0, then

$$m_i = Q_i/(w_0 n\pi_i) \tag{18.16}$$

In practice, the factors $1/(w_0 n\pi_i)$ will be supplied to the enumerators; on reaching any block, the enumerators will list the households and the adults in these households. The total number of adults (Q_i) in the ith block, when multiplied by the factors $1/(w_0 n\pi_i)$, will determine the sample number of households, which is also the sample number of adults, in the block.

The blocks may be selected with probability proportional to size (the most recent census population, for example), or with equal probability. In the latter case, the factor $1/(w_0 n\pi_i) = N/(w_0 n)$ will be a constant for all the blocks.

Further reading

Hansen *et al.* (1953), Vol. I, sections 7.12 and 9.11; Murthy (1967), section 12.3; Singh and Chaudhary, section 9.9; Som (1958-59) and (1959); Sukhatme *et al.* (1984), section 9.8.

PART IV

STRATIFIED MULTI-STAGE SAMPLING

CHAPTER 19

Stratified Multi-stage Sampling: Introduction

19.1 Introduction

In this part, we deal with stratified multi-stage sample designs - a synthesis of what has gone earlier. In such a design, the universe is divided into a number of sub-universes (strata) and sampling is conducted in stages separately within each stratum.

The general theorems relating to stratified multi-stage sampling will be considered in this chapter; subsequent chapters will deal with different types and aspects of stratified multi-stage sampling: simple random sample; varying probability sample; size of sample and allocation to different strata and stages; and self-weighting designs.

19.2 Reasons for stratified multi-stage sampling

Stratified multi-stage designs are the commonest of all types of sample designs. This is because they combine, as regards costs and efficiency, the advantages of both stratification and multi-stage sampling.

In stratified sampling, ideally the strata should be formed so as to be internally homogeneous and heterogeneous with respect to one another (with respect to the values of study variables in srs or with respect to the ratios of the values of study to the size-variable in pps sampling): this, as has been observed before, results in increased efficiency. On the other hand, in a multi-stage design the first-stage units should be internally heterogeneous and homogeneous with respect to one another. In a stratified multi-stage design, all the strata have of course to be covered, but when the strata are made internally homogeneous and the first-stage units internally homogeneous, a small number of first-stage units need be selected from each stratum in order to provide an efficient sample.

Almost all nation-wide sample surveys, including those mentioned in section 1.1, use stratified multi-stage designs. For example, for the National Market Research Survey in Great Britain (1952), where the universe was the adult civilian population aged sixteen and over, the sample was stratified by factors such as geographical region, urban/rural areas, industrialization index and zoning, and was spread over four stages: administrative districts, polling districts, households, and individuals (Moser and Kalton, Chapter 8). In the monthly Current Population Survey in the U.S., the sampling design is stratified three-stage; heterogeneous first-stage units are defined, comprising individual counties (or sometimes two or more adjoining counties), and grouped into a number of strata; the first-stage units are sampled with varying probabilities; relatively small areal units of about two hundred households are the second-stage units; and the third-stage units are clusters of approximately six households, called "segments": the sample contains 449 first-stage units, about 900 segments and about 50,000 households (Hansen *et al.* (1953), Vol. I, Chapter 12; Hansen and Tepping, 1969). In the Indian National Sample Survey (1964-5), in the rural sector there were 353 strata, comprising administrative units, and each covering an average of 1.2 million persons; for the socioeconomic inquiries, villages and households were the two stages of sampling, and for crop yield surveys, the four stages of sampling were respectively villages, clusters of plots, crop-plots, and circular cuts of radius $2'\ 3''$ or $4'$ (see Appendix V). In the Sample Survey on Goods Traffic Movement in Zambia (1967), the design was stratified two-stage, with proprietors of transport vehicles constituting the first-stage and the vehicles the second-stage units.

Furthermore, almost all multi-country survey programs, including those considered in Appendix VI, adopt stratified multi-stage designs.

19.3 Fundamental theorems in stratified multi-stage sampling

The fundamental theorems for the estimation of universe parameters in stratified multi-stage sampling follow from those for a stratified single-stage sample (section 9.3) and for an unstratified multi-stage sample (section 14.4) taken in combination.

1. The universe is sub-divided into L mutually exclusive and exhaustive strata. In the hth stratum ($h = 1, 2, \ldots, L$), of the total N_h first-stage units, n_h (≥ 2) are selected with replacement (and with equal or varying probabilities); if t_{hi} ($h = 1, 2, \ldots, L$; $i = 1, 2, \ldots, n_h$) are the n_h independent and unbiased estimators of the universe parameter T_h, obtained from the ith selected first-stage unit in the hth stratum,

then the combined unbiased estimator of T_h is the arithmetic mean

$$\bar{t}_h = \sum^{n_h} t_{hi}/n_h \tag{19.1}$$

An unbiased estimator of the variance of $\bar{t}_h$ is

$$s^2_{\bar{t}_h} = \sum^{n_h}(t_{hi} - \bar{t}_h)^2/n_h(n_h - 1) = SSt_{hi}/n_h(n_h - 1) \tag{19.2}$$

An unbiased estimator of the universe parameter for all the strata combined

$$T = \sum_{h=1}^{L} T_h \tag{19.3}$$

is obtained on summing up the L unbiased stratum estimators $\bar{t}_h$ (given in equation (19.1)), namely

$$t = \sum^{L} \bar{t}_h \tag{19.4}$$

An unbiased estimator of the variance of t is obtained on summing the L unbiased estimators of the stratum variances $s^2_{\bar{t}_h}$ (sampling in one stratum is independent of sampling in another, so that the stratum estimators $\bar{t}_h$ are mutually independent),

$$s^2_t = \sum^{L} s^2_{\bar{t}_h} \tag{19.5}$$

We shall see later in this section how to compute the t_{hi} values for a stratified multi-stage design.

For another study variable, similarly defined, an unbiased estimator of the universe parameter

$$U = \sum^{L} U_h \tag{19.6}$$

is the sum of the L unbiased stratum estimators $\bar{u}_h$ of U_h

$$u = \sum^{L} \bar{u}_h \tag{19.7}$$

and an unbiased estimator of the variance of u is the sum of L unbiased estimators of the stratum variances $s^2_{\bar{u}_h}$,

$$s^2_u = \sum^{L} s^2_{\bar{u}_h} \tag{19.8}$$

2. An unbiased estimator of the covariance of $\bar{t}_h$ and $\bar{u}_h$ for the hth stratum is

$$\begin{aligned} s_{\bar{t}_h\bar{u}_h} &= \sum^{n_h}(t_{hi}-\bar{t}_h)(u_{hi}-\bar{u}_h)/n_h(n_h-1) \\ &= SPt_{hi}u_{hi}/n_h(n_h-1) \end{aligned} \qquad (19.9)$$

For all the strata combined, an unbiased estimator of the covariance between t and u is the sum of the stratum estimators $s_{\bar{t}_h\bar{u}_h}$, i.e.

$$s_{tu} = \sum^{L} s_{\bar{t}_h\bar{u}_h} \qquad (19.10)$$

3. A consistent but generally biased estimator of the ratio of two universe parameters in the hth stratum $R_h = T_h/U_h$ is the ratio of the sample estimators $\bar{t}_h$ and $\bar{u}_h$ in the stratum,

$$r_h = \bar{t}_h/\bar{u}_h \qquad (19.11)$$

with a variance estimator

$$s^2_{r_h} = (s^2_{\bar{t}_h} + r^2_h s^2_{\bar{u}_h} - 2r_h s_{\bar{t}_h\bar{u}_h})/\bar{u}^2_h \qquad (19.12)$$

For all the strata combined, a consistent but generally biased estimator of the ratio of two universe parameters $R = T/U$ is the ratio of the respective sample estimators t and u,

$$r = t/u \qquad (19.13)$$

with a variance estimator

$$s^2_r = (s^2_t + r^2 s^2_u - 2rs_{tu})/u^2 \qquad (19.14)$$

4. A consistent but generally biased estimator of the correlation coefficient ρ_h between the two study variables for the hth stratum is

$$\hat{\rho}_h = s_{\bar{t}_h\bar{u}_h}/s_{\bar{t}_h}s_{\bar{u}_h} \qquad (19.15)$$

For all the strata combined, an estimator of the correlation coefficient ρ is

$$\hat{\rho} = s_{tu}/s_t s_u \qquad (19.16)$$

5. ***Unbiased estimators of stratum universe totals obtained from the sample units.*** Here we change the notations and also deal with the totals of a study variable (see section 14.4.5).

In a stratified multi-stage design with L strata and u stages ($u > 1$), in the hth stratum ($h = 1, 2, \ldots, L$), and tth stage ($t = 1, 2, \ldots, u$), let $n_{n12\ldots(u-1)}$ sample units be selected with replacement out of the total $N_{12\ldots(t-1)}$ units and let the (initial) probability of selection of the $(h12\ldots t)$th unit be $\pi_{h12\ldots t}$. We define

$$f_{ht} \equiv n_{h12\ldots(t-1)}\ \pi_{h12\ldots t} \tag{19.17}$$

An unbiased estimator of the universe total Y_h of the study variable in the hth stratum is

$$\begin{aligned} y_h &= \sum_{\substack{\text{stratum}\\\text{sample}}} \left(\prod_{t=1}^{u} \frac{1}{f_{ht}}\right) y_{h12\ldots u} \\ &= \sum_{\substack{\text{stratum}\\\text{sample}}} w_{h12\ldots u} y_{12\ldots u} \end{aligned} \tag{19.18}$$

where $y_{h12\ldots u}$ is the value of the study variable in the $(h12\ldots u)$th ultimate-stage sample units and the factor

$$\prod_{t=1}^{u} \frac{1}{f_{ht}} = w_{h12\ldots u} \tag{19.19}$$

is the multiplier (or the weighting-factor) for the $(h12\ldots u)$th ultimate-stage sample units; the sum of products of these multipliers and the values of the study variable for all the ultimate-stage sample units in the stratum provides the unbiased linear estimator of the universe total Y_h in equation (19.18).

An unbiased estimator of the variance of the sample estimator y_h in the general estimating equation (19.18) is

$$s_{y_h}^2 = \frac{\sum^{n_h} \left[n_h \sum_{1,2,\ldots,u} w_{h12\ldots u} y_{h12\ldots u} - \sum_{1,2,\ldots,u} w_{h12\ldots u} y_{h12\ldots u}\right]^2}{n_h(n_h - 1)} \tag{19.20}$$

where n_h (≥ 2) is the number of the first-stage units selected with replacement in the hth stratum, the summation outside the square brackets being over these first-stage units.

The formulae can be put in simpler forms. Using a slightly different notation, an unbiased estimator of the universe total Y_h of the study variable obtained from the ith first-stage unit ($i = 1, 2, \ldots, n_k$) in the hth stratum is

$$y^*_{hi} = n_h \sum_{2,3,\ldots,u} w_{h12\ldots u} y_{h12\ldots u} \tag{19.21}$$

the summation being over the second- and subsequent-stage sample units.

The combined unbiased estimator of the universe total Y_h from all the n_h sample first-stage units in the hth stratum is the arithmetic mean

$$y^*_{h0} = \sum^{n_h} y^*_{hi}/n_h \tag{19.22}$$

which is the same as the estimator y_h in equation (19.18), the summation in equation (19.22) being over the n_h sample first-stage units in the hth stratum.

An unbiased estimator of the variance of y^*_{h0} is, from equation (19.2),

$$s^2_{y^*_{h0}} = \sum^{n_h} (y^*_{hi} - y^*_{h0})^2/n_h(n_h - 1) \tag{19.23}$$

For all the strata combined, an unbiased estimator of the universe total $Y = \sum^L Y_h$ is the sum of the stratum estimators

$$y = \sum^L y^*_{h0} \tag{19.24}$$

An unbiased estimator of the variance of y is, from equation (19.5),

$$s^2_y = \sum^L s^2_{y^*_{h0}} \tag{19.25}$$

Notes

1. ***Setting of confidence limits.*** The results for stratified single-stage sampling, given in section 9.4, also hold for stratified multi-stage sampling.

2. *Selection of two first-stage units from each stratum.* As for stratified single-stage sampling (section 9.5), so also for stratified multi-stage sampling, the computation of estimates is simplified considerably when in each stratum two first-stage units are selected with replacement out of the total number of units. The results of section 9.5 also hold here, t_{h1} and t_{h2} denoting the unbiased estimators of the universe parameter T_h in the hth stratum, obtained from the two sample first-stage units; and similarly for u_{h1} and u_{h2}.

3. For other notes, see section 9.3, notes (2)-(4) and section 14.4, notes (1)-(5).

Further reading

Foreman, section 8.6.

CHAPTER 20

Stratified Multi-stage Simple Random Sampling

20.1 Introduction

In this chapter we will consider the estimating method for totals, means, and ratios of the values of study variables and their variances in a stratified multi-stage sample with simple random sampling at each stage: the methods for two- and three-stage designs are illustrated in some detail, and the procedures for four and higher stages indicated. The methods of estimation of proportion of units in a category and the use of ratio estimators will also be considered.

20.2 Stratified two-stage srs: Estimation of totals, means, ratios and their variances

20.2.1 Universe totals and means

The universe is divided on some criteria into L mutually exclusive and exhaustive strata. Let the number of first-stage units (fsu's) in the hth stratum ($h = 1, 2, \ldots, L$) be the N_h and the total number of fsu's in all the strata taken together $N = \sum^L N_h$; the number of second-stage units (ssu's) in the ith fsu ($i = 1, 2, \ldots, N_h$) in the hth stratum is denoted by M_{hi}, the total number of ssu's in the hth stratum and in all strata taken together being respectively $\sum_{i=1}^{N_h} M_{hi}$ and $\sum_{h=1}^{L} \sum_{i=1}^{M_h} M_{hi}$. Denoting by Y_{hij} the value of the study variable of the jth ssu ($j = 1, 2, \ldots, M_{hi}$) in the ith fsu in the hth stratum, we define the following universe totals and means:

For the hth stratum

Total of the values of the study variable in the ith fsu:

$$Y_{hi} = \sum_{j=1}^{M_{hi}} Y_{hij} \tag{20.1}$$

Mean for the ith fsu:

$$\overline{Y}_{hi} = Y_{hi}/M_{hi} \tag{20.2}$$

Total for the hth stratum:

$$Y_h = \sum^{N_h} \sum^{M_{hi}} Y_{hij} = \sum^{N_h} Y_{hi} = \sum^{N_h} M_{hi}\overline{Y}_{hi} \tag{20.3}$$

Mean per fsu:

$$\overline{Y}_h = Y_h/N_h = \sum^{N_h} M_{hi}\overline{Y}_{hi}/N_h \tag{20.4}$$

Mean per ssu:

$$\overline{\overline{Y}}_h = Y_h \Big/ \left(\sum^{N_h} M_{hi}\right) = \sum^{N_h} M_{hi}\overline{Y}_{hi} \Big/ \left(\sum^{N_h} M_{hi}\right) \tag{20.5}$$

For all strata combined

Total of the values of the study variable:

$$Y = \sum^{L} Y_h = \sum^{L} \sum^{N_h} Y_{hi} = \sum^{L} \sum^{N_h} \sum^{M_{hi}} Y_{hij} \tag{20.6}$$

Mean per fsu:

$$\overline{Y} = Y/N \tag{20.7}$$

Mean per ssu:

$$\overline{\overline{Y}} = Y \Big/ \left(\sum^{L} \sum^{N_h} M_{hi}\right) \tag{20.8}$$

Table 20.1: Sampling plan for a stratified two-stage simple random sample with replacement. In the hth stratum ($h = 1, 2, \ldots, L$)

Stage (t)	Unit	No. in universe	No. in sample	Selection method	Selection probability	f_t
1	First-stage	N_h	n_h	srswr	Equal = $1/N_h$	n_h/N_h
2	Second-stage	M_{hi}	m_{hi}	srswr	Equal = $1/M_{hi}$	m_{hi}/M_{hi}

20.2.2 Structure of a stratified two-stage srs

In the hth stratum, out of the N_h fsu's, n_h are selected; and out of the M_{hi} ssu's in the ith selected fsu ($i = 1, 2, \ldots, n_h$), m_{hi} are selected: sampling at both stages is srs with replacement. The total number of sample ssu's in the hth stratum is $\sum_{i=1}^{n_h} m_{hi}$, and in all the strata taken together is $\sum_{h=1}^{L} \sum_{i=1}^{n_h} m_{hi}$: the latter is the total sample size. The sampling plan is shown in summary form in Table 20.1.

20.2.3 Estimation of the totals of a study variable

Unbiased estimators of the stratum totals Y_h and the overall total Y of a study variable and their variance estimators could be obtained from the general estimating equations of section 19.3(5), or by extending the methods of sections 15.2.3(a) and 10.2 for srs (with replacement).

Let y_{hij} ($h = 1, 2, \ldots, L$; $i = 1, 2, \ldots, n_h$; $j = 1, 2, \ldots, m_{hi}$) denote the value of the study variable in the jth selected ssu in the ith selected fsu in the hth stratum. For the hth stratum, the unbiased estimator of the stratum total Y_h, obtained from the ith fsu, is (from equation (15.12))

$$y_{hi}^* = N_h \frac{M_{hi}}{m_{hi}} \sum_{j=1}^{m_{hi}} y_{hij} = N_h M_{hi} \bar{y}_{hi} \tag{20.9}$$

where

$$\bar{y}_{hi} = \sum_{j=1}^{m_{hi}} y_{hij}/m_{hi} \tag{20.10}$$

The combined unbiased estimator of Y_h from all the n_h sample fsu's is (from equation (15.13))

$$y_{h0}^* = \frac{1}{n_h} \sum^{n_h} y_{hi}^* \tag{20.11}$$

For all the strata combined, the unbiased estimator of the total Y is (from equation (19.4)) the sum of unbiased estimators of the stratum totals Y_h,

$$y = \sum^{L} y^*_{h0} = \sum^{L} \frac{N_h}{n_h} \sum^{n_h} \frac{M_{hi}}{m_{hi}} \sum_{j=1}^{m_{hi}} y_{hij} \tag{20.12}$$

20.2.4 Sampling variances of the estimators y^*_{h0} and y

From section 15.2.4, the sampling variance of y^*_{h0} in srswr at both stages is

$$\sigma^2_{y^*_{h0}} = \frac{N_h^2 \sigma^2_{hb}}{n_h} + \frac{N_h}{n_h} \frac{M^2_{hi} \sigma^2_{hwi}}{m_{hi}} \tag{20.13}$$

where

$$\sigma^2_{hb} = \sum_{i=1}^{N_h} (Y_{hi} - \overline{Y}_h)^2 / N \tag{20.14}$$

is the variance between the fsu's in the hth stratum; and

$$\sigma^2_{hwi} = \sum_{j=1}^{M_{hi}} (Y_{hij} - \overline{Y}_{hi})^2 / M_{hi} \tag{20.15}$$

is the variance between the ssu's within the ith fsu in the hth stratum.

The sampling variance of $y = \sum^L y^*_{h0}$ is

$$\sigma^2_y = \sum^{L} \sigma^2_{y^*_{h0}} \tag{20.16}$$

Note: The sampling variances and their unbiased estimators in srs, without replacement at both stages and with replacement at the first stage and without replacement at the second stage, follow from section 17.2.4 for any stratum; for the sampling variance y of all the strata taken together, the stratum sampling variances are summed up and an unbiased estimator is obtained on summing up the unbiased stratum variance estimators.

20.2.5 Estimation of the variances of sample estimators of totals

An unbiased estimator of $\sigma^2_{y^*_{h0}}$, the sampling variance of the stratum estimator y^*_{h0} is, from equation (19.23) or (15.14),

$$\begin{aligned} s^2_{y^*_{h0}} &= \sum^{n_h} (y^*_{hi} - y^*_{h0})^2 / n_h(n_h - 1) \\ &= SSy^*_{hi} / n_h(n_h - 1) \end{aligned} \tag{20.17}$$

An unbiased estimator of $\sigma_y^2 = \sum^L \sigma^2_{y^*_{ho}}$, the variance of the overall estimator y is, from equation (19.25), the sum of the stratum unbiased estimators of variances

$$s_y^2 = \sum^L s^2_{y^*_{ho}} \tag{20.18}$$

20.2.6 Estimation of the ratio of the totals of two study variables

For another study variable, the unbiased estimators of the stratum total X_h are similarly defined.

From the hith fsu:

$$x^*_{hi} = N_h \frac{M_{hi}}{m_{hi}} \sum^{m_{hi}} x_{hij} \tag{20.19}$$

Combined from n_h fsu's:

$$x^*_{h0} = \sum^{n_h} x^*_{hi}/n_h \tag{20.20}$$

$s^2_{x^*_{ho}}$, an unbiased estimator of variance of x^*_{h0}, is defined as for $s^2_{y^*_{ho}}$, given in equation (20.17).

For the whole universe, the unbiased estimator of X is (from equation (20.12))

$$x = \sum^L x^*_{h0} \tag{20.21}$$

and s_x^2, the unbiased estimator of the variance of x, is defined as for s_y^2, given in equation (20.18).

A consistent but generally biased estimator of the ratio of two universe totals at the stratum level $R_h = Y_h/X_h$ is the ratio of the sample estimators,

$$r_h = y^*_{h0}/x^*_{h0} \tag{20.22}$$

with a variance estimator (from equation (19.12))

$$s^2_{r_h} = (s^2_{y^*_{ho}} + r_h^2 s^2_{x^*_{ho}} - 2r_h s_{y^*_{h0}x^*_{h0}})/{x^*_{h0}}^2 \tag{20.23}$$

where

$$s_{y^*_{h0}x^*_{h0}} = \sum_{i=1}^{n_h} (y^*_{hi} - y^*_{h0})(x^*_{hi} - x^*_{h0})/n_h(n_h - 1) \tag{20.24}$$

is an unbiased covariance estimator of y^*_{h0} and x^*_{h0} (from equation (19.9)).

A consistent but generally biased estimator of two universe totals for all the strata combined, $R = Y/X$, is the ratio of the sample estimators,

$$r = y/x \tag{20.25}$$

with a variance estimator (from equation (19.14))

$$s_r^2 = (s_y^2 + r^2 s_x^2 - 2r s_{yx})/x^2 \tag{20.26}$$

where

$$s_{yx} = \sum_{h=1}^{L} s_{y^*_{h0} x^*_{h0}} \tag{20.27}$$

is an unbiased covariance estimator of y and x (from equation (19.10)).

20.2.7 Estimation of the means of a study variable

For the estimation of the means of a study variable in a two-stage srs in any stratum, we use the methods of section 15.2.6 adding the subscript h for the hth stratum. Here we consider the estimation of the overall mean, i.e. the mean for all the strata combined.

(a) *Estimation of the mean per first-stage unit.* The unbiased estimator of the overall mean per fsu is

$$\overline{y} = y/N \tag{20.28}$$

with an unbiased variance estimator

$$s_{\overline{y}}^2 = s_y^2/N^2 \tag{20.29}$$

The unbiased estimator y of Y, is defined by equation (20.12), and s_y^2, the unbiased variance estimator of y, is defined by equation (20.18).

(b) *Estimation of the mean per second-stage unit.* Two situations will arise: (i) M_{hi} (the total number of ssu's) is known for all the fsu's; and (ii) M_{hi} is known for the n_h sample fsu's only.

(i) *Unbiased estimator.* When M_{hi} is known for all the fsu's, the unbiased estimator of mean per ssu is

$$\overline{\overline{y}} = y\Big/\left(\sum^{L}\sum^{N_h} M_{hi}\right) \tag{20.30}$$

with an unbiased variance estimator

$$s_y^2 \Big/ \left(\sum^L \sum^{N_h} M_{hi} \right)^2 \tag{20.31}$$

See the notes to section 15.2.6(b-i).

(ii) *Ratio estimator.* When M_{hi} is known for only the sample fsu's, an unbiased estimator of the total number of ssu's in the universe, $\sum^L \sum^{N_h} M_{hi}$, is given by an estimating equation of the type (20.12) by putting $y_{hij} = 1$ for the sample ssu's, i.e. by

$$m = \sum^L m_{h0}^* \tag{20.32}$$

with an unbiased variance estimator

$$s_m^2 = \sum^L s_{m_{h0}^*}^2 = \sum^L \sum^{n_h} (m_{hi}^* - m_{h0}^*)^2 / n_h(n_h - 1) \tag{20.33}$$

where

$$m_{h0}^* = \sum^{n_h} m_{hi}^* / n_h \tag{20.34}$$

and

$$m_{hi}^* = N_h m_{hi} \tag{20.35}$$

A consistent but generally biased estimator of the universe mean $\overline{\overline{Y}}$ per ssu is (from equation (20.25))

$$r = y/m \tag{20.36}$$

with variance estimator

$$s_r = (s_y^2 + r^2 s_m^2 - 2r s_{ym})/m^2 \tag{20.37}$$

where

$$\begin{aligned} s_{ym} &= \sum^L s_{y_{h0}^* m_{h0}^*} \\ &= \sum^L \sum^{n_h} (y_{hi}^* - y_{h0}^*)(m_{hi}^* - m_{h0}^*)/n_h(n_h - 1) \end{aligned} \tag{20.38}$$

is an unbiased covariance estimator of y and m.

See notes to section 15.2.6(b-ii).

(iii) *Unweighted mean of means.* An estimator of the mean per ssu in the universe is the simple (unweighted) mean of means.

$$\sum^{L}\sum^{n_h}\overline{y}_{hi}/n \tag{20.39}$$

where $\overline{y}_{hi}$ is the mean of the y_{hij} values over the m_{hi} sample ssu's in the hith sample fsu (equation (20.10)), and $n = \sum^L n_h$. This estimator is neither unbiased nor consistent: it will be unbiased and consistent when the design is self-weighting.

Note: For a comparison of the different types of estimators, see section 15.2.6(b-iv).

20.2.8 Estimation for sub-universes

Here as for stratified uni-stage and unstratified multi-stage sampling, unbiased estimators are obtained for the total value of a study variable in the sub-universe Y_h' in the hth stratum or for all the strata taken together Y' by defining

$$\begin{aligned} y'_{hij} &= y_{hij} \text{ if the sample ssu belongs to the sub-universe} \\ y'_{hij} &= 0 \text{ otherwise.} \end{aligned}$$

Unbiased estimators of the variances of the totals for the sub-universe are similarly given.

20.2.9 Estimation of proportion of units

Here we combine the methods of section 10.3 for a stratified uni-stage srs and section 15.4 for an unstratified two-stage srs. Let M'_{hi} be the total number of second-stage units possessing a certain attribute in the ith first-stage unit ($i = 1, 2, \ldots, N_h$) in the hth stratum ($h = 1, 2, \ldots, L$); the total number of such ssu's in the hth stratum is

$$\sum^{N_h} M'_{hi} \tag{20.40}$$

The universe number of ssu's with the attribute for all the strata combined is

$$\sum^{L}\sum^{N_h} M'_{hi} \tag{20.41}$$

The proportion of such ssu's with the attribute in the ith fsu in the hth stratum is

$$P_{hi} = M'_{hi}/M_{hi} \tag{20.42}$$

and for all the N_h fsu's in the stratum

$$P_h = \sum^{N_h} M'_{hi} \Big/ \sum^{N_h} M_{hi} = \sum^{N_h} M_{hi}P_{hi} \Big/ \sum^{N_h} M_{hi} \tag{20.43}$$

The proportion for all the strata combined is

$$P = \sum^{L}\sum^{N_h} M'_{hi} \Big/ \sum^{L}\sum^{N_h} M_{hi} \tag{20.44}$$

As for unstratified two-stage srs (section 15.4), so also here, unbiased estimators of the total number of ssu's in a stratum and for all strata combined are obtained respectively from the estimating equations (20.11) and (20.12) by putting $y_{hij} = 1$ if the sample ssu has the attribute, and 0 otherwise, and so, also, for the unbiased variance estimators.

For the estimators for any stratum, we use the methods of section 15.4 adding the subscript h for the hth stratum. Here we consider the estimation of the overall universe proportion, i.e. for all the strata combined.

An unbiased estimator of $M' = \sum^L M'_h$ is the sum of the stratum estimators m'^*_{h0}

$$m' = \sum^{L} m'^*_{h0} \tag{20.45}$$

where

$$m'^*_{h0} = \sum^{n_h} m'^*_{hi}/n_h \tag{20.46}$$

from equation (15.54),

$$m'^*_{hi} = N_h M_{hi} p_{hi} \tag{20.47}$$

from equation (15.53), and

$$p_{hi} = m'_{hi}/m_{hi} \tag{20.48}$$

from equation (15.52), m'_{hi} of the m_{hi} sample fsu's in an srs having the attribute in the hth stratum.

An unbiased variance estimator of m' is

$$s^2_{m'} = \sum^{L} s^2_{m'^*_{h0}} = \sum^{L}\sum^{n_h}(m'^*_{hi} - m'^*_{h0})^2/n_h(n_h - 1) \tag{20.49}$$

If all the M_{hi} values are known for the universe, then an unbiased estimator of the universe proportion P is

$$m' \Big/ \sum^{L}\sum^{N_h} M_{hi} \tag{20.50}$$

with an unbiased variance estimator

$$s_{m'}^2 \Big/ \left(\sum^L \sum^{N_h} M_{hi}\right)^2 \tag{20.51}$$

If $\sum^L \sum^{N_h} M_{hi}$ is not known, it is estimated unbiasedly by m in equation (20.32) with an unbiased variance estimator given by equation (20.33).

A consistent but generally biased estimator of the universe proportion P is then

$$p = m'/m \tag{20.52}$$

with a variance estimator

$$s_p^2 = (s_{m'}^2 + p^2 s_m^2 - 2ps_{mm'})/m^2 \tag{20.53}$$

where

$$\begin{aligned} s_{mm'} &= \sum^L s_{m^*_{h0} m'^*_{h0}} \\ &= \sum^L \sum^{n_h} (m^*_{hi} - m^*_{h0})(m'^*_{hi} - m'^*_{h0})/n_h(n_h - 1) \end{aligned} \tag{20.54}$$

is an unbiased covariance estimator of m and m'.

See notes to sections 10.3 and 15.4.

Example 20.1

In each of the simple random samples of two villages, each from the three stages selected in Example 10.2 from the list of villages in Appendix IV, select five households at random, given the listing of households in the sample villages. On the basis of collected data for the sample households on the size, number of adults, and possession of TV satellite dishes, estimate the total number of persons, the average household size, the proportion of adults, and the proportion of households with TV satellite dishes for the three states separately and also combined, along with their standard errors. The sample data are given in Table 20.2. The use of the last column will be illustrated in Example 20.2.

As there are two sample first-stage units, namely villages, in each stratum, we follow the simplified procedure mentioned in note 2 of section 19.3, illustrated also in Example 10.2 for a stratified single-stage srs. We denote the household's size by y_{hij}, the number of adults by x_{hij}, the possession of TV satellite dishes by y'_{hij} (= 1 for yes; and = 0 for no) in the sample households ($h = 1, 2, 3$ for the states; $i = 1, 2$ for the sample villages; and $j = 1, 2, 3, 4, 5$ for the sample households). The respective means per sample household in the sample villages are denoted by

Table 20.2: Size, number of adults (18 years or over), possession of TV satellite dishes, and agreement by the selected adult to an increase of state taxes for education in the sample households in each of the two sample villages in the three states of Example 10.2: Stratified two-stage simple random sample

Household serial no.	Household size (y_{hij})	Number of adults (x_{hij})	Possession of TV satellite dishes ($y'_{hij} = 1$, for Yes; $= 0$, for No)	Agreement by the selected adult to an increase of state taxes for education ($q'_{hij} = 1$, for Yes; $= 0$, for No)
State I: Village serial no. 2; total number of households = 18				
7	6	4	1	0
11	4	2	0	0
12	4	2	0	0
15	3	1	0	1
18	4	2	1	1
State I: Village serial no. 10; total number of households = 22				
1	4	1	0	0
4	5	2	0	0
6	4	2	1	1
9	3	2	1	0
17	4	3	0	0
State II: Village serial no. 5; total number of households = 20				
5	6	3	1	0
7	7	4	1	0
8	5	3	0	1
12	6	4	0	0
16	6	2	1	0
State II: Village serial no. 11; total number of households = 18				
4	4	2	0	1
7	5	3	0	0
9	5	3	1	0
12	7	3	1	0
16	5	3	0	0
State III: Village serial no. 2; total number of households = 21				
7	6	3	0	0
10	5	3	0	0
17	5	2	1	0
19	5	3	0	0
20	7	4	1	1
State III: Village serial no. 3; total number of households = 18				
6	6	4	0	0
10	6	3	0	1
11	6	4	0	0
16	5	3	0	0
18	4	2	1	0

$\bar{y}_{hi}$, $\bar{z}_{hi}$, and $\bar{m}'_{hi}$, defined by equations of the type (20.10). Unbiased estimates of the total number of households in the three zones separately and combined have already been obtained in Example 10.2. These along with the other required computations, are shown in Tables 20.3-20.5.

The final estimates are given in Table 20.6. The coefficients of variation of the estimation proportion of households with TV satellite dishes are rather high. To reduce standard errors, information on items such as this, that do not require detailed probing, could be collected for all the households in the sample villages when a list of households is prepared before a sample of households could be drawn for collecting other information. The present computations are merely illustrative.

20.2.10 Ratio method of estimation

(a) *Estimation procedures for totals.* As for unstratified single- or multi-stage and stratified single-stage, so also for stratified multi-stage sampling, the ratio method of estimation may increase the efficiency of estimators under favorable conditions if information on ancillary variables is available for the second-stage units in the universe.

For the hth stratum

A combined unbiased estimator of the stratum total of the ancillary variable $Z_h = \sum^{N_h} \sum^{M_{hi}} z_{hij}$ is first obtained from equation (20.11)

$$z^*_{h0} = \sum^{n_h} z^*_{hi}/n_h \tag{20.55}$$

with an unbiased variance estimator

$$s^2_{z^*_{h0}} = \sum^{n_h} (z^*_{hi} - z^*_{h0})^2/n_h(n_h - 1) \tag{20.56}$$

where

$$z^*_{hi} = N_h \frac{M_{hi}}{m_{hi}} \sum^{m_{hi}} z_{hij} \tag{20.57}$$

The ratio estimator of the total Y_h, using the ancillary information, is, from equation (15.66),

$$y^*_{hR} = Z_h y^*_{h0}/z^*_{h0} = Z_h r_{1h} \tag{20.58}$$

where

$$r_{1h} = y^*_{h0}/z^*_{h0} \tag{20.59}$$

Table 20.3: Computation of the estimated total numbers of persons, adults and households with TV satellite dishes: data of Table 20.2

State	Number of villages		Sample village	Number of households		
	Total	Sample		Total	Sample	Stratum estimate†
h	N_h	n_h	i	M_{hi}	m_{hi}	$m^*_{hi} = N_h M_{hi}$
(1)	(2)	(3)	(4)	(5)	(6)	(7)
I	10	2	1	18	5	180
			2	22	5	220
					Total	400
					Mean (= stratum estimate of total)	200(m^*_{h0})
					Difference	−40 (d_{mh})
					$\frac{1}{2}$ |Difference| (= estimated standard error)	20 ($s_{m^*_{h0}}$)
II	11	2	1	20	5	220
			2	18	5	198
					Total	418
					Mean (= stratum estimate of total)	209(m^*_{h0})
					Difference	22 (d_{mh})
					$\frac{1}{2}$ |Difference| (= estimated standard error)	11 ($s_{m^*_{h0}}$)
III	9	2	1	21	5	189
			2	18	5	162
					Total	351
					Mean (= stratum estimate of total)	175.5(m^*_{h0})
					Difference	27 (d_{mh})
					$\frac{1}{2}$ |Difference| (= estimated standard error)	13.5 ($s_{m^*_{h0}}$)

(continued)

† The estimated number of households were obtained in Table 10.4, denoted by x^*_{hi}. In this example, x^*_{hi} denotes the estimated numbers of adults, and m^*_{hi} those of households.

Table 20.3 continued

State	Sample village	Number of persons		
		Sample number	Sample mean	Stratum estimate
h	i	$\sum_{j=1}^{m_{hi}} y_{hij}$	$\bar{y}_{hi}$	$y^*_{hi} = N_h M_{hi} \bar{y}_{hi}$
(1)	(4)	(8)	(9)	(10)
I	1	21	4.2	756
	2	20	4.0	880
			Total	1636
			Mean (= stratum estimate of total)	818(y^*_{h0})
			Difference	−156 (d_{yh})
			$\frac{1}{2}$ |Difference| (= estimated standard error)	78 ($s_{y^*_{h0}}$)
II	1	30	6.0	1320
	2	26	5.2	1029.6
			Total	2349.6
			Mean (= stratum estimate of total)	1174.8(y^*_{h0})
			Difference	290.4 (d_{yh})
			$\frac{1}{2}$ |Difference| (= estimated standard error)	145.2 ($s_{y^*_{h0}}$)
III	1	28	5.6	1058.4
	2	27	5.4	874.8
			Total	1933.2
			Mean (= stratum estimate of total)	966.6(y^*_{h0})
			Difference	183.6 (d_{yh})
			$\frac{1}{2}$ |Difference| (= estimated standard error)	91.8 ($s_{y^*_{h0}}$)

(continued)

Table 20.3 continued

State	Sample village	Number of adults		
		Sample number	Sample mean	Stratum estimate
h	i	$\sum_{j=1}^{m_{hi}} x_{hij}$	$\overline{x}_{hi}$	$x^*_{hi} = N_h M_{hi} \overline{x}_{hi}$
(1)	(4)	(11)	(12)	(13)
I	1	11	2.2	396
	2	10	2.0	440
			Total	836
			Mean (= stratum estimate of total)	418(x^*_{h0})
			Difference	−44 (d_{xh})
			$\frac{1}{2}$ |Difference| (= estimated standard error)	22 ($s_{x^*_{h0}}$)
II	1	16	3.2	704
	2	14	2.8	554.4
			Total	1258.4
			Mean (= stratum estimate of total)	629.2(x^*_{h0})
			Difference	149.6 (d_{xh})
			$\frac{1}{2}$ |Difference| (= estimated standard error)	74.8 ($s_{x^*_{h0}}$)
III	1	15	3.0	567
	2	16	3.2	518.4
			Total	1085.4
			Mean (= stratum estimate of total)	542.7(x^*_{h0})
			Difference	48.6 (d_{xh})
			$\frac{1}{2}$ |Difference| (= estimated standard error)	24.3 ($s_{x^*_{h0}}$)

(continued)

Table 20.3 continued

State	Sample village	Number of households with TV satellite dishes		
		Sample number	Sample mean	Stratum estimate
h	i	$\sum_{j=1}^{m_{hi}} y'_{hij}$	$\overline{m}'_{hi}$	$m'^{*}_{hi} = N_h M_{hi} \overline{m}'_{hi}$
(1)	(4)	(14)	(15)	(16)
I	1	1	0.2	36
	2	2	0.4	88
			Total	124
			Mean (= stratum estimate of total)	62(m'^{*}_{h0})
			Difference	−52 ($d_{m'h}$)
			$\frac{1}{2}$ |Difference| (= estimated standard error)	26 ($s_{m'^{*}_{h0}}$)
II	1	3	0.6	132
	2	2	0.4	79.2
			Total	211.2
			Mean (= stratum estimate of total)	105.6(m'^{*}_{h0})
			Difference	52.8 ($d_{m'h}$)
			$\frac{1}{2}$ |Difference| (= estimated standard error)	26.4 ($s_{m'^{*}_{h0}}$)
III	1	2	0.4	75.6
	2	1	0.2	32.4
			Total	108.0
			Mean (= stratum estimate of total)	54.0(m'^{*}_{h0})
			Difference	43.2 ($d_{m'h}$)
			$\frac{1}{2}$ |Difference| (= estimated standard error)	21.6 ($s_{m'^{*}_{h0}}$)

Table 20.4: Computation of estimated total numbers of households, persons, adults, and households with TV satellite dishes: data of Tables 20.2 and 20.3

State	Number of households†			Number of persons		
	Estimate	Standard error	Variance	Estimate	Standard error	Variance
h	m^*_{h0}	$s_{m^*_{h0}}$	$s^2_{m^*_{h0}}$	y^*_{h0}	$s_{y^*_{h0}}$	$s^2_{y^*_{h0}}$
I	200	20	400	818	78	6084
II	209	11	121	1174.8	145.2	21083.04
III	175.5	13.5	182.25	966.6	91.8	8427.24
All states combined	584.5	26.52	703.25	2959.4	188.66	35594.28

State	Number of adults			Number of households with TV satellite dishes		
	Estimate	Standard error	Variance	Estimate	Standard error	Variance
h	x^*_{h0}	$s_{x^*_{h0}}$	$s^2_{x^*_{h0}}$	m'^*_{h0}	$s_{m'^*_{h0}}$	$s^2_{m'^*_{h0}}$
I	418	22	484	62	26	676
II	629.2	74.8	5595.04	105.6	26.4	696.96
III	542.7	24.3	590.49	54	21.6	466.56
All states combined	1589.9	81.67	6669.53	221.6	42.89	1839.52

† Also from Example 10.2.

Table 20.5: Computation of estimated ratios and variances: data of Table 20.3

State	For the average household size			
	$r_h = y^*_{h0}/m^*_{h0}$	$s_{y^*_{h0} m^*_{h0}} = \frac{1}{4} d_{yh} d_{mh}$	$s^2_{y^*_{h0}} + r_h^2 s^2_{m^*_{h0}} - 2 r_h s_{y^*_{h0} m^*_{h0}}$	$s^2_{r_h} =$ col. (4)$/m^{*2}_{h0}$
(1)	(2)	(3)	(4)	(5)
I	4.09	1560.0	14.4400	0.000361
II	5.62	1597.2	6952.2244	0.149159
III	5.51	1239.3	303.2822	0.009847
All states combined	5.06 (r)	4396.5 (s_{ym})	9107.4317 ($s_y^2 + r^2 s_m^2 - 2r s_{ym}$)	0.026658 (s_r^2)

States	For the proportion of adults			
	$p_h = x^*_{h0}/y^*_{h0}$	$s_{y^*_{h0} x^*_{h0}} = \frac{1}{4} d_{yh} d_{xh}$	$s^2_{x^*_{h0}} + p_h^2 s^2_{y^*_{h0}} - 2 p_h s_{y^*_{h} x^*_{h0}}$	$s^2_{p_h} =$ col. (8)$/y^{*2}_{h0}$
(1)	(6)	(7)	(8)	(9)
I	0.5110	1716.00	318.9082	0.00047661
II	0.5356	10 860.96	8.8081	0.00000638
III	0.5615	2230.74	742.3261	0.00079451
All states combined	0.5372 (p)	14 807.70 (s_{yx})	1032.0768 ($s_x^2 + p^2 s_y^2 - 2p s_{yx}$)	0.00011784 (s_p^2)

States	For the proportion of households with TV satellite dishes			
	$p'_h = m'^*_{h0}/m^*_{h0}$	$s_{m^*_{h0} m'^*_{h0}} = \frac{1}{4} d_{mh} d_{m'h}$	$s^2_{m'^*_{h0}} + p'^2_h s^2_{m^*_{h0}} - 2p'_h s_{m^*_{h} m'^*_{h0}}$	$s^2_{p'_h} =$ col. (12)$/m^{*2}_{h0}$
(1)	(10)	(11)	(12)	(13)
I	0.3100	520.0	392.0400	0.00980100
II	0.5053	290.4	434.3764	0.00994429
III	0.3077	291.6	304.3646	0.00988189
All states combined	0.3791 (p')	1102.0 ($s_{mm'}$)	1105.0526 ($s_m^2 + p'^2 s_m^2 - 2p' s_{mm'}$)	0.00323455 ($s^2_{p'}$)

Table 20.6: Estimates and standard errors computed from the data of stratified two-stage srs in Table 20.2

	Item	State I	State II	State III	All states combined
1.	*Total number of persons*				
	(a) Estimate	818	1174.8	966.6	2959.4
	(b) Standard error	78	145.2	91.8	188.66
	(c) Coefficient of variation (%)	9.54	12.36	9.50	6.37
2.	*Average household size*				
	(a) Estimate	4.09	5.62	5.51	5.06
	(b) Standard error	0.019	0.399	0.099	0.163
	(c) Coefficient of variation (%)	0.46	7.10	1.80	3.22
3.	*Proportion of adults*				
	(a) Estimate	0.5110	0.5356	0.5615	0.5372
	(b) Standard error	0.02183	0.00253	0.02819	0.01086
	(c) Coefficient of variation (%)	4.27	0.47	5.02	2.02
4.	*Proportion of households with TV satellite dishes*				
	(a) Estimate	0.3100	0.5053	0.3077	0.3791
	(b) Standard error	0.0990	0.0997	0.0994	0.0569
	(c) Coefficient of variation (%)	31.94	19.73	32.30	15.01

A consistent but generally biased estimator of the variance of y^*_{hR} is (from equation (15.68))

$$s^2_{y^*_{hR}} = Z^2_h s^2_{1h} = (s^2_{y^*_{ho}} + r^2_{1h} s^2_{z^*_{ho}} - 2r_{1h} s_{y^*_{h0} z^*_{h0}}) \tag{20.60}$$

where

$$s_{y^*_{h0} z^*_{h0}} = \sum^{n_h} (y^*_{hi} - y^*_{h0})(z^*_{hi} - z^*_{h0})/n_h(n_h - 1) \tag{20.61}$$

is an unbiased covariance estimator of y^*_{h0} and z^*_{h0}.

For all strata combined (see also section 10.4)

For the estimation of universe total $Y = \sum^L Y_h$, two types of ratio estimators can be used:

(i) *Separate ratio estimator.* The stratum ratio estimators y^*_{hR} are summed up:

$$y_{RS} = \sum^L y^*_{hR} \tag{20.62}$$

the stratum ratio estimators y^*_{hR} being given by equation (20.58), the additional subscript S in y_{RS} standing for *separate* ratio estimator.

The variance estimator of Y_{RS} is the sum of the stratum variance estimators $s^2_{y^*_{hR}}$, given by equation (20.60):

$$s^2_{y_{RS}} = \sum^{L} s^2_{y^*_{hR}} \tag{20.63}$$

(ii) *Combined ratio estimator.* The ratio method is applied to the estimators of the overall totals, thus

$$y_{RC} = Zy/z = Zr \tag{20.64}$$

the additional subscript C in y_{RC} standing for *combined* ratio estimator; y is an unbiased estimator of the universe total Y and is given by equation (20.12), z is an unbiased estimator of $Z = \sum^L Z_h$ given by

$$z = \sum^{L} z^*_{h0} \tag{20.65}$$

and

$$r = y/z \tag{20.66}$$

A variance estimator of y_{RC} is

$$s^2_{y_{RC}} = Z^2 s^2_r = (s^2_y + r^2 s^2_z - 2r s_{yz}) \tag{20.67}$$

where

$$s_{yz} = \sum^{L} s_{y^*_{h0} z^*_{h0}} \tag{20.68}$$

is the unbiased covariance estimator of y and z.

See notes to sections 10.4 and 15.5.

Example 20.2

For the data of Example 20.1, given the additional information on the previous census population of all the villages as in Example 10.5, obtain ratio estimates of the total population for the three states separately and also combined, along with their standard errors.

Here the ratio method of estimation is applied at the level of the first-stage units. Unbiased estimates of the previous census population in the three states along with their standard errors have already been obtained in Example 10.5. The rest of the computations are shown in Table 20.7 and the final results given in Table 20.8.

Table 20.7: Computation of ratio estimates of total population for a stratified two-stage srs: data of Tables 20.2 and 10.12

State	Z_h	$r_{1h} = y^*_{h0}/z^*_{h0}$	$y^*_{hR} = Z_h r_{1h}$	$s_{y^*_{h0} z^*_{h0}} = \frac{1}{4} d_{yh} d_{zh}$	$s^2_{y^*_{hR}} = s^2_{y^*_{h0}} + r^2_{1h} s^2_{z^*_{h0}} - 2r_{1h} s_{y^*_{h0} z^*_{h0}}$	$s_{y^*_{hR}}$
(1)	(2)	(3)	(4)	(5)	(6)	(7)
I	863	1.0161	876.89	390.0	6902.369475	83.08
II	1010	1.1609	1172.51	9583.2	4788.032436	69.20
III	942	1.0278	968.19	6196.5	502.844801	22.42
Total	2815		3017.67 (y_{RS})	15389.7	12 193.246712 ($s^2_{y_{RS}}$)	110.42 ($s_{y_{RS}}$)
Combined	2815 (Z)	1.0732 ($r_1 = y/z$)	3021.06 ($y_{RC} = Zr_1$)	15389.7 (s_{yz})	12 855.394980 ($s^2_{y_{RC}} = s^2_y + r^2_1 s^2_z - 2r_1 s_{yz}$)	113.38 ($s_{y_{RC}}$)

Notes

1. The ratio estimates of the total numbers of households have already been obtained in Example 10.2.

2. The ratio estimates of population are more efficient than the unbiased estimates, as shown by Table 20.6. No marked difference appears between the separate and combined ratio estimates for the whole universe.

(b) *Ratio of ratio estimators of totals of two study variables.* Ratio estimators of totals of two study variables may be used in estimating their ratios. However, as noted in section 10.5 for stratified single-stage sampling, for any stratum (and also for all strata if the combined ratio estimators are used), the ratio of the ratio estimators of totals of two study variables becomes the same as the ratio of the corresponding unbiased estimators. For the separate ratio estimators of totals of two study variables, the ratio of the estimators and its variance estimators have been defined in section 10.5 for stratified single-stage srs. The same methods will apply for stratified multi-stage designs, noting that the appropriate estimating formulae have to be used. For an example, see exercise 3 of this chapter.

Table 20.8: Ratio estimates of population and their standard errors computed from the data of the stratified two-stage srs in Table 20.2, using the previous census population of the sample villages (data of Table 20.7)

State	Ratio estimate	Standard error	CV (%)
I	876.9	83.1	9.48
II	1172.5	69.2	5.90
III	968.2	22.4	2.31
All states			
Separate ratio estimate	3017.7	110.4	3.66
Combined ratio estimate	3021.1	113.4	3.75

20.3 Stratified three-stage srs

20.3.1 Universe totals and means

Following on the formulation in section 20.2.1, let the number of third-stage units (tsu's) in the jth second-stage unit (ssu) in the ith first-stage unit (fsu) of the hth stratum be denoted by Q_{hij} ($h = 1, 2, \ldots, L$; $i = 1, 2, \ldots, N_h$; $= 1, 2, \ldots, M_{hi}$). The total number of tsu's in the ith fsu of the hth stratum is $\sum_{j=1}^{M_{hi}} Q_{hij}$. The total number of tsu's in the hth stratum for all the N_h fsu's is $\sum_{i=1}^{N_h} \sum_{j=1}^{M_{hi}} Q_{hij}$, and in the whole universe $\sum_{h=1}^{L} \sum_{i=1}^{N_h} \sum_{j=1}^{M_{hi}} Q_{hij}$. Denoting by Y_{hijk} the value of the study variable of the kth tsu in the jth ssu in the ith fsu in the hth stratum, we define the following universe totals and means:

For the hth stratum

Total of the values of the study variable in the jth ssu in the ith fsu in the hth stratum:

$$Y_{hij} = \sum_{k=1}^{Q_{hij}} Y_{hijk} \tag{20.69}$$

Mean for the jth ssu in the ith fsu:

$$\overline{Y}_{hij} = Y_{hij}/Q_{hij} \tag{20.70}$$

Total for the ith fsu:

$$Y_{hi} = \sum_{j=1}^{M_{hi}} Y_{hij} \tag{20.71}$$

Mean for the ith fsu per ssu:

$$\overline{\overline{Y}}_{hi} = Y_{hi}/M_{hi} \tag{20.72}$$

Total for the hth stratum:

$$\begin{aligned} Y_h &= \sum^{N_h}\sum^{M_{hi}}\sum^{Q_{hij}} Y_{hijk} = \sum^{N_h}\sum^{M_{hi}} Y_{hij} = \sum^{N_h} Y_{hi} \\ &= \sum^{N_h}\sum^{M_{hi}} Q_{hij}\overline{Y}_{hij} = \sum^{N_h} M_{hi}\overline{Y}_{hi} \end{aligned} \tag{20.73}$$

Mean per fsu in the hth stratum:

$$\overline{Y}_h = Y_h/N_h = \sum^{N_h} M_{hi}\overline{Y}_{hi}/N_h \tag{20.74}$$

Mean per ssu:

$$\overline{\overline{Y}}_h = Y \Big/ \left(\sum^{N_h} M_{hi}\right) = \sum^{N_h} M_{hi}\overline{Y}_{hi} \Big/ \left(\sum^{N_h} M_{hi}\right) \tag{20.75}$$

Mean per tsu:

$$\begin{aligned} \overline{\overline{\overline{Y}}}_h &= Y \Big/ \left(\sum^{N_h}\sum^{M_{hi}} Q_{hij}\right) \\ &= \sum^{N_h}\sum^{M_{hi}} Q_{hij}\overline{Y}_{hij} \Big/ \left(\sum^{N_h}\sum^{M_{hi}} Q_{hij}\right) \end{aligned} \tag{20.76}$$

For all strata combined

Total of the values of the study variable:

$$\begin{aligned} Y &= \sum^{L} Y_h = \sum^{L}\sum^{N_h} Y_{hi} = \sum^{L}\sum^{N_h}\sum^{M_{hi}} Y_{hij} \\ &= \sum^{L}\sum^{N_h}\sum^{M_{hi}}\sum^{Q_{hij}} Y_{hijk} \end{aligned} \tag{20.77}$$

Table 20.9: Sampling plan for a stratified three-stage simple random sample with replacement. In the hth stratum $(h = 1, 2, \ldots, L)$

Stage (t)	Unit	No. in universe	No. in sample	Selection method	Selection probability	f_t
1	First-stage	N_h	n_h	srswr	Equal = $1/N_h$	n_h/N_h
2	Second-stage	M_{hi}	m_{hi}	srswr	Equal = $1/M_{hi}$	m_{hi}/M_{hi}
3	Third-stage	Q_{hij}	q_{hij}	srswr	Equal = $1/Q_{hij}$	q_{hij}/Q_{hij}

Mean per fsu:

$$\overline{Y} = Y/N \tag{20.78}$$

Mean per ssu:

$$\overline{\overline{Y}} = Y\Big/\left(\sum^{L}\sum^{N_h} M_{hi}\right) \tag{20.79}$$

Mean per tsu:

$$\overline{\overline{\overline{Y}}} = Y\Big/\left(\sum^{L}\sum^{N_h}\sum^{M_{hi}} Q_{hij}\right) \tag{20.80}$$

20.3.2 Structure of a stratified three-stage srs

In the hth stratum, out of the N_h fsu's, n_h are selected; out of the M_{hi} ssu's in the ith selected fsu $(i = 1, 2, \ldots, n_h)$, m_{hi} are selected; and finally, out of the Q_{hij} tsu's in the jth selected ssu $(j = 1, 2, \ldots, m_{hi})$, q_{hij} are selected; sampling at the three stages in srs with replacement. The total number of sample tsu's in the hth stratum is $\sum_{i=1}^{n_h}\sum_{j=1}^{m_{hi}} q_{hij}$, and in all the strata taken together is $\sum^{L}\sum^{n_h}\sum^{m_{hi}} q_{hij}$: the latter is the total sample size. The sampling plan is shown in summary form in Table 20.9.

20.3.3 Estimation of the totals of a study variable

Unbiased estimators of the stratum totals Y_h and the overall total Y of a study variable and their variance estimators could be obtained from the general estimating equations of section 19.3(5) or by extending the methods of sections 15.6.3 and 10.2 for srs with replacement.

Let y_{hijk} ($h = 1, 2, \ldots, L$; $i = 1, 2, \ldots, n_h$; $j = 1, 2, \ldots, m_{hi}$; $k = 1, 2, \ldots, q_{hij}$) denote the value of the study variable in the kth selected tsu in the jth selected ssu in the ith selected fsu in the hth stratum.

For the hth stratum, an unbiased estimator of Y_h from the ith selected fsu is (from section 15.6.3)

$$\begin{aligned} y^*_{hi} &= N_h \frac{M_{hi}}{m_{hi}} \sum_{j=1}^{m_{hi}} \frac{Q_{hij}}{q_{hij}} \sum_{k=1}^{q_{hij}} y_{hijk} \\ &= N_h \frac{M_{hi}}{m_{hi}} \sum_{j=1}^{m_{hi}} Q_{hij} \overline{y}_{hij} \end{aligned} \tag{20.81}$$

where

$$\overline{y}_{hij} = \sum_{k=1}^{q_{hij}} y_{hijk}/q_{hij} \tag{20.82}$$

is the simple arithmetic mean of the y_{hijk} values in the hijth ssu.

A combined unbiased estimator of Y_h is the arithmetic mean

$$\begin{aligned} y^*_{h0} &= \frac{1}{n_h} \sum_{i=1}^{n_h} y^*_{hi} \\ &= \frac{N_h}{n_h} \sum_{i=1}^{n_h} \frac{M_{hi}}{m_{hi}} \sum_{j=1}^{m_{hi}} \frac{Q_{hij}}{q_{hij}} \sum_{k=1}^{q_{hij}} y_{hijk} \end{aligned} \tag{20.83}$$

For all the strata combined, an unbiased estimator of the total Y is the sum of the estimators of the stratum totals

$$y = \sum_{h=1}^{L} y^*_{h0} = \sum_{h=1}^{L} \frac{N_h}{n_h} \sum_{i=1}^{n_h} \frac{M_{hi}}{m_{hi}} \sum_{j=1}^{m_{hi}} \frac{Q_{hij}}{q_{hij}} \sum_{k=1}^{q_{hij}} y_{hijk} \tag{20.84}$$

20.3.4 Estimation of the variances of the sample estimators of totals

An unbiased variance estimator of the stratum total estimator y^*_{h0} is

$$s^2_{y^*_{h0}} = \sum^{n_h} (y^*_{hi} - y^*_{h0})^2 / n_h(n_h - 1) \tag{20.85}$$

and that of the universe total estimator y

$$s^2_y = \sum^{L} s^2_{y^*_{h0}} \tag{20.86}$$

20.3.5 Estimation of the ratio of the totals of two study variables

For another study variable, the stratum and universe totals may be defined and estimated similarly. An unbiased estimator x^*_{h0} of the stratum total X_h is obtained as for y^*_{h0} (equation (20.83)), as also its unbiased variance estimator $s^2_{x^*_{h0}}$ (by equation (20.85)). For the whole universe, an unbiased estimator x is obtained in the same manner as for y (equation (20.84)), as also its unbiased variance estimator (equation (20.86)).

A consistent but generally biased estimator of the ratio of the stratum totals $R_h = Y_h/X_h$ is the ratio of the sample estimators

$$r_h = y^*_{h0}/x^*_{h0} \tag{20.87}$$

with a variance estimator

$$s^2_{r_h} = (s^2_{y^*_{h0}} + r^2_h s^2_{x^*_{h0}} - 2r_h s_{y^*_{h0}x^*_{h0}})/x^{*2}_{h0} \tag{20.88}$$

where

$$s_{y^*_{h0}x^*_{h0}} = \sum^{n_h}(y^*_{hi} - y^*_{h0})(x^*_{hi} - x^*_{h0})/n_h(n_h - 1) \tag{20.89}$$

is an unbiased covariance estimator of y^*_{h0} and x^*_{h0}.

For the whole universe, a consistent but generally unbiased estimator of the ratio of totals $R = Y/X$ is the ratio of the sample estimators

$$r = y/x \tag{20.90}$$

with a variance estimator

$$s^2_r = (s^2_y + r^2 s^2_x - 2r s_{yx})/x^2 \tag{20.91}$$

where

$$s_{yx} = \sum^{L} s_{y^*_{h0}x^*_{h0}} \tag{20.92}$$

is an unbiased covariance estimator of y and x.

20.3.6 Estimation of the means of a study variable

For the estimation of the means of a study variable in a three-stage srs in any stratum, see section 15.6.5, and for the estimation of the means per fsu and ssu in all the strata combined, see sections 20.2.7(a) and 20.2.7(b). We consider here the estimating method for the mean per tsu for the whole universe.

Two situations will arise: (i) the total number of the tsu's Q_{hij} is known for all the ssu's in the universe; and (ii) Q_{hij} is known only for the sample ssu's.

(i) *Unbiased estimator.* When Q_{hij} is known for all the ssu's in the universe, an unbiased estimator of the universe mean per tsu $\overline{\overline{\overline{Y}}}$ is

$$\overline{\overline{\overline{y}}} = y \Big/ \left(\sum^{L} \sum^{N_h} \sum^{M_{hi}} Q_{hij} \right) \tag{20.93}$$

with an unbiased variance estimator

$$s_y^2 \Big/ \left(\sum^{L} \sum^{N_h} \sum^{M_{hi}} Q_{hij} \right)^2 \tag{20.94}$$

An unbiased estimator y of Y is defined by equation (20.84), and its unbiased variance estimator s_y^2 by equation (20.86).

(ii) *Ratio estimator.* When Q_{hij} is known only for the sample ssu's, an unbiased estimator of the total number of tsu's in the universe, namely, $\sum^{L} \sum^{N_h} \sum^{M_{hi}} Q_{hij}$, is given by an estimation equation of the type (20.84) by putting $y_{hijk} = 1$ for the sample tsu's, i.e. by

$$q = \sum^{L} q_{h0}^* \tag{20.95}$$

with an unbiased variance estimator

$$s_q^2 = \sum^{L} s_{q_{h0}^*}^2 \tag{20.96}$$

where

$$q_{h0}^* = \sum^{n_h} q_{hi}^* / n_h \tag{20.97}$$

$$q_{hi}^* = N_h \frac{M_{hi}}{m_{hi}} \sum^{m_{hi}} Q_{hij} \tag{20.98}$$

and

$$s_{q_{h0}^*}^2 = \sum^{n_h} (q_{hi}^* - q_{h0}^*)^2 / n_h(n_h - 1) \tag{20.99}$$

A consistent but generally biased estimator of $\overline{\overline{\overline{Y}}}$ is the ratio

$$r = y/q \tag{20.100}$$

with a variance estimator

$$s_r^2 = (s_y^2 + r^2 s_q^2 - 2r s_{yq})/q^2 \tag{20.101}$$

where

$$\begin{aligned} s_{yq} &= \sum^{L} s_{y^*_{h0} q^*_{h0}} \\ &= \sum^{L} \sum^{n_h} (y^*_{hi} - y^*_{h0})(q^*_{hi} - q^*_{h0})/n_h(n_h - 1) \end{aligned} \tag{20.102}$$

(iii) *Unweighted mean of means.* An estimator of $\overline{\overline{\overline{Y}}}$ is the simple (unweighted) mean of means,

$$\sum^{L} \sum^{n_h} \sum^{m_{hi}} \overline{y}_{hij} \bigg/ \left(\sum^{L} \sum^{n_h} m_{hi} \right) \tag{20.103}$$

where $\overline{y}_{hij}$ is the sample mean in the hijth ssu (equation (20.82)).

This estimator will be both biased and inconsistent: it will be unbiased and consistent when the design is self-weighting.

For a comparison of the estimators and other observations, see notes to section 15.6.5.

20.3.7 Estimation for sub-universes

Here also, unbiased estimators are obtained for the total values of a study variable in the sub-universe Y'_h in the hth stratum or for all the strata taken together Y' by defining

$$\begin{aligned} y'_{hijk} &= y_{hijk} \text{ if the } hijk\text{th third-stage sample unit belongs to the sub-universe} \\ y'_{hijk} &= 0 \text{ otherwise.} \end{aligned}$$

Unbiased variance estimators of the totals are similarly obtained.

20.3.8 Estimation of proportion of units

Here we combine the methods of section 10.3 for a stratified single-stage srs and section 15.8 for an unstratified three-stage srs. Let Q'_{hij} be the total number of third-stage units possessing a certain attribute in the jth ssu in the ith fsu in the hth stratum ($h = 1, 2, \ldots, L$; $i = 1, 2, \ldots, N_h$; $j = 1, 2, \ldots, M_{hi}$). The total number of such units in the hith fsu is

$$\sum^{M_{hi}} Q'_{hij} \tag{20.104}$$

in the hth stratum

$$\sum^{N_h} \sum^{M_{hi}} Q'_{hij} \tag{20.105}$$

and in the whole universe

$$\sum^{L} \sum^{N_h} \sum^{M_{hi}} Q'_{hij} \tag{20.106}$$

The proportion of such tsu's with the attribute in the hith fsu is

$$P_{hi} = \sum^{M_{hi}} Q'_{hi} \Big/ \sum^{M_{hi}} Q_{hij} \tag{20.107}$$

in the hth stratum

$$P_h = \sum^{N_h} \sum^{M_{hi}} Q'_{hij} \Big/ \sum^{N_h} \sum_{M_{hi}} Q_{hij} \tag{20.108}$$

and in the whole universe

$$P = \sum^{L} \sum^{N_h} \sum^{M_{hi}} Q'_{hij} \Big/ \sum^{L} \sum^{N_h} \sum^{M_{hi}} Q_{hij} \tag{20.109}$$

Here again, as in section 15.6 for an unstratified three-stage srs, unbiased estimators of the total number of tsu's in the hth stratum and in the whole universe can be obtained respectively from equations (20.83) and (20.84) by putting $y_{hijk} = 1$ if the third-stage sample unit has the attribute, and 0 otherwise. And so also for their unbiased variance estimators.

For estimating the proportion of the third-stage units possessing the attribute in any stratum, we use the method of section 15.8 for an unstratified three-stage srs, adding the subscript h for the hth stratum. For all the strata combined, an unbiased estimator of the total number of such units is

$$q' = \sum^{L} q'^{*}_{h0} \tag{20.110}$$

with an unbiased variance estimator

$$s_{q'}^2 = \sum^{L} s_{q'^*_{h0}}^2 = \sum^{L}\sum^{n_h}(q'^*_{hi} - q'^*_{h0})^2/n_h(n_h - 1) \tag{20.111}$$

where

$$q'^*_{h0} = \sum^{n_h} q'^*_{hi}/n_h \tag{20.112}$$

and

$$q'^*_{hi} = N_h \frac{M_{hi}}{m_{hi}} \sum^{m_{hi}} \frac{Q_{hij}}{q_{hij}} q'_{hij} \tag{20.113}$$

q'_{hij} being the number of tsu's possessing the attribute out of the sample number q_{hij} of tsu's in the jth ssu in the ith fsu in the hth stratum.

If all the Q_{hij} values are known for the universe, then the unbiased estimator of the universe proportion P (equation (20.109)) is

$$q\Big/\left(\sum^{L}\sum^{N_h}\sum^{M_{hi}} Q_{hij}\right) \tag{20.114}$$

with an unbiased variance estimator

$$s_q^2\Big/\left(\sum^{L}\sum^{N_h}\sum^{M_{hi}} Q_{hij}\right)^2 \tag{20.115}$$

If $\sum^{L}\sum^{N_h}\sum^{M_{hi}} Q_{hij}$ is not known, it is estimated unbiasedly by q (equation (20.95)), with an unbiased variance estimator given by equation (20.96).

A consistent but generally biased estimator of the universe proportion P is the ratio

$$p = q'/q \tag{20.116}$$

with a variance estimator

$$s_p^2 = (s_{q'}^2 + p^2 s_q^2 - 2ps_{q'q})/q^2 \tag{20.117}$$

where

$$\begin{aligned} s_{qq'} &= \sum^{L} s_{q^*_{h0}q'^*_{h0}} \\ &= \sum^{L}\sum^{n_h}(q^*_{hi} - q^*_{h0})(q'^*_{hi} - q'^*_{h0})/n_h(n_h - 1) \end{aligned} \tag{20.118}$$

is an unbiased covariance estimator of q and q'.

See notes to section 10.3 and 15.8.

Example 20.3

In Table 20.2 are given the number of adults in the sample households in each of the two sample villages, selected from the three states. In each selected household, one adult member (18 years or over) was further selected at random from the total number of adults in the household, and asked if he/she agrees to an increase in state taxes for education. The information is given also in Table 20.2. Estimate the total number of adults who agree to an increase in state taxes for education and the proportion they constitute of the total number of adults in each of the three states and also in all the states combined, along with their standard errors.

Extending the notation of Example 20.1, let Q_{hij} ($= x_{hij}$ in Example 20.1) denote the total number of adults in the ijth sample household in the hth stratum. The number of adults selected for the interview (i.e. the number of third-stage sample units) is $q_{hij} = 1$. We can therefore dispense with the subscript k in y_{hijk} by which we had denoted the value of the study variable for the ijkth third-stage sample unit in the hth stratum.

Following the method of section 20.3.8, we put $q'_{hij} = 1$ if the selected adult in the ijth sample household in the hth stratum agrees to an increase in state taxes for education, and $q'_{hij} = 0$ otherwise.

From estimating equation (20.113), we have an unbiased estimator of the stratum total of the number of adults who agree to an increase in state taxes for education

$$q'^{*}_{hi} = N_h \frac{M_{hi}}{m_{hi}} \sum_{j=1}^{m_{hi}} Q_{hij} q'_{hij}$$

from the ith first-stage unit (here a village) in the hth stratum. The required computations follow the methods of section 20.3.8, and are shown in Tables 20.10-20.12, and the final results in Table 20.13.

20.3.9 Ratio method of estimation

As for a stratified two-stage srs (section 20.2.10) so also for a stratified three-stage srs, the ratio method of estimation may, under favorable conditions, improve the efficiency of estimators.

The procedures are the same as for stratified two-stage srs, given in section 20.2.10, except that the unbiased estimator of the stratum total Z_h of the ancillary variable will be computed on the basis of the appropriate estimating equation, namely, an equation of the type (20.83), thus:

From the ith fsu:

$$z^{*}_{hi} = N_h \frac{M_{hi}}{m_{hi}} \sum^{m_{hi}} \frac{Q_{hij}}{q_{hij}} \sum^{q_{hij}} z_{hijk} \tag{20.119}$$

Table 20.10: Computation of the estimated total number of adults who agree to an increase in state taxes for education from a stratified three-stage srs with two sample first-stage units (villages) in each stratum: data of Table 20.2

State	Number of villages		Sample village	For the number of adults who agree to an increase in state taxes for education		
h	Total N_h	Sample n_h	i	$\sum_{j=1}^{m_{hi}} Q_{hij} q'_{hij}$	$\frac{1}{5}\sum_{j=1}^{m_{hi}} Q_{hij} q'_{hij}$	$q'^{*}_{hi} = N_h M_{hi} \cdot Col.(6)$
(1)	(2)	(3)	(4)	(5)	(6)	(7)
I	10	2	1	3	0.6	108
			2	2	0.4	88
					Total	196
					Mean (= stratum estimate of total)	98(q'^{*}_{h0})
					Difference	20($d_{q'h}$)
					$\frac{1}{2}$ |Difference| (= estimated standard error)	10($s_{q'^{*}_{h0}}$)
II	11	2	1	3	0.6	132
			2	2	0.4	79.2
					Total	211.1
					Mean (= stratum estimate of total)	105.6(q'^{*}_{h0})
					Difference	52.8($d_{q'h}$)
					$\frac{1}{2}$ |Difference| (= estimated standard error)	26.4($s_{q'^{*}_{h0}}$)
III	9	2	1	4	0.8	151.2
			2	3	0.6	97.2
					Total	284.4
					Mean (= stratum estimate of total)	124.2(q'^{*}_{h0})
					Difference	54.0($d_{q'h}$)
					$\frac{1}{2}$ |Difference| (= estimated standard error)	27.0($s_{q'^{*}_{h0}}$)

Table 20.11: Computation of the estimated total numbers of adults who agree to an increase in state taxes for education and their variances: data of Tables 20.2 and 20.10

State	Number of adults who agree to an increase in states taxes for education		
	Estimate	Standard Error	Variance
h	q'^{*}_{h0}	$s_{q'^{*}_{h0}}$	$s^2_{q'^{*}_{ho}}$
I	98.0	10.0	100.00
II	105.6	26.4	696.96
III	124.2	27.0	729.00
All states	327.8 (q')	39.05 ($s_{q'}$)	1525.96 ($s^2_{q'}$)

Table 20.12: Computation of the estimated proportion of adults who agree to an increase in state taxes for education and their variances

State h	$p'_h = q'^{*}_{h0}/x^{*}_{h0}$	$s_{x^{*}_{h0}q'^{*}_{h0}} = \frac{1}{4} d_{xh} d_{q'h}$	$s^2_{q'^{*}_{ho}} + p_h^2 s^2_{x^{*}_{ho}} - 2p_h \cdot s_{x^{*}_{h0}q'^{*}_{h0}}$	$s^2_{p_h} =$ col. $(4)/x^{*2}_{ho}$
(1)	(2)	(3)	(4)	(5)
I	0.2344	-220.00	229.728896	0.00131481
II	0.1678	1974.72	191.783509	0.00048433
III	0.2289	656.10	459.576144	0.00156047
Total	0.2062 (p)	2410.82 ($s_{xq'}$)	815.312908 ($s^2 + p^2 s_x^2 - 2ps_{xq'}$)	0.00032254 $s_p^2 =$ col.$(4)/x^2$

Table 20.13: Estimated numbers and proportions of adults who agree to an increase in state taxes for education, computed from the data of a stratified three-stage srs in Table 20.2

Adults who agree to an increase in state taxes for education	State I	State II	State III	All states combined
(a) *Number*				
Estimate	98.0	105.6	124.2	327.8
Standard error	10.0	26.4	27.0	39.05
Coefficient of variation (%)	11.36	25.00	21.74	11.91
(b) *Proportion to total adults*				
Estimate	0.2344	0.1678	0.2289	0.2062
Standard error	0.03626	0.02201	0.03950	0.01796
Coefficient of variation (%)	15.47	13.12	9.62	8.71

From all the n_h fsu's:

$$z_{h0}^{*} = \sum^{n_h} z_{hi}^{*}/n_h \qquad (20.120)$$

where z_{hijk} is the value of the ancillary variable in the $hijk$th selected third-stage unit.

An unbiased variance estimator of z_{h0}^{*} will be given by an estimating equation of the type (20.85).

Example 20.4

For the data of Example 20.3, given the additional information on the previous census population of all the villages as in Example 20.2, obtain ratio estimates of the number of adults who agree to an increase in state taxes for education in the three states separately and also combined, along with their standard errors.

Here the ratio method is applied, as in Example 20.2, at the level of the first-stage units. The unbiased estimates of the previous census population in the three states have already been obtained in Example 10.2, along with their standard errors.

We define

$$r_{3h} = q_{h0}^{\prime *}/z_{h0}^{*}$$

as the ratio of the combined unbiased estimator of the stratum number of adults who agree to an increase in states taxes for education $\sum^{N_h} \sum^{M_{hi}} Q'_{hij}$ to that of the stratum estimator of the previous census population Z_h. The computations follow the methods of sections 20.3.10 and 20.2.11 and are shown in Table 20.14. The final results are given in Table 20.15.

Note: The ratio estimates obtained are more efficient than the unbiased estimates of the number of adults who agree to an increase in state taxes for education in Table 20.13. There is no marked difference between the separate and combined ratio estimates for the whole universe.

20.4 Miscellaneous Notes

1. *Stratified four- and higher-stage srs.* Estimating procedures can be derived for stratified four- and higher-stage srs either directly from the general estimating procedures given in section 19.3 or by extending the procedures for a stratified three-stage design given in section 20.3. All that is required are the unbiased estimators of the stratum totals of the study variables, from which estimators of ratios, means, variances etc. follow.

2. *Gain due to stratification.* As for stratified srs (section 10.6), so also for stratified multi-stage srs, it is possible to estimate the gain, if any, due to stratification, as compared with an unstratified srs with the same total number of ultimate-stage units. For this, see Sukhatme *et al.* (1984), section 8.13.

3. *Stratification after sampling.* The same principles as for a single-stage srs, given in section 10.7, hold for multi-stage srs. If the sample number of first-stage units is large, say over twenty in each stratum, and the errors in weights negligible, the method of post-stratification can give results almost as precise as proportional stratified sampling.

4. *Stratification at lower stages of samples.* As observed in note 2 of section 9.2, stratification may be introduced at different stages of sampling. In household budgets and consumption surveys, for example, households are often stratified on the basis of size or a rough measure of the total income (which information can be readily obtained while listing all the households in the selected penultimate stages), and the required number of sample households selected from the different strata with a different sampling fraction in each. The same principles for estimation outlined earlier will apply in such cases.

Further reading

Foreman, section 8.6; Hansen *et al.* (1953), Vols. I and II, chapters 7-10; Kish (1965), sections 5.6, 6.4, and 6.5; Singh and Chaudhary, section 9.5; Sukhatme *et al.* (1984), sections 8.11 and 8.13.

Table 20.14: Computation of ratio estimates of the total number of adults who agree to an increase in state taxes for education for a stratified three-stage srs: data of Tables 20.2, 10.8, and 10.9

State	Z_h	$r_{3h} = q'^{*}_{h0}/z^{*}_{h0}$	$q'^{*}_{hR} = Z_h r_{3h}$	$s_{q'^{*}_{h0} z^{*}_{h0}} = \frac{1}{4} d_{q'h} d_{zh}$	$s^2_{q'^{*}_{hR}} = s^2_{q'^{*}_{h0}} + r^2_{3h} s^2_{z^{*}_{h0}} - 2r_{3h} s_{q'^{*}_{h0} z^{*}_{h0}}$	$s_{q'^{*}_{hR}}$
(1)	(2)	(3)	(4)	(5)	(6)	(7)
I	863	0.1217	105.03	50.0	88.200275	9.39
II	1010	0.1043	105.34	1676.4	394.221928	19.86
III	942	0.1321	124.44	1822.5	327.002062	18.08
Total	2815		334.81 (q'^{*}_{RS})	3548.9	809.424265 ($s^2_{q'_{RS}}$)	28.45 ($s_{q'_{RS}}$)
Combined	2815 (Z)	0.1164 ($r_3 = q'/z$)	327.67 ($q'_{RC} = Zr_3$)	3548.9 ($s_{q'z}$)	920.866880 ($s^2_{q'_{RC}} = s^2_{q'} + r^2_3 s^2_z - 2r_3 s_{q'z}$)	30.35 ($s_{q'_{RC}}$)

Table 20.15: Ratio estimates of number of adults who agree to an increase in state taxes for education

State	Ratio estimate	Standard error	$CV(\%)$
I	105.0	9.39	8.94
II	105.3	19.86	18.86
III	124.4	18.08	14.53
All states:			
Separate ratio estimate	334.8	28.45	8.50
Combined ratio estimate	327.7	30.35	9.26

Exercises

1. For a survey on the yield of corn in a district, the villages were grouped into 10 strata, and from each stratum two villages were selected at random. From each selected village, two fields were selected at random from all fields on which corn was grown. In each selected field, a rectangular plot of area 1/160 acre was located at random, and the yield of corn in the plot was measured in ounces. The sample data are given in Table 20.16, which also shows for each stratum the area under corn. Estimate for the district the total yield of corn and its standard error (Chakravarti *et al.*, Exercise 3.6, adapted).

Table 20.16: Yield of corn in sample plots

Stratum no.	Area under corn (acres)	Yield of corn (oz. from plot of $\frac{1}{160}$ acre)			
		Sample Village 1		Sample Village 2	
		Plot 1	Plot 2	Plot 1	Plot 2
1	2767	90	102	346	166
2	1577	87	69	85	99
3	1778	82	195	40	236
4	6669	206	176	63	110
5	368	60	72	58	67
6	2875	28	241	120	34
7	5305	367	378	339	328
8	7219	48	236	208	180
9	3782	198	241	195	149
10	1457	32	148	160	112

Source: Chakravarti *et al.*, Exercise 3.6 (adapted).

2. For the same data of Example 20.1, estimate for all the states combined the average number of adults per household and its standard error.

3. For the data of Example 20.2, estimate the average household size from the separate ratio estimates of number of persons and households and its standard error.

[illegible]

CHAPTER 21

Stratified Multi-stage Varying Probability Sampling

21.1 Introduction

In this chapter, we will consider the estimating procedures for a stratified multi-stage sample with varying probabilities of selection: the methods combine those for a stratified single-stage varying probability sample (Chapter 11) and for an unstratified multi-stage varying probability sample (Chapter 16). The estimating methods follow from the fundamental theorems of section 19.3 for a stratified multi-stage varying probability sample.

The general estimating procedures for stratified two- and three-stage designs will be considered first, followed by stratified two- and three-stage designs with pps at the first stage or the first two stages and srs at other stages. The special cases of crop survey will also be mentioned.

The method of ratio estimation and the estimating procedures for proportion of units and sub-universe etc. can be derived on similar lines as those for stratified multi-stage srs.

21.2 Stratified two- and three-stage pps designs

21.2.1 General

The general estimating equations for totals of study variables and the ratio of two totals, as well as those for their respective variances, follow from the general estimating equations in section 19.3. The cases of stratified two- and three-stage pps designs are mentioned briefly.

Table 21.1: Sampling plan for a stratified two-stage pps sample design with replacement. In the hth stratum ($h = 1, 2, \ldots, L$)

Stage (t)	Unit	No. in universe	No. in sample	Selection method	Selection probability	f_t
1	First-stage	N_h	n_h	ppswr	$\pi_{hi} = z_{hi}/Z_h$	$n_h \pi_{hi}$
2	Second-stage	M_{hi}	m_{hi}	ppswr	$\pi_{hij} = w_{hij}/W_{hi}$	$m_{hi}\pi_{hij}$

21.2.2 Two-stage pps design

The structure of a stratified two-stage design is given in Table 21.1. The universe is sub-divided into L strata, and in the hth stratum ($h = 1, 2, \ldots, L$), a sample of n_h first-stage units is selected out of the total N_h units with pps (and replacement), the (initial) probability of selection of the ith first-stage unit ($i = 1, 2, \ldots, N_h$) at any draw being

$$\pi_{hi} = z_{hi}/Z_h \tag{21.1}$$

where z_{hi} is the value ("size") of an ancillary variable $Z_h = \sum_{i=1}^{N_h} z_{hi}$ being known. In each of the n_h first-stage units thus selected, a sample m_{hi} ($i = 1, 2, \ldots, n_h$) of second-stage units is selected with pps (and replacement) from the total M_{hi} units, the initial probability of selection of the jth second-stage unit ($j = 1, 2, \ldots, N_h$) being

$$\pi_{hij} = w_{hij}/W_{hi} \tag{21.2}$$

where w_{hij} is the value of another ancillary variable, and $W_{hi} = \sum_{j=1}^{M_{hi}} w_{hij}$.

The universe totals and means are defined as for a stratified two-stage srs in section 20.2.1.

If y_{hij} is the value of the study variable in the jth selected ssu in the ith selected fsu in the hth stratum ($i = 1, 2, \ldots, n_h$; $j = 1, 2, \ldots, m_{hi}$), then from equation (16.3) or (19.21), an unbiased estimator of the stratum total Y_h (as defined by equation (20.3)) from the ith selected fsu is

$$y^*_{hi} = \frac{1}{\pi_{hi} m_{hi}} \sum_{j=1}^{m_{hi}} \frac{y_{hij}}{\pi_{hij}} = \frac{Z_h W_{hi}}{z_h m_{hi}} \sum_{j=1}^{m_{hi}} \frac{y_{hij}}{w_{hij}} \tag{21.3}$$

and a combined unbiased estimator of Y_h from all the n_h fsu's is the arithmetic mean

$$y^*_{h0} = \frac{1}{n_h} \sum^{n_h} y^*_{hi} \tag{21.4}$$

with an unbiased variance estimator

$$s^2_{y^*_{h0}} = \sum^{n_h}(y^*_{hi} - y^*_{h0})^2/n_h(n_h - 1) \tag{21.5}$$

For the whole universe, an unbiased estimator of the total $Y = \sum^L Y_h$ is

$$y = \sum^L y^*_{h0} \tag{21.6}$$

with an unbiased variance estimator

$$s^2_y = \sum^L s^2_{y^*_{h0}} \tag{21.7}$$

Estimators of the ratios of totals of two study variables and their variances follow directly from the fundamental theorems of section 19.3.

Notes

1. *Sampling variance of the estimator* y. From equation (16.6) the sampling variance y^*_{h0} in ppswr at both stages is

$$\begin{aligned}\sigma^2_{y^*_{h0}} &= \frac{1}{n_h}\left(\sum_{i=1}^{N_h}\frac{Y^2_{hi}}{\pi_{hi}} - Y^2_h\right)\\ &+\frac{1}{n_h}\sum_{i=1}^{N_h}\frac{1}{\pi_{hi}m_{hi}}\left(\sum_{j=1}^{M_{hi}}\frac{Y^2_{hij}}{\pi_{hij}} - Y^2_{hi}\right)\end{aligned} \tag{21.8}$$

   which is estimated unbiasedly by $s^2_{y^*_{h0}}$, defined by equation (21.5).

   The sampling variance of y is

$$\sigma^2_y = \sum^L \sigma^2_{y^*_{h0}} \tag{21.9}$$

   which is estimated unbiasedly by s^2_y, defined by equation (21.7).

2. As noted in section 16.2.1, great simplifications result if the same size variable is used for pps sampling in both the stages.

21.2.3 Three-stage pps design

The sampling plan for a stratified three-stage pps design is given in summary form in Table 21.2. If y_{hijk} is the value of the study variable in the kth selected third stage unit of the jth selected second-stage unit of the ith selected first-stage unit in the hth stratum ($h = 1, 2, \ldots, L$; $i = 1, 2, \ldots, n_h$;

Table 21.2: Sampling plan for a stratified three-stage pps sample design with replacement. In the hth stratum ($h = 1, 2, \ldots, L$)

Stage (t)	Unit	No. in universe	No. in sample	Selection method	Selection probability	f_t
1	First-stage	N_h	n_h	ppswr	$\pi_{hi} = z_{hi}/Z_h$	$n_h\pi_{hi}$
2	Second-stage	M_{hi}	m_{hi}	ppswr	$\pi_{hij} = w_{hij}/W_{hi}$	$m_{hi}\pi_{hij}$
3	Third-stage	Q_{hij}	q_{hij}	ppswr	$\pi_{hijk} = v_{hijk}/V_{hij}$	$q_{hij}\pi_{hijk}$

$j = 1, 2, \ldots, m_{hi}$; $k = 1, 2, \ldots, q_{hij}$), then from the general estimating equation (19.21) or from equation (16.12), an unbiased estimator of the stratum total Y_h, obtained from the ith selected fsu is

$$\begin{aligned} y^*_{hi} &= \frac{1}{\pi_{hi} m_{hi}} \sum_{j=1}^{m_{hi}} \frac{1}{\pi_{hij} q_{hij}} \sum_{k=1}^{q_{hij}} \frac{y_{hijk}}{v_{hijk}} \\ &= \frac{Z_h W_{hi}}{z_{hi} m_{hi}} \sum_{j=1}^{m_{hi}} \frac{V_{hij}}{w_{hij} q_{hij}} \sum_{k=1}^{q_{hij}} \frac{y_{hijk}}{v_{hijk}} \end{aligned} \tag{21.10}$$

and a combined unbiased estimator is

$$y^*_{ho} = \sum^{n_h} y^*_{hi}/n_h \tag{21.11}$$

with an unbiased variance estimator

$$s^2_{y^*_{ho}} = \sum_{h=1}^{n_h} (y^*_{hi} - y^*_{h0})^2/n_h(n_h - 1) \tag{21.12}$$

where $Z_h = \sum_{i=1}^{N_h} z_{hi}$; $W_{hi} = \sum_{j=1}^{M_{hi}} w_{hij}$; and $V_{hij} = \sum_{k=1}^{Q_{hij}} v_{hijk}$.

For the whole universe, an unbiased estimator of the total Y is

$$y = \sum^{L} y^*_{h0} \tag{21.13}$$

with an unbiased variance estimator

$$s^2_y = \sum^{L} s^2_{y^*_{ho}} \tag{21.14}$$

Estimators of the ratios of totals of two study variables and their variances follow directly from the fundamental theorems of section 19.3.

Note: As noted in section 16.2.2, great simplifications result if the same size variable is used for pps sampling in all the three stages.

Table 21.3: Sampling plan for a stratified two-stage design with pps sampling at the first-stage and srs at the second-stage. In the hth stratum $(h = 1, 2, \ldots, L)$

Stage (t)	Unit	No. in universe	No. in sample	Selection method	Selection probability	f_t
1	First-stage	N_h	n_h	ppswr	$\pi_{hi} = z_{hi}/Z_h$	$n_h \pi_{hi}$
2	Second-stage	M_{hi}	m_{hi}	srs	Equal $= 1/M_{hi}$	m_{hi}/M_{hi}

21.3 Stratified two-stage design with pps and srs

21.3.1 General case

We illustrate this with a sampling plan for a rural sample survey, where in each stratum into which the universe is divided, villages are the first-stage units selected with ppswr (the "size" being population in a previous census for a demographic inquiry and area for a crop survey) and households or fields the second-stage units selected as an srs or systematically (Table 21.3). This is a very common type of nation-wide sample inquiries.

An unbiased estimator of the hth stratum total Y_h $(h = 1, 2, \ldots, L)$, obtained from the ith fsu $(i = 1, 2, \ldots, n_h)$ is (from equation (21.3))

$$y^*_{hi} = \frac{M_{hi}}{\pi_{hi} m_{hi}} \sum_{j=1}^{m_{hi}} y_{hij} = \frac{M_{hi}}{\pi_{hi}} \bar{y}_{hi} \tag{21.15}$$

where y_{hij} is the value of the study variable in the jth selected ssu (household or field) or the ith selected fsu (village) in the hth stratum, and

$$\bar{y}_{hi} = \sum_{j=1}^{m_{hi}} y_{hij}/m_{hi} \tag{21.16}$$

is the mean of the y_{hij} values in the hith sample fsu.

The combined unbiased estimator of Y_h is

$$y^*_{h0} = \sum^{n_h} y^*_{hi}/n_h \tag{21.17}$$

and an unbiased estimator of the overall total Y is

$$y = \sum^{L} y^*_{h0} \tag{21.18}$$

Unbiased variance estimators of the stratum and the universe estimators of totals are given by equations (19.23) and (19.25) respectively: estimation of ratios of two totals and their variances follow from the methods of section 19.3. These will be illustrated by Examples 21.1 and 21.2.

21.3.2 First-stage units selected with probability proportional to the number of second-stage units

If M_{hi}, the number of ssu's, is known beforehand for all the N_h fsu's, and the n_h fsu's in the hth stratum are selected with probability proportional to M_{hi}, then, as shown in section 16.3.2 for one stratum, the estimating procedure becomes simpler.

Example 21.1

In each of the two sample villages selected in each of the three states in Example 11.1 from the list of villages in Appendix IV with probability proportional to previous census population and with replacement, select five households at random, given the listing of the households in the sample villages. On the basis of the collected data for the sample households on the size and number of adults (persons aged eighteen years or over), estimate the total number of persons, the number and the proportion of adults for three states separately and also combined, along with their standard errors. The sample data are given in Table 21.4. The use of the last column will be illustrated in Example 21.3.

As there are two sample first-stage units, i.e. villages, in each stratum, we follow, as in Example 20.1, the simplified procedures mentioned in note 2 of section 19.3, illustrated also in Example 11.1 for the stratified single-stage pps sampling. As in Example 20.1, we denote by y_{hij} the household size and by x_{hij} the number of adults in the sample households ($h = 1, 2, 3$ for the states; $i = 1, 2$ for the sample villages; and $j = 1, 2, 3, 4, 5$ for the sample households): note that in Example 11.1 the total number of households in a sample village was denoted by x_{hi}, which we denote here by M_{hi}, and the sample number by m_{hi}. Unbiased estimates of the total number of households in the three states separately and combined have already been obtained in Example 11.1, along with their standard errors. These, along with the other required computations, are shown in Tables 21.5-21.7, and the final estimates in Table 21.8.

Table 21.4: Size, number of adults (18 years or over), and agreement of the selected adult to an increase in state taxes for education

Household serial no.	Household size (y_{hij})	Number of adults (x_{hij})	Agreement of the selected adult to an increase in state taxes for education (q'_{hij} = 1 for Yes; = 0 for No)
State I: Village serial no. 5; total number of households = 24			
1	5	2	0
6	5	2	0
16	4	2	1
18	6	3	0
22	3	2	0
State I: Village serial no. 7; total number of households = 20			
4	5	3	1
8	4	2	0
13	5	3	0
16	4	2	0
18	4	2	0
State II: Village serial no. 3; total number of households = 17			
3	5	3	0
5	6	3	0
7	3	2	0
14	5	2	1
15	3	2	0
State II: Village serial no. 7; total number of households = 19			
4	3	1	0
12	6	4	0
14	7	4	0
16	5	3	1
18	5	2	0
State III: Village serial no. 2; total number of households = 21			
3	5	3	0
5	6	3	0
11	6	4	1
14	5	2	0
15	8	5	0
State III: Village serial no. 4; total number of households = 21			
7	5	2	0
8	6	3	1
12	5	3	0
13	7	3	0
15	8	5	0

Table 21.5: Computation of the estimated total number of persons and adults: data of Tables 11.2 and 21.9

State	Sample village	Reciprocal of probability $1/\pi_{hi}$	Total number of households M_{hi}	$m^*_{hi} = M_{hi}/\pi_{hi}$ †	For number of persons $\sum_{j=1}^{5} y_{hij}$	$\bar{y}_{hi} = \frac{1}{5}\sum_{j=1}^{5} y_{hij}$	$y^*_{hi} = (M_{hi}/\pi_{hi})\bar{y}_{hi}$
(1)	(2)	(3)	(4)	(5)	(6)	(7)	(8)
I	1	9.3804	24	225.1296	23	4.6	1035.60
	2	11.9861	20	239.7220	22	4.4	1054.78
			Total	464.8516			2090.38
			Mean (= Stratum estimate of total)	232.45 (m^*_{h0})			1045.19 (y^*_{h0})
			Difference	-14.59 (d_{mh})			-19.18 (d_{yh})
			$\frac{1}{2}$ |Difference|(= Estimated standard error)	7.295 ($s_{m^*_{h0}}$)			9.59 ($s_{y^*_{h0}}$)
II	1	13.8356	17	235.2052	22	4.4	1034.90
	2	11.8824	19	225.7656	26	5.2	1173.98
			Total	460.9708			2208.88
			Mean (= Stratum estimate of total)	230.49 (m^*_{h0})			1104.44 (y^*_{h0})
			Difference	9.44 (d_{mh})			-139.08 (d_{yh})
			$\frac{1}{2}$ |Difference|(= Estimated standard error)	4.72 ($s_{m^*_{h0}}$)			69.54 ($s_{y^*_{h0}}$)
III	1	8.4107	21	176.6247	30	6.0	1059.75
	2	8.0513	21	169.0773	31	6.2	1048.28
			Total	345.7020			2108.03
			Mean (= Stratum estimate of total)	172.85 (m^*_{h0})			1054.02 (y^*_{h0})
			Difference	7.55 (d_{mh})			11.73 (d_{yh})
			$\frac{1}{2}$ |Difference|(= Estimated standard error)	3.775 ($s_{m^*_{h0}}$)			5.865 ($s_{y^*_{h0}}$)

(continued)

† These are the estimated number of households in Tables 11.2, denoted by x^*_{hi}. In the present example, x^*_{hi} denotes the estimated numbers of adults, and m^*_{hi} those of households.

Table 21.5 continued

State	Sample village	For number of adults		
		$\sum_{j=1}^{5} x_{hij}$	$\bar{x}_{hi} = \frac{1}{5}\sum_{j=1}^{5} x_{hij}$	$x^*_{hi} = (M_{hi}/\pi_{hi})\bar{x}_{hi}$
(1)	(2)	(9)	(10)	(11)
I	1	11	2.2	495.29
	2	12	2.4	575.33
			Total	1070.62
			Mean (= Stratum estimate of total)	535.31 (x^*_{h0})
			Difference	−80.04 (d_{xh})
			$\frac{1}{2}$ |Difference|(= Estimated standard error)	40.02 ($s_{x^*_{h0}}$)
II	1	12	2.4	564.49
	2	14	2.8	620.26
			Total	1184.75
			Mean (= Stratum estimate of total)	592.38 (x^*_{h0})
			Difference	−55.47 (d_{xh})
			$\frac{1}{2}$ |Difference|(= Estimated standard error)	27.885 ($s_{x^*_{h0}}$)
III	1	17	3.4	600.52
	2	16	3.2	541.05
			Total	1141.57
			Mean (= Stratum estimate of total)	570.78 (x^*_{h0})
			Difference	59.47 (d_{xh})
			$\frac{1}{2}$ |Difference|(= Estimated standard error)	29.735 ($s_{x^*_{h0}}$)

Table 21.6: Computation of estimated numbers of households, persons and adults: data of Tables 21.4 and 21.5

State h	Number of households†			Number of persons			Number of adults		
	Estimate m^*_{h0}	S.E. $s_{m^*_{h0}}$	Variance $s^2_{m^*_{h0}}$	Estimate y^*_{h0}	S.E. $s_{y^*_{h0}}$	Variance $s^2_{y^*_{h0}}$	Estimate x^*_{h0}	S.E. $s_{x^*_{h0}}$	Variance $s^2_{x^*_{h0}}$
I	232.43	7.295	53.2170	1045.19	9.59	91.9681	535.31	40.020	1601.6004
II	230.49	4.720	22.2784	1104.44	69.54	4835.8116	592.38	27.885	777.5732
III	172.85	3.775	14.2506	1054.02	5.865	34.3982	570.78	29.735	884.1702
All states combined	635.77 (m)	9.473 (s_m)	89.7460 (s^2_m)	3203.65 (y)	70.443 (s_y)	4962.1779 (s^2_y)	1698.47 (x)	57.126 (s_x) (s^2_x)	3263.3438

† Also from Example 11.1.

Table 21.7: Computation of estimated ratios and variances: data of Table 21.5

State	For the household size			
	$r_h = y^*_{h0}/m^*_{h0}$	$s_{y^*_{h0}m^*_{h0}} = \frac{1}{4} d_{yh} d_{mh}$	$s^2_{y^*_{h0}} + r^2 s^2_{m^*_{h0}} - 2r_h s_{y^*_{h0}m^*_{h0}}$	$s^2_{r_h} =$ col.(4)$/m^{*2}_{h0}$
(1)	(2)	(3)	(4)	(5)
I	4.497	70.0070	538.5325	0.009971
II	4.791	-328.2288	8493.3551	0.159859
III	6.100	22.1116	541.1492	0.018123
All states combined	5.039 (r)	-255.0108 (s_{ym})	9811.2035 ($s^2_y + r^2 s^2_m - 2r s_{ym}$)	0.024278 (s^2_r)

State	For the proportion of adults			
	$p_h = x^*_{h0}/y^*_{h0}$	$s_{y^*_{h0}m^*_{h0}} = \frac{1}{4} d_{yh} d_{xh}$	$s^2_{x^*_{h0}} + p^2_h s^2_{y^*_{h0}} - 2p_h s_{y^*_{h0}x^*_{h0}}$	$s^2_{p_h} =$ col.(8)$/y^{*2}_{h0}$
(1)	(6)	(7)	(8)	(9)
I	0.5122	383.8398	1232.9229	0.00112859
II	0.5363	139.9123	2018.3693	0.00165480
III	0.5416	174.3958	705.3547	0.00063493
All states combined	0.5302 (p)	698.1479 (s_{yx})	3918.3563 ($s^2_x + p^2 s^2_y - 2p s_{yx}$)	0.00038179 (s^2_p)

Table 21.8: Estimates and standard errors computed from the data of stratified two-stage (pps and srs) design in Table 21.4

Item	State I	State II	State III	All states combined
1. *Number of households*†				
(a) Estimate	232.4	230.5	172.8	635.8
(b) Standard error	7.30	4.72	3.78	9.47
(c) Coefficient of variation (%)	3.14	2.05	2.18	1.49
2. *Number of persons*				
(a) Estimate	1045.2	1104.4	1054.0	3203.6
(b) Standard error	9.59	69.54	5.86	70.44
(c) Coefficient of variation (%)	0.92	6.30	0.56	2.20
3. *Number of adults*				
(a) Estimate	535.3	592.4	570.8	1698.5
(b) Standard error	40.02	27.88	29.74	57.13
(c) Coefficient of variation (%)	7.48	4.71	5.21	3.36
4. *Average household size*				
(a) Estimate	4.50	4.79	6.10	5.04
(b) Standard error	0.0999	0.3998	0.1346	0.1558
(c) Coefficient of variation (%)	2.22	8.34	2.21	2.61
5. *Proportion of adults*				
(a) Estimate	0.5122	0.5363	0.5416	0.5302
(b) Standard error	0.0336	0.0407	0.0252	0.0195
(c) Coefficient of variation (%)	6.56	7.59	4.65	3.69

† Also from Example 11.1.

Example 21.2

For estimating the total yield of paddy in a district, a stratified two-stage sample design was adopted, where four villages were selected from each stratum, with ppswr, the "size" being the geographical area, and four plots were selected from each sample village circular systematically for ascertaining the yield of paddy. Using the data given in Table 21.9, estimate unbiasedly the total yield of paddy in the district and its standard error (Murthy (1967), Problem 9.7).

The required computations are given in Tables 21.10 and 21.11 denoting by y_{hij} the yield of paddy (in kilograms) in the jth sample plot of the ith sample village of the hth stratum ($h = 1, 2, 3$, for the strata, $i = 1, 2, 3, 4$ for the sample villages, and $j = 1, 2, 3, 4$ for the sample plots).

The estimated total yield of paddy in the district is 3976 tonnes (1 tonne = 1000 kg). The estimated standard error of estimate of this total is $\sqrt{(118,352)} =$ 344 tonnes, i.e. a CV of 8.65 per cent.

Table 21.9: Yield of paddy in sample plots, selected circular systematically in the sample villages, and selected with probability proportional to geographical area in each of the three strata of a district

Stratum h	Sample village i	Reciprocal of probability $1/\pi_{hi}$	Total number of plots M_{hi}	Yield of paddy (in kg) y_{hij} 1	2	3	4
I	1	440.21	28	104	182	148	87
	2	660.43	14	108	64	132	156
	3	31.50	240	100	115	50	172
	4	113.38	76	346	350	157	119
II	1	21.00	256	124	111	135	216
	2	16.80	288	123	177	106	138
	3	24.76	222	264	78	144	55
	4	49.99	69	300	114	68	111
III	1	67.68	189	110	281	120	114
	2	339.14	42	80	61	118	124
	3	100.00	134	121	212	174	106
	4	68.07	161	243	116	314	129

Source: Murthy (1969), Problem 9.7.

21.4 Stratified three-stage design with pps, srs, and srs

The example of section 21.3 will be extended in this section by taking a simple random sample of q_{hij} third-stage units out of the total number Q_{hij} of such units in the selected jth ssu of the selected ith fsu in the hth stratum ($h = 1, 2, \ldots, L$; $i = 1, 2, \ldots, n_h$; $j = 1, 2, \ldots, m_{hi}$). The sampling plan is shown in summary form in Table 21.12.

An unbiased estimator of the stratum total Y_h (as defined by equation (20.3)), obtained from the ith sample fsu is, from equation (16.25) or (19.21) or (21.10),

$$\begin{aligned} y^*_{hi} &= \frac{M_{hi}}{\pi_{hi} m_{hi}} \sum_{j=1}^{m_{hi}} \frac{Q_{hij}}{q_{hij}} \sum_{k=1}^{q_{hij}} y_{hijk} \\ &= \frac{M_{hi}}{\pi_{hi} m_{hi}} \sum_{j=1}^{m_{hi}} Q_{hij} \bar{y}_{hij} \end{aligned} \tag{21.19}$$

where y_{hijk} is the value of the study variable in the kth selected tsu of the jth selected ssu of the ith selected fsu in the hth stratum ($h = 1, 2, \ldots, L$;

Table 21.10: Computation of the estimated total yield of paddy: data of Table 21.9

Stratum h	Sample village i	(M_{hi}/π_{hi})	$\sum_{j=1}^{4} y_{hij}$ (kg)	$\bar{y}_{hi} = \frac{1}{4}\sum_{j=1}^{4} y_{hij}$ (kg)	$y_{hi}^{*} = (M_{hi}/\pi_{hi})\bar{y}_{hi}$ (tonnes; 1 tonne = 1000 kg)
(1)	(2)	(3)	(4)	(5)	(6)
I	1	12 325.88	521	130.25	1605
	2	9246.03	460	115.00	1063
	3	7560.00	437	109.25	826
	4	8616.88	972	243.00	2094
Total					5588
Mean (= stratum estimate of total)					1397 (y_{h0}^{*})
II	1	5376.00	586	146.50	788
	2	4838.40	544	136.00	658
	3	5496.72	541	135.25	743
	4	3449.31	593	148.25	511
Total					2700
Mean (= stratum estimate of total)					675 (y_{h0}^{*})
III	1	12 791.52	625	156.25	1999
	2	14 243.88	383	95.75	1364
	3	13 400.00	613	153.25	2054
	4	10 959.27	802	200.50	2197
Total					7614
Mean (= stratum estimate of total)					1903.5 (y_{h0}^{*})

Table 21.11: Computation of the estimated variance of the estimated total yield of paddy: data of Tables 21.9 and 21.10

Stratum h	y^*_{h0}	$\sum_{i=1}^{4} y^{*2}_{hi}$	$\frac{1}{4}\left(\sum_{i=1}^{4} y^*_{hi}\right)^2$	SSy^*_{hi} = col. (3) - col. (4)	$s^2_{y^*_{h0}} = SSy^*_{hi}/12$
(1)	(2)	(3)	(4)	(5)	(6)
I	1397.0	8773.106	7806.436	966.670	80 555.83
II	675.0	1867.078	1822.500	44.578	3714.83
III	1903.5	14 902.222	14 493.249	408.973	34 081.08
Total	3975.5 (y)				118 351.74 (s^2_y)

Table 21.12: Sampling plan for a three-stage design with pps sampling at the first stage, and srs at the second and third stages. In the hth stratum ($h = 1, 2, \ldots, L$)

Stage (t)	Unit	No. in universe	No. in sample	Selection method	Selection probability	f_t
1	First-stage (village)	N_h	n_h	ppswr	$\pi_{hi} = z_{hi}/Z_h$	$n_h \pi_{hi}$
2	Second-stage (household or field)	M_{hi}	m_{hi}	srswr	Equal = $1/M_{hi}$	m_{hi}/M_{hi}
3	Third-stage (person or plot)	Q_{hij}	q_{hij}	srswr	Equal = $1/Q_{hij}$	q_{hij}/Q_{hij}

$i = 1, 2, \ldots, n_h$; $j = 1, 2, \ldots, m_{hi}$; $k = 1, 2, \ldots, q_{hij}$) and

$$\overline{y}_{hij} = \sum_{k=1}^{q_{hij}} y_{hijk}/q_{hij} \tag{21.20}$$

is the arithmetic mean of the y_{hijk} values in the hijth ssu.

A combined unbiased estimator of Y_h is

$$y^*_{h0} = \sum^{n_h} y^*_{hi}/n_h \tag{21.21}$$

and an unbiased estimator of the universe total Y is

$$y = \sum^{L} y^*_{h0} \tag{21.22}$$

Unbiased variance estimators of y^*_{h0} and y are given respectively by estimating equations (19.23) and (19.25). An estimator of the ratio of the totals of two study variables and its variance estimator follow from estimating equations (19.11)-(19.14). These will be illustrated with Examples 21.3 and 21.4.

Example 21.3

In Table 21.3 are given the number of adults in the sample households, randomly selected out of the total number of households in each of the two sample villages, selected with pps from the three states. In each selected household, one adult member (eighteen years or over) was further selected at random from the total number of adults in the households and asked if he/she agrees to an increase in state taxes for education. The information is given also in Table 21.4. Estimate the total number of adults who agree to an increase in state taxes for education and also the proportion they constitute of the total number of adults in each of the three states and in all the states combined, along with their standard errors.

Extending the notation of Example 21.1, let Q_{hij} (= x_{hij} in Example 21.1) denote the total number of adults in the ijth sample household in the hth stratum. The number of adults selected for interview (i.e. the number of third-stage sample units) is $q_{hij} = 1$. We can therefore dispense with the subscript k in y_{hijk} by which we had denoted the value of the study variable for the ijkth third-stage sample unit in the hth stratum.

Similar to the procedures adopted in Examples 16.2 and 20.3, we put $q'_{hij} = 1$ if the selected adult in the ijth sample household in the hth stratum agrees to an increase in state taxes for education, and 0 otherwise.

From estimating equation (21.19), we have an unbiased estimator of the stratum total Q'_{hi} of the number of adults with knowledge of development plans

$$q'^*_{hi} = \frac{M_{hi}}{\pi_{hi} m_{hi}} \sum_{j=1}^{m_{hi}} Q_{hij} q'_{hij}$$

from the ith first-stage unit (i.e. village) in the hth stratum. The required computations are shown in Tables 21.13-21.15, and the final results in Table 21.16.

Example 21.4

For the household inquiries in the Indian National Sample Survey (1953-4) in the rural sector, the design was stratified three-stage. The total number of 2522 *tehsils* were grouped into 240 strata on the basis of consumer expenditure (as estimated in earlier surveys) and geographical continuity, such that each stratum contained approximately equal population, as in the census of 1951. In each stratum, two *tehsils* and in each selected *tehsil*, two sample villages, were selected, sampling at both stages being with probability proportional to the 1951 Census population or area and with replacement; within each selected village, nine households on an average were selected by the enumerators systematically with a random start from the lists of households in the village, which they had prepared on reaching the villages. The sample comprised 8235 households and 49,177 persons.

With the previous notations, the reader will see that an unbiased estimator of the stratum total Y_h in the hth stratum ($h = 1, 2, \ldots, 240$), obtained from the ith ($i = 1, 2$) first-stage unit is (from equation (21.19))

$$h^{\bullet}_{hi} = \frac{1}{2} \cdot \frac{1}{\pi_{hi}} \sum_{j=1}^{2} \frac{Q_{hij}}{\pi_{hij} q_{hij}} \sum_{k=1}^{q_{hij}} y_{hjik}$$

where y_{hijk} is the value of the study variable in the kth selected third-stage unit of the jth selected ssu of the ith selected fsu in the hth stratum. A combined unbiased estimator of Y_h is

$$y^{\bullet}_{h0} = \frac{1}{2}\left(y^{\bullet}_{h1} + y^{\bullet}_{h2}\right)$$

and the unbiased estimator of the universe total $Y = \sum_{h=1}^{240} Y_h$ is

$$y = \sum_{h=1}^{240} y^{\bullet}_{h0}$$

Unbiased variance estimators of the stratum and the universe estimators of totals are given by equations (19.23) and (19.25) respectively: estimators of ratios of two totals and their variance estimators follow from section 19.3. As two fsu's are selected with replacement in each stratum, the simplified procedure of note 2 to section 19.3 could be used for the estimation of totals and ratios of totals and their variances.

Some results are given in Table 21.17 on the rates of births, deaths, marriage, and sickness. The coefficients of variations of the estimates are rather large, but the sample was not designed specifically to provide demographic estimates: for

Table 21.13: Computation of the estiamted total number of adults who agree to an increase in state taxes for education from a stratified three-stage design with two pps first-stage units (villages) in each stratum, srs of second-stage units (households), and srs of one adult in the selected households: data of Tables 11.2, 21.4, and 21.5

State h	Sample village i	$\sum_{j=1}^{5} Q_{hij'} q'_{hij}$	$\frac{1}{5}\sum_{j=1}^{5} Q_{hij'} q'_{hij}$	$q'^{*}_{hi} = (M_{hi'}/\pi_{hi})\times$ col. (4)
(1)	(2)	(3)	(4)	(5)
I	1	2	0.4	90.05
	2	3	0.6	143.83
			Total	233.88
			Mean (= Stratum estimate of total)	116.94 (q'^{*}_{h0})
			Difference	-53.78
			$\frac{1}{2}$ |Difference| (= Estimated standard error)	26.89 ($s_{q'^{*}_{h0}}$)
II	1	2	0.4	94.08
	2	3	0.6	135.46
			Total	299.54
			Mean (= Stratum estimate of total)	114.77 (q'^{*}_{h0})
			Difference	-41.38
			$\frac{1}{2}$ |Difference| (= Estimated standard error)	20.69 ($s_{q'^{*}_{h0}}$)
III	1	4	0.8	141.30
	2	3	0.6	101.45
			Total	242.75
			Mean (= Stratum estimate of total)	121.38 (q'^{*}_{h0})
			Difference	39.85
			$\frac{1}{2}$ |Difference| (= Estimated standard error)	19.925 ($s_{q'^{*}_{h0}}$)

Table 21.14: Computation of the estimated total number of adults who agree to an increase in state taxes: data of Tables 21.4 and 21.13

State h	Estimate q'^{*}_{h0}	Standard error $s_{q'^{*}_{h0}}$	Variance $s^2_{q'^{*}_{h0}}$
I	116.94	26.89	723.0721
II	114.77	20.69	428.0761
III	121.38	19.92	397.0056
All states combined	353.09 (q')	39.347 ($s_{q'}$)	1548.1538 ($s^2_{q'}$)

Table 21.15: Computation of the estimated proportions of adults who agree to an increase in state taxes for education and their variances

State	$p'_h = q'^{*}_{h0}/x^{*}_{h0}$	$s_{x^{*}_{h0}q'^{*}_{h0}} = \frac{1}{4} d_{xh} d_{q'h}$	$s^2_{q'_{h0}} + s'^2_h s^2_{x^{*}_{h0}} - 2p'_h s_{x^{*}_{h0}q'^{*}_{h0}}$	$s^2_{p'_h} =$ col. (4)$/x^{*2}_{h0}$
(1)	(2)	(3)	(4)	(5)
I	0.2185	1076.1378	329.2635	0.00114903
II	0.1937	576.9406	233.7438	0.00066610
III	0.2127	592.4699	184.9696	0.00056776
All states combined	0.2079 (p')	2254.5483 ($s_{xq'}$)	755.4598 ($s^2_{q'} + p'^2 s^2_x - 2p' s_{xq'}$)	0.00026189 ($s^2_{p'}$)

Table 21.16: Estimated numbers and proportions of adults who agree to an increase in state taxes for education, computed from the data of stratified three-stage (pps, srs, and srs) design in Table 21.9

Adults who agree to an increase in states taxes for education	State I	State II	State III	All states combined
1. *Number*				
(a) Estimate	116.9	114.8	121.4	353.1
(b) Standard error	26.9	20.7	19.9	39.35
(c) Coefficient of variation (%)	2.99	18.03	16.42	11.14
2. *Proportion of total adults*				
(a) Estimate	0.2185	0.1937	0.2127	0.2079
(b) Standard error	0.0339	0.0258	0.0238	0.0162
(c) Coefficient of variation (%)	15.51	13.31	11.19	7.79

Table 21.17: Estimates of vital rates per 1000 persons: Indian National Sample Survey, rural sector, 1953-4

	Estimate	Standard error	CV (%)
Births (annual)	34.6	1.0	2.9
Deaths (annual)	16.6	1.1	6.6
Marriages (annual)	7.1	0.5	7.0
Prevalence of sickness (monthly)	64.8	4.4	6.8

Source: Ganguly and Som (1958).

results of a later survey, see Example 10.3. The other point to note is that the reported birth, death, and sickness rates were obvious under-estimates, as could be seen from available external evidence. One method of obtaining adjusted birth and death rates from such defective data is described briefly in section 25.6.5.

21.5 Special cases of crop surveys

21.5.1 Introduction

The estimating procedures in crop surveys are greatly simplified if the sampling units at each stage in a multi-stage design are selected with probability proportional to area (section 16.5). The extension to stratified multi-stage design is straightforward and will be illustrated with a stratified two-stage design.

21.5.2 Area surveys of crops

In a stratified design with L strata, in each stratum the fsu's (villages) are selected with probability proportional to their (geographical) areas and in the selected fsu's, the ssu's (fields) are selected with probability proportional to their areas. We assume that the same number n_0 of fsu's is selected in each stratum and the same number m_0 of ssu's selected in the selected fsu's.

If a_{hij} is the total area, y_{hij} the area under the crop, and $p_{hij} = y_{hij}/a_{hij}$ the proportion of the area under the crop in the hijth sample field ($h = 1, 2, \ldots, L$; $i = 1, 2, \ldots, n_0$; $j = 1, 2, \ldots, m_0$), then an unbiased estimator of proportion of area under the crop in the hth stratum is (from equation (16.31))

$$\bar{p}_h = \sum_{i=1}^{n_0} \sum_{j=1}^{m_0} p_{hij}/n_0 m_0 \tag{21.23}$$

and an unbiased estimator of the total area under the crop in the hth stratum is (from equation (16.29))

$$y_{h0}^* = A_h \bar{p}_h \tag{21.24}$$

(where A_h is the total area of the hth stratum) and in the whole universe

$$y = \sum^{L} y_{h0}^* \tag{21.25}$$

Unbiased variance estimators of $\bar{p}_h$ and y_{h0}^* are given respectively by equations of the type (16.32) and (16.33), and an unbiased variance estimator of y is given by equation (21.7).

21.5.3 Yield surveys of crops

If x_{hij} is the yield of a crop and $r_{hij} = x_{hij}/y_{hij}$ is the average yield per unit area in the hijth sample field, then an unbiased estimator of the total yield in the hth stratum is (from equation (16.34))

$$x^*_{h0} = A_h \bar{r}_h \tag{21.26}$$

where

$$\bar{r}_h = \sum_{i=1}^{n_0} \sum_{j=1}^{m_0} r_{hij}/n_0 m_0 \tag{21.27}$$

is an unbiased estimator of the average yield per unit area in the stratum.

An unbiased estimator of the total yield in the whole universe is

$$x = \sum^{L} x^*_{h0} \tag{21.28}$$

Unbiased variance estimators of x^*_{h0} and $\bar{r}_h$ are given respectively by estimating equations of the type (16.33) and (16.32), and an unbiased variance estimator of x is given by an estimating equation of the type (21.7).

See notes to section 16.5.

Further reading

Foreman, section 8.6; Hansen *et al.* (1953), Vols I and II, chapters 7-10; Kish (1965), chapter 7; Sukhatme *et al.* (1984), sections 9.7, 9.9-9.10.

Exercises

1. The sampling design for a socio-economic survey was stratified two-stage. In a stratum, two villages were selected with ppswr, and households within a selected village were selected with equal probability and with replacement. Table 21.18 gives the data collected on household size and monthly consumption of cereals in the sample households. Estimate the total and *per capita* monthly expenditures on cereals and the average household size in the region covered by the strata, along with their standard errors (Chakravarti *et al.*, Exercise 3.10, adapted).
2. For the data of Example 21.1, estimate for the three states combined, the average number of adults per household and its standard error.

Table 21.18: Summary data on household size and per household consumption of cereals in sample villages

Stratum	Sample village	Selection probability	Number of household	Average per household	
				Number of persons	Monthly expenditures on cereals (Ind.Rs.)
h	i	π_{hi}	M_{hi}	$\bar{y}_{hi}$	$\bar{x}_{hi}$
1	1	0.0537	996	3.23	37.18
	2	0.0423	761	5.21	81.05
2	1	0.0634	1165	6.43	63.00
	2	0.0990	3108	4.26	108.04
3	1	0.0157	802	7.13	29.46
	2	0.0482	2324	5.68	59.54
4	1	0.0646	3085	6.28	42.74
	2	0.1285	7981	4.39	81.34
5	1	0.2092	8592	7.55	34.86
	2	0.0965	5921	3.68	60.37
6	1	0.0785	1119	4.10	70.48
	2	0.0167	1742	5.54	60.37
7	1	0.0268	1356	7.76	47.45
	2	0.0712	4098	5.09	43.76
8	1	0.1159	8976	6.07	45.12
	2	0.0297	1387	7.02	83.89

Source: Chakravarti *et al.*, Exercise 3.10 (adapted).

CHAPTER 22

Size of Sample and Allocation to Different Strata and Stages

22.1 Introduction

The procedure for the allocation of the total sample to the different strata and stages in a stratified multi-stage design follow from those for a stratified design (Chapter 12) and a multi-stage design (Chapter 17). The total sample size is fixed in general by the availability of resources, especially the available number of trained enumerators and the number of sample units that can be surveyed by the enumerators during the survey period.

22.2 Optimum allocations

Extending the notations in sections 12.3 and 17.2, for a stratified two-stage design (and sampling with replacement), and a fixed number m_{0h} of sample ssu's in each selected fsu, the variance of the unbiased estimator of a universe total can be expressed as

$$\sum_h \frac{V_{1h}}{n_h} + \sum_h \frac{V_{2h}}{n_h m_{0h}} \tag{22.1}$$

where V_{1h} is the variability between the fsu's V_{2h} that between the ssu's within the fsu's and n_h the sample number of fsu's in the hth stratum $(h = 1, 2, \ldots, L)$.

If the cost of travel between the fsu's within a stratum is small and is not taken into account, the following cost function may be adopted:

$$\sum_h n_h c_{1h} + \sum_h n_h m_{0h} c_{2h} \tag{22.2}$$

where c_{1h} is the cost per fsu and c_{2h} the cost per ssu in the hth stratum.

The optimum sample number of ssu's in the hth stratum is obtained on minimizing the variance for a fixed cost or *vice versa*, and is given by

$$m_0 = \sqrt{(V_{2h}c_{1h}/V_{1h}c_{2h})} \tag{22.3}$$

which is of the same form as expression (17.11) for an unstratified two-stage design. The optimum sample number of fsu's is similarly given by

$$n_h \propto N_h\sqrt{(V_{1h}/c_{1h})} \tag{22.4}$$

which again is of a similar structure as expression (12.6) for a stratified single-stage design when the cost is fixed.

The above formulae indicate the following rules for allocating the sample:

1. Select more fsu's in a stratum where (a) the number of fsu's is large, and (b) the ssu's within fsu's are heterogeneous.

2. Select more ssu's in a stratum when (a) the cost of selecting fsu's is greater than that of selecting the ssu's, (b) the average number of ssu's per fsu is large, and (c) the variability within ssu's is larger than that between fsu's.

22.3 Gain due to stratification

It can be shown that a stratified two-stage srs will be more efficient than an unstratified two-stage srs, with the same total sample size, when the variation of the stratum means is large.

To combine the efficiencies of both stratification and multi-stage sampling, the strata have to be made heterogeneous with respect to each other, with the fsu's within a stratum internally homogeneous (see sections 12.3.2 and 17.2.3). A small number of sample fsu's in each stratum will then provide an efficient sample. When strata are formed to contain an approximately equal sum of the sizes of the fsu's in each, two additional advantages ensue – first, an equal allocation of sample size per stratum, and second, achievement of a self-weighting design.

Further reading

Cochran, section 10.10; Foreman, section 8.6; Hansen *et al.* (1953), Vols. I and II, chapters 7 and 9; Kish (1965), chapter 8; Singh and Chaudhary, section 9.5; Sukhatme *et al.* (1984), section 8.12.

CHAPTER 23

Self-weighting Designs in Stratified Multi-stage Sampling

23.1 Introduction

The problem of how to make a stratified multi-stage design self-weighting will be considered in this chapter. The results follow from a combination of those of Chapter 13 for stratified self-weighting designs and Chapter 18 for multi-stage self-weighting designs.

23.2 General case

Taking the stratified multi-stage design of section 19.3(5), an unbiased estimator of the universe total Y of a study variable is

$$y = \sum_{\text{sample}} \left(\prod_{t=1}^{u} \frac{1}{f_{ht}} \right) y_{h12\ldots u} \tag{23.1}$$

(see also equation (19.18)) where $y_{h12\ldots u}$ is the value of the study variable in the $(12\ldots u)$th ultimate stage unit in the hth stratum, and the factor

$$\prod_{t=1}^{u} \frac{1}{f_{ht}} = w_{h12\ldots u} \tag{23.2}$$

(see also (19.19)) is the multiplier for the $(12\ldots u)$th ultimate stage sample unit in the hth stratum, and $f_{ht} \equiv n_{12\ldots(t-1)}\pi_{h12\ldots t}$, there being $n_{h12\ldots(t-1)}$ sample units each selected with probability $\pi_{h12\ldots t}$ out of the total $N_{h12\ldots(t-1)}$ units at the tth stage.

The design will be self-weighting with respect to y when the multiplier $w_{h12...u}$ defined above is a pre-determined constant, w_0, for all the ultimate-stage sample units. It will be so when

$$f_{hu} \propto \prod_{t=1}^{u-1} \frac{1}{f_{ht}} \tag{23.3}$$

In particular, a stratified multi-stage srs will be self-weighting with proportional allocation for the first-stage units, and for each of the subsequent stages, a constant sampling fraction for selecting the next-stage units. That is, the first-stage sampling fraction n_{h1}/N_{h1} should be a constant, the second-stage sampling fraction n_{h12}/N_{h12} a constant, and so on.

If, further, a fixed number of ultimate-stage units, n_0, is to be selected in each of the selected penultimate-stage units, then the same procedure as for an unstratified multi-stage sample (section 18.2) will apply in each stratum.

See the notes to section 18.2.

23.3 Stratified two-stage design with pps and srs

The procedure for an unstratified two-stage design with pps selection at the first stage and srs (or systematic selection) at the second stage, given in section 18.3, can be extended readily to a stratified design. The multiplier for the ijth sample ssu in the hth stratum is

$$w_{hij} = \frac{1}{f_{h1} f_{h2}} = \frac{M_{hi}}{n_h \pi_{hi} m_{hi}} \tag{23.4}$$

and the design will be self-weighting with a constant multiplier w_0, when

$$m_{hi} = M_{hi}/(w_0 n_h \pi_{hi}) \tag{23.5}$$

In a household inquiry, if the number of households to be sampled in each village is to remain a constant m_0, then with a self-weighting design,

$$m_0 = M_{hi}/(n_h \pi_{hi} w_0) \tag{23.6}$$

or

$$\pi_{hi} = M_{hi}/(w_0 n_h m_0) = M_{hi}/M_h \tag{23.7}$$

where $m_h = \sum^{N_h} M_{hi}$ is the total number of households in the hth stratum, i.e. in any stratum the villages are to be selected with probability proportional to the total number of households; also

$$n_h = M_h/(w_0 m_0) \tag{23.8}$$

i.e. the number of sample villages is to be allocated to the different strata in proportion to the total number of households. It will also be seen that

$$\begin{aligned} w_0 &= M/(nm_0) \\ &= \text{total number of households/number of sample households} \\ &= 1/\text{sampling fraction} \qquad (23.9) \end{aligned}$$

where $n = \sum^L n_h$ is the total number of sample villages and $M = \sum^L M_h$ is the total number of households in the universe.

As the total number of households existing at the time of actual enumeration will not in general be known unless the inquiry is conducted simultaneously with a census, the following operational steps may be prescribed in order to restrict the variation of the number of sample households in a sample village:

1. Make a first estimate (M'') of the present total number of households in the universe, and fix the overall sampling fraction, which is the ratio of the required number of households to be sampled (nm_0) to the first estimate of the total number of households in the universe (M''). The reciprocal of this will be the constant overall multiplier w_0.

2. Divide the whole universe into L strata by using some suitable criteria.

3. Allocate the number of sample villages to the different strata in proportion to the number of households in them as per the latest census.

4. Select the sample villages in any stratum with probability proportional to the number of census households.

5. Compute the estimated increase in the number of households in the universe from the latest census, $I = M''/M'$; and compute $m'_0 = m_0/I$, the expected number of sample households per village, where M' is the total number of households in the universe as per the census.

6. Give to the enumerators the values m'_0 and M'_{hi} (the number of households in the villages in the census) with instruction to select at random (or systematically) $m_{hi} = m'_0 M_{hi}/M'_{hi}$ households out of the M_{hi} households listed in the ith village in the hth stratum.

See the notes to section 18.3.

23.4 Opinion and marketing research

Extending the procedures of section 18.4, it can be seen that with the selection of one individual from a household, the sampling design can be made self-weighting if the households are selected with probability proportional to the respective sizes.

Expanding the example given in section 18.4, if households are selected with probability proportional to size (number of adults in this case), the multiplier for the adult in the ijth sample household in the hth stratum becomes

$$Q_{hi}/(n_h \pi_{hi} m_{hi}) \tag{23.10}$$

If this is to be a constant, then

$$m_{hi} = Q_{hi}/(w_0 n_0 \pi_{hi}) \tag{23.11}$$

In practice, the factors $1/(w_0 n_h \pi_{hi})$ will be given to the enumerators, who will select the required number m_{hi} of households which is the product of this factor and Q_{hi} the total number of adults listed in the ith block in the hth stratum.

If the blocks are selected with equal probability, then the factor $1/(w_0 n_h \pi_{hi}) = N_h/(w_0 n_h)$ will be a constant in the hth stratum. If, in addition, a constant sampling fraction for the first-stage units is taken in all the strata, then the factor $N_h/(w_0 n_h)$ becomes a constant for all strata. In this case, the number of households to be selected in any block will bear a constant ratio to the current number of adults of the block.

23.5 Self-weighting versus optimum design

With simple random sampling, and assuming that $M_{hi} = M_{h0}$, and $m_{hi} = m_{h0}$, it can be shown that the design will be self-weighting when $n_h m_{h0}/N_h M_{h0} = f_0$ is a constant in all strata. On the other hand, the optimum allocation gives

$$f_{0h} = n_h m_{h0}/N_h M_{h0} \propto \sqrt{(V_{2h}/c_{2h})}$$

While the cost per ssu (c_{2h}) may be approximately equal in fsu's of different sizes (i.e. with different total number of ssu's), the variation among the ssu's (as measured by V_{2h}) may be greater in the larger fsu's. A self-weighting design will not thus be necessarily optimum. But as the theoretically optimum design is not generally attainable while the optimum values are generally broad, a self-weighting design will often be as efficient as the optimum.

As the above formulation assumes the same field cost function for both the optimum and the self-weighting designs, while the latter would entail considerably less cost for tabulation, there is an added justification in trying to achieve a self-weighting design.

Further reading

Cochran, section 10.10; Foreman, section 8.6; Hansen *et al.* (1953), Vol. I, sections 7.12 and 9.11; Murthy (1967), section 12.3; Singh and Chaudhary, section 9.9; Som, 1958-59, 1959; Sukhatme *et al.* (1984), section 9.8.

PART V

MISCELLANEOUS TOPICS

CHAPTER 24

Miscellaneous Sampling Topics

24.1 Introduction

In this chapter will be considered miscellaneous topics such as multi-phase sampling; sampling on successive occasions and panel sampling; estimation of mobile populations; inverse sampling; interpenetrating networks of sub-samples; simplified methods of variance computations; bootstrap and jackknife methods of variance estimation; method of collapsed strata; controlled selection; the use of Bayes's theorem in sampling; the use of focus groups to collect data; the use of randomized response technique of estimation; and surveys on disability, nutritional status of children, women's status, and HIV/AIDS infection.

24.2 Multi-phase sampling

24.2.1 Reasons for multi-phase sampling

It is sometimes convenient and economical to collect certain items of information from all units of a sample, and other items of information from a sub-sample of the units which constitute the original sample. This plan is termed two-phase sampling or double sampling: when extended to three or more phases, it is termed multi-phase sampling. Multi-phase sampling can be used with stratified multi-stage designs, but its use will be illustrated for double sampling with srs (with replacement) at both phases in an unstratified design.

Multi-phase sampling can be used for various reasons:

1. When the number of units needed to give the required precision on different items is widely different, either owing to the fact that the variabilities of the variables are different or because the precisions required are different. If no use is made of the relations between the

different variables, such multi-phase sampling is equivalent to taking samples of different sizes for the different items. Thus in a household budget inquiry, information on irregular purchases, such as of clothing and furniture, could be collected, because of their larger variability, from a large sample, and information on regular purchases, such as of foodstuffs, from a sub-sample.

2. When it is comparatively difficult or costly to collect information on some items, this can be done at the second-phase sample, while the first-phase sample can be canvassed for simpler items only. In a crop survey, it is usual to select a large sample of fields (or farms or plots), generally in a stratified multi-stage design; the acreage of the crop is determined from the whole sample of fields (or farms or plots) and yield rates are determined from a sub-sample.

3. First-phase information may also be used as ancillary information, in order to improve the efficiency of second-phase information by the ratio and regression methods, where ancillary information on the whole universe is not available. If the first-phase information is collected prior to the second-phase information, it may be used as a basis for the sampling plan, e.g. for stratification of the first-phase sample units (for the selection of the second-phase units) or for pps selection of the second-phase units or for both.

Notes

1. For selecting the second-phase sample, the first-phase sample serves as the universe, but since the first-phase information is based on a sample, it is itself subject to sampling errors, and this must be taken into account in the estimating procedure.
2. Multi-phase sampling is structurally different from multi-stage sampling: in the former the same sampling units are used throughout, but in the latter a hierarchy of sampling units is used. The two may of course be combined.
3. An example of double sampling is found in the survey of level of living among rural Africans in Cameroon, 1961-5. For the demographic sample survey, which was conducted first, the design was stratified single-stage, a systematic sample of villages (fsu's) being taken in each stratum. For the socio-economic inquiry, a sub-sample of the original sample fsu's was taken probability proportional to the existing number of households, and in each such selected unit a fixed number of households (second-stage units) were selected with equal probability and without replacement. For obtaining estimates of totals and means, the ratio method of estimation was used.

24.2.2 Double Sampling

We shall consider simple random sampling with replacement at both the phases, a sub-sample of n' ($< n$) being taken of the original sample of n units from a universe of N units. In the second phase, if no account is taken of the information collected at the first phase, then an unbiased estimator of the universe mean

$$\overline{Y} = \sum^{N} Y_i/N$$

is the sample mean

$$\overline{y}' = \sum^{n'} y_i'/n' \tag{24.1}$$

with sampling variance (from equation (2.17))

$$\sigma^2_{\overline{y}'} = \sigma^2_y/n' \tag{24.2}$$

an unbiased estimator of which is (from equation (2.20))

$$s^2_{\overline{y}'} = \sum^{n'} (y_i' - \overline{y}')^2/n'(n'-1) \tag{24.3}$$

However, account may profitably be taken of the first-phase information by using the regression or the ratio method of estimation. If the regression method is used, then analogous to the estimating equation (3.24), which applies in single-phase sampling, we have the regression estimator of the universe mean

$$y'_{Reg} = \overline{y}_i + \hat{\beta}(\overline{x} - \overline{x}') \tag{24.4}$$

where $\overline{x}$ and $\overline{x}'$ are the sample means of the ancillary variable, obtained respectively from the first phase of n units and the second phase of n' units, and $\hat{\beta}'$ is an estimator of β, the regression coefficient of y and x, based on the second phase sample.

The variance of this regression estimator is approximately

$$\sigma^2_{\overline{y}_{Reg}} = \frac{\sigma^2_y}{n} + \frac{\sigma^2_y(1-\rho^2)}{n'} = \frac{\sigma^2_y}{n'}\left[1 - \rho^2\left(1 - \frac{n'}{n}\right)\right] \tag{24.5}$$

where ρ is the universe correlation coefficient between the study and the ancillary variables.

Clearly, the variance of the regression estimator will be less than that of the unbiased estimator based on an srs of n' units (namely, σ^2_y/n'), unless

$\rho = 0$. When $\rho = 0$, the two become equal, and there is no advantage in double sampling.

Optimum sizes. Considering the cost aspect, it is necessary to determine the optimum sizes of the original sample (n) and that of the sub-sample (n') which will provide estimators with the minimum error at a given cost or *vice versa*.

Taking the simple cost function

$$C = nc_1 + n'c_2 \tag{24.6}$$

where c_1 and c_2 are the respective cost per unit of the original sample and the sub-sample. Then for a fixed cost C, the optimum values of n and n' are respectively

$$n = C\left[c_1 + c_2\sqrt{\left(\frac{1-\rho^2}{\rho^2}\cdot\frac{c_1}{c_2}\right)}\right]^{-1} \tag{24.7}$$

and

$$n' = (C - nc_1)/c_2 = n\sqrt{\left[\frac{1-\rho^2}{\rho^2}\cdot\frac{c_1}{c_2}\right]} \tag{24.8}$$

The minimum variance in this case is

$$(\sigma_y^2\{\sqrt{[(1-\rho^2)c_2]} + \rho\sqrt{c_1}\}^2)/C \tag{24.9}$$

where $\rho > 0$.

In the comparison of single-phase and double sampling, it may be noted that if a single-phase sample has to provide the same information as a double sample, the cost per unit in the former would also be c_2 , so that with the given total cost C, the size of such a single-phase sample is

$$n = \frac{C}{c_2} = n\,\frac{c_1}{c_2} + n' \tag{24.10}$$

and the value of ρ at which the variance of the optimum double sample becomes equal to that of an equivalent single-phase sample is

$$\sqrt{\left[\frac{4c_1c_2}{(c_1+c_2)^2}\right]} \tag{24.11}$$

ρ must be considerably large for the survey design to benefit from a double sample. For example, if $c_2/c_1 = 10$, i.e. the unit cost of collecting data on y is ten times that for observing x, then the two variances are equal where $\rho = 0.58$; when $\rho = 0.8$, the variance in double sampling is about three-fourths, and when $\rho = 0.9$, the variance is about one-half, that in an equivalent single sample.

Estimating equations. In the regression estimator in double sampling in srs (equation (24.4)), β' is estimated by (cf. equation (3.23))

$$\hat{\beta}' = SPy'_i x'_i / SSx'_i \tag{24.12}$$

and a variance estimator of $\overline{y}'_{Reg}$ is

$$\frac{s^2_{y'}\hat{\rho}'^2}{n} + \frac{s^2_{y'}(1-\rho'^2)}{n'} \tag{24.13}$$

where $s_{y'}$, $s_{x'}$, and $\hat{\rho}' = \hat{\beta}' s_{x'}/s_{y'}$ are computed from the sub-sample.

Example 24.1

A simple random sample of 75,308 ($= n$) farms out of a total 1,200,000 ($= N$) farms gave the estimated average area per farm as 31.25 acres ($= \overline{x}$). A sub-sample of 2,055 ($= n'$) farms gave the following data (y'_i denoting the number of cattle and x'_i the area of the ith farm):

$$\sum^{n'} y'_i = 25,751; \sum^{n'} y'^2_i = 597,737; \sum^{n'} x'_i = 62,989;$$

$$\sum^{n'} x'^2_i = 2,937,851; \text{ and } \sum^{n'} y'_i x'_i = 1,146,391.$$

Estimate the average number of cattle per farm and its standard error using the second-phase sample data (United Nations, *Manual*, Example 30; also see Exercise 3 in Chapter 3 of this book).

Here $\overline{y}' = 25,751/2,055 = 12.5309$; $\overline{x}' = 62,989/2,055 = 30.6516$ acres; $s'_y = 11.551$; $s'_x = 22.143$; and $\hat{\beta}' = 0.354551$.

The regression estimate of the average number of cattle per farm in double sampling is (from equation (24.4)) , $\overline{y}'_{Reg} = 12.7431$.

As $\hat{\rho}' = \hat{\beta}' s_{x'}/s_{y'} = 0.679666$, the variance estimator of $\overline{y}'_{Reg}$ is (from equation (24.13)) 0.03573, so that the estimated standard error of $\overline{y}'_{Reg}$ is 0.1891, with estimated CV of 1.48 percent.

If no account is taken of the information in the original sample, then the unbiased estimate of the average number of cattle per farm is $\overline{y}' = 12.5309$, with an estimated standard error of 0.2548 and CV of 2.03 percent.

Note: In double sampling, the ratio method of estimation may also be used, i.e.

$$\overline{y}'_R = \overline{x}\,\overline{y}'/\overline{x}' \tag{24.14}$$

but this estimator does not have any general superiority over the regression estimator, except that it is simpler to compute; it may however be better than the unbiased estimator.

See also the notes to sections 3.2 and 3.3.

Further reading

Chaudhuri and Stenger, Chapter 6; Cochran, sections 12.1-12.8; Foreman, sections 5.4 and 9.2; Hansen *et al.* (1953), Vols. I and II, Chapter 11; Hedayat and Sinha, section 6.9; Kish (1965), sections 12.1 and 12.2; Murthy (1967), sections 9.12 and 11.5; Singh and Chaudhary, Chapter 10; Sukhatme *et al.* (1984), section 6.6; Yates, section 3.12.

24.3 Sampling on successive occasions and panel sampling

24.3.1 Introduction

Sampling inquiries are often carried out at successive intervals of time on a continued basis covering the same universe, such as in the Current Population Survey of the U.S. and in the National Sample Survey of India. Various methods are employed for this purpose, e.g. by the selection of a new sample on each occasion, by the partial replacement of the sample, and by sub-sampling the initial sample. When the same sample of respondents in a mail, personal or telephone interview is selected to provide information on more than one occasion, the sampling plan is termed "**panel sampling**". In such surveys, interest centers on measuring one or more of the following:

1. the change in a universe parameter, such as the mean value of a study variable

2. the average of the mean values for all the occasions combined

3. the mean value of the most recent occasion

The methods of estimation for the above will be illustrated for sampling on two occasions with srs. Owing to the generally high positive correlation between the values on two occasions, it is better to retain the same sample to estimate the change in the mean values, and to draw a new sample to estimate the average for both the occasions combined. For estimating the mean for the second occasion, the same initial sample or a new sample of the same size would give equally precise estimators, but a more efficient scheme would be to replace a part of the sample on the second occasion and to use the double sampling method with the regression estimator.

24.3.2 Estimating the change in the mean value

If $\overline{y}_1$ and $\overline{y}_2$ are the sample means (which are unbiased estimators of the respective universe means $\overline{Y}_1$ and $\overline{Y}_2$ in srs) on two occasions (the subscripts referring to the occasions), both being based on samples of size n, the change in the universe mean values $(\overline{Y}_2 - \overline{Y}_1)$ is estimated unbiasedly by the change in the sample means, namely,

$$\overline{y}_2 - \overline{y}_1 \tag{24.15}$$

If the samples are independent, then the sampling variance of the estimator $(\overline{y}_2 - \overline{y}_1)$ in equation (24.15) is

$$(\sigma_1^2 + \sigma_2^2)/n \tag{24.16}$$

If, however, the same sample is used on both the occasions, the sampling variance of the estimator in equation (24.15) is

$$(\sigma_1^2 + \sigma_2^2 - 2\rho\sigma_1\sigma_2)/n \tag{24.17}$$

As ρ, the correlation coefficient between the values on the same units on successive occasions, is likely to be positive and very high, the variance in expression (24.17) will be less than the variance in expression (24.16), and it is therefore better to retain the same sample to estimate more efficiently the change in the mean values.

24.3.3 Estimating the mean for both the occasions combined

The estimator of the overall mean is $1/2(y_1 + y_2)$. Its sampling variance is

$$(\sigma_1^2 + \sigma_2^2 + 2\rho\sigma_1\sigma_2)/4n \tag{24.18}$$

When ρ is positive – as it likely to be – it is better to draw a new sample in order to estimate the overall mean more efficiently.

24.3.4 Estimating the mean for the second occasion

With the reasonable assumption that $\sigma_1^2 = \sigma_2^2 = \sigma^2$, the means obtained from the initial sample and from an independent second sample (of the same size n) will have the same sampling variance σ^2/n, i.e. will be equally efficient.

Let us, however, consider that the sample is replaced in part, with m out of the initial n units matching (i.e. retained) and a fresh sample of

$n - m = u$ unmatching units (i.e. replaced) taken on the second occasion. Unbiased estimators of the universe mean $\overline{Y}_2$ for the second occasion are

$$\begin{aligned} &\text{from the unmatched } u \text{ units: } \overline{y}_{2u}, \text{ with variance } \mathrm{var}(\overline{y}_{2u}) = \sigma^2/n \\ &\text{from the matched } m \text{ units: } \overline{y}_{2m}, \text{ with variance } \mathrm{var}(\overline{y}_{2m}) = \sigma^2/m \end{aligned} \tag{24.19}$$

However, for the matched part, a more efficient estimator than the unbiased estimator $\overline{y}_{2m}$ is obtained by using the double sampling method with the regression estimator, i.e. from equation (24.4),

$$\overline{y}_{2Reg} = \overline{y}_{2m} + \hat{\beta}(\overline{y}_1 - \overline{y}_{1m}) \tag{24.20}$$

where $\overline{y}_1$ is the sample mean for all the n units and $\overline{y}_{1m}$ that for the m matched units on the first occasion. From equation (24.5), the sampling variance of $\overline{y}_{2Reg}$ is approximately

$$\mathrm{var}(\overline{y}_{2Reg}) = \frac{\sigma^2\rho^2}{n} + \frac{\sigma^2(1-\rho^2)}{m} \tag{24.21}$$

A still better estimator of $\overline{Y}_2$ is obtained on weighting the two independent estimators $\overline{y}_{2u}$ and $\overline{y}_{2Reg}$ inversely as their variances, i.e.

$$\overline{y}_{2Comb} = \frac{\overline{y}_{2u}\,\mathrm{var}(\overline{y}_{2Reg}) + \overline{y}_{2Reg}\,\mathrm{var}(\overline{y}_{2u})}{\mathrm{var}(\overline{y}_{2u}) + \mathrm{var}(\overline{y}_{2Reg})} \tag{24.22}$$

The variance of this combined estimator is approximately

$$\frac{\sigma^2(n - u\rho^2)}{n^2 - u^2\rho^2} \tag{24.23}$$

The optimum matching fraction (not taking the cost and other factors into consideration) is

$$\frac{m}{n} = \frac{\sqrt{(1-\rho^2)}}{1+\sqrt{(1-\rho^2)}} \tag{24.24}$$

and the minimum variance in this case is

$$V_{Min} = \frac{\sigma^2}{2n}\left[1+\sqrt{(1-\rho^2)}\right] \tag{24.25}$$

The optimum percentage to match never exceeds 50 percent and decreases steadily as $|\rho|$ increases. Considering more than one item in a survey, a good matching fraction will be one-third or one-fourth. Unless ρ is very high, the gain from using the above procedure is modest.

Notes

1. In sampling on more than two occasions, the optimum matching fraction tends to 1/2, irrespective of the value of ρ.
2. Sampling the same units on successive occasions reduces the field cost. But in surveys of human populations, the co-operation of the respondents may diminish after a time and the process of repetitive surveys itself may condition response and behavior; it is better, therefore, to rotate the sample by partial replacement of the units according to a specific pattern. In the Current Population Survey in the U.S., and the Labour Force surveys in Israel, an individual household remains in the sample during four consecutive months, and after an eight-month interval is brought back for another four months. In the Micro-Census of the Federal Republic of Germany, where the sample design is stratified single-stage (see Exercise 5, Chapter 10), one-third of the sample enumeration districts are replaced every year. Partial replacement of sampling units is also made in the labor force surveys in Canada and Egypt.
3. A relevant question in sampling on successive occasions is the choice of the sampling unit. Problems, often intractable, arise from the addition and depletion of units, such as buildings, dwelling units, households, and families. In a household or demographic inquiry, for example, over a period of time households change their composition (due to births, deaths, and migration), re-form or dissolve; taking a specific geographical area, some households move out of it and some new households move in. Similar problems arise in agricultural inquiries where the farmers are the ultimate-stage sampling units. To overcome these problems, it is preferable to use areal units as the ultimate-stage sampling units and enumerate completely the elementary (or recording) units in the sample areal units: areal sampling units have the other advantage of reducing response errors mentioned in section 25.7. At the same time, to avoid the bias arising from repeated visits of the same recording units (note 2 above), a compromise would be either to have a partial replacement of the areal units as in the Micro-Census of the Federal Republic of Germany, or to retain the same penultimate stage areal units (such as villages and urban-blocks) with a partial rotation of the elementary units within them.

Further reading

Binder and Hudiroglou; Cochran, sections 12.9-12.12; Chaudhuri and Stenger, section 6.3; Foreman, section 9.3; Hansen *et al.* (1953), Vols. I and II, Chapter 11; Kish (1965), sections 12.4-12.7; Murthy (1967), sections 9.12 and 11.5; Singh and Chaudhary, Chapter 10; Sukhatme *et al.* (1984), section 6.7; Yates, sections 6.21, 6.22 and 7.19.

Additional references for panel sampling

Panel surveys by Kasprzyk *et al.*; Kish (1965), section 12.5; Moser and Kalton, section 6.5; and Sudman, sections 9.11-9.17.

24.4 Estimation of mobile populations

The problem of estimation of the number and characteristics of mobile populations, such as animals, fish and insects, requires the adoption of special sampling techniques. One is the capture-recapture method.

24.4.1 Capture-recapture method

Let N_A birds be captured, marked (banded), and released. If afterwards n birds are recaptured, of which n_h are found to be marked, from simple consideration of probability an estimator of the total number of birds in the universe is

$$N^* = N_A\, n/n_h \tag{24.26}$$

This estimator is known as the "**Lincoln Index**" after F.C. Lincoln (1930) who had used it for estimating the total number of waterfowl on the basis of banding returns in the U.S. in 1920-6. It is a maximum likelihood estimator and consistent but biased.

The estimator (24.26) can also be derived from the ratio estimator of the total of a study variable from a simple random sample (without replacement) of n out of N units (Hájek and Dupač). Denoting by y and z the sample totals of the study and the ratio variables respectively and Z the total of the ratio variables for the N units, the ratio estimator of the total Y is

$$y_R^* = Zy/z \tag{24.27}$$

If $Y_i = 1$ for all i and $z_i = 1$ for the marked units and $z_i = 0$ for the other units, then $Y = N$; $y = n$; $Z = N_h$; $z = n_h$, and from formula (24.27) we obtain as an estimator of the total number of units in the universe

$$N^* = Zn/z = N_h n/n_h \tag{24.28}$$

A variance estimator of N^* is

$$\begin{aligned} s_{N^*}^2 &= \frac{Z}{z}\left(\frac{Z}{z}-1\right)\frac{n^2}{n-1}\sum^{n}\left(1-\frac{n}{z}z_i\right)^2 \\ &= \frac{n^2}{n-1}\,\frac{N_A}{n_A}\left(\frac{N_A}{n_A}-1\right)\left(\frac{n}{n_A}-1\right) \end{aligned} \tag{24.29}$$

The assumption in this method is that the "capture and recapture" method represents a simple random sample without replacement.

24.4.2 Enumeration of nomads

For estimating mobile human population groups, such as nomads, experiences in some African and West Asian countries show the hierarchical approach through the tribal chiefs and administration to have considerable promise; use of water holes as sampling units presents certain difficulties and requires progressive modification of the sample design, while aerial photography is much less successful, besides being very costly. The use of the capture-recapture method for this purpose has yet to be explored.

Further reading

Bailey; Boswell *et al.*; Hájek and Dupač, section VIII.8; Hammersley; Leslie; Pollock *et al.*; Ramsey *et al.*; Sen (1971, 1972, 1973 and 1991); Thompson, Chapter 18; United Nations (1993b).

Exercise

1. In an area of 4500 m^2, 156 running beetles were captured, marked by punching their wing sheaths by special forceps, and then released; after 24 hours, 191 unmarked and 32 marked beetles were found in a number of earth traps. Estimate the total number of beetles in the area (data of Skuhravý, quoted by Hájek and Dupač).

24.5 Inverse sampling

In inverse sampling, sampling is continued until certain specified conditions, dependent on the results of the sampling, have been fulfilled. We shall consider inverse sampling for proportions and for continuous data.

24.5.1 Inverse sampling for proportions

When the universe proportion P $(= N'/N)$ of the number of units possessing an attribute (N') is very small, i.e. when the attribute is rare, an estimate of P obtained from even a very large (simple random) sample by the method of section 2.13, may not be reasonably precise. And a bad guess of P, required to estimate the sample size to provide an estimate with a given precision, may result in considerable under- or over-sampling.

With inverse sampling for an attribute in a simple random sample without replacement, the sample size n is not fixed in advance, but sampling is continued until a predetermined number n' possessing the attribute are returned. An unbiased estimator of P is

$$p = \frac{n' - 1}{n - 1} \tag{24.30}$$

where n is the total sample size with the n' required units.

An unbiased variance estimator of p is

$$s_p^2 = \frac{p(1-p)}{n-2}\left(1 - \frac{n-1}{N}\right) \qquad (24.31)$$

When N is large, the variance estimator of P can be taken as

$$s_p^2 = \frac{p(1-p)}{n-2} \qquad (24.32)$$

Further, with a small p, the sample size n will be fairly large, so that the variance estimator s_p^2 in equation (24.32) could be approximated by $p(1-n)/(n-1)$, which is the form of the unbiased variance estimator of $p = n'/n$ in srs with replacement (equation (2.74)).

The estimation equations in inverse sampling are attributed to J.B.S. Haldane.

24.5.2 Inverse sampling of continuous data

Suppose that samples are taken with equal or varying probabilities (and with replacement), until a predetermined number $(r+1)$ of distinct units are included, and this takes $(n+1)$ samples; the last unit, which is the $(r+1)$th distinct unit, is discarded. The sample, now of size n with r distinct units, may then be analyzed as if it were a sample deliberately chosen by the method of section 2.9 for srs or of section 5.3 for pps, both with replacement.

The (combined) unbiased estimator of the universe total Y is, as before,

$$y_0^* = \sum^{N} y_i^*/n \qquad (24.33)$$

where

$$y_i^* = N\,y_i \text{ in srs} \qquad (24.34)$$

$$y_i^* = y_i/\pi_i \text{ in pps} \qquad (24.35)$$

where N is the total number of units, y_i the value of the study variable of ith selected unit $(i = 1, 2, \ldots, n)$, and π_i the probability of selection of the ith unit $(i = 1, 2, \ldots, N$ for the universe) in pps sampling.

An unbiased variance estimator of y_0^* is, as before,

$$s_{y_0^*}^2 = \sum^{n}(y_i^* - y_0^*)^2/n(n-1) \qquad (24.36)$$

But $s^2_{y_0^*}$, as defined in equation (24.36), is not an unbiased estimator of the sample variance

$$\sigma^2_{y_0^*} = \sum^N (Y_i - \overline{Y})^2/nN \text{ in srswr} \quad (24.37)$$

$$\sigma^2_{y_0^*} = \frac{1}{n}\sum^N \left(\frac{Y_i}{\pi_i} - Y\right)^2 \pi_i \text{ in ppswr} \quad (24.38)$$

(see also equations (2.17) and (5.4)). The sampling variance of y_0^* in inverse sampling is approximately

$$\sigma^2_{y_0^*}\left(1 - \frac{n}{2N}\right) = \sigma^2_{y_0^*}\left(1 - \frac{1}{2}f\right) \quad (24.39)$$

where $\sigma^2_{y_0^*}$ is given by equation (24.37) for srswr and by equation (24.38) for ppswr, and $f = n/N$ is the sampling fraction.

The sampling variance in inverse sampling is thus less than that of $\sigma^2_{y_0^*}$.

Further reading

Cochran, section 4.5; Hedayat and Sinha, section 12.1; Levy and Lemeshow, sections 14.5 and 14.6; Sampford, sections 7.10 and 12.5; Sukhatme *et al.* (1984), section 2.13.

24.6 Estimation of rare events

Rare events, such as the incidence of a disease that affects, say, less than 10 per cent of a population, require a rather large sample to provide reliable estimates. Three options are available in such cases the use of the "capture-recapture" techniques; inverse sampling of proportions; and network sampling.

The first two options have been outlined in the preceding two sections. We describe network sampling briefly with reference to a relatively rare disease. If a patient (an "element") with such a disease is treated at more than one health center (one "enumeration unit"), then "the counting rule that allows an element to be linked to more than one enumeration is called Multiplicity counting rule" and "sampling designs that use multiplicity counting are called network sampling" (Levy and Lemeshaw, 1991, section 14.5). For a fuller introduction to network sampling, see Levy and Lemeshaw, *op.cit.*

Further reading

Kalton and Anderson (1986); Levy and Lemeshow, sections 14.5 and 14.6; Sudman, Chapter 9; Sudman, Sirken, and Cowan; U.N., 1993b.

24.7 Interpenetrating network of sub-samples

When two or more sub-samples are taken from the same universe by the same sampling plan so that each sub-sample covers the universe and provides estimators of the universe parameters on application of the same estimating procedures, the sub-samples are known as interpenetrating networks of samples. The sub-samples may or may not be drawn independently and there may be different levels of interpenetration corresponding to different stages in a multi-stage sample scheme. The sub-samples may also be distinguished by differences in the survey procedures or of processing features. These are sometimes known as **replicated samples**. The technique was developed by P.C. Mahalanobis in 1936; a variant of this is sometimes called the **Tukey plan** (Deming, 1960, pp. 186-7, footnote).

This technique enables one to:

(a) examine the factors causing variation, e.g. enumerators, field schedules, different methods of data collection and processing;

(b) compute the sampling error from the first-stage units if these comprise one level of interpenetration (both by the standard method as also by a non-parametric test);

(c) provide control in data collection and processing;

(d) supply advanced estimates on the basis of one or more sub-samples and provide estimates based on one or more sub-samples when the total sample cannot be covered due to some emergency;

(e) provide the basis of analytical studies by the method of fractile graphical analysis.

The technique may be incorporated as an integral part of the standard sample designs and has been used in a number of sample surveys in India, including the Indian National Sample Survey, Peru, Zimbabwe, the Philippines, and the U.S.A.

The first three uses of the technique are illustrated in section 26.6.5. In this section we shall briefly deal with the other two uses of the technique.

24.7.1 Advanced estimates from sub-samples

Provisional, advanced estimates can be obtained as required on earlier processing of one or more of the sub-samples, as was done in crop surveys in Bengal (India) and in the 1961 Census of Agriculture in Peru. In emergencies also, one or more sub-samples can be jettisoned and the other(s) used.

24.7.2 Fractile graphical analysis

A simple graphical representation can supply a visual (or geometric) means of assessing and controlling errors and also a measure of the margin of error (or of uncertainty), especially in the case of an interpenetrating network of independent sub-samples. The observed sample units in each sub-sample are ranked in ascending order of any suitable variable (or even in order of time of observation) and the whole sub-sample is divided into a suitable number of fractile groups of equal size (i.e. consisting of equal estimated number of universe units). For each sub-sample the average value, or median, or some other estimate of any variable for each fractile group, is plotted on paper and the successive plotted points are connected by straight lines; these polygonal lines are called the fractile graphs of the sub-samples. A fractile graph for the whole sample can be drawn in the same way, retaining the same number of fractile groups, as in the case of the sub-samples. The area between the fractile graphs for the sub-samples provides a visual and geometric estimate of the error associated with the fractile graph of the whole sample. This method can be used not only to assess the significance of observed deviation from a statistical hypothesis relating to the results of the survey for any particular part or the whole range of observed data, but also for comparisons between results based on two or more surveys carried out in different regions or at different periods of time in the same region using the fractile graphical error for each survey. This method has been extensively used in the Indian National Sample Survey (see Mahalanobis, 1960 and 1966).

Further reading

Cochran, sections 13.15 and 13.16; Deming (1960, Chapter 11; and 1963); Kish (1965), section 4.4; Koop (1960 and 1988); Lahiri; Mahalanobis (1946a, 1958, and 1961); Murthy (1967), sections 2.12 and 13.10g; Rao, C.R. (1993); Som (1965); Sukhatme *et al.* (1984), section 11.10; U.N. (1964a, section V18, and 1972, Part I, section 11); Yates, sections 3.17, 5.4 5.6, 7.24 7.26; Zarkovich (1963, section 10.8).

24.8 Simplified methods of variance computation

As has been described before, the standard method of estimating the sampling variance of the estimator of a universe total is to base it on the estimators of the total obtained from each of the sample first-stage units selected with replacement. If a fixed number of sample fsu's is selected in each stratum, the computation of variance is simplified, the more so if two sample fsu's are selected in each stratum (section 9.5 and note 2 of section

19.3). It has also been noted earlier that the computation of estimates and their variances become very simple if the design is made self-weighting, either at the field or at the tabulating stage. The use of interpenetrating networks of samples in providing variance estimators will be discussed in section 25.6.5.

24.8.1 Method of random groups

As noted in section 2.14, if the sample units in an srs without replacement are divided into a number of groups with an equal number of sub-sample units in each, estimates of sampling variances could be computed from the group means.

A variation of this method, followed in the Agricultural Survey in the U.S. is to assign, in each of the eight regions covered, a random group number, 1 to 10, to a schedule before it is punched, and in tabulating to sort the cards first according to the tabulation requirements and then by the random group number, obtaining successive cumulative estimates relating to the random groups. Estimates of variance were obtained by "decumulating" the data from the auxiliary tapes and performing the appropriate computations.

24.8.2 Selection of a random pair of units from each stratum

Another simplified procedure would be to select a random pair of sample units from each of a large number of strata with approximately equal sizes and to estimate the variance from the random pairs.

24.8.3 Use of "error graphs"

A sample survey usually contains many items of information, and the process of computation of variance estimators for each of the items and their cross-classifications becomes laborious and expensive. If a relation could be found for a survey between the estimator of Y and its standard error that holds substantially for a large number of items, then a curve (the "error graph"), fitted on a suitable scale, may be used for estimating the standard errors of less important item totals. Such error graphs have been used by Yates (section 7.20), and in surveys conducted by the U.S. Bureau of the Census (Hansen *et al.* (1953), Vol. II, Section 12.B), and the U.S. National Center for Health Statistics.

Note: Whenever a sub-sample is drawn from the original sample for computing the variance estimate of an estimate obtained from the large, original sample, caution must be exercised in using the appropriate estimating equations. Suppose, for example, that from an srs with replacement of n out of N units, the unbiased estimator y_0^* of $\overline{Y}$ is computed by the method of section 2.9, i.e., $y_0^* = N\sum^n y_i/n$. If a sub-sample of n' units is selected at random from the n units, then an unbiased estimator of σ^2, computed from the sub-sample of n' units, will be

$$s'^2 = \sum^{n'} (y_i' - \overline{y}')^2/(n'-1) \tag{24.40}$$

where $\overline{y}' = \sum^{n'} y_i'/n'$, and y_i' $(i = 1, 2, \ldots, n')$ are the values of the study variable in the sub-sample. To estimate the sampling variance of y_0^*, defined above, this value of s'^2 is substituted for s^2 in estimating equation (2.44), which then becomes

$$s_{y_0^*}^2 = N^2s^2/n = N^2s'^2/n = N^2SSy_i'/[n(n'-1)] \tag{24.41}$$

Further reading

Hansen *et al.* (1953), Vol. I, section 10.16, Vol. II, section 10.4; Kish (1965), sections 8.6C and 14.3.

24.9 Other methods of estimating sampling variance

In the last two decades, a number of methods have emerged that are applicable to non-linear functions of the sample data (and, of course, as a special case, to linear functions) and do not involve the assumption of normality. We discuss here two such methods – the bootstrap method and the jackknife method: both require the re-sampling of the sample data.

24.9.1 Bootstrap

The bootstrap method was developed by Efron (1979) to provide estimates of sampling variances for complex statistics.

Suppose, in a universe comprising N units, interest centers on a universe parameter R which is a non-linear function of the universe variate(s) and that a sample of size n has been drawn according to a specified sample design. In the bootstrap method, the sample of size n is considered as the universe, from which repeated samples are drawn (with replacement), each of size n. Suppose, for example, that the sample size is $n = 10$. The sample is re-sampled (with replacement), using a computer, to provide constructs of $n = 10$ artificial data sets each. Then M (100 or more) such constructs are obtained from the computer, in each of which the original ten numbers would be selected by chance and some might not be selected at all. If there are M re-samples, for each of which a copy r is made of universe parameter

R, which gives M bootstrap estimators r_j ($j = 1, 2, \ldots, M$). Then, the (bootstrap) estimator of sampling variance is given by

$$s_{\bar{r}}^2 = \sum^{M} (r_j - \bar{r})^2 / M(M-1) \tag{24.42}$$

where

$$\bar{r} = \sum^{M} r_j / M \tag{24.43}$$

Computer software programs, such as Resampling Stats for IBM or MAC personal computers, are now available for the large-scale re-sampling required by this technique (Goldstein); Willeman has shown how to implement the bootstrap in a simple spreadsheet of the Lotus 1-2-3 type.

Efron calls his method "the bootstrap" as, in a sense, the sample data pull themselves up by their own bootstraps by generating new data sets through which their reliability can be determined.

Mammon has set down conditions under which bootstrap can be expected to work satisfactorily. Efron has applied it in a study on the decay of tau particles and a study of the effects of the cholesterol-reducing drug cholestyramine and found it to give more plausible results.

It has been argued that the bootstrap method "has been examined for its applicability to sample surveys, but the results have been somewhat mixed so far. ... At present, the use of the bootstrap as a general solution to computing variances is not very promising" (Brick), a major problem being computational, particularly in surveys with many estimates.

Further reading

New York Times of 9 November 1988 and the London *Economist* of 19 November 1988 gave popular accounts of the bootstrap method. For a detailed introduction to the method, see J.N.K. Rao's article on variance estimation (1988). For a full treatment of the topic, see the book by Efron and Tibshirani (1993), and Mammon (1992), and articles by Efron (1979, 1982, and 1989), Efron and Gong (1983), and J.N.K. Rao and Wu (1988) and the papers for the 1990 Conference on Exploring the Limits of Bootstrap, edited by LePage and Billard.

24.9.2 Jackknife

A method proposed by Quenouille (1956) for reducing the bias in ratio estimators is used, in a modified form suggested by Tukey (1958), in variance estimation of non-linear functions; Durbin (1959) and Schucany, Gray and Owen (1971), among others, have further developed the method. It is a handy, utility method analogous to the jackknife which is a strong clasp-knife for the pocket, hence the naming of the method.

The jackknife method is based upon the concept that if r_n is an estimator biased to order $1/n$, and r_{n-1} is the mean of the n estimators derived by omitting one unit of the sample, then

$$nr_n - (n-1)r_{n-1} \tag{24.44}$$

is biased to order $1/n^2$. Suppose, interest centers on estimating a universe ratio $R = Y/X$, and from a sample of size n, unbiased estimators y and x respectively of Y and X, obtained from a sample of size n. The sample estimator $r = y/x$ is known to be biased generally (see section 3.2, Notes). Then, n samples, each of size $n-1$, are obtained in the following manner. First, randomly select one sample unit, delete it and obtain a new estimate of R, based on the sample of size $n-1$; call this new estimator r_1. Then replace the deleted unit and randomly select another unit, delete it and obtain the second estimator r_2. Continue the process (noting that some of the units may be deleted, and replaced, more than once) until n new estimators are obtained, each based on the sample of size $n-1$; call these new estimators r_j. Pseudovalues

$$\tilde{r}_j = r + (n-1)(r - r_j) \tag{24.45}$$

are then computed, and the revised, jackknifed estimator of R is given by

$$\tilde{r} = \sum \tilde{r}_j / n \tag{24.46}$$

The estimated MSE of the jackknife estimator (24.46) is given by

$$MSE_{\tilde{r}} = \sum (\tilde{r}_j - \tilde{r})^2 / n(n-1) \tag{24.47}$$

Further reading

For a more detailed introduction to the jackknife method, see J.N.K. Rao's (1988) article on variance estimation. For full details, see the books by Efron (1982), and Gray and Schucany (1972), and articles by R.G. Miller (1974), and Schucany, Gray, and Owen (1991).

24.10 Method of collapsed strata

When the universe is stratified taking into account the characteristics of more than one study variable, stratification may have to be carried to a degree at which there is only one first-stage sample unit in each stratum. Estimators of variances of totals, means, etc. cannot then be computed by the standard method. In such a situation, the strata may be combined into a smaller number of groups such that each group contains at least two first-stage sample units. Estimators of variances can then be computed from the sample fsu's in the newly formed groups. This is known as the method of collapsed strata.

The variance estimator, computed from such collapsed strata, over-estimates the sampling variance. A very satisfactory approximation may, however, be obtained when the number of strata is large and the strata are combined into groups such that for each group the strata are about equal in size (total Y). On the other hand, the variance estimator will be a serious under-estimate if the groups are constructed by seeing the sample results and making them differ as little as possible.

Further reading

Cochran, sections 5A.12 and 8.12; Hansen *et al.* (1953), Vols I and II, Chapter 10; Kish, sections 4.3, 7.4D, and 8.6B; Singh and Chaudhary, section 3.12; Sukhatme *et al.* (1984), section 4.15.

24.11 Controlled selection

Controlled selection is any process of selection in which, while maintaining the assigned probability for each unit, the probabilities of selection for some or all preferred combinations of n out of N units are larger than in stratified srs (and correspondingly the probabilities of selection for at least some non-preferred combinations are smaller than in stratified srs).

To illustrate from an actual survey situation described by Hess, Riedel, and Fitzpatrick (1961), suppose a universe of nine hospitals is divided into two strata, one containing four large hospitals A, B, C and D, and the other five small hospitals a, b, c, d and e. If it is decided to select one hospital each from each stratum with equal probability, the probability of selection of a hospital from stratum I is 0.25 and that from stratum II is 0.20 and all the $4 \times 5 = 20$ combinations of one hospital each from a stratum has the same probability, 0.05. If, however, hospitals A, B, a, and b are state owned and the others are privately owned, then of the twenty combinations ten will comprise one state and one private hospital. Although each of the twenty combinations will provide an unbiased estimator of a universe total, it is preferable to have in the sample one state hospital and one private hospital; but the chance of selection of such preferred samples is 10/20, i.e. 0.50.

One method of ensuring higher probabilities of selecting such preferred samples is due to Goodman and Kish (1950), which in the above example would increase the probability of selection of a preferred sample from 0.50 to 0.90 in stratified srs.

Controlled sampling is mainly practicable and useful when the sample is made up of a few large first-stage units. It then enables additional control to be introduced beyond what is possible by stratification alone, this being limited by the fact that the number of strata cannot exceed the number of units in the sample. However, in the words of one of the originators of the method, "it has several drawbacks ... It may be safer for non-specialists to avoid any controlled selection in favor of simpler and more standard probability methods" (Kish (1965), section 12.8). The drawbacks relate mainly to the difficulties of application that require skill and care, arbitrariness involved in selection, and the estimation of the sampling variance.

Further reading

Cochran, section 5A.5; Chaudhuri and Stenger, section 1.1.4; Foreman, section 11.8; Kish (1965), section 12.8; Singh and Chaudhary, section 3.15; Sukhatme *et al.*, (1984), section 4.16.

24.12 Use of Bayes' theorem in sampling

In section 2.8 and elsewhere, probabilistic statements concerning an unknown universe parameter have been made from considerations of confidence intervals. Another approach involves the use of Bayes' theorem, which plays an important role in the treatment of the foundations of probability on which the theory of survey sampling is based. The application of Bayes' theorem in sampling is somewhat controversial.

Consider k events $H_1, \ldots, H_k$, which are mutually exclusive and exhaustive (i.e. they exclude each other two-by-two and one necessarily occurs). An event A is observed and is known to have occurred in conjunction with or as a consequence of the event H_j. The joint probability of occurrence of the events A and H_j is

$$P(A \mid H_j) = P(A \mid H_j)P(H_j) = P(H_j \mid A)P(A) \tag{24.48}$$

where $P(A \mid H_j)$ is the conditional probability of A given H_j and $P(H_j \mid A)$ that of H_j given A. As the event can occur only in conjunction with some H_i ($i = 1, 2, \ldots, k$), and the AH_i are mutually exclusive, the probability of occurrence of A is

$$P(A) = \sum^{k} P(AH_i) = \sum^{k} P(A \mid H_i)P(H_i) \tag{24.49}$$

from equation (24.48). Also, from equation (24.48), the probability of occurrence of H_j, given A, is

$$P(H_j \mid A) = \frac{P(AH_j)}{P(A)} = \frac{P(A \mid H_j)P(H_j)}{\sum^k P(A \mid H_i)P(H_i)} \tag{24.50}$$

That is, the probability of H_j, given A, is proportional to the probability of A, given H_j, multiplied by the probability of H_j. This is Bayes' theorem or rule for the probability of causes. $P(H_j \mid A)$ is called the posterior probability, $P(A \mid H_j)$ the likelihood, and $P(H_j)$ the prior probability; $H_1, H_2, \ldots, H_k$ are called the ***hypotheses or causes*** of A.

The theorem has many statistical applications but requires a knowledge of the prior probabilities. Suppose a human population is divided into k strata on the criteria of race, and let p_i $(i = 1, 2, \ldots, k)$ be the probability that an individual chosen at random belongs to stratum H_i. Let the event A denote the possession of blood group A, and let p'_j denote the conditional probability $P(A \mid H_j)$ that a person belonging to race H_j also has blood group A. The probability that an individual chosen at random has blood group A is, from equation (24.49), $\sum^k p_i p'_i$. Given that a person has blood group A, what is the probability of his belonging to race H?

From Bayes' Rule (equation 24.50), the answer is

$$P(H_j \mid A) = \frac{P(AH_j)}{P(A)} = \frac{P(H_j)P(A \mid H_j)}{\sum^k P(H_i)P(A \mid H_i)} = \frac{p_j p'_j}{\sum^k p_j p'_j}$$

A Bayesian approach has also been applied to determine the optimal design in stratified sampling (Ericson, 1969); and further work has been done on the optimal cluster size etc.

Unlike the usual Bayesian posterior inference problem, where the primary focus is on the posterior distribution of the hypothesis H_i, in sample surveys with the finite universe model it is the predictive distribution of the unobserved components of H_i (or a future sample of $N_h - n_h$ units) which is primarily of interest.

Further reading

Chaudhuri and Stenger, section 3.2.1; Ericson (1969) and other relevant articles in Johnson and Smith (1969); Ericson (1988); Feller, sections V.1 and V.2; Lindley (1965 and 1972); Schmitt.

24.13 The use of focus groups

One technique of qualitative research is focus group interviews that also goes by the name of group interviews or group discussions. This technique has found wide applications in consumer market research and, more recently, in political and opinion research. Focussed group interviews should be "conceived of as an ecumenical research tool for use in every domain of social life; it is not at all confined to market research as many evidently assume is the case. For sound research results, it must be supplemented by more systematic and typically quantitative research" (personal communication from Robert K. Merton).

Group interviews have been used for over 35 years in the fields of psychotherapy, public opinion surveys and market research (Slavson, 1979, and Higginbotham and Cox, 1979, cited by Scherr, 1980). Following the pioneering work of Merton and his colleagues in the 1940s at the Columbia Office of Radio Research during World War II and later at the Columbia Bureau of Applied Social Research (into which the Office of Radio Research had evolved), providing a theoretical underpinning of the basic concept of *focus* (Merton and Kendall, 1946, Merton, Fiske, and Kendall, 1956), focus groups have gained a huge currency, particularly in market research, all over the world. In the United States alone, 100,000 focus groups were conducted in consumer market research in 1992, with more than 1 million participants (Impulse Research Corp.). Focus groups are now increasingly being used in fields such as social security administration, population censuses and family planning surveys.

In focus group interviews, a number of respondents or "participants" are brought together for half an hour to two hours under the direction of a "group leader" or "moderator"; the size of the group varies according to the topic of discussion and the type of participants, generally ranging from five to twelve. In contrast to individual interviews, the prime concern in focus group research is group interaction, and thus the suggested criterion for the success of a focus group: in a productive focus group, the participants talk to each other more than they talk to the moderator.

It has been argued that unlike individual interviews that are based on structured questionnaires, a focus group discussion allows greater in-depth probing of a topic and is, therefore, expected to provide more representative data from the public point of view. The contrary argument is that groups can inhibit individual articulation: private interviews are used, e.g. in KAP studies, to encourage responses that are normally inhibited by the presence of others (Stycos). As Merton has indicated, group and private focussed interviews are complementary.

Focus group sessions are commonly tape recorded, sometimes videotaped, and are usually conducted in facilities equipped with one-way glass to view participants without inhibiting them.

Focus groups have been found to be an appropriate tool for developing insights and hypotheses and for exploring the range of pertinent attitudes, opinions, concerns, experiences, and suggestions of participants and can thus be a helpful preliminary step in developing the conceptual framework, data specification, and question wording of a survey or evaluating draft questionnaires for a quantitative survey that will use structured questionnaires in a probability sample or in administration of or after a survey to collect information on non-response. Thus, a legitimate role of focus groups is as a supplement to, and not as a replacement for, (quantitative) survey sampling.

The U.S. Social Security Administration used focus groups in 1978 in Washington D.C., Maryland, and California to evaluate proposed revisions to an administrative form for collecting race and ethnic data (Scherr; U.S. Office of Management and Budget, 1983). The U.S. Census Bureau used focus groups in the 1986 Test Census in Mississippi and Los Angeles on motivational messages and themes that would encourage the hard-to-enumerate populations, such as people with low incomes. It was found that the test censuses were not well covered by the media; the groups were not aware that the census was mandatory, let alone that a penalty existed; the groups suggested greater publicity of the census with endorsements by church leaders and national celebrities (such as Bill Cosby, Jerry Lewis, Danny Thomas, and Richard Pryor) and not by local officials (Bush; Freeman). These focus groups have led to the improvement of the Bureau's census and survey methodology.

At the request of the Office of Child Support Enforcement at the U.S. Department of Health and Human Services to assist in revising the Child Support Supplement to the Current Population Survey, the Census Bureau conducted in 1992 focus groups with previous respondents to the Supplement.

In epidemiological and KAP-type studies, focus groups have been used recently by the U.S. National Cancer Institute to pre-test public education programs and materials for cancer prevention, by the U.S. National Heart, Lung and Blood Institute to design a public information program on high blood pressure, by the UNICEF to gain a deeper understanding of people's behavior related to breast feeding in developing countries, by the Population Council in a similar undertaking in four developing countries, by a large public sector research program in Mexico to provide information to develop a commercial contraceptive marketing program, in Indonesia to gather information for survey research on family planning (Schearer; Folch-Lyon and Frost; Folch-Lyon *et al.*; and Suyono *et al.*), and in Bangladesh to understand the determinants of contraceptive use (Nag and Duza).

Focus groups are also used to identify the desired criteria of potential jurors, e.g. in the trial of O.J. Simpson.

Further reading

For the use of focus groups and other methods of qualitative and motivation research in general, consult the books by Merton, Fiske and Kendall (1990) and by Morgan (1988) and articles by Merton and Kendall (1946), Merton (1987) and Forsyth and Lessler (1991). For the use of focus groups in consumer market research, consult articles by Calder (1977), Cox *et al.* (1976), Fern (1982), Gage (1978), Grik, Parker, and Hetegikamana (1987), McDaniel (1979), Reynolds and Johnson (1978), Sampson (1986), and Szybillo and Berger (1978); for its use in population, family planning, and community health studies, consult the references cited in the text, the special issue on focus group research of *Studies in Family Planning* (12(12), 1981), and articles by Grik *et al.* (1987), Knodel (1990), and Sikes (1993); for the U.S. Census Bureau's use of focus groups, see Market Dynamics's report, Alder (1985), Bush (1985), Freeman (1985) and U.S. Census Bureau (1993a).

24.14 Randomized response technique

In socio-economic and epidemiological studies, accurate information is difficult to obtain on topics that are sensitive in nature, due either to personal or socially imbedded reasons – such as the practice of family planning or regular visits to a church, mosque or temple – or due to legal reasons, such as not reporting, or incorrectly reporting, income or consumption of alcohol on illegal substance abuse (such as cocaine), or use of abortion (in countries where abortion is illegal under certain conditions). It is quite possible that the practice of regularly visiting a house of worship would be over-reported and that of illegal substance abuse under-reported. Warner (1965) suggested a technique to obtain such information by giving the respondents a choice of answering a question on the topic under study or a related or unrelated question. The technique, as modified by Horvitz, Shah and Simmons (1967), is briefly described below.

Suppose, in a family planning survey, an objective is to estimate the number (and the proportion) of women aged 15-49 years practicing family planning. Two questions are formulated:

1. Are you now using any family planning method?

2. Does the last digit of your age (in years) lie between 0 and 4?

The first question is related directly to the topic under study; the second question is an unrelated one, the universe proportion of the response to the latter – let us call it θ – being known, or could be estimated fairly easily, in the example above by 0.5: (θ should be so chosen as to lie between 0.5 and < 1). The respondent is given a choice (determined by herself) to answer question 1 or 2, depending on the outcome of a drawing, the universe probability of which – let us call it P_2 – is known beforehand; for example, the respondent might be given a coin and asked to toss it, away from the interviewer, the actual outcome of which – head or tail – is known only to the respondent (but the universe probability, 0.5, is known to the survey designer), and the respondent is asked to answer ("Yes"/"No") to question 1 if the result of the toss is a head to answer ("Yes"/"No") to question 2 otherwise.

It is thus assumed that the composition of the randomization process (e.g. the coin tossed resulting in a head) and the universe probability of answering "Yes" to the unrelated question are known, the two probabilities being designated respectively by θ and P_2. Suppose, in an srswr of n individuals, n_1 answer "Yes" (but it is not known how many have answered "Yes" to question 1 and how many answered "Yes" to question 2). The problem is to estimate the "true" proportion of individuals answering "Yes" to the relevant question (question 1). The RRT estimate of this proportion is

$$p_1^* = \frac{p_1 - P_2(1-\theta)}{\theta} \tag{24.51}$$

where $p_1 = n_1/n$.

An unbiased estimator of the variance of p_1^* is given by

$$\text{var}(p_1^*) = \frac{p_1(1-p_1)}{(n-1)(1-\theta)^2} \tag{24.52}$$

Suppose in a sample of 600 women, each women is asked first to toss a coin and to answer the first question if the toss results in a head and to answer question 2 otherwise. Suppose further that 180 women answer "Yes", but it is not known how many have answered "Yes" to question 1 and how many have answered "Yes" to question 2. Here $n_1 = 180$, $p_1 = 180/600 = 0.3$. And, since $\theta = 0.5$, the RRT estimate of the proportion of women using any family planning method is, from equation (24.51): $[0.3 - 0.5(1 - 0.5)]/0.5 = 0.1$, or 10 per cent.

This can also be seen from simple mathematical considerations. Half of the 600 women, i.e. 300, are expected answer the second question, and again half of them, i.e. 150, are expected to answer "Yes" to the second question. Thus, the remaining (180 – 150 =) 30 women answering "Yes" must have responded to the first question, the total number of women responding to the first question being 300. Hence the estimated proportion of women answering "Yes" to the first question, i.e. practicing family planning is $30/300 = 0.1$, or 10 per cent.

The estimated standard error of this estimate is, from equation (24.52) 0.0374, and the percentage error is 37.44 per cent.

The techniques sometimes returns inadmissible estimates – proportions that are negative or greater than 1.

The putative advantage of RRT is that it is impossible to tell which question the individual answered and therefore it is impossible to tell their standing on the controversial item, so that the respondent can feel anonymous. But respondents are often suspicious that they have been tricked and lose their confidentiality, so they still might not answer honestly.

Randomized response technique has been used in studies, among others, of induced abortion in urban North Carolina (Abernathy, Greenberg, and Horvitz, 1970), and outcomes of pregnancies in Taiwan (I-Cheng, Chow, and Rider, 1972). RRT is also being used in AIDS research (*Discover*, July 1987, p. 12, cited by Scheaffer *et al.*).

In addition to the limitations of the randomized response technique mentioned earlier, another reason why the technique is not being used more is that cross-tabulations of the sensitive topic responses by independent variables are not possible. The above considerations call for a most judicious application of this technique.

Further reading

For the admissibility criteria of RRT estimates, see Bourke and Dalenius (1974). For a full treatment of the subject, see the references cited in the text and the book by Chaudhuri and Mukherjee; consult also the papers by Horvitz, Greenberg and Abernathy (1975). Droitcour *et al.* (1991) have discussed a number of randomized response techniques. They also reviewed an experiment using the item count method, due to J. Miller (1984), where each respondent is provided with a short list of items describing behaviors and asked to count and report the total number (not the names) of behaviors engaged in. In the simplest case, a probability sub-sample (Sub-sample A) of respondents are shown a four-list item that includes the socially disapproved behavior item; the remaining respondents (Sub-sample B) are shown a list of three items (the original four-item list minus the disapproved item). By comparing responses from the two sub-samples, an estimate is obtained.

24.15 Line sampling and lattice sampling

In **line sampling**, lines are drawn across a geographical area and all universe units falling on the line, or intersected by it, are included in the sample. If the lines are straight parallel equally spaced across the area, the sampling becomes one variant of systematic sampling (Chapter 4).

If instead of all intercepts on the line, a series of evenly spaced points are chosen on each line, the sampling is equivalent to choosing the points on a lattice and may also be regarded as two-stage sampling.

In **lattice sampling**, sub-strata are selected for the sampling of individual units according to some pattern analogous to the allocation of treatments on a lattice experimental design. Lattice sampling can be considered a variant of "deep stratification."

An example of **line-intercept sampling** would be in estimating the total quantity of berries of a certain plant species in a study area, when a random sample of n lines, each of length l, is selected and drawn on a map of the study area; in the field survey, whenever a line is seen to intersect a bush to the species, the berries of that bush are collected and their quantity measured (Thompson, Chapter 19, who also provides estimating equations in line sampling.)

Further reading

Thompson, Chapter 19; Yates, sections 3.15 and 3.18.

24.16 Surveys on disability

One of the concerns in today's world is disability, whether congenital or disease- or trauma-induced. But a major problem of the international comparability of data on disability still arises from the different definitions adopted by national organizations that are different from the International Classification of Impairments, Disabilities and Handicaps (WHO, 1980).

The decennial population census that is *de rigueur* in most countries has been the traditional source of obtaining data on disability, as in the 1990 Population Census in the U.S. and the 1986 Census of Canada. However, the decennial census suffers from two major limitations: first, it is conducted only every ten years, and, second, with the competing demands that are put upon it, it cannot provide in-depth data on disability.

Special-purpose surveys are the other major source of disability data. While utilizing the census frames, such disability surveys are conducted either in connection with the population census, in the form of post-census surveys as in Canada (in 1986) or Egypt (in 1960) or otherwise independent from the population census as in Egypt (in 1978-1983), India, the Netherlands, and the United States.

In the 1986 Census of Population in Canada, the survey strategy was to ask 20 per cent of households to complete a long questionnaire that included one question on disability ((a) "Are you limited in the kind or amount of activity that you can do because of a long-term physical condition, mental condition or health problems – at home; at school or at work; in other activities, e.g., transportation to or from work, leisure activities?" ; (b) "Do you have any long-term disabilities or handicaps?") and 23 other additional questions on ethnic origin, labor force participation, income and education. The question on disability was aimed at providing a sampling frame for the conduct of a post-censal survey entitled the Health and Activity Limitation Survey (Furrie, 1989). Such post-census disability surveys could be a recommended strategy due to lowered cost (the 1986 survey in Canada obtained five times the sample at twice the cost; see Statistics Canada, 1992) and simplified sampling.

Another strategy that has recently gained currency is to piggy-back a disability survey, by attaching special modules, on national surveys, as was done in the Labor Force Survey of Canada in 1983-84, the Health Interview Survey of Netherlands in 1986-88, and the Health Interview Survey of the United States in 1993-94 (Statistics Canada, 1992). A variant of this would be to combine a disability survey with a survey on aging, since disability is often a function of age.

Screening questions could be on general disability, as in the 1986 Population Census of Canada (see above), or comprise a checklist of specific disabilities, as in the U.S. Population Census of 1990, and New Zealand Population of 1990 (Chamie, 1989): the choice would obviously depend on other additional questions asked.

As with most large-scale sample surveys, the sample design should be stratified multi-stage, with strata being formed taking into account the prevalence of certain forms of disability (such as river blindness (onchocerciasis) which is specific to southern Mexico, Guatemala, Venezuela, and Central Africa) and also some population groups with higher than the national prevalence of disabilities.

In the 1986 Health and Activity Limitations Survey of Canada, two major strata were formed – Indian reserves and other areas. To estimate "false positives", a sub-sample was taken from the sample of persons who had responded "Yes" to the Census screening question and to estimate "false negatives", another sub-sample was taken from those who responded "No" (see section 25.3.2(i) for discussion of the terms "false positive" and "false negative.") For the institutional population, the institutions – orphanages and children's homes; special care homes and institutions for the elderly and chronically ill; general hospitals; psychiatric hospitals; and treatment centers and institutions for the physically handicapped – were grouped in three strata according to size (number of permanent residents), and within each stratum, a sample of residents was selected and personally interviewed, or if the selected individual was unable to respond for himself or herself, an interview was conducted with a staff member or next-of-kin (Furrie, 1989;

for details, see Statistics Canada, 1986).

Further reading

For details of the Health and Activity Survey in Canada in 1983-84, see Statistics Canada (1986) and for the 1991 survey, see Statistics Canada (1991 and 1994); for details of the Netherlands Health Interview Survey in 1981-91, see Netherlands, Central Bureau of Statistics (1992); for the U.S. Health Interview Survey in 1984-88, see Ries and Brown (1991); and for the 1983-85 Survey, see LaPlante (1988). Also see Chamie (1989) for survey design strategies; for case studies of the development of disability statistics in Egypt, Iraq, Jordan, Lebanon, and Syria, see U.N. (1986) and for development of statistical concepts and methods for household surveys, see U.N. (1988a).

24.17 Surveys on nutritional status of children

The U.N. World Summit for Children, held in September 1990, established, in line with the Rights of the Child, a series of specific goals to be achieved by the year 2000: the major goals being reducing infant and child mortality by one-third or to 50 and 70 per 1000 live births respectively, whichever is less; halving of the maternal mortality rate; halving the prevalence of severe and moderate malnutrition among children; providing universal access to safe drinking water and to sanitary means of excreta disposal; and improving protection of children in especially difficult circumstances.

One of the instruments for achieving these goals is the monitoring of the nutritional status of children, for child malnutrition is an accepted indicator of social development that has been seen to be endemic not only in developing countries, but also in a number of developed countries, particularly among the urban poor and the "inner cities" in the United States. The three recommended nutritional status indicators bounding the early childhood period are: birth weight, weight-for-age, and height-for-age (WHO, 1962; FAO, 1984; Carlson, 1987).

The principal objective of a nutritional study of children in this context is to assess the frequency and degree of underweight (indicator: weight-for-age), stunting or chronic malnutrition (indicator: height-for-age), and wasting or acute malnutrition (indicator: weight-for-height).

A sample survey is the only feasible data source for such information. A linked nutritional status survey module that combines a survey on the nutritional status of children with another compatible survey could be the proper strategy for a field study (U.N., 1990d; Carlson and Jaworski, 1994).

In such a linked survey, the nutritional module is self-standing but with links to the relevant identification and characteristics of the sample households, so that studies of interrelationships of the nutritional status of children with, say, household income and environments could be undertaken later on. This was the approach used in the Child Nutritional Status Survey in Bangladesh in 1989-90 (Carlson and Jaworski, 1994; Bangladesh, 1991). In the Gulf Child Health Survey conducted in seven member countries of the Gulf Co-operation Council and in the Pan Arab Project for Child Development in 14 Arab countries, information on the nutrition of children was collected for the sample households (see Appendix VI, Sections A6.3.7 and A6.3.8). As a general rule, the sample design should be stratified multi-stage. Typically, the sample size in these surveys ranges from 5,000 to 10,000 sample households, yielding, on an average, respectively 3,400 and 6,800 children between 3 months and 5 years of age in most developing countries (Carlson and Jaworski, 1994). The Pan Arab Child Development Programme has revealed the importance of paying particular attention to the urban poor (Jemiai and Khalifa, 1993), e.g. those in "the inner cities" in the U.S.: it may thus be advisable to have a separate stratum for them in order to estimate the nutritional status of their children, the vulnerable group, distinct from the urban affluent.

In view of the seasonal character of the nutritional status of children (Carlson and Jaworski, 1994), the sample should be spread over a year and each month a fresh sample, one-twelfth the total size, should be enumerated so as to eliminate the seasonal effect (Som, 1973, Chapter 2).

For **software for nutritional surveillance**, see Fichner *et al.* (1989) and Sullivan *et al.* (1990). Other statistical software programs and packages that could be considered are ANTHRO-Software for Calculating Pediatric Anthropometry (Sullivan and Gorstein, 1990) and EPI-Info (Dean *et al.*, 1990; also see section 27.10.4 of this book). Of the available software packages for data processing, the Integrated Microcomputer Processing System (IMPS), developed by the United States Bureau of the Census (see sec. 27.10.1), appears to be the most flexible and is recommended.

24.18 Surveys on the status of women

The collection and analysis of accurate and comprehensive statistics on the situation of women has become a top priority in the United Nations system: the United Nations has published a data base on the situation of women called, WISTAT (U.N., 1992).

Women's activities cover a wide range, arguably wider than that of men. Studies of the social and economic roles of women in the family and the household would require data, among others, on the composition of the family/household, including headship rates, i.e. the number and proportion of households headed by women, age and marital status of the household head, female property inheritance, widow inheritance, the nature of labor migration, the impact of male migration, the participation of women both in domestic activities and informal sector activities that had, until recently, been "invisible" (U.N., 1984a, 1988b, 1990b, and 1993c).

The informal sector includes non-monetary production (in the sense that the output is retained by the producers for their own use), monetary production, and other activities that lead to production that constitutes "an important contribution to the total consumption of the household" (U.N., 1990b).

The population census has been the traditional source of collecting general demographic and social data, from which some required indicators, such as headship rate by sex and age, could be calculated. Recommendations have been made, among others by the U.S. Agency for International Development and the United Nations for using the population census for collecting data and preparing reports on women's status in national development (USAID, 1981; U.N., 1990a).

But most other data on the status of women would have to be collected through household or labor force surveys (particularly their activities in the informal sector) (U.N., 1988b; ILO, 1990), time-use surveys (for detailed information on the amount and intensity of labor inputs), and division of labor modules (U.N. 1988b and 1993c).

As men and women are fairly equal in number in the population and are distributed generally equally among households, no new sampling technique would be required in a study of women's status. Consideration should, however, be given to forming separate strata for "inner city" populations with large proportions of woman-headed households and for rural and semi-urban areas with a preponderance of non-formal economic activities and to have these sectors adequately represented in the sample.

The survey strategy could be the piggy-backing of, or linking with, a survey on women's status and other data collections, such as a population census, or a household or demographic (particularly, fertility) survey or a labor force survey. The women's survey would then constitute a sample of the census or a sub-sample of the household or demographic or labor force survey: this would have the further advantage of the potential of the analysis of interrelationships between the status of women with household or demographic or labor force characteristics.

24.19 Estimation of HIV/AIDS infection

During the last decade, the spread of Human Immunodeficiency Virus (HIV) infection/AIDS (Acquired Immunodeficiency Syndrome) has become a world-wide public health problem affecting vulnerable population groups – intravenous drug users, homosexual males, prison population (males), female partners of HIV-infected males, commercial sex workers and infants born to HIV-infected women. Several governments and international organizations, particularly, the World Health Organization, are attaching great importance to this problem.

Before any action programs to contain HIV infection could be taken, it would be necessary to estimate its prevalence among different population groups. The U.S. Centers for Diseases Control implemented a family of HIV Seroprevalence Surveys in the United States in 1987-89 in sexually transmitted disease clinics, in drug treatment centers, among women in women's health clinics, among patients of tuberculosis clinics, at U.S. sentinel hospitals, among childbearing women, and among blood donors. The survey design (with options for local variations) was blinded and unlinked, where "there is no interaction or intervention with eligible people for the purpose of the survey, and HIV serological test results cannot be linked to identifiable persons. ... Only demographic and behavioral information already collected for other purposes is used in the survey" (Pappaioanou *et al.*, 1990a). This approach obviates the need for informed consent, and thus eliminates the self-selection bias in most voluntary HIV seropositive surveys.

In the survey among childbearing women, for the first survey periods, minimum sample sizes for estimating statewide seroprevalence with 50 per cent relative error were calculated by using seroprevalence rates observed in civilian female applicants for military service (Pappaioanou *et al.*, 1990b).

The Center for International Research of the U.S. Bureau of the Census has studied the recent HIV seroprevalence levels by country and the trends and patterns of HIV/Infection in selected developing countries (U.S. Bureau of the Census, 1993b and 1993c; also see Way and Stanecki, 1993) and has published an HIV/AIDS Surveillance Database (U.S. Bureau of the Census, 1993d).

In designing a national survey to estimate the prevalence of HIV infection and the characteristics of the infected, it should be useful (a) to stratify the universe according to the generally known prevalence of HIV infection, and (b) to take samples in proportion to the rough estimates of the number of the infected. Thus it may be necessary to put cities such as San Francisco and New York in the U.S., Bombay in India, and Bangkok in Thailand, and prison populations (males) with sub-strata for specific urban localities. Lacking the informed consent of the general population to be tested for HIV infection, data would *par force* have to be collected from hospital and clinic records. However, informed consent would have to be obtained for testing, on humans, potential vaccines against AIDS.

To test the hypothesis of association of HIV and BIV (Bovine Immunodeficiency Virus), it would be necessary to survey separately women who live in rural areas (with exposure to cattle) but commute daily to the cities for commercial sex activities (Bhattacharya).

Further reading

In addition to the references cited in the text, see the articles in *Public Health Reports*, Vol. 105, No. 2, March April 1990, "Special Section: The Sentinel HIV Seroprevalence Surveys."

CHAPTER 25

Errors and Biases in Data and Estimates

25.1 Introduction

In the previous chapters, we had assumed implicitly that the basic data collected either through a complete enumeration or a sample are free from any error or bias and that the (unbiased) estimates obtained from a sample are subject only to sampling errors. This is not so in actual survey conditions. From the stage of collection to that of preparation of final tables, the data are generally subject to different types of errors and biases. In this chapter will be reviewed the various types of errors and biases in data and in estimates derived from them and the methods of measurement and control of these errors and biases.

The importance of giving proper attention to errors and biases in data and estimates can best be illustrated by the telling words of W. Edwards Deming (1950, page 25):

> *"For what profiteth a statistician to design a beautiful sample when the questionnaire will not elicit the information desired, or if the universe has not been satisfactorily defined, or the field-force is so badly organized that the results will not be worth tabulating? And again, what is accomplished if a well-designed questionnaire and well-disciplined field-force are used with a biased sampling procedure?"*

This observation applies equally to the design of censuses.

25.2 Coverage and content errors

It will be convenient at this stage to distinguish two types of errors – errors of coverage and errors of content. If in a survey for a specified area, some households, persons, etc. are missed, i.e. not enumerated at all or enumerated more than once, that is a **coverage error**: complete villages in remote areas being missed in some African censuses are an extreme case in point. If a particular unit is covered in the survey, but there is a mistake in recording its relevant characteristics (the age of a person, for example, who may not, for whatever reasons, know or report his exact age), that is a **content** (or **classification**) **error**. There may be some balancing out of these two types of errors.

25.3 Classification of errors in data and estimates

The different types of errors in data and estimates can be classified first according to the source: (a) errors having their origin in sampling and (b) errors which are common to both censuses and samples.

25.3.1 Errors having their origin in sampling

These are: (i) sampling errors and (ii) sampling biases. Errors other than sampling errors are called non-sampling errors and biases.

(a) *Sampling errors.* A measure of the degree to which the sample estimate differs from the expected survey value (which is obtained on repeated applications of the same survey procedure) is given by the **standard error**, the square of which is called **sampling variance**.

(b) *Sampling biases.* In a sample survey, in addition to sampling errors, sampling biases may arise from inadequate or faulty conduct of the specified probability sample or from faulty methods of estimation of the universe values. The former includes defects in frames (in both the available frames and those prepared by the enumerators in a multi-stage design), wrong selection procedures, and partial or incomplete enumeration of the selected units.

Biases of the estimating procedure may either be deliberate, due to uses of a biased estimation procedure itself, or may be due to the inadvertent use of a wrong formula for estimation. An example of the unavoidable bias in a particular estimate, even though the components may be unbiased, is the ratio of the unbiased estimates of two totals (see sections 2.7, 3.2 etc.).

Three examples will be cited of bias due to methods of estimation not based on the sampling plan. First, biased estimates will result if the sample is treated for simplicity as simple random or self-weighting when it is neither of these. In an inquiry when it is not feasible to construct a host of estimates by applying proper multipliers, unweighted estimates of ratios, which may be subject to much smaller biases than the estimates of the numerators and denominators, may be obtained for the less important items. A comparison should in any case be made of the estimates based on proper weights and simplified weights for some important items in order to obtain an idea of the magnitude and direction of the bias in simpler estimates. Results from a number of demographic surveys in India have shown that such a procedure could be used with proper caution (Som, 1973, section 1.3.2(ii)b1).

Second, biased estimates will result when estimates of variances and covariances are computed from results at higher levels of aggregation. Results (such as birth rate and proportion of childless women) are not generally published for the ultimate strata but only for the higher levels of aggregation, such as states and provinces: the components of these results (estimated numbers of births and total population for the birth rate and estimated numbers of childless women and total women for the proportion of childless women) are almost never published. Yet, in a study of a demographic survey in Zaire in 1955-1957, the correlation coefficient between (the estimated) birth rate and (the estimated) proportion of childless women was computed from the published results of birth rate and proportion of childless women for the 28 districts in the country: this is a wrong method and the desired estimates should be obtained with a design-based procedure by building up estimates of variances, covariances etc. from the ultimate strata in the sample.

As a last point, biases will arise when the sample is treated as if it had come from a stratified design when the design is either unstratified or stratified with respect to another variable (see section 10.7).

25.3.2 Errors common to both censuses and samples

Errors common to both censuses and samples will be classified under seven broad headings: (i) errors due to inadequate preparation; (ii) errors of nonresponse; (iii) response, ascertainment or observational errors; (iv) processing errors; (v) errors in constructing substitute estimates; (vi) errors in interpretation; and (vii) errors in publication. These categories are not strictly mutually exclusive, and there may be some amount of overlap, especially for the first three.

(i) *Errors due to inadequate preparation*. These might be due to the failure to state carefully the objectives of the inquiry or to decide on the required statistical information; drawbacks of the questionnaire, including lack of clarity of concepts, definitions and instructions; failure to define the universe with enough precision or to provide precise instructions and definitions; and careless and disorganized field procedure, including faulty methods of selection of enumerators or supervisors and faulty or insufficient training.

One common source of such errors is the faulty demarcation of the boundaries of the geographical units which leads to wrong coverage, mainly under-coverage; available maps do not often contain the details required and such maps cannot take into account new premises that are built between the time the map was prepared and the inquiry is conducted.

In a demographic inquiry, if the question of current births is asked only of the mothers who are alive on the survey date, those occurring to mothers who had died by the survey date would be omitted; similarly, deaths to all the members of a household, or to persons after whose death the household dissolves or re-forms in more than one part, may not be obtained unless specific procedures are laid down to collect such data.

In an inquiry designed to estimate the size and characteristics of a specific population group, selected after a preliminary screening (where the mesh cannot perforce be fine) of the general population, bias may be introduced at the stage of screening by the failure to take precautions against the occurrences of (a) "**false negatives**", i.e. persons actually belonging to the specific population group but not classified as such, and (b) "**false positives**", i.e. persons not actually belonging to the population group but classified as such. For example, inquiries on the prevalence of pulmonary tuberculosis were generally conducted by first screening, by means of miniature X-rays, the general population (excluding children under age five years) in the selected areas (or households) and then bacteriologically examining sputum and laryngeal swabs of only those showing evidence of any pathology in the X-ray films. But a study made in the U.S. in 1947 showed that in a single reading of X-rays, 54 per cent of cases showing a moderate degree of tuberculosis and 0.8 per cent of cases with no evidence of tuberculosis were diagnosed as positive (Yerushalmy and Neyman). Although in such a procedure the number of "false positives" could be estimated, that of "false negatives" (and, therefore, the true total number of persons with pulmonary tuberculosis) could not be estimated, unless a sample of persons from those not showing evidence of any pathology in the X-ray films were also covered for the bacteriological examinations of sputum and laryngeal swabs. The latter procedure is now reportedly being followed, and recommended, by the World Health Organization. Estimates from both the groups should be combined to provide estimates for the specific population group under study.

For the 1986 Health and Activity Limitations Survey of Canada, screening questions were introduced in the national population census of Canada in 1986 to define a sample of individuals likely to have a disability; the survey included sub-samples from both those who answered "Yes" to the census screening questions and those who answered "no": the proportion of persons with some disability was estimated at 13 per cent, of whom "false negatives" – i.e. the census screening missed – constituted 40 per cent and the ratio of "false positives" – i.e. those falsely identified by the census screening as having some disability when they had none – to the number with actual disability was 1/16 (U.N., 1994; for details, see Statistics Canada, 1992).

(ii) *Errors of non-response or incomplete sample.* Non-response could be due to the non-coverage of areal units (coverage errors) because of incomplete listing or inaccuracy, difficulties of the terrain, the absence from home of respondents when the dwelling units are visited ("not at homes"), the inability to answer some questions, and the refusals to answer a question. Non-response increases both the sampling error (by decreasing the effective sample size) and the non-sampling error.

An example of a response error is the deficiency in the reported number of two-member households in a morbidity inquiry in Syracuse, New York, in 1930-31, which was attributed to the failure of the enumerators to revisit missed households in which childless married women working away from home are likely to predominate (Kiser, 1934; cited by Brookes and Dick, section 47b.7 and Yates, 1981, section 2.4).

In other situations, where information is required on physical characteristics of some items by adopting objective techniques of measurement, the necessary information may be lacking due to the failure to reach the survey units at the right time to collect the data. These difficulties arise mostly in agricultural surveys designed to estimate crop production, because of failure of the enumerators to visit the sample fields selected for crop-cutting experiments before harvesting.

(iii) *Response or ascertainment or observational errors.* These refer to the differences between the individual survey values and the corresponding true values (elaborated in section 25.4).

The magnitude and direction of response errors depend on the survey conditions and procedures – the skill of the enumerator, the time given for interview, the inclusion of fully detailed probes in the schedule, the acceptance of proxy interviews, and the understanding and co-operation of the respondents, the general climate of attitudes and opinions etc. The response error may be unintentional or it may be deliberate on the part of the respondents: a person may not know his exact age, or he may know his exact age but make an assertion more pleasing to his personal vanity ("**the vanity effect**," vide Kendall and Buckland).

The fear of the "**evil eye**" is known in some cultures in reporting births, especially of males (Das Gupta, 1958), farm produce (Sukhatme *et al.* (1984), section 11.19), etc.

In the 1951 Census of England and Wales, for an appreciable number of women married more than once whose earlier marriages were not reported and who were thus included among the "married once only", the total number of children in all the marriages was associated with the current marriages; at marriage ages 35-39 years, where this error was concentrated, the total fertility might have been overstated by as much as 10 per cent (Benjamin, 1955).

The enumerator may also have conscious or unconscious biases. For example, he may like to record numerical answers in round figures, or he may over- or under-record some figures for particulars groups of units. In crop surveys, eye estimates are always subject to bias, which would vary from enumerator to enumerator, and the net effect can often be substantial. In inquiries on crop yields, the use of too small a "cut" would give rise to biases when the demarcation of the boundaries becomes of importance, as the size of the unit is decreased, and also because the possibility of influencing the results by small changes in location, e.g. so as to include particularly good plants, is greater the smaller the unit areas (Mahalanobis, 1946b; Mahalanobis and Sengupta, 1951; Sukhatme, 1946a, 1946b, 1947a, and 1947b; and Yates, section 2.5).

One major source of response error is that of recall. Many of the items in an inquiry relate to events that occurred in the near or distant past, and the problem is both to remember these events and to place them in the correct time periods. Such a recall error is called recall lapse when it can be expressed as a function of the recall period (i.e. the period that elapsed between the occurrence of the event and the inquiry). Recall lapse may relate to the reporting of births and deaths, reporting of the births of a preferential sex, under-reporting of infant deaths, reporting of sickness, etc., when these can be seen to be a function of the recall period (Som, 1973). Recall lapse has been observed in both developed and developing countries, particularly related to nondurable consumer goods, spells of minor sickness, etc. (Bailar and Biemer (1984)).

(iv) *Processing errors.* Such errors may be introduced at different stages of data processing, from editing to final presentation in the form of summary tables. However, these errors are generally easier to control on the basis of checking and sample re-verification, including the insertion of dummy entries (Mahalanobis, 1946a). The possibility of occurrence of simple computation errors is not generally admitted explicitly, a refreshing exception being the study of the registration data in Uttar Pradesh, one of the larger states in India: "The published figures do not necessarily conform with the actual registration in the villages. This vitiates all calculations regarding corrected birth rates" (Government of India, 1955). For the population census data of Yugoslavia in 1953, errors made during the individual phases of manual processing were found to be quite considerable when compared with the totals of corrected figures (Macura and Balaban, 1961).

(v) *Errors in constructing substitute estimates.* When the inquiry, either a census or a sample, fails to provide accurate estimates, substitute measures are often devised. These are sometimes based on assumptions which cannot be validated or do not apply in a particular situation. Thus, a method of estimating the current fertility level based on the assumption that fertility has not changed in the recent past cannot be applied to provide estimates aimed at measuring fertility changes.

(vi) *Errors in interpretation.* These relate to misleading interpretations and conclusions that do not follow from the data. Take, for example, the crude death rates for single and married males in Portugal in 1950 (10.5 and 12.6 per 1,000 single and married males respectively) and in Chile in 1952 (10.5 and 12.8 per 1,000 respectively). From these, it would be wrong to conclude that in these countries a male could increase his chance of survival by remaining single: for, in each age group, the death rate for the single is higher than that for the married (United Nations, *Demographic Yearbook, 1956*, Table 21). The paradox stems from the different age distributions of the single and the married.

(vii) *Errors in publication.* These are mainly mechanical in nature and relate to proof reading etc.; they are not generally as serious as the other types of errors.

This inventory and description of errors and biases should not lead one to suppose that all inquiries are worthless because all have errors. *The errors are of varying types and degrees, and can occasionally be measured and subtracted out; this could be the main aim in research on errors and biases (Deming, 1950). On the other hand, a survey in which the magnitude and direction of different types of errors are not evaluated, or at least indicated, may give a false sense of accuracy in the collected data and constructed estimates.*

25.4 Simple statistical model

25.4.1 Individual value, true value, and expected survey value

The true value of ith individual item will be designated by z_i. The survey may be a census with the total number of units N, in which case i runs from 1 to N; or it may be a sample of n units with i running from 1 to n. Thus, the study variable may refer to the number of births in the reference period in a particular household, which may be zero (no birth), $1, 2, \ldots$; or it may refer to the age in years last birthday of a household member, which may have any of the values $0, 1, 2, \ldots$; or it may refer to the size of the household, which may be $1, 2, \ldots$; and so on. The recorded survey value, obtained either from a census or a sample, will be designated by y_i. The error or bias in the individual ith unit is defined as

$$b_i = y_i - z_i$$

Thus, if in a particular household no birth is reported to have occurred during the reference period ($y_i = 0$), when in fact a birth had occurred ($z_i = 1$), the error in the survey value is $b_i = y_i - z_i = 0 - 1 = -1$; and so on.

When the survey value coincides, perhaps fortuitously, with the true value, i.e. when $y_i = z_i$, i.e. $b_i = 0$, we say that the survey value is *accurate*; if $y_i \neq z_i$, i.e. $b_i \neq 0$, the survey value is said to be *inaccurate*.

The value which is obtained on repeated applications of the same survey procedures is called the *expected survey value*, and designated by $E(y)$. Note that the expected survey value would differ from survey to survey and from a sample to a census. Under the term "same survey procedure" are included the skill and training of the enumerators, the quality of the questionnaires etc.

In most cases the true individual value can be defined, but it may not be known, e.g. the age of a person. In some cases, however, the true value is difficult to define, e.g. attitudes and opinions, but is nevertheless useful conceptually.

Note also that the true values of individual and derived items are generally unknown. What are obtained in a survey (census or sample) are the survey values.

25.4.2 Census

First we deal with a census or complete enumeration. As before, let the individual true value of an item under study be designated by z_i and the recorded survey value by y_i ($i = 1, 2, \ldots, N$). The errors in the individual survey values are $b_i = y_i - z_i$, from which

$$y_i = z_i + b_i \tag{25.1}$$

The relation between the survey value and the true value of the universe total is given by summing both sides of equation (25.1),

$$\sum^{N} y_i = \sum^{N} z_i + \sum^{N} b_i$$

or

$$Y = Z + B \tag{25.2}$$

and that between the means by dividing both sides of equation (25.2) by N

$$\overline{Y} = \overline{Z} + \overline{B} \tag{25.3}$$

where B is the error of the survey total (Y) and ($\overline{B}$) that of the survey mean ($\overline{Y}$).

When $Y > Z$, i.e. B, the error in the survey total, has a positive sign, Y gives an overestimate of Z; and when $Y < Z$, i.e. B has a negative sign, Y gives an underestimate of Z. And similarly for the means.

From equation (25.1) the variance of an individual survey value is given by

$$\sigma_y^2 = \sigma_z^2 + \sigma_b^2 + 2\sigma_{zb} \tag{25.4}$$

where σ_y^2, σ_z^2, and σ_b^2 are the variances (per unit) of y, z, and b respectively and σ_{zb} is the covariance of z and b, defined respectively by

$$\begin{aligned} \sigma_y^2 &= \sum^{N}(y_i - \overline{Y})^2/N; \ \sigma_z^2 = \sum^{N}(z_i - \overline{Z})^2/N; \\ \sigma_b^2 &= \sum^{N}(b_i - \overline{B})^2/N; \text{ and } \sigma_{zb} = \sum^{N}(z_i - \overline{Z})(b_i - \overline{B})/N \end{aligned} \tag{25.5}$$

The covariance σ_{zb} can also be expressed as $\rho_{zb}\sigma_z\sigma_b$, where ρ_{zb} is the correlation coefficient between z and b, given by

$$\rho_{zb} = \sum^{N}(z_i - \overline{Z})(b_i - \overline{B})/N\,\sigma_z\sigma_b$$

Apart from the general case where individual errors are present ($b_i \neq 0$), and do not cancel out ($B \neq 0$), so that the survey total, survey mean and the variance of survey value are given respectively by equations (25.2), (25.3), and (25.4), we may have the following special cases:

(i) *Individual errors absent, data accurate.* Here $b_i = 0$, and $B = 0$, so that

$$Y = Z; \ \overline{Y} = \overline{Z}; \text{ and } \sigma_y^2 = \sigma_z^2 \text{ (Here } \sigma_b^2 = 0 = \sigma_{zb}).$$

(ii) *Individual errors present, but total error absent.* Here $b_i \neq 0$, but $B = 0$, so that

$$Y = Z;\ \overline{Y} = \overline{Z};\ \text{but } \sigma_b^2 \neq 0;\ \text{and } \sigma_y^2 = \sigma_y^2 + \sigma_b^2 + 2\sigma_{zb}$$

Thus in this case, the survey total or mean would be free from error, but the variance will be affected; even if $\sigma_{zb} = 0$, i.e. even if there is no correlation between z and b, but $\sigma_y^2 > \sigma_z^2$.

(iii) *Constant individual error.* Here $b_i = b_0$, a constant, so that $Y = Z + Nb_0, \overline{Y} = \overline{Z} + b_0$, but $\sigma_y^2 = \sigma_z^2$ (since $\sigma_b^2 = 0 = \sigma_{zb}$). In this case, the survey total, mean etc. will be affected, but not their variances.

The above results show that *even in a census, the results (total, mean etc.) may be misleading unless some idea is obtained about the errors of the data.*

25.4.3 Sample

Here we assume that a simple random sample of n units is drawn with replacement from the total of N units. The individual survey values are

$$y_i = z_i + b_i \tag{25.6}$$

The relation between the totals is

$$\sum^n y_i = \sum^n z_i + \sum^n b_i \tag{25.7}$$

and that between the means (dividing both sides of equation (25.7) by n) is

$$\overline{y} = \overline{z} + \overline{b} \tag{25.8}$$

where $\overline{y}$ is the survey mean, $\overline{z}$ the true mean and $\overline{b}$ the error of the survey mean, for the n sample units.

When $\overline{y} > \overline{z}$, i.e. $\overline{b}$ has a positive sign, $\overline{y}$ will give an overestimate of $\overline{z}$; and when $\overline{y} < \overline{z}$, i.e. $\overline{b}$ has a negative sign, $\overline{y}$ will give an underestimate of $\overline{z}$.

The expected value of the survey mean $\overline{y}$ is obtained on taking the means of all the possible values of $\overline{y}$ from different samples all of size n under the same survey procedure and is given by

$$E(\overline{y}) = E(\overline{z}) + E(\overline{b}) = \overline{Z} + \overline{B} \tag{25.9}$$

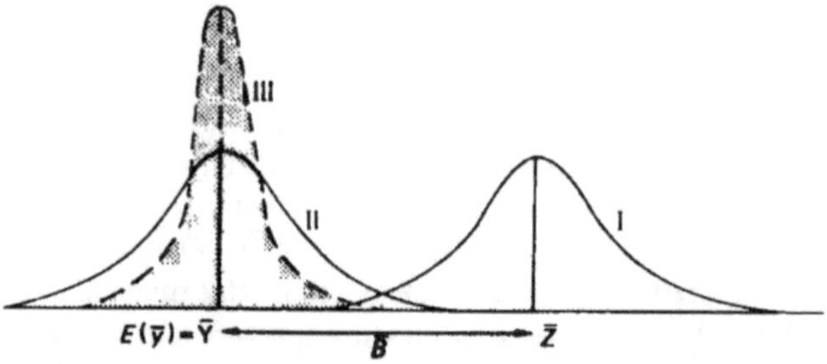

Figure 25.1: Distribution of true and survey individual values and of survey mean (adapted from Zarkovich (1963), Figure 1).

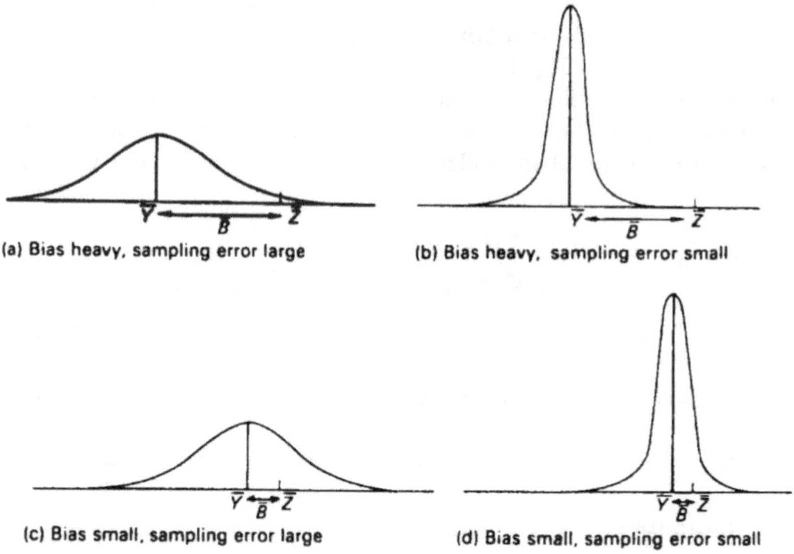

Figure 25.2: Distribution of the sample estimate ($\overline{y}$) in relation to the true value ($\overline{Z}$).

and the sampling variance of $\overline{y}$ is

$$\sigma_{\overline{y}}^2 = E[\overline{y} - E(\overline{y})]^2 = E(\overline{y} - \overline{Y})^2 = (\sigma_z^2 + \sigma_b^2 + 2\,\sigma_{zb})/n \qquad (25.10)$$

In Figure 25.1, the distribution (I) of the true values z_i around the true mean $\overline{Z}$ has been shown by Curve I; the variance of the distribution is σ_z^2. The distribution (II) of the survey values y_i around the survey mean $\overline{Y}$ is shown by Curve II; the variance of the distribution is σ_y^2. If a sample of n units are taken from the distribution II, the estimates of the mean $\overline{y}$ will have the distribution shown by Curve III around the expected survey mean $\overline{Y}$; a measure of this variation is given by $\sigma_{\overline{y}}^2$, as defined by equation (25.10). This does not, however, give any indication of the expected behavior of the distribution of the sample estimate $\overline{y}$ around the true mean $\overline{Z}$, which is the basic aim in any survey. That is measured by the mean square error defined below, which, in addition to the sampling error, takes account of the expected value of the bias, measured by $\overline{B} = \overline{Y} - \overline{Z}$.

Consider also the four special cases given in Figure 25.2(a)-(d). It is clear that in the two cases (a) and (b), where the bias is heavy, the values of both the estimate and its (estimated) variance would be misleading in setting any confidence limits to the universe value: the ideal situation is, of course, given by case (d) with both the bias and the sampling error small (see also Deming, 1950, Fig. 1).

Mean square error. The variability of the survey mean around the true value is measured by the mean square error, which is defined as the expectation of the square of the difference of the survey mean and the true mean:

$$\begin{aligned} MSE_{\overline{y}} &= E(\overline{y} - \overline{Z})^2 = E[(\overline{y} - \overline{Y}) + (\overline{Y} - \overline{Z})]^2 \\ &= E(\overline{y} - \overline{Y})^2 + (\overline{Y} - \overline{Z})^2 + 2(\overline{Y} - \overline{Z})E(\overline{y} - \overline{Y}) \\ &= \sigma_{\overline{y}}^2 + \overline{B}^2 \qquad (25.11) \end{aligned}$$

as $(\overline{Y} - \overline{Z}) = \overline{B}$ is a constant (for the same survey procedure) and $E(\overline{y} - \overline{Y}) = 0$.

Some special cases of the above equations are given in Table 25.1.

25.4.4 Response errors and biases

Response errors and biases have been considered jointly here, although a distinction could be made between them. Those response errors which have expectation zero may be called **response errors** (consider in the preceding sub-section the case where $b_i \neq 0$ but $\overline{b} = 0$); thus, without affecting the estimates except in a statistical sense, these increase the variability or the

Table 25.1: Effects of individual and total errors on the expected value, the sampling variance, and the mean square error of the survey mean

Individual error b_i	Mean error $\bar{b}$	Expected value survey mean $E(\bar{y}) = \bar{Y}$	Sampling variance of survey mean $\sigma_{\bar{y}}^2$	Mean square error of survey mean $MSE_{\bar{y}}$
General case: Individual and total errors present				
$\neq 0$	$\neq 0$	$\bar{Z} + \bar{B}$	$\frac{1}{n}(\sigma_z^2 + \sigma_b^2 + 2\sigma_{zb})$	$\frac{1}{n}(\sigma_z^2 + \sigma_b^2 + 2\sigma_{zb}) + \bar{B}^2$
Estimate, sampling variance, and mean square error affected.				
Special cases:				
(i) Individual errors absent, data accurate				
$= 0$	$= 0$	$\bar{Z}$	σ_z^2/n	σ_z^2/n
Estimate, sampling variance, mean square error unaffected. (Note the classical formula for the sampling variance of the mean for smple random samples.)				
(ii) Individual errors present, but mean error absent				
$\neq 0$	$= 0$	$\bar{Z}$	$\frac{1}{n}(\sigma_z^2 + \sigma_b^2 + 2\sigma_{zb})$	$\frac{1}{n}(\sigma_z^2 + \sigma_b^2 + 2\sigma_{zb}) + \bar{B}^2$
Estimate unaffected, sampling variance and mean square error affected.				
(iii) Constant individual error				
$= b_0$ (constant)	b_0	$\bar{Z} + B_0$	σ_z^2/n	$\frac{1}{n}\sigma_z^2 + \bar{B}^2$
Estimate affected, sampling variance unaffected, mean square error affected.				

Source: Som, 1973, section 1.4.3.

sampling error of the estimates. **Response biases** are the other type of response errors that do not cancel out, and affect the estimates themselves, while the variability of the estimates may or may not be affected (consider in the preceding subsection the cases where $\bar{b} = 0$; the sampling errors will not be affected if b_i is a constant.) Thus, in a census or a sample, if the differential effects of the enumerators cancel out, but there is a net bias common to all the enumerators, these may be stated to constitute a response bias.

25.4.5 Reliability and accuracy

It is important to clarify the difference between reliability and accuracy of data and estimates. The difference between the result obtained from a sample and that obtained from a complete enumeration, conducted under the same survey conditions (controlled or uncontrolled), or on the repeated application of the sampling procedure, gives **the reliability or precision of the sample estimate** and **is measured by the sampling error**; the difference between the result from an inquiry, either a sample or census, and the true value (which, in general, would be unknown) is known as **the accuracy of the estimate** and **is measured by the root mean square error**. We have seen in Equation (25.11)

$$\begin{aligned} \textit{Mean square error} &= \textit{Sampling variance} + (\textit{Bias})^2 \\ \textit{Root mean square error} &= \sqrt{(\textit{Sampling error})^2 + (\textit{Bias})^2} \end{aligned}$$

The above relation may be expressed by a Pythagorean right-angled triangle, where the hypotenuse is the root mean square error and the two sides the sampling error and the bias respectively (Fig. 25.3).

In most cases, we can measure only the precision of an estimate by the sampling error, although our main interest lies in measuring the accuracy of the estimate.

In a census with complete enumeration of all the units, the sampling variance of the survey mean and the census mean are the same; but the mean square error of the (survey) mean, obtained on taking the expectation of the square of the difference of the (survey) mean from the "true" mean, would be present unless $b_i = 0$, individually, even if $\overline{B}$ (i.e., even if the errors cancel out): see the special case (i) in Table 25.1.

Furthermore, the bias in a census is likely to be of a different type and degree than that of a sample. This is because, although the "true" value remains the same, the expected survey value would be different. It is also generally difficult to conduct a census under the same general conditions as that of a sample. The scale of operations being comparatively small,

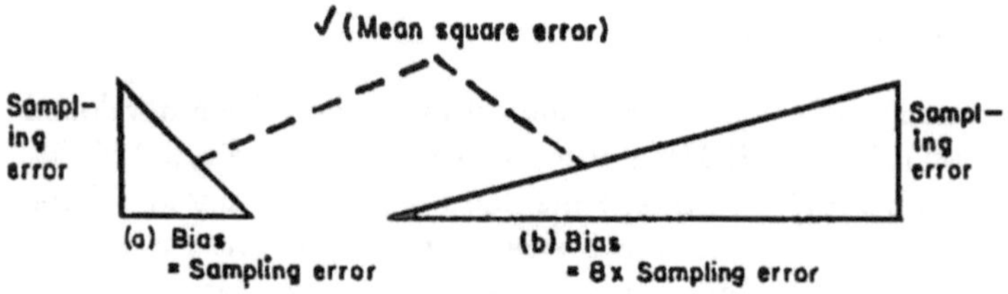

Figure 25.3: Geometrical representation of total error, sampling error, and bias.

sample surveys make it possible to exercise better control over the collection and processing of data by employing superior enumerators, giving them intensive training, and requiring interviews in depth (see section 1.3).

The aim in an inquiry, be it a census or a sample, should be to obtain estimates with the greatest possible accuracy by controlling the total error (as measured by the root mean square error) and not merely the sampling error.

25.4.6 Effect of bias on errors of estimation

Evidently some error will be involved if the sampling error is used to set confidence limits of the true value. Assuming that the estimator $\overline{y}$ is distributed normally about the expected value $E(\overline{y}) = \overline{Y}$, which is at a distance $\overline{B}$ from the true value $\overline{Z}$ (expression (25.9)), it is possible to estimate the effects of the bias on errors of estimate (Cochran, section 1.7).

If the bias is one-tenth of the standard or less, its effect on the total probability of an absolute error, i.e. $|\overline{y} - \overline{Z}|$ of more than $1.96\,\sigma_{\overline{y}}$ is negligible. But when the bias is of the same order as the standard error, the total probability becomes 0.1700, instead of the presumed 0.05 at $\overline{B} = 0$. For a bias which is four times the standard error, the total probability is 0.9793, more than nineteen times the presumed value. It has further been shown that, in the above technique of setting confidence limits with the standard error, when we presume, with a negative bias, that we shall be wrong in the direction of underestimate only in 2.5 per cent of the cases, we are more likely to be wrong in 98 per cent of the cases if the bias is four times the standard error (Som, 1973, section 1.4.6).

25.5 More complex models

More complex models of errors in data and estimates have been developed by Cochran (1977); Fellegi (1964); Hansen and his colleagues (1951; 1953, Vol. II, Chapter 12; 1961), and Sukhatme and Seth (1952), among others; more recently developed models are reviewed in *Measurement Errors in Surveys* (edited by Biemer *et al.*, 1991), in particular by Biemer and Stokes (1991). Following in general the formulation by Cochran, and elaborating the model in section 25.4, we express y_{ij}, the value obtained on the ith unit ($i = 1, 2, \ldots, n$) in the jth independent repetition, as

$$y_{ij} = z_i + b_0 + (b_i - b_0) + d_{ij} = z_i' + d_{ij} \tag{25.12}$$

where z_i is the true value; b_0 the constant bias over all the units; $(b_i - b_0)$ the variable component of the bias which follows some frequency distribution with mean zero as i varies and may be correlated with the correct value z_i; $d_{ij} = y_{ij} - (z_i + b_i)$ is the fluctuating component which follows some frequency distribution with mean zero and variance σ_i^2 as j varies for fixed i; and $z_i' = z_i + b_i$.

The mean $\overline{y}$ of an srs would be unbiased with variance σ^2/n (ignoring the finite multiplier) if all the measurements were accurate. However, because of the errors of measurement, the mean may be subject to a bias $\overline{B} = b_0$ and its mean square error is

$$MSE(\overline{y}_j) = \frac{1}{n}\,[\sigma_{z'}^2 + \sigma_d^2\{1 + (n-1)\rho_w\}] + b_0^2 \tag{25.13}$$

where $\sigma_d^2 \doteq \sum^N \sigma_i^2/N$, and ρ_w is the intrasample correlation of measurements within the same sample defined by analogy with cluster sampling by the equation

$$E(d_{ij}d_{i'j}) = \rho_w \sigma_d^2$$

It can be shown that when errors of measurements are uncorrelated within the sample and cancel out over the whole universe, the formulae given in the preceding chapters for estimating sampling variances remain valid for stratified and multi-stage sampling, although such errors decrease the efficiency of the estimators. On the other hand, when ρ_w is present (in measurements and in processing, particularly if subjective judgement is involved), it is likely to be positive, and the standard formulae for estimating the variance of $\overline{y}_j$ usually give an underestimate.

25.6 Measurement of response errors

Some types of response errors may be immediately evident in the data. The misreporting of age is a well-known case in point, reflected in the heaping of ages at some individual years (ending mostly in zeroes and fives) and the corresponding deficiency at some others (mostly ending in odd digits other than 5): very high or very low birth and death rates reported in some demographic surveys and civil registration constitute another case.

Some methods of measuring response errors will be considered: external record checks; re-surveys; interpenetrating networks of sub-samples; internal consistency checks; and analysis by recall periods.

25.6.1 External record checks

Such checks can be made only if accurate external data are available for each unit of the universe. An example is the "Reverse Record Checks" made for coverage errors in the 1960 Census of Population and Housing in the U.S., where probability samples of persons were drawn from different sources of records, namely persons enumerated in the previous census, registered aliens, children born in the intervening period, etc., or of special groups, such as the aged social security beneficiaries and students enrolled in colleges and universities: these were checked against the census returns (U.S. Bureau of the Census, 1963b). In countries with good hospital records, data on hospitalization as reported in household interview surveys can be compared with hospital records: this was done in the U.K. and the U.S. In a large number of countries of the world, this method obviously either cannot be used or would be of very limited value.

Where feasible, a related procedure is to monitor responses which are objective in nature. This is illustrated by the checking of the validity of the telephone coincidental method for determining the in-home radio station rating. In this method, telephone calls are made at random times during each quarter hour period of the day to a random sample of individuals in listed telephone households; the selected person is asked if the radio is on, and if it is, to identify the station that was tuned in. In a survey conducted the late 1960s in the New York Metropolitan area, this method was tested as follows. Each interviewer who did the calling had at her disposal an electronic device whereby she could transmit, over the telephone she was using, the broadcast currently coming from any of the leading twenty A.M. stations in the area by pressing any of the twenty buttons. In the telephone interview, when the radio was reportedly on and the program was identified by the respondent, the interviewer pressed the test button corresponding to the station reported. If then the respondent reported that what was coming over the telephone was the same as what was on his radio, the original response was considered to be correct. Of the total 854 responses validated in the test, 91 per cent of the responses were found correct by this procedure (Frankel, 1969). Groves *et al.* (1989) have described more recent techniques of validation of telephone interviews.

25.6.2 Re-survey

A standard method of measuring and adjusting for response errors in a survey, whether a census or a sample, is the re-survey of a sub-sample of units in the original survey, preferably using a more detailed schedule, and, in a personal interview, with better staff (either the supervisors or the better set of interviewers). For population and housing censuses, the ultimate sampling unit in re-surveys should be compact areal units. In the form of a **post-enumeration check**, such re-surveys are now almost universal with population censuses.

A re-survey can form an integrated part of the original survey, as in the Demographic Sample Survey of Guinea, in 1954-5, where in addition, whenever infant deaths were considered to be grossly under-reported in a village, a medical team was sent to re-interview the females (Ministère de la France d'Outre-Mer, 1956). Unless conducted simultaneously with the original survey or immediately after it, the re-survey introduces some operational and technical difficulties.

The conducting of a re-survey, in the form of a **post-enumeration check**, i.e. a **Post-Enumeration Survey (PES)**, to evaluate a census is a relatively recent phenomenon. For four countries with a long tradition of census-taking, the year when the first population census was conducted and the year when a PES was first introduced in the census are: the U.S. - 1790 and 1950; the U.K. - 1801 and 1961; Canada - 1871 and 1951; and India - 1872 and 1951.

Unitary checks with one-to-one matching (e.g. of persons in a population census and the PES) is considered as the essential, integral part of post-enumerative checks of censuses to check both **gross and net errors**; however, the difficulty of such matching with the available resources may be so enormous in some countries that checks at the aggregate levels (e.g. of the total number of persons in the selected areal units) may be considered practical to check only the net errors.

The statistical model in a re-survey is

$$E(\bar{y}') = \bar{Z}$$

on the assumption that the bias element $\bar{B}' = 0$, the letters with primes denoting the value obtained from the re-survey. The difference $(\bar{y} - \bar{y}')$ therefore gives an estimator of the bias element $\bar{B}$ in the original survey. Adjustments may also be made by regression and ratio estimators.

Some estimates of the under-enumeration of the total population (i.e. the coverage error) in a number of censuses are as follows: 1/2 per cent in the U.S.S.R. in 1959; 1.1 per cent in Canada in 1956; 1.1 per cent (and 3.5 per cent by analytical methods) in the U.S. in 1950; 3.5 per cent in Sierra Leone in 1963; and 8.0 per cent in Swaziland in 1956. For the U.K., an estimate at census date in 1951 exceeded the final count by 0.3 per cent; and in 1961 a careful retrawl of sample areas and interviews with a small sub-sample of households, carried out immediately after the census, gave no evidence of significant under- or over-enumeration (Benjamin, 1968). Note that the use of demographic-analytical techniques generally provides a higher degree of under-count in the census, e.g. 3.6 percent (Coale, 1955) as compared to 1.4 percent given by the PES in the 1950 population census of the U.S.

In agricultural inquiries too, under-enumeration can be sizable. In an experiment conducted in Greece, the farmers were found to have under-reported by 36 per cent the number of parcels operated by them, and in the 1979 Census of Agriculture in the U.S., the total number of farms were under-counted by 8.4 per cent and the area of farms in lands by 6 per cent, whereas the coefficients of variation of the estimated totals were of the order of 1 per cent only (Sukhatme *et al.* (1984), section 11.19).

In crop surveys, whenever **eye estimates** are used, the results should be calibrated by comparison with the physical measurements of a sub-sample: this assumes a high positive correlation between the two sets of figures (Yates, section 4.25, examples 6.12b and 7.15b, and sections 6.15 and 7.17).

Note: The **net error** is the resultant of positive and negative error, obtained on a one-to-one matching of the orginal survey, e.g. a census, and the re-survey; and the **gross error** is the absolute total of both positve and negative errors, ignoring their sign. The net error can be considered to be a measure of the non-sampling bias in the original survey and the gross error as a measure of the non-sampling variation, i.e., the *response variance* (U.S. Bureau of the Census, 1963a).

25.6.3 Supervision on a probability basis

If supervision of the field work done by the primary enumerators could be arranged on a probability basis, the results of the field checks by the supervisors could be utilized to measure and control the response biases in the work of the enumerators (Sukhatme *et al.* (1984), Section 11.10). However, this is not generally done on various theoretical and practical grounds. When the comparably weak enumerators receive a proportionately greater amount of supervision, the overall quality of the basic data is undoubtedly improved, but the final results will contain an unknown element of correction consequent on supervision.

In the U.S. Current Population Surveys, the re-interview survey is used as a quality control operation for the enumerators as well as an evaluation and research tool (see Hansen, Hurwitz, and Bershad, 1961, and Hansen, Hurwitz and Pritzker, 1964). Tentative consideration given to the possibility of adjustment of the results of the original survey on the basis of the re-interview caused some doubt that it would be worthwhile, because, on the whole, the differences were not large. It could not be done currently, month by month, with the present sampling design, because the re-interview sample was too small. The introduction of the re-interview results by a double sampling procedure would, on the other hand, call for a considerable change in sampling design. None of these seemed to have sufficient merit to be justified but would have caused additional work and cost (Personal communication in 1961 from Morris H. Hansen, then Assistant Director for Research and Development, U.S. Bureau of the Census).

In the Indian National Sample survey also, although there is on an average one supervisor to four enumerators, supervision is not arranged on a probability basis, primarily because there is reason to believe that the quality of work of the primary enumerators in the sub-sample for supervision would be different from that in the rest of the sample, because in actual practice, it is extremely difficult, if not impossible, to keep the enumerators ignorant of the supervisor's sub-sample: the scheme itself then might encourage negligence in the sample falling outside the supervisor's sub-sample. Moreover, the supervisor himself might be interested in demonstrating that the enumerators working under him are doing a good job, otherwise his competence as a supervisor might be called into question (Personal observations made to the author in 1962 by D.B. Lahiri, then Advisor to the National Sample Survey Department, Indian Statistical Institute).

25.6.4 Internal consistency checks

A simple check of consistency of data is to examine items with implausibly high or low figures. A number of checks may also be introduced in the schedules to ascertain whether consistent replies are obtained from two sets of questions, the first a broad question or item of information in a schedule, and the second with **cross-checks and fully detailed probes**. The total number of children borne by a woman can be asked directly and compared with the total of the number of children at present living in the same household, the number living away, and the number who have died. Information on the sexual life within marriage has been asked in the U.S. and that on the use of contraceptives from both husband and wife in a large number of countries, sometimes with startlingly different results. Some of these checks are negative in the sense that they would not reveal any systematic bias.

Another check of internal consistency is the comparison of differential reporting in **proxy** and **self-interviews**. In the U.S. National Health Survey, for example, about one-fifth of the minor non-chronic conditions were missed due to the acceptance of proxy interviews (Nisselson and Woolsey, 1959).

In crop-yield surveys, internal consistency checks are provided by measurements of yields from cuts of concentric circles: a comparison could also be made of the cuts of the same area but with different shapes such as a circle, triangle, rectangle, square, etc., where a circle with the minimum perimeter would be least subject to the **border bias**. In any case, the final checks will be made with the result of harvesting the whole sample field, working under the same survey conditions (see the references cited in section 25.3.2iii).

In demographic inquiries, with repeated visits of the same sample of households after an interval of time, such as six months or a year, information on pregnancies collected in a visit can be checked with their outcomes in the subsequent visit(s): this method was used with considerable success in Pakistan and Turkey. Such repeated visits can also provide generally accurate estimates of births, deaths, and migration in the intervening period as they did in India in 1958-9; these are, however, more expensive than one-time surveys.

25.6.5 Interpenetrating network of sub-samples

This technique, introduced by P.C. Mahalanobis in sample surveys in India in the 1940s, consists of subdividing the total number of units in the survey (sample of census) into a number of parallel, random groups and permits the testing by the analysis of variance the differential effects of enumerators and of other factors, e.g. variation in the field schedules, and methods of collection, etc. (section 24.7).

A simple tabulation may often reveal differences in the estimates obtained by different enumerators covering the same universe. In the models discussed in sections 25.4 and 25.5, the bias element ($\overline{B}_e$), where e refers to the enumerator, may be taken to consist of two parts, a constant bias ($\overline{B}'$) that affects all the enumerators alike, and another component ($\overline{B}'_e$) that affects the eth enumerator. The F-ratio of "between enumerator" variance/the relevant "error" variance tests the hypothesis $\overline{B}'_e$ = constant, i.e. that the biases of the enumerators do not differ significantly.

It is also possible, by the use of the Studentized smallest Chi-square, to test the significance of the smallest of the difference between two enumerators (Ramachandran, 1959): a significant smallest variance ratio would lead one to suspect collusion between the two enumerators.

The net bias common to all enumerators ($\overline{B}'$) passes undetected by this technique. It has to be controlled by improvement in survey methodology, comparisons with independently obtained estimates, and adoption of special analytical techniques.

Note: As mentioned earlier, this technique can be used to test the differential bias of the primary enumerators versus that of the supervisors, or, in a health survey, the lay enumerators versus medical or para-medical staff, or to test the differences of estimates obtained by different types of schedules (one with direct questions and another with fully detailed probes), or by different methods of inquiry (retrospective versus periodic observations), and these different sources of variation could be taken up in combination in different sub-samples. As it is very unlikely that there will be a constant bias running through all these combinations, one could meet one of the commonest objections to the use of this technique (Mahalanobis, 1958, Preface; Som, 1965, Appendix 2).

Another formulation would be that due to Cochran (section 14.15). A random sample of n units is divided at random into k sub-samples, each containing $m = n/k$ units, and each of the k enumerators is assigned a different sub-sample. We assume that the correlation we have to deal with is due solely to the biases of the enumerators and that there is no correlation between errors of measurement for different enumerators.

The previous statistical model (25.12) is rewritten as

$$y_{ei} = z_{ei} + d_{ei}$$

where e denotes the sub-sample (enumerator) and i the member (unit) within the sub-sample. The finite multiplier is ignored.

The variance of the mean of the eth random sub-sample is, by equation (25.13)

$$V(\overline{y}_e) = \frac{1}{m}\ [\sigma_{z'}^2 + \sigma_d^2\{1 + (m-1)\rho_w\}]$$

where ρ_w is the correlation between the d_{ei} values obtained by the same enumerator.

As the errors are assumed to be independent in the different sub-samples,

$$V(\overline{y}) = \frac{1}{k}\ V(\overline{y}_e) = \frac{1}{n}\ [\sigma_{z'}^2 + \sigma_d^2\{1 + (m-1)\rho_w\}] \tag{25.14}$$

From the sample data, analysis of variance could be computed for the variations due to "between enumerators (sub-samples)" and "within enumerators" with respective mean squares

$$\begin{aligned} s_b^2 &= m\sum^k(\overline{y}_e - \overline{y})^2/(k-1) \\ s_w^2 &= \sum^k \sum^m (y_{ei} - \overline{y}_e)^2/k(m-1) \end{aligned}$$

As s_b^2/n is an unbiased estimator of $V(\bar{y})$, the technique provides an estimate of error that takes account of the enumerator bias; and the F-ratio s_b^2/s_w^2 tests the null hypothesis $\rho_w = 0$.

The technique is readily extended to stratified multi-stage sampling by simply ensuring that the sample consists of a number of sub-samples of the same structure in which errors of measurements are independent in different sub-samples. Strictly speaking, the process of interpenetration should go deeper than the field, with not only different enumerator-teams, but also of supervisors, data processors etc. used in different sub-samples: in epidemiological surveys, which may require laboratory testing and X-ray reading by independent observers.

An unbiased estimator of $V(\bar{y})$ is provided by $\sum^k(\bar{y}_e - \bar{y})^2/k(k-1)$ with $(k-1)$ degrees of freedom. If $k = 2$ in each stratum, each stratum provides one degree of freedom and the computation of estimates is simplified, as noted in section 9.5.

This method was followed in computing the variance estimates in the Indian National Survey quoted in this book (Examples 10.2 and 21.4). Conformance to the fundamental formulae in stratified multi-stage sampling was established by assigning in a stratum each of two or more enumerators, each sub-sample consisting of the same number of first-stage sample units.

Similar estimates of errors (but excluding the enumerator bias) can in many cases be obtained from a sample – which was not planned as an interpenetrating network – by sub-dividing the sample into sub-samples in some appropriate random manner (see also section 24.8.1, "Method of random groups").

The use of this method in providing errors of estimates in complex sample designs and computations is particularly to be noted: one example is the computation of price index numbers from a sample (Koop, 1960), and another is the situation when the data pass through a number of adjustment processes. If the inquiry is arranged in the form of a number of interpenetrating sub-samples, the sub-sample estimates would remain comparable when subjected to the same transformation and adjustment procedures.

There is a simple non-parametric way of expressing the margin of uncertainty of the estimates from the sub-samples. Each of the independent sub-samples provides an independent and equally valid estimate of the parameter under study, and therefore gives directly an estimate of the margin of uncertainty involved in the estimate. This is an advantage of particular relevance to surveys in countries where suitable personnel and equipment for estimating the sampling variances of a large number of estimates are scarce.

The probability that the median of the distribution of the sample estimates (which is assumed to be symmetrical) will be contained by the range of k estimates obtained from the k independent, interpenetrating sub-samples is $1-(1/2)^{k-1}$. With two sub-samples, for example, the range between the two sub-sample estimates provides 50 per cent margin of uncertainty for the combined estimate of the two sub-samples; with four sub-samples, the range between the smallest and the largest estimates will provide 87.5 per cent margin of uncertainty for the combined estimate of the four sub-samples. The derivation of such limits, besides being simple, involves a minimum of assumptions regarding the sampling distribution of the estimates (see also, Deming (1960), p. 116). An example is provided by the two sub-sample estimates of the birth rate in India in 1958-9, 38.5 and 38.0, with the combined estimate of 38.3 per 1000 persons (Som *et al.*, 1961).

In regard to testing enumerator differences, one example is provided by the Mysore Population Survey in 1951-2, conducted jointly by the Government of India and the United Nations in the then State of Mysore; only one F-ratio came out statistically significant in one "round" (there were four "rounds" of work) relating to the number of vacant dwelling units in Bangalore City (U.N., 1961, Chapter 31). Another example relates to the 1953 Demographic Sample Survey of the Indigenous African Population of Southern Rhodesia (now Zimbabwe), where the 10 sub-samples adopted to determine the sampling errors were not statistically significant, showing that the survey was under statistical control, but the six districts showed significant variation of birth and death rates (Som, 1965).

A limitation of the technique is an increase in the cost of travels by the enumerators. This could be reduced by stratifying the sample into compact areas and assigning, say, two enumerators in each stratum, each covering one half-sample. In the Indian National Sample Survey, the increase in travel cost due to interpenetration constituted about 3 1/2 per cent of the total cost (Som, 1965).

Small enumerator differences cannot in general be detected by this technique, unless linked samples are used. It should not of course be considered as a substitute for supervision and other control of filed and processing work.

25.6.6 Analysis by recall period

In retrospective inquiries, where questions are asked about events which occurred in the near and distant past, a tabulation by recall periods of varying lengths often reveals the existence of recall lapse, which may otherwise lie hidden in the overall estimates. Recall lapse has been observed or suspected in a number of situations in both the developing the developed countries relating to demographic and household consumption and many other types of inquiry.

The statistical model of recall analysis in current vital data is briefly as follows. Estimates (Y_k) are built up from cumulative recall periods k (= 1 month, 2 months,...,12 months, for example), the mathematical expectations of which are given by $E(Y_k) = Z + B_k$, where Z is the "true" total number of events, and B_k the recall bias, which is a function of k, the recall period.

If there were no recall bias, i.e. $B_k = 0$, then $E(Y_k) = Z$. In practice, the Y_ks, as obtained from the data, are smoothed by a curve $f(k)$, from which $\hat{Y}_0 = f(0)$, corresponding to the recall period "0", can be estimated and taken to be equal to Z on the assumption that $B_0 = 0$. The estimate of the vital rate is then obtained on dividing $\hat{Y}_0$ by the corresponding population estimate. This is in essence the recall analytical technique (*"the Som method"*).

By applying this analytical technique, the reported crude birth and death rates in the Indian National Sample Survey (1953-4), 34.3 and 16.6 per 1000, were adjusted to 40.9 and 24.0 respectively; the reported crude birth rates in Upper Volta (1960-1) and Chad (1964) and the reported crude death rate in Morocco (1973), were 49.6, 45.0 and 10.8 per 1000, and these were adjusted to 52.6, 49.0 and 16.5 respectively (Som, 1973; Nadot, 1966; Hosni, 1975).

Bailar and Biemer (1984) have extended the recall analysis model. A template for the application of this technique, based on Quattro Pro spreadsheet, was prepared by the author in 1990 and distributed by the U.N. Statistical Division/Software Development Project; an update is available from the author.

25.6.7 Other external analytical checks

The external unitary check most relevant in estimating vital rates is the one-to-one matching of the current vital events, reported in a survey, with those registered. The method for adjustment of the data from such a check in demographic inquiries is based on the assumption that the chance of an event being missed in either list (survey or registration) is independent of its chance of being missed in the other (Chandra Sekar and Deming, 1949). This is the same as the Lincoln Index, used in estimating the size of mobile populations such as birds and animals by the capture and recapture method (section 24.4). This technique has found applications in a series of nation-wide demographic surveys known as **Population Growth Estimation (PGE) Studies** – conducted, with initial technical support from the Population Council, in Liberia, Malawi, Pakistan, Thailand, Turkey, and a few other countries in the 1960s (Marks, Seltzer, and Krotki, 1976;

Krotki, 1966 and 1978). For the experiences of "Dual Record System", operated by the Laboratories for Population Statistics, University of North Carolina at Chapel Hill, see Adlaka *et al.* (1977) and Myers (1976).

Assuming that the under-reporting in the survey and under-registration operate independently, the adjustment (multiplying) factor for the total number of events reported by either or both the agencies is $1/(1-p_1-p_2)$, where p_1 is the probability of an event not being reported in the survey,

$$p_1 = y_{\bar{s}r}/(y_{sr} + y_{\bar{s}r})$$

and p_2 is the probability of an event not being registered,

$$p_2 = y_{s\bar{r}}/(y_{sr} + y_{s\bar{r}})$$

the subscripts s and $\bar{s}$ denoting whether the event was reported in the survey or not; and similarly for r and $\bar{r}$. For the total number of events, the estimator ("*the Chandra Sekar - Deming formula*") is

$$N^* = (y_{sr} + y_{\bar{s}r} + y_{s\bar{r}})/(1-p_1-p_2) = (y_{sr} + y_{s\bar{r}})(y_{sr} + y_{\bar{s}r})/y_{sr}$$

It is a consistent, maximum likelihood estimator but is generally biased. A variance estimator of N^* is

$$N^*(1-p_1)(1-p_2)/p_1p_2$$

In the Survey of Population Change in Thailand, this method raised the survey-estimated birth rate and death rates of 42.2 and 10.9 respectively to 46.0 and 12.9 per 1000 persons (Lauriat and Chinatakananda, 1966), and in the Pakistan PGE Study in 1961-3, the method provided for an additional 5 per cent of births and 7 per cent of deaths missed by the special registrars and the survey-enumerators (Krotki, 1966).

25.7 Control of non-sampling errors

Response errors can be controlled by the proper selection, training, and supervision of enumerators and the control of enumeration. From expression (25.14) it might appear that, since an increase in the number of enumerators would decrease the intra-enumerator correlation, the contribution to the total variance due to the variability between enumerators would also decrease. But when a very large number of enumerators have to be employed, one has to accept a lower level of staff, training and supervision, resulting in a change in survey conditions that would increase the response errors.

Response errors can also be controlled by introducing internal consistency checks in the questionnaires and by asking **screening questions and probes**. In the U.S. National Health Survey, for example, only half the chronic conditions were reported in response to the initial questions and the checklist questions were necessary to pick up the other half; for sicknesses without medical care and restricted activity, the proportion of cases missed by the initial probes was as high as two-thirds (Nisselson and Woolsey, 1959). In the morbidity studies conducted in the Indian National Survey, 1960-61, the broad questions, "During the last thirty days did you or any member of your household have any sickness or injury?", elicited three-quarters of the total sicknesses, and a reference to specific symptoms and sites included all but less than 4 per cent; additional probing on medical treatment and medicines and on women's complaints reported the remaining sicknesses (Das, 1969).

In the processing of data, **quality control techniques** can also be usefully introduced, For example, the work of each data entry operator can be verified one hundred per cent initially, and then when his error rate falls below a set level, only a sample verification is required of him. This method has been used in practice at the U.S. Bureau of the Census. Owing to the danger of back-sliding, it might not be advisable to withdraw completely verification of the work on data editing and data entry.

When data are processed manually, estimates for the important items may be calculated by two or more independent teams of computing clerks.

Some further consideration is given to the subject in sections 26.11, 26.12, and 26.14-26.17 on methods of data collection, questionnaire preparation, pilot inquiries and pretests, selection and training of enumerators, supervision and processing of data.

25.8 Evaluation and control of sampling versus non-sampling errors

For a given total size of a sample, the sampling error can be controlled and evaluated in a suitably designed survey with considerations of optimization of stratification, allocation of the total sample size into different strata and stages, probabilities of selection, etc., and by using appropriate formulae for estimation, relating all these to the important variables to be studied. The sampling error can be reduced by increasing the sample, but it may introduce additional non-sampling errors in the estimates unless the survey conditions remain the same. Some requirements of the reduction of sampling errors may, however, come in conflict with those of the response errors, e.g. while sampling consideration might call for a widespread sampling with little clustering, a concern for the response errors might lead the survey designer to confine the sample to a sample of large clusters in which

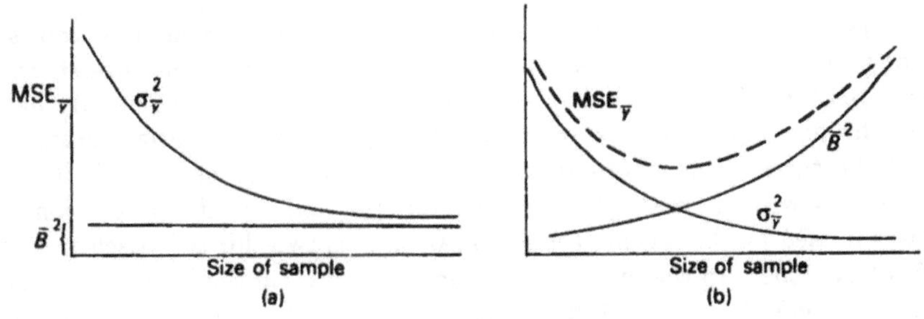

Figure 25.4: Effect of sample size on mean square error.

supervisory inspection and control can be more fully exercised (Hansen, Hurwitz, and Pritzker, 1964). In addition, complete enumeration of the areal units may have other advantages, e.g. cross-check of information of the neighboring households in a social or demographic inquiry, and the psychological effect that the inquiry relates to all persons in the selected area rather than those in some households. The latter factor is of importance both in developing and developed countries, and is one of the factors determining the choice of enumeration districts as the sampling units in the Micro-Census of the Federal Republic of Germany (see Exercise 5, Chapter 10).

The effect of increasing sample size may also be represented by Figure 25.4(a). This may, however, be somewhat misleading, as the magnitude of the bias has been assumed to be constant, independent of the sample size. A better way of representing the effect of sample size on the sampling error and the bias would be given in Figure 25.4(b); for a particular sample size, the total error, as measured by the root mean square, may be minimum; beyond that, increasing sample size would, no doubt, decrease the sampling error but would also increase the non-sampling errors, which might more than offset the decrease in sampling error.

This is also seen from equation (25.13). In the formula for the mean square error, the two terms $\sigma_{z'}^2/n$ and $\sigma_d^2(1 - \rho_w)/n$ will decrease with increasing n; but it cannot be assumed that the other two terms, σ_d^2 and b_0^2, although independent of n, will not be affected by the survey procedures due to a change in the size of the sample, and that the intrasample correlation between errors and the constant bias will remain the same; i.e., σ_d^2 and b_0^2 might well increase with sample size. In large samples, the mean square error will be dominated by these terms and the ordinary sampling variance will be a poor guide to the accuracy of the results, unless the non-sampling errors are controlled substantially.

To try to reduce sampling errors while a bias several times as large is allowed to creep in is not only pointless but also a waste of resources (Sukhatme and Sukhatme (1960), p. 543).

Reference has been made in section 1.3 to the study made at the U.S. Bureau of the Census, which showed that for many of the more difficult items in a census, such as occupation, industry, work status, income and education, the enumerator variability is approximately the same as the sampling variance for a 25 per cent sample of households; the census results were further seen to be subject to a bias which varied from one item to another and was 6 per cent on average. The census and the sample will have approximately the same bias if the census enumerators collect the data for the sample of 25 per cent of households as part of the regular census. Under these assumptions, the root mean square errors that can be expected for the complete enumeration, and the sample for various percentages of units possessing a certain attribute, are given in Table 25.2 for areas with different population sizes. While the root mean square errors for the complete enumeration and the sample are appreciably different in areas with small populations, they converge with increasing population size and become almost identical for areas with 50,000 persons. Thus, when the major census results are published for areas with 50,000 or more persons, it is more advantageous to take a sample for these particular items. This procedure has been followed since the 1960 census of population in the United States (U.S. Bureau of the Census, 1960).

25.9 Evaluation and adjustment for non-response

25.9.1 Effect of non-response

We assume that the universe of N units can be divided into two strata, the first consisting of N_1 units for which measurements would be obtained and the second of N_2 for which no measurement would be obtained either in a census or a sample. Let $\overline{Y}_1$ and $\overline{Y}_2$ denote the two stratum-means. When the field work is completed for an srs of n units, consisting of n_1 units from the first stratum and n_2 units from the second stratum, we would have measurement only from the n_1 sample units, giving the sample mean $\overline{y}_1$. Assuming that n_1 and n_2 are random samples from the two strata, the

Table 25.2: Magnitude of root mean square of the estimated percentages of population in complete enumeration (census) and a sample survey of 25 per cent of households for areas of specific size: Experimental Study of the U.S. Bureau of the Census

Percentage of population to be estimated	Root mean square (in percentage) for an area with a population of					
	2,500		10,000		50,000	
	Census	Sample	Census	Sample	Census	Sample
0.5	0.3	0.4	0.1	0.1	0.1	0.1
2.0	0.6	0.8	0.3	0.4	0.2	0.2
5.0	0.9	1.2	0.5	0.7	0.4	0.4
20.0	2.0	2.5	1.4	1.6	1.2	1.3
50.0	3.6	4.1	3.2	3.3	3.0	3.1

Source: U.S. Bureau of the Census (1960).

amount of bias in the sample mean is

$$E(\bar{y}_1) - \bar{Y} = \bar{Y}_1 - \bar{Y} = \bar{Y}_1 - (N_1\bar{Y}_1 + N_2\bar{Y}_2)/N = (N_2/N)(\bar{Y}_1 - \bar{Y}_2)$$

The amount of bias is thus the product of the proportion of non-responding units and the difference between the two stratum-means. Any sizeable proportion of non-response or difference between the two stratum-means would produce a substantial bias. For example, a 5 per cent non-response observed in some Africa and Indian fertility surveys may not substantially affect fertility rates, but the same percentage will produce a large bias in surveys of incomes, sales etc., if the non-responding units are those with very high incomes or sales.

25.9.2 Adjustment for non-response

Three methods of adjustment for non-response will be considered: (i) selecting random substitutes from the responding units; (ii) "call-back" or selecting a random sub-sample of the non-responding units; and (iii) adjustment for bias without call-backs.

(i) *Selecting random substitutes from the responding units.* When the proportion of non-response is low and the difference between the two stratum-means is believed not to be substantial, a simple practical method is to replace the non-responding units by a random sub-sample of the responding units. This is relevant particularly in self-weighting designs in order to keep the multiplier a constant.

(ii) *"Call-backs" or selecting a random sub-sample of the non-responding units (Hansen-Hurwitz method).* Of the total number n_2 of the non-responding sample units, suppose a sub-sample n_2' is chosen for call-backs (or re-mailing of the questionnaire in a mail inquiry), all of whom respond as result of special efforts. Then an unbiased estimator of the universe mean $\overline{Y}$ is $(n_1\overline{y}_1 + n_2\overline{y}_2)/n$, where $\overline{y}_2$ is the sample mean from n_2' call-backs.

Assume the cost function $C = nc_1 + nPc_2 + n(1-P)c_3/k$, where c_1 is the cost per unit of making the first attempt, c_2 the cost per unit of processing the data of the first attempt, c_3 the cost per unit of collecting and processing the data of the second attempt, $P(= N_1/N)$ is the proportion of universe units that would have responded at the first attempt, and $k = n_2/n_2'$. Under some broad assumptions, the optimum value of k is given by

$$k = \sqrt{[Pc_3/(c_1 + Pc_2)]} \tag{25.15}$$

whence

$$n = n'[1 + (k-1)(1-P)] \tag{25.16}$$

Note that although we have assumed that all the n_2' units respond at the second attempt, they may not in fact do so. The process of call-backs could be continued until all but a very insignificant proportion of the units respond: the Hansen-Hurwitz method has been extended by El Badry (1956).

Successive calls may help to diminish the bias and a graph (Clausen and Ford, 1947) or a regression curve fitted to the estimates and the cumulated rates of responses at successive attempts when extrapolated to the hypothetical case of 100 per cent response could give a better approximation to the true value (Hendricks, 1949; 1956, Chapter XI). Some results of the inquiry of fruit growers in North Carolina, U.S., are given in Table 25.3; the initial response gave a very high average number of fruit trees per farm (456) in North Carolina because a farmer's interest in a fruit survey could be expected to be positively associated with his scale of operation, but a regression curve of the form $y = ax^b$, where y represents the total number of trees picked by a sample when the (cumulative) percentage return is x (and a and b are estimated from the data), when extrapolated to $x = 100$ per cent, gave an estimate of 344 trees per farm, as against the true value of 329, which was known (Hendricks, 1956).

Table 25.3: Results of three repeated mailings to a list of 3,241 North Carolina fruit growers

Mailing	Percentage of Questionnaires returned	Average trees per farm for reporting farms	Cumulative	
			Percentage of questionaires returned	Average trees per farm for reporting farms
1	9.3	456	9.3	456
2	16.7	382	26.0	406
3	13.4	340	39.4	385
Estimated from fitted regression			100	346
True value			100	329

Source: Hendricks (1956).

In the 1946 Family Census of the U.K., the initial response gave a very high birth rate due to the fact that the majority of the initial non-respondents were women with few or no children: of the 230,000 initial non-respondents (who constituted 17 per cent of the total sample), 50,000 responded to the follow-up appeal and the replies of the first 12,000 of them, when combined with the remainder of the sample with a weight of 230/12, gave an adjusted birth rate which corresponded to that already known from other sources (Glass and Grebenik, 1954; cited by Yates, section 5.22, and by Moser & Kalton, section 7.4).

(iii) *Adjustment for bias without call-backs (Hartley-Politz-Simmons method).* For dealing with the not-at-home cases, a procedure that saves the cost of call-backs may be adopted which consists of ascertaining from the respondents the chance of their being at home at a particular point of time and weighting the results by the reciprocal of this probability. Suppose the enumerators make calls on households during the evening on six nights of the week. The household members are asked whether they were at home at the time of the interview on each of the five preceding evenings. The households may then be classified according to r, the number of evenings the household members were at home out of five, and the ratio $(r+1)/6$

taken as an estimate of the probability of the household members being at home during the enumeration hours. If n_r is the number of interviews obtained in the group r and $\overline{y}_r$ is the group mean, then the estimator of the universe mean is

$$\overline{y} = \frac{\sum_{r=0}^{5} 6n_r\overline{y}_r/(r+1)}{\sum_{r=0}^{5} 6n_r/(r+1)} \tag{25.17}$$

The method is obviously biased as no allowance is made for persons who are away from home during the enumeration hours for all the six evenings: the bias is however negligible if such persons are relatively few. The variance formula for $\overline{y}$ is rather complicated.

25.10 Concluding remarks

Sampling inquiries have assumed a crucial rôle in providing data in areas with defective or non-existent registration systems and also in enabling in-depth studies of socioeconomic variables and their interrelations. A sample survey also makes it possible to ensure quality checks and incorporate special techniques to adjust for response biases.

To that end, however, there can be no routine procedure. Methods suitable to the particular data have to be applied as critically as possible. Assessment should be made with existing knowledge on the characteristics of the universe and the field procedures that could be obtained. The greater the detail recorded and tabulated in the survey, the more potent should the checks for accuracy and internal consistency could be, and hence the greater the likelihood that important discrepancies could be detected and adjusted for in estimates.

A combination of several methods for evaluating response errors is naturally likely to give better results than the use of a single method. For these methods to be effective, they should be built into the survey design so as to permit the required analysis: one example is the staggering of the total sample over the survey period in a demographic inquiry in order to eliminate seasonality in vital events in recall analysis. This should, of course, be supplemented by *post hoc* techniques, wherever possible. But these too, in order to be effective and theoretically valid, should be integrated into the overall survey design.

Further reading

Two books – *Nonsampling Error in Surveys* by Lessler and Kalsbeek (1992) and *Measurement Errors in Surveys* (edited by Biemer *et al.*, 1991) – describe recent developments on the topics.

For classification and examples of non-sampling errors: Deming (1950; 1960); Mahalanobis (1946a); Som (1973), Chapter 1.

For evaluation of non-sampling errors (especially response and non-response errors), including incomplete data: Bailar and Biemer (1984); Chaudhuri and Stenger, Chapter 11; Cochran, Chapter 13; Deming (1960), Chapter 5, (1966), Chapter 2, and (1963); Foreman, Chapter 11; Groves; Kalton (1983b); Kish (1965), Chapter 13; Madow *et al.*, three volume on incomplete data in sample surveys; Martin (1993); Moser and Kalton, Chapter 15; Murthy (1967), Chapter 12; Neter and Waksberg (1965) for response errors in household expenditure data; Rubin; Singh and Chaudhary, Chapter 12; Som (1973), Chapter 1; Sogunro; Sukhatme *et al.* (1984), Chapter 11; U.N. (1972).

For evaluation of non-sampling errors in specific subject-fields, see the references given for Chapter 26.

For other topics covered in this chapter see the references cited in the text.

Exercises

1. In spite of the deficient reporting of two-member households, the sample for the morbidity study in Syracuse in 1930-31, cited in section 25.3.2 (ii), was considered satisfactory, as is indicated by the statement that the workers "had been primarily concerned with securing a sample, representative of the area in regard to prevalence of sickness rather than size of household." Comment on this statement.

   (*Hint*: "Actually such a biased sample can scarcely be regarded as wholly satisfactory for even a morbidity study, since sickness rates are likely to vary with the size and composition of the family" (Yates, 1981, section 2.4)).

2. Check for possible inconsistency the following results obtained from a demographic sample survey of Kinsenso (a suburb of Kinshasa, Zaire) in mid-1965; the reference period for births and deaths was 1 July 1964-30 June 1965 (Knoop, 1966, Tables IV and VIII): Total population under one year of age: 790; Total births: 708; and Deaths of infants under one year of age: 51.

   (*Hints*: of the total births, numbering 708, during the reference period, some died within the reference period itself (of course, before attaining one year of age). Such infant deaths, i.e., the number of infants born and dying within the same reference of one year, represent in general two-thirds of the total infant deaths during the reference period (i.e., the number of infants born and dying within the reference year plus the number of infants born before the reference year but dying within the reference year before attaining one year of age). The total number of infant deaths during the reference year is given as 51. Even taking the absurd assumption that the deaths of infants under one year age, numbering 51, related only to infants born during the reference year, then subtracting this number (51) from the total births during the reference year (708), one would get the number of surviving infants, under one year of age, which would normally equal the number of total population under one year of age at the time of the survey. The latter number is given as 790, higher than the total births during the year. The normal relationship may, however, be upset in the unusual situation of a very heavy immigration of infants to the area selected for the survey, who are enumerated in the survey but whose births are not covered in the survey owing to adopted definitions (Som, 1973, Exercise 1.10.4)).

CHAPTER 26

Planning, Execution and Analysis of Surveys

26.1 Introduction

W. Edwards Deming (1969, page 666) has stated the importance of the preparation of survey procedures in the following forceful words:

> *"Statistical teaching centres might well include in the curriculum some small amount of time on preparation of procedures. How many students with doctor's degrees have had the experience of writing a set of procedures for a study, complete with statistical controls to detect departures from the procedure described, and of learning the hard way that he may have to write these instructions twelve times before he learns that if there is any possible way to misconstrue an instruction, someone will do it?"*

Some aspects of the planning, execution, and analysis of statistical surveys are considered in this chapter. Most of the considerations excepting those relating to the choice of the sample design would apply also to complete enumerations.

The considerations will of necessity be brief here. References to survey methods in some selected subject fields are given at the end of the chapter. A case study dealing with the survey methods and sample design of the Indian National Sample Survey (1964-5) is given in Appendix V. References to some selected case studies are given under "Further reading" in Appendix V.

26.2 Objectives of the survey

The Objectives of the survey should be clearly specified to include the following:

1. A clear statement of the desired information in statistical terms, such as estimation of total personal incomes and expenditures on food and other items in a family budget inquiry, or the area and the yield of a certain crop in an agricultural inquiry, or the rates of birth, death and natural increase in a demographic inquiry, or the number of persons with pulmonary tuberculosis in an epidemiological inquiry.

2. A description of the coverage and the definition of the universe such as the geographic region or branch of the economic or social group or other categories constituting parts of the universe covered by the survey. In a survey of human population, for example, it is necessary to specify whether hotel residents, persons in institutions (e.g. boarding houses and sanatoria), homeless people, persons without fixed abode, military personnel, etc. are to be included and, where possible, to indicate the order of magnitude of the categories omitted; in an agricultural survey, it will be necessary to indicate whether all agricultural holdings have been included.

3. The geographical and classificatory breakdowns by which results are to be tabulated.

4. The degree of precision desired, specifying the sampling error that can be tolerated, as also an indication of the possible inaccuracies in the final estimates.

5. The total cost of the survey.

26.3 Legal basis

Surveys, especially field enumeration, should be conducted with the legal authority obtained in a country. The confidentiality of individual information should be clearly established and guaranteed by adequate sanctions so as to form a basis for the confident cooperation of the respondents.

26.4 Publicity and cooperation of respondents

The objective of publicity in the form of educational campaigns is twofold: first, to allay any anxiety regarding the purposes of the surveys concerning, for example, taxation, conscription, and repatriation of immigrant labor and forced labor in some countries, and, second, to explain the reasons for the various questions to be asked. The publicity media may include films, articles for the press, booklets, talks for radio, television and general press release.

On the other hand, in some surveys, such as the Social Survey in the U.K., publicity was never used for it was believed better to take the respondents by surprise: if they had had time to think out the matter they may have decided to refuse or to present some socially acceptable response. But faced with an actual human appeal from an interviewer on the doorstep, it may be more difficult to take this decision.

26.5 Budget and cost control

Preparation of a preliminary budget estimate is a priority activity that should be planned and executed at an early stage. The budget will depend on the survey design, including the levels of precision desired for various estimates, as well as on the geographical and other classifications for presentation of results, and the operational conditions prevailing in the region. The preliminary budget will have to be re-examined and revised as the survey activities progress. This will inculcate cost consciousness into those responsible for the survey and indicate the existence of any inefficiencies, often acting as an impetus to devise economies and innovations.

As practices vary widely, there does not seem to be a standard way of preparing survey budgets. The budget for the demographic sample surveys in West Cameroon in 1965 is given as example in Table 26.1.

On the basis of this and other available survey budgets for African demographic censuses and samples, the overall cost per person enumerated around 1970 for the United Nations sponsored African Census Program which consists of a simple head-count of the total population by sex and geographical divisions and sample surveys (with about 100,000 sample persons) for other demographic and associated items was estimated at U.S. 10 cents for the simple head-count and U.S. $3.50 for the sampling inquiry covering all the items. These are, of course, notional estimates, prepared in order to arrive at an overall figure for the program covering several countries: for the actual operation in any country, the budget is worked out on the basis of local conditions; the mapping cost in particular would vary widely in different countries.

Thus for an African country with an estimated 10 million persons, the cost of the simple head-count could be estimated at U.S. $1 million, and of the sample inquiry (with 100,000 sample persons) for all the demographic and associated items at U.S. $350,000 per year in the mid-1970s.

Table 26.1: Budget for demographic sample survey, West Cameroon, 1965: Sample of 170,000 with single-round team-enumeration

Item	Cost estimate In CFA Francs ('000)
1 *Expatriate personnel*	
(a) *Outside France*	
1 Survey Director for 14 months at Fr.F. 4,800 per month	3,360
1 Field organizer for 10 months at Fr.F. 3,200 per month	1,600
Social Security contributions 40%	1,984
Subsistence: 2 months at Fr.F. 60 per day; 22 months at Fr.F. 30 per day	1,170
Total outside France	8,114
(b) *In France*	
Preparation of mission:	
1 Survey Director for 15 days at Fr.F. 2,400 per month	60
1 Field Organizer for 15 days at Fr.F. 1,600 per month	40
Social Security contributions 40%	40
Total in France	140
(c) *Miscellaneous charges*: 15% of {(a) + (b)}	1,238
Total Item 1	9,492

(continued)

Table 26.1 (continued)

Item	Cost estimate In CFA Francs ('000)
2 *Local Personnel*	
(a) *Selection and training:*	
80 persons for a month at CFA.F. 12,500 per month	1,000
(b) *Field work and data processing:*	
1 Assistant Field Organizer for 13 months at CFA.F. 40,000 per month	520
48 Enumerators for 7 months at CFA.F. 25,000 per month	8,400
12 Field Supervisors for 7 months at CFA.F. 40,000 per month	3,360
59 months drivers at CFA.F. 25,000 per month	1,475
1 Messenger for 10 months at CFA.F. 8,000 per month	80
1 Secretary/Accounts Clerk for 10 months at CFA.F. 30,000 per month	300
15 Processing Clerks and Coders for 6 months at CFA.F. 25,000 per month	2,250
Miscellaneous allowances	390
(a) + (b)	17,775
(c) *Miscellaneous charges* 15% of {(a) + (b)}	2,666
Total Item 2	20,441

(continued)

Table 26.1 (continued)

	Item	Cost estimate In CFA Francs ('000)
3	*Equipment and local transport:*	
	3 Vans (1,400 kg)	3,000
	1 Small van	750
	Running costs, maintenance, repairs, insurance	5,250
	Printing of questionnaire and data processing forms	1,500
	Office equipment and furniture	300
	Camping equipment for Field Organizers and Supervisors	250
	Enumerators' equipment	300
	Bicycle allowance	270
	Local transport costs	250
	Rent of offices and accommodation	2,000
	Total Item 3	13,870
4	*Air transport:* 3 RT Paris-Yaoundé	459
5	*Miscellaneous:*	
	Mechanical data processing	2,500
	Publication of provisional report	800
	Total Item 5	3,300
6	*Contingency:* Approximately 5% of total	2,438
	Grand total (approximately)	50,000

1 CFA Franc = French Franc 0.02 (or US $0.004 or UK £0.0017), at the then rates of exchange.

Source: U.N. Economic Commission for Africa (1971).

26.6 Project management

One important element of project management is the preparation of a provisional calendar or timetable, indicating the sequence and estimated durations of the various component operations of the survey. It is essential in survey planning and execution; such a provisional calendar should be made final as early as possible and kept constantly under review. In addition to a detailed checklist of operations, the calendar may sometimes take the form of a chart or graph.

Furthermore, it is important to keep track of the sequence of the different items of input, e.g. ordering of equipment, such as computers, site-preparation, and the training of staff in their use, and to take needed measures to keep the project on track. All these call for installing **project management software programs**, such as TimeLine DOS or WIN (by Symantec Corporation), Project/WIN (by Microsoft Corporation), with their built-in Gantt chart (named after its creator, the behavioral psychologist Dr. William Gantt), or **PERT (Project Evaluation and Review Technique) charting**.

26.7 Administrative organization

The final responsibility for survey planning should rest with the executive agency. The permanence of a survey organization, within the framework of the statistical system of a country, has definite advantages in furthering the development of specialized and experienced personnel who could undertake the continuing research needed for evolving increasingly efficient survey designs, and also for improving the quality of survey data by assessing and controlling non-sampling errors and biases. A permanent survey organization can also undertake urgent *ad hoc* surveys at short notice.

26.8 Coordination with other inquiries

Needed cooperation may often have to be obtained from different bodies by coordinating the survey with other compatible inquiries, using, for example, the same sampling frame, and also trying to accommodate the needs of other users, but without overloading the questionnaires.

26.9 Cartographic (mapping) work

In surveys using areal sampling units at some (or all) stages, the detailed subdivision of the universe into identifiable areal units is one of the basic important operations in a survey. Censuses of population, housing and agriculture are the most common source of such information. However, maps have to be prepared for selected areas when these are not available with the required details.

26.10 Tabulation programs

At the very early stages of survey planning, the tabulation program should be drawn up so that the procedures and costs involved may be investigated thoroughly and the questionnaire tested to indicate whether it is possible to gather the required information.

26.11 Methods of data collection

Data may be collected:

1. by direct investigation comprising (a) physical observation which may involve subjective methods (e.g. eye estimation of the areas under a crop in sample fields) or objective methods (e.g. physical measurements, such as the yield of a crop in the sample fields) or both, and (b) personal interview;

2. by mail inquiry;

3. by telephone;

4. from registration;

5. from transcription of records.

The objectives of the survey, the nature of the items of information, the operational feasibility and cost will often determine the method of data collection. Mail inquiries are practicable in predominantly literate cultures where respondent cooperation is ensured; the last three methods have limited applications in most developing countries.

Physical observations are required in inquiries involving measurements such as crop yields, housing conditions, and anthropometric studies.

The method of personal interview is widely used in economic and social surveys. Some items of information may be collected only through personal interview, which has the added advantage that the concepts and definitions can be explained to the respondents, thus reducing response errors and non-response. Direct person-to-person investigation is, however, generally more expensive than other methods.

In a household inquiry, after a sample household is contacted the required data may be obtained by interviewing one or more members (the

canvasser or **enumeration method**) or the questionnaire may be left with the household for it to be filled by the members (the "**householder method**"), and later collected by the enumerator.

Recent innovations in the use of computers in data collection include Computer Assisted Personal Interview (CAPI), Computer Assisted Telephone Interview (CATI), and Computerized Self-administered Questionnaire (CSAQ): some of these techniques have been reviewed in the next chapter.

26.12 Questionnaire preparation

26.12.1 Questionnaire design

The type of questionnaire, its content, format, and the exact wording and arrangement of the questions determine to a great extent the quality of the responses. Among the many factors which should be taken into consideration in designing the questionnaire are the method of enumeration, the type of inquiry, the data to be collected, the most suitable form of the questions and their arrangement, and the processing techniques to be employed.

Although we have sometimes used the terms **questionnaire** and **schedule** interchangeably, a distinction might be made between the two: a schedule contains a list of items on which information is required but the exact form of the questions to be asked is not standardized but left to the judgment of the enumerators; a questionnaire contains a list of the actual questions that the enumerators would put to the respondents verbatim in a specified order. The choice between the two would be generally determined by the nature of respondent biases, the type of inquiry, the number of items, the skill and training of the enumerators, and the costs involved.

In general, a questionnaire is more suitable for inquiries where the number of items, quantitative answers and cross-tabulations are not too many and especially in surveys on opinions and attitudes. On the other hand, a schedule is more compact and flexible: with a schedule a well-trained enumerator is generally in a better position to frame questions to elicit the required information and to meet the exigencies of the situation.

The framing of the questions and the exact wording should be embedded in specific socio-cultural contexts. For example, in a number of African countries, the term "married" includes those consensually living together; "race" is another term subject to different interpretations. Effective communication requires the process of meaning-making on the part of the interviewee; in linguistics, this is known as "**transderivational search**" . The U.S. Bureau of the Census has used focus groups to sharpen concepts and definitions for use in population censuses (see section 24.13, "The use of focus groups").

26.12.2 Details of the information to be collected

The statement of the desired information and the form in which final estimates are required would determine the details of the items of information to be collected (see, in this connection, Appendix VI, section A6.2.4, "Inter-country data comparability"). The problems naturally differ in different subject fields and it is not generally possible to give a formulation which will apply equally in all situations. A few general observations may, however, be made.

Some **derived information** may be obtained from the items of information collected during the survey. Examples are family composition, including family nucleus, and the socio-economic status of persons. For accurate reporting, some items may require separate recording of their components or inclusion of supplementary and supporting items of information. For example, data on the total number of children borne by a woman cannot often be collected accurately, unless account is taken separately for each marriage of a woman, of the children who were born alive but since died and, of those living, both those living in the household (or family) and those living separately: note in this connection the British experience (section 25.3.2(iii)).

A common mistake in questionnaire design is to restrict oneself to the topics or items on which final estimates are required (e.g. infant deaths), and not to include their components (e.g. deaths to male and female infants) or those which are required to elicit the desired information. The inclusion of such auxiliary items at a marginal cost has two advantages: first, such information acts as a check on that of the principal items (e.g. computation separately of male and female infant deaths provides a check of the overall infant mortality and of differential infant mortality between male and female births); second, these may be standardized and not left to the enumerators to formulate. Similar considerations apply to the total number of children borne by a woman.

Some **difficult or sensitive items** such as income, savings, indebtedness, opinions and attitudes may better be put at the end of the schedule.

Some items of information require the inclusion of others in order to ensure proper interpretation. For example, it is necessary to ask the actual programs which the respondents listen to in a survey of preference of radio listeners and the actual books and periodicals read in a survey of readership preferences.

The framing and arrangements of the questions or items of information should be done in such a manner as to facilitate accurate data collection and tabulation. The schedule or the questionnaire should not be too lengthy or tedious.

26.12.3 Time-reference period for data

The data collected should relate to a well-defined reference period. The time-reference period need not be identical for all items of information.

With a fixed length, the reference period can be either fixed or moving; the former is a period whose end-points are fixed, for example the week 10-16 January; the latter moves with the date of inquiry, such as the seven days preceding it.

The choice of the length and type (**fixed** or **moving**) of the **reference period** depends on the frequency of occurrence of the items, the manner of accounting, the memory factor, and the type of survey. In an integrated household inquiry, for example, the reference period could be a year for infrequent purchases of durable goods such as furniture, and a week or month for food consumption: those for household income and production would depend on the receipt of incomes and production cycles, such as agricultural seasons.

For most of the data in a demographic inquiry, the reference period will be the enumeration day; for economic characteristics, it may be a brief period prior to enumeration, preferably not longer than a week, but when this short period is unrepresentative of seasonal changes in the level of activity and regular periodic sample surveys are not conducted, supplementary information on "usual" economic characteristics over a longer period may also be collected; for information on current fertility and mortality, the reference period is generally taken as twelve months preceding the enumeration, but for data on fertility history it should obviously cover the whole reproductive span of the sample women.

In continuing surveys with common ultimate stage sample units (see sections 24.3 and 26.13.1), the time-reference period for most items will be the interval since the last survey.

26.13 Survey design

26.13.1 Types of surveys

The objectives of the survey will determine the choice of sample design and the type of survey. The following types of surveys have been distinguished by the Sub-Commission on Statistical Sampling of the United Nations Statistical Commission (U.N., 1947):

(i) *Integrated survey.* In an integrated survey, data on several subjects (or items, or topics) are collected for the same set of sampling units for studying the relationship among items belonging to different subject fields. Such surveys are of special importance in developing countries where the related activities are frequently undertaken in an integrated manner in the household.

(ii) *Multi-subject survey.* When in a single survey operation several subjects, not necessarily very closely related, are simultaneously investigated for the sake of economy and convenience, the investigation may be called a multi-subject survey. The data on different subjects need not necessarily be obtained for the same set of sampling units, or even for the same type of units (e.g. households, fields, schools, etc.).

(iii) *Continuing surveys.* The most usual example of these surveys occurs where a permanent sampling staff conducts a series of repetitive surveys which frequently include questions on the same topics in order to provide continuous series deemed of special importance to a country. Questions on the continued topics can frequently be supplemented by questions on other topics, depending upon the needs of the country.

(iv) *Ad hoc survey.* This is a survey without any plan for repetition.

(v) *Multi-purpose survey.* This term is sometimes used in connection with sampling organizations which conduct surveys in various fields of interest to several departments or parties, keeping in view their diverse purposes. This permits economies in the technical and other resources and their more effective use, particularly in developing countries.

(vi) *Specialized or special-purpose survey.* This may be defined as an investigation focusing on a single set of objectives which, because of their nature or complexity, requires specialized knowledge or the use of special equipment by a technical staff with training in the subject field of enquiry.

Multi-subject integrated surveys are generally more economical than a series of single-subject non-integrated surveys due to savings in overhead costs and in survey cost (travel to the penultimate-stage sample units such as villages, enumeration districts etc., camp setting and listing and contacting the ultimate-stage units). The economies become considerable with a permanent survey organization, when multi-stage and multi-phase sampling designs can be adopted. Further, grouping of different subjects of inquiry in the same sample of first-stage units helps in increasing the sample size at the first stage, which generally improves the efficiency of the survey estimates. In many circumstances, therefore, the efficiency attainable by separate single-subject surveys may be achieved by incurring a much smaller expenditure in an integrated multi-subject survey.

It will be noted again that only an integrated survey makes it possible to study the inter-relations of different subject fields. Further, if integration is carried out up to the penultimate-stage units (e.g. locality or a village), infrastructure information such as the distance from the nearest school, health and communication centers, and irrigation system and existence of anti-malarial operations, etc., need be collected only once (often available from routine administrative sources) and analyzed in respect of the relevant items of inquiry.

Recent trends have therefore been toward adopting multi-subject integrated surveys both in established survey organizations such as the U.S. Bureau of the Census and the Indian National Sample Survey and in a number of developing countries which have embarked on sampling inquiries. The United Nations-sponsored National Household Capability Program also advocated collection of integrated data on the household and its members through household surveys (see Appendix VI, section A6.3.4).

A multi-subject integrated survey is generally more suitable for household inquiries. It may be impractical or inefficient for combining inquiries that require very different types of designs: there are also the problems of overloading the informant, the training of enumerators and the data processing facilities.

Often it is also found impractical or inefficient to collect some types of agricultural statistics through household sample surveys. These relate mainly to the time element involved in crop cutting (i.e. at the right time for harvesting) and the difference between the so-called "**biological yield**" (i.e. the yield without any loss) and the "**economic yield**" (i.e. the part of the yield available to farmers for all uses). The important point here is that to be realistic, yield surveys must be conducted under conditions of farm practices, and this is not generally possible in a multi-subject integrated inquiry (see also Zarkovich (1962)).

Some specialized inquiries cannot be fitted into schemes of an integrated multi-subject survey. Thus, while generally accurate information on the demographic and socio-economic aspects of morbidity such as sex, age, days lost due to sickness, expenditure on medicines etc., could be collected through an integrated household survey, a separate diagnostic survey with a fully equipped team is required for clinical and laboratory examinations to supply information on causes of sickness: this method is currently being followed in the U.S. National Health Survey. Other examples are surveys of industrial establishments, fisheries surveys, etc.

26.13.2 Choice of sample design

Stratified multi-stage designs constitute the commonest type of designs used in practice. They combine the advantages (including economies) of stratification and clustering when the strata are made internally homogeneous (but heterogeneous with respect to one another) and the first-stage units internally homogeneous, so that a small number of first-stage units need be selected from each stratum in order to provide an efficient sample (see section 22.3). In practice, geographically compact strata make field enumeration easier.

Usually, the primary strata are compact geographical areas. Within these, further stratification may be done on the basis of available supplementary information on the universe. The ultimate strata may be smaller homogeneous areas or groups of units having similar characteristics. Stratification could also be used at second and subsequent stages in a multi-stage design.

For a given total cost, multi-stage sampling will often give more efficient estimates than unrestricted simple random sampling, because of the usual economies of clustering and better sampling plans for the second and subsequent stages. Also, the advantages of stratification are generally considerable and will almost outweigh the slight increase in the complexity of analysis. For a highly variable universe, the very large units may be put in a separate stratum and completely enumerated: this is the general practice in surveys of establishments.

Sample selection in each stratum should be according to the most efficient method applicable and need not be uniform in all strata. For the stages for which samples can be drawn in the central offices, varying probability selection may be used, but for the ultimate and sometimes the penultimate stages, for which the samples have to be drawn by the field enumerators, simpler methods are required such as simple random or systematic selection.

The criteria of stratification and allocation of the sample size in the different strata and stages have already been outlined in Chapters 12, 17 and 22. In multi-subject inquiries, these evidently have to be decided on the basis of which variables are the most important: different sampling fractions for the ultimate units could be called for (e.g. for demographic vis-à-vis employment data), or different and appropriate samples of second-stages units within the same primary units can be selected using different criteria of size (e.g. selection of households may be based on expenditure or income criteria, whereas those for agricultural holdings may be selected on land criteria).

For computing estimates, available supplementary information may be used to obtain ratio or regression estimates in order to improve the efficiency of the estimators. The ratio and regression methods of estimation can be used in different stages and also at different levels of aggregation, but it may be desirable to have some idea of their efficiency and bias before they are used on a wide scale. Furthermore, one would need to consider the facilities available for processing when these methods are used, as they involve greater computational work.

There are considerable advantages in adopting a self-weighting design, mentioned in section 23.5, and the technique of interpenetrating subsamples mentioned in sections 24.7 and 25.6.5.

26.13.3 Sampling frames

The frame for sampling consists of previously available descriptions of the objects or material related to the physical field in the form of maps, lists, directories, etc., from which sampling units may be constructed and a set of sampling units selected. Generally a frame may consist either of area units or of list units covering the items being investigated in a survey. A frame becomes more valuable if it contains some supplementary information which can be used to improve sampling and estimating procedures; also, information on communications, transport, types of crops etc. may be of value in improving the design for the choice of sampling units, in programming the field work and in the formation of strata. The sampling frame should be accurate, free from omissions and duplication, adequate and up-to-date and the units should be identifiable without ambiguity.

Unless the sample survey is conducted simultaneously with a census – of population, housing, agriculture, etc. – the initially available frame will often require amendments for updating. And even when the sample survey is conducted along with the census, the question of under- or over-enumeration in the census would remain to be tackled.

The problems arise particularly in the latter stages of multi-stage sampling, and often a frame may require to be constructed *ab initio*. Thus a frame of higher-stage units, such as provinces, counties, towns and villages, and enumeration areas, may be available from a census, but for the selection of households, farms etc., a complete listing of such units in the selected penultimate stages will have to be made. In addition to the requirement of making the frame of ultimate units up-to-date and complete, such a listing has the advantage that useful supplementary information may be collected with little more effort, which can be used in stratification or estimation.

Supplementary information, whether available from the frame or obtained during the listing of all the ultimate stage units in the selected penultimate stages, may also be highly useful in improving the efficiency of estimates obtained from the survey, through suitable stratification, pps or systematic selection, ratio method of estimation etc. In repeated surveys, information collected in earlier surveys may also serve to improve the frame for later surveys.

In some parts of Africa, villages are sometimes abandoned altogether and the whole population shifts elsewhere, or a village might be split into two or more parts which take on the names of the new chiefs: these make difficult the updating of a frame in the absence of an adequate administrative machinery.

Note: The following method of estimating the number of common units in two lists is due to Deming and Glasser (1959):

Suppose there are two lists, one having M units and the other N units with an unknown number D common to both. Samples of m and n units are selected with srswor from the two lists, and d units are found to be common to the two samples. Then an unbiased estimator of D is MNd/mn, and its unbiased variance estimator is

$$(MN/mn)[(MN/mn) - 1]\, d$$

26.13.4 Survey periods

In countries where the economy is subject to pronounced seasonal fluctuations within a year, socio-economic inquiries might preferably be conducted for a year or multiples of a year with the work of data collection spread evenly over the survey period. This would enable not only the study of the seasonal elements but also a meaningful annual average. In a permanent survey organization where the trained field personnel are small in number, such a staggering of the survey is the only recourse.

In such cases, to minimize recall lapse and to represent all parts of the year it is necessary to adopt moving rather than fixed reference periods; for example, the reference period for the collection of information on some items of household consumption could be the seven days preceding the date of enumeration rather than a fixed calendar week.

In addition to the survey period considered above (such as one year, an agricultural season etc.), the periodicity of surveys – annual, biennial etc. – would also have to be considered.

26.14 Pilot inquiries and pre-tests

In undertaking a large-scale survey, particularly of unexplored material, it should be the general rule to conduct pre-tests and pilot and exploratory inquiries to test and improve field procedures and schedules and to train field enumerators, and also to obtain information on the variability of the data and on the cost-pattern which will enable the sample design to be planned more efficiently and to finalize the tabulation program. For example, the results of a pilot inquiry may be used to estimate the first- and second-stage components of variance relevant to a two-stage sample design which is envisaged for the main survey, and also the relevant components of cost from which it is possible to determine the optimum intensity of sampling at each of these two stages. Pilot inquiries also provide information required to determine the most effective type and size of sampling units.

Pilot inquiries required to test different methods of data collection, types of schedules and enumerators etc. may be organized in the form of interpenetrating networks of sub-samples that will enable valid comparisons to be made of the factors under study.

The pilot inquiry, which can be considered a "dry run" for the survey proper, may be preceded by a series of small pre-tests on isolated problems of the design or enumeration procedures; the pre-test need not be based on probability samples, and even 50 to 100 households or persons covered in a pre-test could provide valuable information. One important component of pre-test in household sampling should be the practice of house-listing in some representative areas in order to train the enumerators so that the coverage errors in the survey proper are minimized.

26.15 Selection and training of enumerators

The problem of selection of enumerators depends on whether the sample survey is conducted on an *ad hoc* basis or is conducted over time as an integrated process of data collection. For the first type, part-time staff, such as teachers and housewives, can be used sometimes; but for the latter, full-time enumerators often have to be selected and incentive provided so that their performance is satisfactory and the turnover is not high. In countries with different linguistic and ethnic groups, it is necessary to ensure that the enumerators are accepted by the population being surveyed: female enumerators may, for example, be employed to interview female respondents in some surveys. The enumerators should have nearly equal workload, and each should cover a manageable geographical area.

It is often necessary to decide whether to employ locally based staff confined to a few sampling units, or mobile or itinerant staff covering a wider area. The choice between the two types of staff would again depend mainly on the type of survey being undertaken.

In a small-scale survey, enumerators can be trained in the central office. But in a large-scale survey, this may not be practicable and the training will be organized in two phases: first the supervisors will be trained by the technical staff, and, second, the supervisors in turn will go to their field offices and train the enumerators under them. It would be beneficial to organize refresher courses for both the supervisors and the enumerators, not only for the field procedures but also on the objectives and methodology of the survey. The training method would normally comprise classroom teaching, provision of manuals, demonstration enumeration, field work under supervision, scrutiny of filled-in schedules, and oral and written tests. The manual for the enumerators may contain the following topics: the survey objectives and program; preliminary work in the sample areas; how and whom to interview; detailed instructions on the questions, including concepts, definitions and examples; checking completed questionnaires; preparation of summary results; and return of documents and equipment.

26.16 Supervision

Accurate field work can only result from thorough training of efficient enumerators. Nevertheless, adequate supervision must be an integral part of the field work.

The ratio of supervisors to enumerators is 1:6 in the U.S. Current Population Survey and 1:4 in the Indian National Sample Survey; in the 1969 National Survey of Family Income and Expenditure in Japan it was 1:4. A supervisor should undertake some field work, either independently or as a check on the work of the enumerators, and this information can be used to adjust the data collected by the enumerators.

It is preferable to have a copy of the filled-in forms made before they are passed to the central offices for tabulation and analysis. The enumerators should themselves carry out simple numerical calculations and consistency checks of the data. When a copy of the enumerators' filled-in schedules is retained in the supervisor's office, a percentage of both the original and the copy should be scrutinized for copying and other mistakes; and, whenever possible, gross mistakes should be referred back to.

Any correction made by the supervisor on the enumerators' schedules should not be done by erasing the enumerator's entry but should be made in such a way that both can be used for future checks and analysis.

The enumerators should also be required to keep records of the times spent on the different aspects of the field work such as journey, identification and contact of sampling units, listing, enumeration, etc., which may be useful for designing subsequent inquiries.

26.17 Processing of data

After the completion of the enumeration, the schedules should be checked as carefully as possible and those entries which are not in the form of numerical quantities should be converted into numerical form; e.g. the sexes recorded as "male" and "female" may be converted to 1 and 2 respectively. The first of these steps is called **editing** and the second, **coding**.

While a precoded questionnaire can speed up a survey very greatly, it has several disadvantages, the major being the lack of checks. It is generally desirable to have the coding done in the central office.

The filled-in forms should be scrutinized by the technical staff for consistency and, as far as possible, for accuracy before they are sent on for processing. In a large-scale survey, this can be done on a sub-sample basis separately for each enumerator: if the percentage of inconsistent and inaccurate entries exceeds a certain limit for an enumerator, all the schedules filled in by him should be checked. Forms containing too many mistakes or doubtful entries may have to be rejected.

The importance of preparing a tabulation plan well in advance has been mentioned in section 26.10.

Tabulations can be manual, mechanical, or electronic. The data can be processed into final tabulations either by analyzing direct from the schedules or by transferring the data to cards or tapes for mechanical or electronic processing. In small-scale sample inquiries, the main required estimates can often be obtained manually from the forms and the cross-tabulations made through mechanical or electronic processing. In large-scale inquiries too, estimates that are required quickly can often be obtained from the summaries which the enumerators may be asked to prepare; for example, for the Indian National Sample Survey on population, births and deaths in 1958-9 (with 1.2 million persons in the sample), estimates of birth and death rates were prepared (from the village summary tables done by the enumerators) within a fortnight of the receipt of schedules: further estimates for cross-tabulations and also the computations of standard errors were made later through electronic processing.

Such preliminary results of the main items of a survey often serve important purposes. If the survey is arranged in the form of interpenetrating sub-samples, advance estimates can be obtained on earlier processing of one or more of the sub-samples (section 24.7.1).

During processing, it is advantageous to make provisions for obtaining estimates for the first-stage sample units. From a listing of these, both the final estimates and their standard errors can be computed in relation to totals and ratios: for the computation of standard errors in each stratum, two or more first-stage units should be selected with replacement, or the sampling fraction should be small when they are selected without replacement.

Checks should be introduced at every stage of processing. For data entry, for example, some dummy entries might be made which would ensure the checking of the punching and the verification work.

The computations required for a large-scale survey are made on a different basis from those illustrated in this book. For each ultimate sample unit, the multiplier or the weighting factor is computed and the product of the multiplier and the value of the study variable for the sample unit is obtained. When these are summed up for all the sample units by the machine, the required estimate is obtained and a suitable procedure is adopted for obtaining its standard error.

Suppose in a household inquiry (as in Example 21.1), the sample is stratified two-stage, with villages, the first-stage units, selected with pps and with replacement, and households, the second-stage units, selected at random or systematically. For estimating a stratum total, the multiplier of each household is given by

$$\frac{(\textit{Reciprocal of the probability of selection of the sample village})}{(\textit{Number of sample villages in the stratum})} \times \frac{(\textit{Total number of households in the sample village})}{(\textit{Number of sample households in the sample village})}$$

This multiplier will be computed for each sample household. To estimate the total consumption of food in a stratum, for example, the food consumption in a sample household will be multiplied by its multiplier, and the sum of the products for each stratum would give the estimated total food consumption in it. These stratum estimates when summed up for all the strata will provide an unbiased estimate of the total food consumption in the universe. When the ultimate units in the sample penultimate units are selected with equal probability (as in this example), the multiplier will be the same for the sample ultimate units in a particular selected penultimate unit.

Of course, in a self-weighting design, for estimating the totals the multiplier will be applied only once at the overall level, and it will not be required for estimating the ratio of two totals and its variance estimator.

The advent of electronic computers in recent years has been in many ways a great help in obtaining the required estimates speedily and accurately: for a sample survey, standard errors of a host of estimates can be computed with their aid. These can also be programmed to tasks of editing with consequent improvement of consistency of the primary data and possibly of its accuracy. They make possible some forms of estimation on a scale which would have been impractical with manual and mechanical tabulations. For example, in some cases the use of the regression method of estimation on previous surveys of a series with overlapping samples can substantially improve the efficiency of the estimators. The use of computers makes possible changes in the allocation of resources between the collection of data and processing. Computers can also be used in analysis and studies for interrelations of various factors by multivariate analysis (United Nations, 1964b), for example. "Off-the-shelf" software packages and programs, such as the Integrated Microcomputer Processing Systems (IMPS) of the U.S. Bureau of the Census, are now available for use in personal computers for obtaining not only estimates of totals, means, ratios, etc. of study variables but also their standard errors (more fully discussed in the next chapter, "The Use of Computers in Survey Sampling").

However, an efficient use of computers would be predicated upon the availability of trained programmers, especially the local staff, continuous service facilities, and intensive utilization of computer time.

A record of time, kept on the different processing operations, would help to improve future operations.

26.18 Preparation of reports

In addition to the presentation of the numerical results, the report of a survey, either complete enumeration or a sample, should contain some discussion and interpretation of the results.

For the preparation of the report of a sample survey, the United Nations Sub-Commission on Statistical Sampling has made certain recommendations (1964a); these are summarized below.

26.18.1 General Report

The General Report should contain a general description of the survey for the use of those who are primarily interested in the final results and broad findings rather than in the technical and statistical aspects of the sample design, execution, and analysis. The general description should include information on the following points:

1. Statement of the purposes of the survey;
2. description of the scope and coverage;
3. items of information and method of data collection;
4. date and duration - the survey and time-reference periods;
5. repetition - whether the survey is an isolated one or is one of a series of similar surveys;

6. numerical results;

7. main findings;

8. accuracy - a general indication of the accuracy achieved, distinguishing between sampling errors and non-sampling errors;

9. costs - an indication of the cost of the survey, under such headings as preliminary work, enumeration, analysis etc.;

10. assessment - the extent to which the purposes of the survey were fulfilled;

11. responsibility - the names of the organizations sponsoring and conducting the survey;

12. references - references to any available reports or papers relating to the survey.

26.18.2 Technical report

Technical reports should be prepared for surveys of particular importance and those using new techniques and procedures of special interest. In addition to covering such fundamental points as the purposes of the survey, stipulated margins of error, conditions to be fulfilled and resources available for the survey, the report should deal in detail with technical and statistical aspects of the sampling design, execution and analysis; the operations and other special aspects should also be fully covered. These aspects may be considered under the following headings:

1. *Design of the survey.* The sample design should be described in detail, sampling frames (their specification, accuracy, and adequateness), including types of sampling units, sampling fractions, particulars of stratification, sampling procedures for different stages, etc.

2. *Personnel and equipment.* It is desirable to give an account of the organization of the personnel employed in planning, collecting, and processing and tabulating the primary data, together with information regarding their previous training and experience. Arrangements for recruitment, training, inspection and supervision of the staff should be explained, as also should methods of checking the accuracy of the primary data at the point of collection. A brief mention of the equipment used is frequently of value to readers of the report.

3. *Statistical analysis and computational procedure.* The statistical methods followed in the estimation and in compilation of the final summary tables from the primary data should be described, the relevant formulae being reproduced where necessary. If any elaborate processes of estimation have been used, the methods followed should be explained with reasons.

4. *Reliability and accuracy of the survey.* The difference between reliability and accuracy has been explained in Section 25.4.5. This present heading should include:

   (a) Reliability (or efficiency) as indicated by the standard errors deducible from the survey;

   (b) degree of agreement observed by the independent enumerators covering the same material when interpenetrating samples are used;

   (c) study of accuracy of the data and estimates.

   In addition, methods adopted for evaluating the accuracy of the estimates and the method of controlling and adjusting for non-sampling errors and biases should be described.

5. *Technical analysis and interpretation of survey results.* These would be of value to answer some questions or to test some hypotheses; in a demographic inquiry, for example, the fertility and mortality levels and trends can be measured and the socio-economic differentials tested.

6. *Comparisons with other sources of information.* Every reasonable effort should be made to compare the survey results with other independent sources of information. The object of this is not to throw light on the sampling error, since a well-designed survey provides adequate internal estimates of such errors, but rather to gain knowledge of biases and other non-random errors.

   Disagreement between results of a sample survey and other independent sources may, of course, sometimes be due, in whole or in part, to differences in coverage, concepts and definitions or to errors in the information from other sources.

7. *Costing analysis.* The sampling method can often supply the required information with greater speed and at lower cost than a complete enumeration. For this reason, information on the costs involved in sample surveys is of particular value for the development of sample surveys within a country and is also of help to other countries.

- Fairly detailed information should be given on costs of sample surveys under such headings as planning (showing separately the cost of pilot studies), travel, enumeration proper, supervision, processing, analysis and overhead costs. In addition, labor costs in person-weeks of different grades of staff, and also the time required for different operations, are worth providing, for these may suggest methods of economizing in the planning of future surveys. However, the preparation of an efficient design involves a knowledge of the various components of costs as well as of the components of variance. The concept of cost in this respect should be regarded broadly in the sense of economic cost and should therefore take account of indirect costs which may not have been charged administratively to the survey.

8. *Efficiency.* The results of a survey often provides information which enables investigations to be made on the efficiency of the sample designs, in relation to other sample designs which might have been used in the survey. The results of any such investigations should be reported. To be fully relevant, the relative costs of the different sampling plans must be taken into account when assessing the relative efficiency of different designs and intensities of sampling. A study of the stage-variances, in a multi-stage design, would help in designing future surveys more efficiently.

9. ***Observations of technicians.*** The critical observations of technicians (in statistics and the specific subject fields) in regard to the survey, or any part of it, should be given. These observations will help improve future operations.

Further reading

Good introductions to the planning, execution and analysis of surveys are given by Casely and Lury (1981) relating to developing countries and by Salant and Dillman (1994) relating to the U.S. Other references include Foreman, Chapters 12-15; Levy and Lemeshaw, Chapters 15 and 16; Mahalanobis (1946a); Murthy (1967), Chapter 14; Murthy and Roy; United Nations, *Recommendations for the Preparation of Sample Survey Reports* (1964a); Yates, Chapters 3-5, and section 7.26.

On survey research in general, see Fienberg and Tanur (1983), Fowler (1988), and Sudman and Bradburn (1982).

In addition to the references given under "Further reading" in Chapter 1, the following are some references in specific subject fields:

Agricultural surveys: Monroe and Finkner; FAO (1969); Panse (1966); Royer; Sanderson; Sukhatme and Sukhatme (1969); Zarkovich (1963, 1965a, 1965b).

Demographic surveys: Blanc; España; Som *et al.*; United Nations (1971b); U.S. Bureau of the Census (1963b).

Fertility and family planning surveys: Population Council; Acsádi, Klinger, and Szabady (1969).

Forest inventory: Schreuder, Gregoire, and Wood (1993).

Health surveys: see the references under "Further reading" in Appendix V.

Household and social surveys: United Nations (1964b).

Sexuality surveys: Bullough; Hite (1979 and 1981); Janus and Janus (1993); Kinsey *et al.* (1948) and (1953); Masters and Johnson; Laumann *et al.*; Michael *et al.*; Tavris and Sadd (the *Redbook* study).

For some methods of utilizing **imperfect frames**, see the papers by Hansen, Hurwitz, and Jabin (1964), and Szameitat and Schaffer (1964) and the discussion at the meeting on "Sampling from Imperfect Frames" at the 34th session of the International Statistical Institute, 1963 (Proceedings, 1964).

For examples of **good technical reports**, see the publications on the World Fertility Survey and the Demographic and Health Surveys (further details on these two survey programs are given in Appendix VI, "Multi-country Survey Programs.")

Exercise

1. Prepare a critique of the sampling aspects of U.S. surveys on human sexuality.

   (*Hints:* Consult the studies on human sexuality listed under *Further reading* above and the definitive statistical critique of Kinsey reports by Cochran, Mosteller, and Tukey (1953). Note that all these studies, excepting the one conducted through the University of Chicago's National Opinion Research Center (NORC), reported by Laumann *et al.* and Michael *et al.*, were based on non-probability samples. Consider the implications of self-selection in such non-probability samples. Note that in the NORC survey, 6 per cent of the interviews were made in the presence of a spouse or a sex partner and that an additional 15 per cent were done in front of the respondent's children. The researchers of the NORC study have stated that this had a significant effect on responses: "When interviewed alone, 17 per cent reported having two or more sex partners in the past years, while only 5 per cent said so when their partners were present during the interview." – cited by John K. Wilson, in Letter to the Editor, *New York Times*, November 1, 1994; see also M.G. Lord's article, "What That Survey Didn't Say," in the *New York Times*, Op-Ed page, October 25, 1994. The homeless and persons in institutions, e.g. in prisons, were not included in the NORC study. Consider the possible effects of such exclusion on the estimation of HIV/AIDS infection. See Section 24.19, "Estimation of HIV/AIDS infection" in this book and the review by Robinson of the NORC study and by Lewontin of Bullough's and NORC studies.)

CHAPTER 27

The Use of Personal Computers in Survey Sampling

27.1 Introduction

Computers have now become an indispensable tool for processing all types of data in public and private sector activities, such as censuses and surveys, financial transactions, telephone listing, payroll preparation, educational testing, and space programs.

In a typical census, there is a fairly large amount of data input, but the data structure and statistical tabulations are relatively simple, comprising first-level tabulations and second- and third-level cross-tabulations; in contrast, sample surveys require far less data input (5 per cent or less than that of a corresponding census), but the statistical operations, such as the estimation of universe totals and their variances, are complex.

Many countries with populations up to 10 million (such as Honduras with 5.1 million and Niger with 7.7 million around 1990) have been increasingly using powerful Personal Computers (PCs), with shared (local or remote) area network systems to process their entire census data (Toro and Chamberlain, 1989). For example, of the 25 African countries that had undertaken the processing of their 1985-91 population censuses, 16 had used only microcomputers, 9 only minicomputers and 3 a combination of both (Dekker, 1991).

And, almost invariably, sample survey data are now being processed on PCs. For example, the data of 48 Demographic and Health Surveys, conducted in different countries of the world during 1985 through 1992, were all processed on PCs, using an integrated software package especially developed for these surveys and the first sets of tabulations were completed within a fortnight of the receipt of data (see Appendix VI, section A6.3.3, and Croft, 1992).

PCs are replacing mainframes (and minicomputers) in census and survey data processing in most countries on consideration of capacity, access, and cost; there has been, simultaneously, nothing short of an explosion of PC-based software packages for office and personal (including recreational) use that has further expanded the appeal of PCs. While PCs do not have the capacity and speed of mainframes, they are improving very fast. PCs are geared toward individual users and software packages are more "user-friendly" and easier to learn than mainframe packages; thus, more people can be trained with less effort – an important consideration in developing countries where training and resources are limited.

In most countries, access to mainframes or minicomputers is easier for a population census than for a sample survey. As a national endeavor, a population census rightfully gets a high priority for its processing, in mainframes or mini, that, in many developing countries, are located in government offices or government-sponsored corporations, doing routine chores such as the preparation of government payrolls and electricity bills. Processing of data of surveys, even when conducted by government departments, may not get a high priority there. On the other hand, PCs are, by their very nature, individualized machines doing only what the user tells them to, but doing it immediately. This control over computing can speed up the processing time considerably. And until recently census and survey data processing could be started only after *ad hoc* computer programs for data editing and tabulation had been prepared; but now a variety of software programs and packages are available for personal computers, more than for mainframes and minicomputers, so that "off-the-shelf" software packages can be easily used without much knowledge of computer programming. The delays, if any, now in data processing are due to problems of management, work organization, and training (Gerland, 1992). And, finally, the number of staff members and the level of training required to operate and maintain computers and updating of software packages are much less for a PC than for a mainframe. The decreasing cost of PCs and their increasing processing power often become the clinching argument for their recommended use in census and survey data processing, particularly in developing countries (U.N., 1990c, Chap. VI).

In this chapter, we shall first consider a number of general issues in using PCs for data processing and then briefly describe the use of computers in sample surveys, starting from sample selection and proceeding through data collection, data editing and imputation, data entry, data tabulation, computation of sampling variance, data analysis, digital mapping, and publishing; we shall mention only those software packages that are widely in use. We shall also mention some new computer technologies that are being evolved in developed countries such as Australia, Canada, the Netherlands,

Sweden, the U.K., and the U.S.

27.2 General issues

The general trend in the developed world has been toward an increasing use of PCs and, for the field, powerful portables, with facilities for wireless or telephone communication. Perhaps countries in the third world could go directly to using PCs (desktops) in the office with a Local Area Network (LAN) and portables in the field without going through the Mainframe → Minicomputer → Desktop + Portable route. Given resources and access to electricity for recharging batteries, computer-assisted personal interview could be adopted in these countries right now.

There is no escaping the computer revolution. Like Alice in *Through the Looking Glass*, if you just run, you will remain in the same place: to advance, you must run faster.

It has been argued that the experiences of developed countries may not be appropriate to the developing countries for several reasons: significant differences in the computerization and introduction of microcomputers; the need for the developing countries to have stand-alone PCs, as compared to linking micros to existing mainframe systems; and PC use in developing countries not having immediate relevance to developed country PC applications, such as in finance and law enforcement in the government sector in the U.S.

For the developing countries, there are two important lessons. First, in developing computers and their software packages, initial emphasis should fall on computer engineering and program development. Later on, emphasis should shift to end-users, both present and potential. Otherwise, computers will continue to be considered esoteric. The time-frame of the history of computers from the first electronic computing machine, the **ENIAC (the Electronic Numeric Integrator and Computer)** in the mid-1940s and the early mainframes (that needed highly skilled mathematicians to be programmers) to the current crop of PCs that are being used by almost everybody (most of whom are not programmers) has to be telescoped in the developing countries. Second, experiences of the developed countries will have to be internalized in the developing countries.

The speed and effectiveness of the transfer of microcomputer technology, or for that matter of any technology, depend on the receptivity of the population – receptivity that is embedded in the social and cultural milieu. In cultures where hands-on machine use is still considered *infra dig.*, the PCs provided could become mere status symbols and office adornments. The provision of equipment does not guarantee its use, as some potential users could continue to be passive toward the technology (Sanwal, 1988).

27.3 Sample selection

For selecting the sample, sampling frames are required. Such frames are normally available at the central sample survey office, at least up to the first few stages (in a multi-stage survey) in the form of a data base, from which a computer program can be written for selecting the (first few stages') sample units. For example, a master sample may contain information on small areal units (either census blocks or census enumeration areas) in the form of a data base. If in a demographic survey, all the extant households in the selected small areal units are to be enumerated, then these units become the ultimate-stage sample units, that is, no further sampling need be done in the field. Households and individuals then become observation or recording units. If, however, a sample of households is to be selected from the selected small areal units, sampling at the ultimate stage must be carried out by the enumerators in the field.

For selecting the sample, the sampling frames are usually put in a data base format, such as dBase III or IV, Fox Pro or Paradox, and the sample drawn following the sample design.

27.4 Data collection

Until recently, the required information in sample surveys had to be collected by personal interview, and in a few cases by mail or telephone interview. Computer-assisted data collection, through personal or telephone interview, makes it possible to combine data collection, data editing and data entry; thus, soon after enumeration, the fully edited data become available in machine-readable form and then easily transferred to the central computers for further processing (see section 27.6).

27.5 Data editing and data entry

The availability of computer power right at the point where data are entered now makes real-time error checking possible. The data entry software packages can be programmed to refuse to accept an entry or to flash a warning should the operator try to register a wrong code or if variables appear to be inconsistent with each other; also, fields that are known to remain unchanged from one record to the next may be duplicated automatically (U.N. ESCAP, 1988).

Among the data processing operations of a census or a survey, data entry is the most constricting bottleneck such that for a population census it can take a year and often more. Data entry software packages have been designed to relieve this constriction.

Two data entry software packages that are in most common use in censuses and surveys in developing countries are **CENTRY**, developed by the U.S. Bureau of the Census, and **PC-EDIT**, developed by the U.N.; both include editing, verification, data modification, and statistics on operator performance. **CONCOR**, a software package also developed by the U.S. Census Bureau for data scrutiny and editing, can be run during data entry with CENTRY.

Computer-assisted Data Entry and Editing (CADE) Program. CADE, also called CADI (for computer-assisted data input), is an interactive system for data entry and data editing of data collected in PAPI (pencil-and-paper interview; see the next section). It checks for consistency being done by a subject-matter specialist either by consulting the forms or calling the supplier of the information. This allows a dynamic checking of errors that is absent in CAPI and CATI (Computer assisted telephone interview; see the next section) (Bethlehem and Hundepool, 1992).

Computer Assisted Coding (CAC). Coding of responses in censuses and surveys, an important part of data processing but very error-prone, can be computerized in two ways: (1) though **automated coding**, codes are assigned automatically by the computer, based on a file with verbal descriptions and a coding program but without any user interference; and (2) **computer-assisted coding**, the verbal description is presented on the screen and additional facilities are available to assist the coder to establish the correct codes (Bethlehem, Hofman and Schuerhoff, 1989). The experience of computer-assisted coding has been studied in the Netherlands, Canada, France, Switzerland, and Sweden, among others, and in general they have shown the superiority of CAC over both manual (clerical) coding and automated coding (Dekker, 1991). In Latin America, Argentina and Brazil are the only two countries that had developed computer software packages to assist in automatic coding of the census data (Ellis, 1991).

In the 1990 U.S. Census, with over 20 million persons providing industry and occupation descriptions, the Bureau of the Census applied a combination of automated and computer-assisted (clerical) coding (Scopp and Tornell, 1991).

Application of computer-assisted coding requires the development of necessary indexes according to some strict syntax rules. One version of the CAC system of the Australian Standard Classification of Occupations uses a simple matching algorithm.

Reference should also be made to automated matching with good administrative records being used at census offices in a number of developed countries such as Canada (*vide* the **Fellegi-Sunter algorithm** in Fellegi and Sunter, 1969), the Netherlands, Sweden, and the U.S.

In adopting computer-assisted coding in data processing, consultation with subject-matter specialists should be maintained because in case of substantive issues regarding coding, the coders would not generally be able to decide the issues.

Image Capture Technique. Optical mark readers (OMRs) have been used in many censuses and surveys for data capture. However, the OMR technique cannot be used to capture written responses, such as on industry and occupation; these can be captured using the **optical character recognition** (OCR) technique for image capture, including the handwriting recognition technique. In Latin American countries, OMR is the second preference for data capture (used in 6 countries) after micros (16 countries), with manual data entry used in 3 countries; the major limitation for using OMRs in Latin America is that the forms have to meet a certain quality and often have to be printed outside the countries, adding to the cost (Ellis, 1991).

27.6 Combined data collection, editing and entry

Computer-based new technologies have emerged to combine data collection, data editing and data entry. They include computer-assisted personal interview (CAPI) and computer-assisted telephone interview (CATI).

Computer-Assisted Personal Interview (CAPI). The traditional interview using pen and paper (**pen-and-paper interview, PAPI**) is replaced by computer programs containing the questionnaires, the sequence of questions to be asked, and the responses checked against the given alternatives. In CAPI, the computer program is loaded into a portable computer, which is taken to the field, and the answers by the respondents entered into the portable.

Computer-Assisted Telephone Interview (CATI). Here too, the traditional PAPI is replaced by computer programs containing the questionnaires, the sequence of questions to be asked, and the responses checked against the given alternatives. In CATI, the interviewer calls the respondents from a central unit, carries out the interview by telephone and enters the data in the computer program in a desktop computer.

These methods eliminate printing, transporting, office editing, and keypunching questionnaires at the central office.

Studies made at the U.S. Bureau of the Census and the Department of Labor and in other developed countries have indicated that CAPI can give better quality data than PAPI (Baker and Bradburn, 1992; Bradburn,

Frankel, *et al.*, 1992; Olsen, 1992; Speizer and Dougherty, 1991). CAPI has thus become a common feature in many surveys in countries such as the U.S. [for example, to collect supplementary data, in the U.S. National Health Interview Survey (Rice *et al.*, 1989) and in a survey to assess nutritional needs of the population (Rotschild and Wilson, 1989)] in Canada, the Netherlands (in the labor force survey, van Bastelaar *et al.*, 1988), Sweden, the U.K. and Australia. CATI is being used, among others, in the U.S. National Crime Victimization Survey and has been adopted by literally hundreds of survey organizations as the preferred method of data collection (Lyberg and Kasprzyk, 1991).

An experiment was conducted in Guatemala for the Demographic and Health Survey there (see Appendix VI, section A6.3.3) by interviewing 300 women once by PAPI and then by CAPI that showed a saving of about 25 per cent in interviewing time in CAPI. (Cantor and Rojas, 1992).

27.7 Data tabulation

There are three major data tabulation software packages in use in many developing countries – CENTS and QUICKTAB, developed by the U.S. Bureau of the Census, and XTable, developed by the U.N. Statistical Division.

27.7.1 CENTS

CENTS has been used successfully in many censuses, surveys and other activities requiring the production of statistical tables. It can produce tables in virtually any format. The user "draws" the tables, then writes instructions on the relationship between the data and table cells. Once tables are defined, they can be produced automatically for up to five geographical levels. CENTS is available for PCs only.

27.7.2 QUICKTAB

QUICKTAB is a menu-driven tabulation package for frequency distributions and cross-tabulations. It produces cross-tabulations of up to three dimensions and processes at a speed of 80,000 records per minute on an average PC.

Tables produced by QUICKTAB are in a pre-specified format. That is, QUICKTAB, unlike CENTS, does not allow for the flexibility of "drawing" tables in any format. QUICKTAB is simple to use and can be learned in a few minutes.

27.7.3 XTable

XTable is an easy-to-use package that produces frequencies and summary tabulations of censuses, surveys, vital and civil registration records, population statistical databases, or any other administrative data.

27.8 Variance estimation

Currently available commercial statistical software packages programs and packages do not allow the computation of estimates and their variances from complex survey designs with several stages and strata and with varying probabilities of selection. The software package, **Integrated System of Survey Analysis (ISSA)**, developed at the DHS (see section A6.3.3), is a powerful tool but it produces estimates and variances of DHS-type surveys only (see section 27.10.3 below). The software package **SUDAAN (for SUrvey DAta ANalysis)**, developed at the Research Triangle Institute (1989), requires the linearization of the estimates.

The variation calculation system, **CENVAR**, developed at the U.S. Census Bureau, is the only software package that is adaptable to complex sample survey designs to estimate universe parameters and their sampling variances (see section 27.10.1) by following the methods described in this book.

27.9 Data dissemination

With the census data and tables being put in databases, data dissemination, particularly between government agencies, has been facilitated by putting the databases on-line. In Japan, the Statistical Bureau and the Statistics Center are jointly operating a statistical database called **SISMAC (Statistical Information System of Management and Coordinating Agency**; Ide and Kawasaki, 1991). In Latin America, on-line access to census data by external users is not widespread except in Brazil; for the region as a whole, micro data of the population censuses will be disseminated using magnetic tapes and CD-ROM diskettes (Ellis, 1991).

27.10 Integrated systems of data processing

Under integrated systems, we shall classify those software packages that are designed for the specific purpose of covering all, or the major, data processing requirements of a survey to avoid the interfaces between separate packages that can do only part of the job. Integrated systems are especially useful for relatively unskilled users working on PCs. We shall review integrated systems of data processing, using PCs, developed by the U.S. Census Bureau, by the Netherlands Central Bureau of Statistics, and by the Institute for Research Development. We shall also mention two other not fully integrated systems developed by the World Health Organization, the U.S. Centers for Disease Control and by the U.N.

27.10.1 IMPS (Integrated Microcomputer Processing System)

The most comprehensive integrated system of survey data processing to date is the **Integrated Microcomputer Processing System (IMPS)**, developed by the U.S. Bureau of the Census. It allows persons with little or no computer experience to contribute to the data processing operation. It has been constructed from packages, some of which have been mentioned earlier, which perform the following processing tasks: data entry, editing, tabulation, operational control, and statistical analysis.

IMPS is available for use with PCs and has been used or is planned to be used in 75 countries for processing their population and housing censuses and in 43 counties for processing their surveys on agriculture, farm incomes; economic statistics - income and expenditure; education and literacy; demography and vital statistics; household statistics; labor force; and Social Dimensions of Adjustment, sponsored by the World Bank (see Appendix VI, section A6.3.6).

The mainframe version of IMPS is not yet available, so mainframe users will have to continue to use separate modules of the programs available to them.

The following are the modules of IMPS version 3.1, issued in December 1993:

Data Entry. CENTRY (CENsus enTRY) is a screen-oriented, menu-driven package for developing data entry applications (see section 27.5).

Editing. CONCOR (CONsistency and CORrection) is a package for rapid identification and correction of invalid and inconsistent data using available and appropriate editing techniques. CONCOR can be used independently or together with CENTRY (see section 27.5).

Tabulation. QUICKTAB (QUICK TABulation) is a menu-driven package for the rapid production of frequency distributions and cross-tabulations. CENTS (CENsus Tabulation System) is a package for tabulating, summarizing, and displaying statistical tables for publication (see Sections 27.7.1 and 27.7.2).

Table Retrieval System. While a census tabulation plan generates a large number of tables, only a relatively small proportion of them is published. To make the unpublished mass of data accessible to staff both in the census offices and outside (e.g. businesses, press, universities, and research institutions), such that data users (even those with no previous computer experience) could examine all tables and accompanying text stored in a computer readable format - whether published or unpublished - Carlos Ellis of the U.N. Statistical Division developed a menu-driven software package that makes it easy to select, retrieve, display and print statistical tables from either the CENTS tabulation component of IMPS or from text files. The Bureau of the Census elaborated the TRS system to form an integral part of IMPS.

Operational Control. CENTRACK (CENsus TRACKing) is a management and control package to help census managers monitor, control, and track the various operations necessary between receipt of questionnaires from the field and data entry.

Variance Calculation. CENVAR (Variance Calculation System) is a statistical package for analyzing data from stratified, multi-stage sample surveys. It provides estimates of totals, means, ratios, etc., and also estimates of their sampling variances. Sub-population estimates can be obtained by specifying classification variables, which can be crossed with each other. In addition, CENVAR output presents for each estimate the coefficient of variation, the number of observations and the "design effect."

For demographic analysis, the Center for International Research of the U.S. Bureau of the Census has developed a software package called "**Population Analysis Spreadsheets**" for analyzing demographic data and preparing projections. The accompanying manual, "Population Analysis Using Microcomputers," presents many useful and accepted methods of demographic analysis.

IMPS and its modules are available free of charge to eligible, noncommercial users; these can also be downloaded from the Internet. For information, contact International Systems Team-CASIC, U.S. Bureau of the Census, Department of Commerce, Washington, D.C. 20233-3102, U.S.A.; phone: 301-457-1453; fax: 301-457-3033; Telex: 62761615; Internet: imps@info.census.gov.

27.10.2 Integrated statistical information processing on microprocessors (the Netherlands)

The Central Bureau of Statistics of the Netherlands has developed a system to make data processing more efficient with production of better quality information by using computer technology. It is based on the principles of standardization of hardware and software packages and integration of the software packages required in the various processing steps: the basis of the integrated environment is the Blaise system, a system designed to describe a survey questionnaire in a formal language, the Blaise (named after the French theologian and mathematician Blaise Pascal, 1623-1662), which in turn is partly based upon the programming language Pascal (named also after Blaise Pascal).

Questionnaire specification, collection of information (through CAPI, CATI, and PAPI), and data input are parts of the BLAISE system. Weighting is done through Bascula, a general weighting program run on PCs in the DOS platform, tabulation through the Abacus tabulation package, which is menu-driven and interactive, and can read files created by the Blaise system and also ASCII files; statistical analysis through commercial software packages SPSS and Stata, after the data file is converted from the Blaise format to ASCII; and finally publication through CBSview, a publication package developed by the Netherlands Central Bureau of Statistics (Bethlehem and Hundepool, 1992).

27.10.3 ISSA (Integrated System of Survey Analysis)

The Integrated System of Survey Analysis (ISSA) was developed to meet the data processing requirements of Demographic and Health Surveys (DHS) (see Appendix VI, Section A6.3.3). It is an integrated and interactive system that can handle complex surveys and data files, consistent with established procedures and compatible with other systems. Its modules can run "stand alone" or a part of an integrated system. The system comprises a designer, a data entry processor, a batch processor and a number of utility modules. Within the ISSA system, there are two basic entities: a data dictionary, describing the structure and contents of a data file; and an application, which instructs the system how the data are processed. It provides estimates of universe totals and their variances and includes two random group estimators: Jackknife (see section 24.9.2) and Balanced Repeated Replications.

ISSA can be applied to process data from surveys of the DHS type. Additional information can be obtained from Macro International/DHS, 11785 Beltsville Drive, Suite 300, Calverton, MD 20705, U.S.A.; phone: 301-572-0200; fax: 301-572-0999.

27.10.4 Epi Info and Epi Map

Epi Info and Epi Map constitute an integrated program of data correction, tabulation and thematic mapping.

Epi Info (now in version 5) is a series of microcomputer programs for handling epidemiological data in questionnaire format and also for organizing study designs and results into text that may form part of written reports. A questionnaire can be set up and processed in a few minutes, but Epi Info can also form the basis of a powerful disease surveillance data base with may files and record types. It includes the features most used by epidemiologists in statistical programs such as SAS or SPSS and data base programs like dBase, combined in a single system that may be freely copied and distributed.

Epi Info is available in English, Spanish, and Arabic from USD Inc., 2075A West Park Place, Stone Mountain, Georgia 30087; phone: 404-469-4098; fax: 404-469-0681. The French language Epi Info is distributed by Editions ENSP, Ecole Nationale de la Santé Publique, Avenue du Professeur Léon Bernard, 35043 Rennes Cedex, France. Chinese, Portugese, Italian, and Czech language versions are being prepared, and an Indonesian language version is planned.

Epi Map has been described briefly in section 27.14.2 under "Geographical Information Systems."

27.10.5 U.N. suite of software packages for data processing

PC-EDIT (see section 27.5) and XTable (see Section 27.7.3), combined with PopMap (see section 27.14.1), developed by the U.N., form a suite of software packages for data editing, data entry, and geographic information services. These are provided free of charge to UNFPA-supported projects, government agencies and academic institutions. By mid-1991, such software packages were delivered in 464 projects or institutions in 126 countries.

For further information, contact Project Coordinator, Computer Software and Support for Population Activities Project, Statistical Division, U.N., Room DC2-1570, New York, NY 10017, U.S.A.; phone: 212-963-4118; fax: 212-963-4116.

27.11 Computer-Assisted Survey Information System (CASIC)

The new technology of Computer-assisted Survey Information System that is now in place, but still evolving, in the U.S., Canada, the Netherlands, the U.K., Sweden, Australia, New Zealand and in other developed countries relates to the use of computers both for data collection and data entry through Computer-Assisted Survey Information Collection (CASIC).

The CASIC integrated system includes the following (Creighton *et al.* (1994)):

Data Collection and Capture:

Computer-assisted personal interview (CAPI)
Computer-assisted telephone interview (CATI)

Computerized self-administered questionnaire (CSAQ)
Touch-tone data entry and voice recognition entry (TDE/VRE)
Standardized technologies assisted mail processing (STAMP)
Electronic data interchange (EDI)
Computer-assisted data entry (CADE)
Imaging
Paperless fax image reporting system (PFIRS)

Data Processing:

Automated coding
Interactive corrections
Batch editing
Imputation
Weighting
Tabulation

Management:

Survey/census design
Authoring
Payroll
Management information system

Some of the processes are still being evolved and many of them will not be applicable to most developed countries at their present level of technology and infrastructure. CAPI, CATI and CADE – three components of CASIC – have been dealt with earlier in section 27.6. Among the newer technologies that are emerging, automated voice recognition, paperless fax image reporting, optical character recognition, and touch-tone data entry, are briefly described below.

Automated voice recognition. A study is underway for possible use in the year 2000 census project with the goal of designing and testing the feasibility of a spoken language system, over the telephone, which will interact with a caller to elicit specific information about the caller (Cole and Novick, 1993).

Paperless fax image reporting system and optical character recognition. In the paperless fax image reporting system (PFIRS), using optical character recognition (OCR), being evaluated currently at the U.S. Census Bureau for possible use in its data collection and processing, the process is: respondents fax questionnaires to the Bureau → the Census Bureau fax-server receives digital image (not paper copy) and the system automatically identifies the questionnaire → OCR software converts survey responses to machine-readable codes (unrecognized data are displayed on the computer screen for manual keying) → the interpreted data are stored directly into a data base (Rowe, Petkunas, Appel, and Cable, 1994).

Touch-tone data entry. Touch-tone data entry (TDE) is an automated data capture technology that allows a respondent, using the keypad of a touch-tone telephone, to reply to computer-generated prompts: the TDE system functions as an interviewer. TDE has been in use for some time in commercial applications outside the survey field, e.g., in banking by telephone, rental car reservations, and college class registration (Lyberg and Kasprzyk, 1991). The U.S. Bureau of Labor Statistics has been using TDE since 1986 (Ponikowski, Copeland and Meily, 1989), many firms in the United States are using it for a variety of applications and the Census Bureau itself currently uses TDE for a number of surveys and automated call routing (Appel, 1992).

The Current Employment Survey, conducted by the U.S. Bureau of Labor Statistics, reports a high positive respondent acceptance of TDE due to its speed and convenience for the respondents. Also TDE offers many advantages to the survey agencies, including the reduction or elimination of costs of interviewing, mail-out, mail-back, check-in, data keying, and CATI (Werking and Clayton, 1991); the limitations of TDE are that it requires respondents to have touch-tone phones and that it is self-initiated by the respondents, the same problem as for mail collection. For an initial technical assessment of TDE, see Appel (1992).

Automated voice recognition, paperless fax image reporting system, and touch-tone data entry are three different forms of **computer-assisted self-interview (CASI)** in which the respondent answers questions without the benefit of an interviewer (U.S. Office of Management and Budget, 1990; also Lyberg and Kasprzyk, 1991).

Prepared data entry. A third example of computer-assisted self-interview is prepared data entry (PDE) that allows respondents with compatible microcomputers or terminals to access and complete a questionnaire on their computers by entering data in response to preprogrammed floppy disks and then mail it either via a carrier (such as the postal service) or electronically. This technique has been applied, among others, by the U.S. Internal Revenue Service for tax payers to submit tax returns electronically and at the U.S. Energy Information Administration for collecting monthly refinery data from petroleum companies (U.S. Office of Management and Budget, 1990, cited by Lyberg and Kasprzyk, 1991). A variant of PDE, called "tele-interview", has been reported by Saris (1989; also cited by Lyberg and Kasprzyk, 1991) where each sample respondent is provided with a computer and a modem; with the help of these and a television and telephone (that are present in almost every household in the United States), an interview is conducted from the central computer to the home computer of the respondent; the respondent answers all the questions on the personal computer and sends the results back to the central computer.

Although not formally a part of the CASIS system, an interesting development for reducing costs and for obtaining better quality of data is the use of an interactive multimedia kiosk.

Interactive multimedia kiosk. Among the variety of technologies assessed by the U.S. Census Bureau in 1993 having the potential of reducing both costs and differential under-count in the Year 2000 Census is the prototype interactive multimedia kiosk that offers a menu, in both English and Spanish, to reach the public with the Census' message, answer questions, vend information products, and collect questionnaire data. This technology, housed in a publicly convenient kiosk (or through in-home access via a computer and modem), is deployed or being considered in many U.S. federal agencies, such as the U.S. Postal Service, the Internal Revenue Service, the Veterans Administration, and the Social Security Administration.

27.12 Statistical analysis

For statistical analysis – for example for fitting regression and studying correlation – commercially available software programs and packages, such as **BMDP, SAS, SPSS (SPSS/PC+** for PCs), and **SYSTAT**, may be used.

27.13 Spreadsheets and graphics

The census and survey tables can be stored in tables created with a spreadsheet software package. Recent versions of the commercially available spreadsheet software packages such as Excel, Lotus 1-2-3 and Quattro Pro include considerable graphic facilities for preparing charts and graphs – bar diagrams, pie charts, and graphs depicting, e.g., the fertility levels in different socioeconomic strata of population – mostly without recourse to graphic software package.

27.14 Geographical information systems

A recent innovative use of PCs is the **digital mapping** of census and other data at different geographical levels. Several geographical information systems (GIS) have been produced for this purpose. In the U.S., software packages for digital mapping and geographic analysis include **Atlas GIS,**

MapInfo (MapInfo Corporation), and **InfoShare** for the city of New York (by the Queens College of the City University of New York).

Atlas GIS (for Windows) can access data from dBase, ASCII, Lotus 1-2-3, Microsoft Excel, and industry-specific marketing information files, from single U.S. city blocks up through the entire world on diskettes or CD-ROM, street maps of the entire U.S. (based on TIGER, see later in this section), Atlas demographic data from the 1990 Census, projected to 1998, business and consumer data from exclusive and syndicated sources, and a library of available marketing information and competitive business locations and data. Atlas GIS is produced by Strategic Mapping Corporate Headquarters, 3135 Kifer Road, Santa Clara, California 95051; phone: 408-970-9800; fax: 408-970-9999.

MapInfo for Windows, version 3.0 includes satellite raster image support to see natural terrain on-screen and chart new roads or other details on old maps; remotely imported data can be displayed in thematic maps or charts in a broad range of colors.

GIS is being used in a rapidly growing number of disciplines. In anthropology, economics, history, sociology, and environmental and urban planning, for example, GIS is providing new insights into geographically-linked data and enabling the spatial display, analysis, manipulation and querrying of data at increasing levels of detail; in other applications, three-dimensional conceptual maps, such as response- or cost-surface, are being created with GIS.

A National Center for Geographic Information and Analysis has been established by the U.S. National Science Foundation to carry out fundamental research into GIS technology and issues central to its effective use. It has covered topics such as utility management in urban areas and earthquake modeling and data keeping.

Censuses of population, agriculture, industrial establishments and others provide a veritable "demographic treasure-trove" for preparing digital mapping for use in war and peace. In the 1991 Gulf War, GIS systems of the American Defense Department helped guide soldiers and missiles to targets (Dataquest, 1991).

The U.S. Census Bureau sells computerized city-block maps of America for around $10,000. For the first time, the 1990 census maps are generated from a computerized data base, the **TIGER (Topographically Integrated Geographic Encoding and Reference System)**.

GIS systems, combining information from censuses and other governmental information, have great potential for developing countries in their national development planning. A thematic mapping software package allows the production, at specified geographical levels, of thematic maps, e.g., maps showing demographic levels, such as population density, fertility and mortality and the available infrastructure, such as schools, hospitals, and family planning clinics.

However, sample survey results, by their very nature, may not be completely amenable for reproduction in a GIS.

Two major thematic mapping software packages are **PopMap** and **Epi Map**.

27.14.1 PopMap

PopMap is an integrated software package, developed by the U.N. Statistical Division in collaboration with the Institute of Computer Science in Viet Nam, which offers graphics and mapping capabilities with a spreadsheet for collecting, storing, tabulating and mapping information. This software package was prepared to support planning and administration of population activities with important geographical or logistic context or to facilitate geographic or graphic expression of population indicators or related data. The system comes with data base modules for declaring the structure and content of the geographical and statistical data base, a map editor for entering map outlines, boundaries, borders, rivers, routes and individual facility locations, and a system for selective retrieval of statistical and facilities data, preparing maps, graphs and spreadsheet tables for on-line study and analysis, and for printing and publication.

For further information, contact Project Coordinator, Computer Software and Support for Population Activities Project, Statistical Division, U.N., Room DC2-1570, New York, NY 10017, U.S.A.; phone: 212-963-4118; fax: 212-963-4116.

27.14.2 Epi Map

Epi Map was jointly developed by the Division of Surveillance and Epidemiologic Studies, Epidemiology Program Office, U.S. Centers for Disease Control and the Global Programme on AIDS of the World Health Organization. It is a microcomputer program for producing thematic maps from geographic boundary files and data values entered from the keyboard or supplied in Epi Info or dBASE files. The data may be counts, rates or other numeric values. Epi Map also produces cartograms, in which the value for each geographic entity is allowed to control the size of the entity; thus, a state or county will be small on the map, but large if the number is large. Base maps may be maps with text and symbols for titles and legends. They may consist of the outlines of a geographical area or they may contain sub-regions or enumeration units; for example, a base map may be a continent by country, a country by provinces, or a country by health district.

27.15 Desktop publishing

Special purpose software packages, such as Aldus Page Maker, have expanded microcomputer use to the business of publishing: they make it possible to produce, right on the PCs, camera-ready reports for the photo offset process, thus bypassing the regular production process. Many reports, studies, and books are now being produced, at reduced costs and more speedily, using desktop publishing. This has been of particular benefit to developing countries, where census and survey reports may languish for years waiting their turn at the government printing offices.

Some word processing software packages, such as **WordPerfect version 6.0** and **Word for Windows**, integrate page layout capabilities with word processing, and are capable of producing all but the most design-intensive projects (Lichty, 1994; Parker, 1994a and 1994b).

27.16 Concluding remarks

No software package should be used without a full understanding of the underlying assumptions of methods involved. This is particularly true of the application of software packages for analysis, whether for estimating universe totals and their variances or cross-tabulations and regression analysis to discern interactions and the so-called proximate determinants of study variables. The adage "garbage in, garbage out" (GIGO) remains true.

Further reading

In a rapidly evolving field such as that of microcomputers, any list of references runs the risk of being out of date soon: that caveat should be sounded in consulting studies prepared before the late 1980s.

For a review of statistical software programs and packages for processing censuses and surveys, see U.N. ESCAP (1988). On the use of PCs in processing census data, see Toro and Chamberlain (1988) and U.N. (1989 and 1990c) and on the use of computers in processing household survey data, see U.N. (1982a). For GIS, see Arbis and Coli (1991), and Maguire, Goodchild, and Rhind (1991); on other topics, see the references cited in the text. For the role of microcomputers in national development, particularly in India, see Sanwal (1988) and Som (1993).

APPENDIX I

List of Notations and Symbols

1. For ***summation and product notations***, see section 1.15.

2. In general, the following ***subscripts*** are used:
h for stratum, $h = 1, 2, \ldots, L$;
i for a first-stage unit, $i = 1, 2, \ldots, N$ for the universe, and $i = 1, 2, \ldots, n$ for the sample;
j for a second-stage unit, $j = 1, 2, \ldots, M_i$ for the universe, and $j = 1, 2, \ldots, m_i$ for the sample;
k for a third-stage unit, $k = 1, 2, \ldots, Q_{ij}$ for the universe, and $k = 1, 2, \ldots, q_{ij}$ for the sample.

3. y denotes a ***study variable*** (also an unbiased estimator of the total Y of the study variable for the universe);
x denotes ***another study variable***; and
z an ***ancillary variable*** either for selection with probability proportional to size or for ratio or regression estimation.

4. The following ***symbols*** are used with additional subscripts, as required:

	Universe parameter	*Sample estimator*
Total of y	Y	y_0^* (or y)
Mean of y (per first-stage unit)	$\overline{Y}$	y_0^*/N
Proportion	P	p
Ratio	R	r
Sampling variance of estimator t	σ_t^2	s_t^2
Standard error of t	σ_t	s_t
Covariance of t and u	σ_{tu}	s_{tu}
Correlation coefficient	ρ	$\hat{\rho}$
Intraclass correlation coefficient	ρ_c	$\hat{\rho}_c$
Regression coefficient	β	$\hat{\beta}$

5. *Other notations and symbols*

ea	enumeration unit
f	sampling fraction (with subscripts added for strata and stages)
fsu	first-stage sample unit
pps sampling	probability proportional to size sampling
ppswr	pps sampling with replacement
srs	simple random sample or simple random sampling
srswor	srs without replacement
srswr	srs with replacement
ssu	second-stage unit
tsu	third-stage unit
$=$	is equal to
$\neq$	is not equal to
$\approx$	is approximately equal to
$\leq$	is less than or equal to
$\geq$	is greater than or equal to
$\propto$	is proportional to
∞	infinity
SSy_i	sum of squares of deviations of y_i from the mean $\overline{y}$ (see section 1.15)
SPy_ix_i	sum of products of deviations of y_i and x_i from their respective means (see section 1.15).

6. *Greek letters used*

α (alpha)	probability point of the t or the normal distribution
β (beta)	regression coefficient
π (pi, small)	probability of selection
$\prod$ (pi, capital)	product notation
ρ (rho)	correlation coefficient
σ (sigma, small)	universe standard deviation
Σ (sigma, capital)	summation notation

APPENDIX II

Elements of Probability and Proofs of Some Theorems in Sampling

A2.1 Introduction

In this appendix we present first the elements of probability and then the theorems themselves required for proving some theorems in sampling.

A2.2 Elements of probability

A2.2.1 Definition of probability

We consider the classical definition of probability.

A number of events are considered equally likely if no one of them can be expected to occur in preference to the others. Suppose there are t total possible outcomes of an experiment conducted under a given set of conditions, and that they are exhaustive (i.e. one of them must necessarily occur), mutually exclusive (i.e. no two of them can occur simultaneously), and equally likely. Let f of these t outcomes represent an event A (i.e. f are favorable to the occurrence of A), so that A happens when one of the f favorable outcomes happens and conversely.

Then the (mathematical) probability of A is

$$P(A) = f/t \qquad \text{(A2.1)}$$

As $f \leq t$,

$$0 \leq P(A) \leq 1 \qquad \text{(A2.2)}$$

$P(A) = 0$ means that the event A is impossible, and $P(A) = 1$ means that the event A is certain to occur.

Example

A six-faced die is thrown. What is the probability that (a) the number 4 turns up? (b) an even number turns up?

The six possible cases are 1, 2, 3, 4, 5, 6, and these are exhaustive and mutually exclusive; they may also be considered equally likely if the die is perfect in shape and homogeneous, and is thrown such that no one face gets any particular preference over others. That is, $t = 6$.

For (a), there is only one favorable case, namely, the figure 4, so that the probability of 4 turning up is $f/t = \frac{1}{6}$; and this holds for each of the other figures.

For (b), the number of favorable cases is $f = 3$ (namely, the even numbers 2, 4, and 6), and therefore the probability that an even number will turn up is $f/t = \frac{3}{6} = \frac{1}{2}$.

A2.2.2 Theorems of total probability

Let $A_1, A_2, \ldots, A_n$ denote n events which may happen as a result of an experiment performed under a given set of conditions. Then $A_1 + A_2 + \cdots + A_n = \sum^n A_i$ will denote the occurrence of at least one of these n events. And $A_1 A_2 \ldots A_n = \prod^n A_i$ will denote the simultaneous occurrence of all these n events.

(a) *Mutually exclusive events.* If $A_1, A_2, \ldots, A_n$ are n mutually exclusive events, then the probability that at least one of these n events will occur is

$$P(A_1 + \cdots + A_n) = P(A_1) + \cdots + P(A_n) = \sum^n P(A_i) \qquad \text{(A2.3)}$$

If in addition the events are exhaustive, i.e. at least one of the events must occur, then

$$P(A_1 + \cdots + A_n) = \sum^n P(A_i) = 1 \qquad \text{(A2.4)}$$

If the non-occurrence of the event A is denoted by $\bar{A}$, then A and $\bar{A}$ being exhaustive and mutually exclusive, i.e., $P(A) + P(\bar{A}) = 1$, so that

$$P(\bar{A}) = 1 - P(A) \qquad \text{(A2.5)}$$

For example, the probability of the figure 4 not turning up in a throw of a perfect die is $P(\bar{A}) = 1 - P(A) = \frac{5}{6}$, where A denotes the turning up of the figure 4.

(b) *Events not mutually exclusive.* If the n events $A_1, A_2, \ldots, A_n$ are not necessarily mutually exclusive, then

$$\begin{aligned} P(A_1 + A_2 + \cdots + A_n) \\ = \sum_{i}^{n} P(A_i) - \sum_{\substack{i,j \\ i<j}} P(A_iA_j) + \sum_{\substack{i,j,k \\ i<j<k}} P(A_iA_jA_k) - \cdots \\ - (-1)^n P(A_1A_2 \ldots A_n) \end{aligned} \quad \text{(A2.6)}$$

A2.2.3 Theorems of compound probability

The conditional probability that A_2 will occur when it is known that A_1 has occurred is denoted by $P(A_2|A_1)$.

If $A_1, A_2, \ldots, A_n$ are n events that may result from an experiment, then the probability of the simultaneous occurrence of all these n events is

$$\begin{aligned} P(A_1A_2 \ldots A_n) = & P(A_1)P(A_2|A_1)P(A_3|A_1A_2) \ldots \\ & \ldots P(A_n \mid A_1A_2 \ldots A_{n-1}) \end{aligned} \quad \text{(A2.7)}$$

provided that each of the factors on the right-hand side is non-zero.

When each one of n events $A_1, A_2, \ldots, A_n$ is independent of the others, the events are said to be ***mutually independent***, and in this case

$$P(A_1A_2 \ldots A_n) = P(A_1)P(A_2) \ldots P(A_n) = \prod^{n} P(A_i) \quad \text{(A2.8)}$$

i.e. the probability of the simultaneous occurrence of the n events is the product of their unconditional probabilities.

For example, when two perfect dice are tossed, the probability of 4 being thrown up by both is, since the throws are independent, $\frac{1}{6} \times \frac{1}{6} = \frac{1}{36}$.

A2.2.4 Mathematical expectation

(a) *Definition.* A variable that may take any of the values of a specified set with a specific relative probability for the different units is called a ***random variable*** or a ***stochastic variable***. For example, with a perfect die the number occurring in its tossing is a random variable which takes the values $1, 2, \ldots, 6$ each with the same associated probability of $\frac{1}{6}$.

Let a random variable Y_i take the values $Y_1, Y_2, \ldots, Y_N$ with respective probabilities $p_1, p_2, \ldots, p_N$, the set $Y_1, Y_2, \ldots, Y_N$ containing all the possible values of Y_i. These values need not all be different, but they are assumed to arise from an exhaustive set of mutually exclusive events $\left(\sum^N p_i = 1\right)$.

The mathematical expectation of Y_i is defined as

$$E(Y_i) = Y_1p_1 + Y_2p_2 + \cdots + Y_Np_N = \sum^{N} Y_ip_i \quad \text{(A2.9)}$$

where the summation is taken over all these possible values of Y_i.

For example, the mathematical expectation of the number being thrown up by the tossing of a perfect die is

$$1 \times \frac{1}{6} + 2 \times \frac{1}{6} + \cdots + 6 \times \frac{1}{6} = 3.5$$

(b) *Some properties of expected values*

(i) The expected value of a constant is the constant itself. For, if $Y_1 = Y_2 = \cdots = Y_N = a$, and the probability is p_i, then the expected value of $Y_i = a$ is, from equation (A2.9),

$$E(Y_i) = \sum^N ap_i = a \sum^N p_i = a \tag{A2.10}$$

since $\sum^N p_i = 1$, as the events are exhaustive.

(ii) $E(aY_i) = a \cdot E(Y_i)$ (A2.11)

(iii) $E(a + Y_i) = a + E(Y_i)$ (A2.12)

(iv) The mathematical expectation of the sum of a number of random variables is the sum of their expectations.

$$E(X_i + Y_i + Z_i + \cdots) = E(X_i) + E(Y_i) + E(Z_i) + \cdots \tag{A2.13}$$

(v) A number of random variables is said to be ***independent*** if the probability of any one of them to take a given value does not depend on the values taken by the other variables. Thus two variables X_i and Y_j which take respective values $X_1, X_2, \ldots, X_N$ and $Y_1, Y_2, \ldots, Y_N$ will be called independent if

$$P(X_i = X_i;\ Y_j = Y_j) = P(X_i = X_i)P(Y_j = Y_j)$$

for all i and j.

(vi) The mathematical expectation of the product of a number of independent random variables is the product of their expectations.

If $X_i, Y_i, Z_i, \ldots$ are the independent random variables, then

$$E(X_iY_iZ_i \ldots) = E(X_i)E(Y_i)E(Z_i) \ldots \tag{A2.14}$$

A2.2.5 Variance and covariance

(a) *Definition of variance.* The variance of a random variable Y_i is defined as

$$V(Y_i) = E(Y_i - EY_i)^2 \quad \text{(A2.15)}$$
$$= \sum^{N}(Y_i - EY_i)^2 p_i$$
$$= E(Y_i^2) - (EY_i)^2 \quad \text{(A2.16)}$$

from the definition of expected values, given in equation (A2.9), EY_i denoting for brevity $E(Y_i)$; $Y_1, Y_2, \ldots, Y_N$ are all the possible values of Y_i and $p_1, p_2, \ldots, p_N$ are the probabilities associated with these possible values.

(b) *Some properties of the variance.* The variance of a constant is zero, i.e.

$$V(a) = 0 \quad \text{(A2.17)}$$

Also

$$V(aY_i) = a^2 V(Y_i) \quad \text{(A2.18)}$$

and

$$V(a + Y_i) = V(Y_i) \quad \text{(A2.19)}$$

(c) *Covariance.* The covariance between two random variables X_i and Y_i is defined as

$$\text{Cov}(X_i, Y_i) = E[(X_i - EX_i)(Y_j - EY_j)] \quad \text{(A2.20)}$$
$$= 0 \quad \text{if } X_i \text{ and } Y_j \text{ are independent} \quad \text{(A2.21)}$$

(d) *Variance of a linear combination of random variables.* By definition,

$$V(X_i + Y_j) = E[(X_i + Y_j) - E(X_i + Y_j)]^2$$
$$= E[(X_i - EX_i) + (Y_j - EY_j)]^2$$
$$= E(X_i - EX_i)^2 + E(Y_j - EY_j)^2$$
$$+2E[(X_i - EX_i)(Y_j - EY_j)]$$
$$= V(X_i) + V(Y_j) + 2\,\text{Cov}(X_i, Y_j) \quad \text{(A2.22)}$$

This result can be generalized to the case of several random variables. Thus for three variables X_i, Y_i, Z_i,

$$V(X_i + Y_i + Z_i) = V(X_i) + V(Z_i) + V(Z_i) + 2\,\text{Cov}(X_i, Y_i)$$
$$+2\,\text{Cov}(X_i, Z_i) + 2\,\text{Cov}(Y_i, Z_i) \quad \text{(A2.23)}$$

When the variables are independent, the covariance terms are zero and

$$V(X_i + Y_i + Z_i + \cdots) = V(X_i) + V(Y_i) + V(Z_i) + \cdots \qquad \text{(A2.24)}$$

Corollary

$$\begin{aligned} V(X_i - Y_j) &= V(X_i) + V(Y_j) - 2\,\mathrm{Cov}(X_i, Y_j) & \text{(A2.25)} \\ &= V(X_i) + V(Y_j) \\ &\qquad \text{when } X_i \text{ and } Y_j \text{ are independent} & \text{(A2.26)} \end{aligned}$$

A2.3 Proofs of some theorems in sampling

A2.3.1 Number of possible simple random samples

(a) There are NC_n or

$$\binom{N}{n} = \frac{N!}{n!(N-n)!}$$

ways of selecting n units of a total of N units without replacement and disregarding the order of selection of the n units, where $n! = 1 \times 2 \times \cdots \times n$ is the product of the first n natural numbers and is known as factorial n. By convention $0! = 1$.

Thus if $N = 6$, $n = 2$, then in sampling without replacement and disregarding the order, the number of possible samples is (see section 2.4.1)

$$\binom{6}{2} = \frac{6!}{4!2!} = 15$$

(b) There are N^n ways of selecting n units out of a total of N units with replacement and considering the order of selection.

Thus if $N = 6$, $n = 2$, then the number of such possible samples is $6^2 = 36$ (see section 2.5.1).

A2.3.2 Expectation and sampling variance of the sample mean

A simple random sample of size n is drawn from the universe of size N. Let Y_i $(i = 1, 2, \ldots, N)$ be the value of the variable for the ith universe unit. The universe mean is by definition

$$\overline{Y} = \frac{1}{N}(Y_1 + Y_2 + \cdots + Y_N) = \frac{1}{N}\sum_{i=1}^{N} Y_i \qquad \text{(A2.27)}$$

and the universe variance per unit

$$\sigma^2 = \frac{1}{N}\sum^{N}(Y_i - \overline{Y})^2 \qquad \text{(A2.28)}$$

Let us denote by y_i $(i = 1, 2, \ldots, n)$ the value of the variable for the ith sample unit (i.e. the unit selected at the ith draw). The sample mean is, by definition,

$$\overline{y} = \frac{1}{n}(y_1 + y_2 + \cdots + y_n) = \frac{1}{n}\sum^{n} y_i \tag{A2.29}$$

We shall obtain the expectation and variance of the sample mean separately for sampling with and without replacement.

(a) *Sampling with replacement.* To prove:

$$E(\overline{y}) = \overline{Y} \tag{A2.30}$$

$$\text{Var}(\overline{y}) = \sigma^2_{\overline{y}} = \frac{\sigma^2}{n} \tag{A2.31}$$

Proof.

$$E(\overline{y}) = E\left(\frac{1}{n}\sum^{n} y_i\right) = \frac{1}{n}E\left(\sum^{n} y_i\right) = \frac{1}{n}\sum^{n} E(y_i)$$

from section A2.2.4.

Now in sampling with replacement, as the successive draws are independent and the universe remains the same throughout the sampling process, y_i $(i = 1, 2, \ldots, n)$ can assume any one of the values $Y_1, Y_2, \ldots, Y_N$ with the same probability $1/N$, so that

$$E(y_i) = \sum^{N} Y_i \cdot P(y_i = Y_i) = \sum^{N} Y_i \frac{1}{N} = \overline{Y}$$

Hence

$$E(\overline{y}) = \frac{1}{n} n\overline{Y} = \overline{Y}$$

For the second equation, by definition the variance of $\overline{y}$ is

$$\begin{aligned}
\text{Var}(\overline{y}) &= \sigma^2_{\overline{y}} = E(\overline{y} - E\,\overline{y})^2 \\
&= E\left[\frac{1}{n}\sum^{n}(y_i - Ey_i)\right]^2 \\
&= \frac{1}{n^2}E\left[\sum^{n}(y_i - Ey_i)\right]^2 \\
&= \frac{1}{n^2}\sum_{i=1}^{n} E(y_i - Ey_i)^2 + \frac{1}{n^2}\sum_{i\neq j}^{n} E(y_i - Ey_i)E(y_j - Ey_j) \\
&= \frac{1}{n^2}\sum^{2}\text{Var}(y_i) + \frac{1}{n^2}\sum_{i\neq j}^{n}\text{Cov}(y_i, y_j)
\end{aligned} \tag{A2.32}$$

Now

$$\begin{aligned}\text{Var}(y_i) &= E(y_i - \overline{Y})^2 \\ &= \sum^N (Y_i - \overline{Y})^2 P(y_i = Y_i) \\ &= \frac{1}{N}\sum^N (Y_i - \overline{Y})^2 = \sigma^2 \end{aligned} \quad \text{(A2.33)}$$

$$\begin{aligned}\text{Cov}(y_i, y_j) &= E(y_i - \overline{Y})(y_j - \overline{Y}) \\ &= \sum_{i,j}^N (Y_i - \overline{Y})(Y_j - \overline{Y}) P(y_i = Y_i; y_j = Y_j) \end{aligned} \quad \text{(A2.34)}$$

In sampling with replacement y_i and y_j are independent, i.e. y_i can take any of the values $Y_1, Y_2, \ldots, Y_N$ irrespective of the value y_j takes, so that

$$P(y_i = Y_i; y_j = Y_j) = P(y_i = Y_i)P(y_j = Y_j) = \frac{1}{N} \cdot \frac{1}{N} = \frac{1}{N^2}$$

Thus

$$\begin{aligned}\text{Cov}(y_i, y_j) &= \frac{1}{N^2}\sum_{i,j}^N (Y_i - \overline{Y})(Y_j - \overline{Y}) \\ &= \frac{1}{N^2}\sum_i^N (Y_i - \overline{Y}) \sum_j^N (Y_j - \overline{Y}) = 0 \end{aligned} \quad \text{(A2.35)}$$

for each i, j $(i \neq j)$, since $\sum^N (Y_i - \overline{Y}) = \sum^N (Y_j - \overline{Y}) = 0$, being the sum of the deviations from the mean.

From equation (A2.32), using equations (A2.33) and (A2.35), we obtain

$$\sigma_{\overline{y}}^2 = \frac{1}{n^2} n\sigma^2 = \frac{\sigma^2}{n}$$

Note: A shorter proof would be as follows.

$$\text{Var}(\overline{y}) = \text{Var}\left[\frac{1}{n}\sum^n y_i\right] = \frac{1}{n^2}\text{Var}\left(\sum^n y_i\right) = \frac{1}{n^2}\sum^n \text{Var}(y_i)$$

since the y_i's are independent (from equation A2.24). Using equation (A2.33), the final result is obtained.

(b) *Sampling without replacement.* To prove:

$$E(\bar{y}) = \bar{Y} \tag{A2.36}$$

$$\begin{aligned} \text{Var}(\bar{y}) &= \sigma_{\bar{y}}^2 = \frac{\sigma^2}{n}\,\frac{N-n}{N-1} \\ &= \frac{S^2}{n}\,(1-f) \end{aligned} \tag{A2.37}$$

where

$$\begin{aligned} S^2 &= \frac{1}{N-1}\sum^N (Y_i - \bar{Y})^2 \\ &= \frac{N}{N-1}\,\sigma^2 \end{aligned} \tag{A2.38}$$

and $f = n/N$ is the sampling fraction.

Proof. As before

$$E(\bar{y}) = \frac{1}{n}\sum^n E(y_i)$$

In sampling without replacement also, y_i can take any one of the values $Y_1, Y_2, \ldots, Y_N$ with the same probability $1/N$ (see note 1 below). Therefore $E(\bar{y}) = \bar{Y}$.

To prove (A2.37), we proceed up to (A2.34) as for srs with replacement. In srs without replacement, however,

$$\begin{aligned} & P(y_i = Y_i;\; y_j = Y_j) \\ &\quad = P(y_i = Y_i)P(y_j = Y_j \mid y_i = Y_i) \\ &\quad = \frac{1}{N}\cdot\frac{1}{N-1} \qquad \text{if } i \neq j \end{aligned}$$

since y_j can take any value excepting Y_i (the value which is already known to have been assumed by y_i) with probability $1/(N-1)$. So

$$\begin{aligned} \text{Cov}(y_i, y_j) &= \frac{1}{N(N-1)}\sum_{i \neq j}^{N}(Y_i - \bar{Y})(Y_j - \bar{Y}) \\ &= \frac{1}{N(N-1)}\sum_{i}^{n}(Y_i - \bar{Y})\left[\sum_{j}^{N}(Y_j - \bar{Y}) - (Y_i - \bar{Y})\right] \\ &= \frac{1}{N(N-1)}\left[\sum_{i}^{N}(Y_i - \bar{Y})\sum_{j}^{N}(Y_j - \bar{Y}) - \sum_{i}^{N}(Y_i - Y)^2\right] \end{aligned}$$

$$= -\frac{1}{N(N-1)}\sum^{N}(Y_i - \overline{Y})^2 \quad \text{as} \sum^{N}(Y_j - \overline{Y}) = 0$$

$$= -\frac{\sigma^2}{N-1} \quad \text{by definition} \qquad \text{(A2.39)}$$

As there are $n(n-1)$ covariance terms in (y_i, y_j) for $i \neq j$,

$$\sum_{i \neq j}^{n} \text{Cov}(y_i, y_j) = -\frac{n(n-1)\sigma^2}{N-1}$$

From equations (A2.32), (A2.33), and (A2.39),

$$\sigma_{\bar{y}}^2 = \frac{n\sigma^2}{n^2} - \frac{n(n-1)}{n^2(N-1)}\sigma^2 = \frac{\sigma^2}{n}\left(1 - \frac{n-1}{N-1}\right)$$

$$= \frac{\sigma^2}{n}\frac{N-n}{N-1} = \frac{S^2}{n}(1-f)$$

Notes

1. In srs with replacement, the probability that the ith universe unit is selected on any draw is clearly $1/N$, since the units are returned after selection so that the universe remains the same throughout the sampling process.

   In srs without replacement, the probability that a specific universe unit is selected on the first draw is $1/N$. The probability that it is selected on the second draw is given by the probability that it is not selected in the first draw multiplied by the conditional probability that it is selected on the second draw, i.e.

   $$\frac{N-1}{N} \cdot \frac{1}{N-1} = \frac{1}{N}$$

   Similarly, the probability that it is selected on the third draw is

   $$\frac{N-1}{N} \cdot \frac{N-2}{N-1} \cdot \frac{1}{N-2} = \frac{1}{N}$$

   and so on. Hence the result.

2. A specific universe unit may be included in the sample of n units at any of the n draws; the events that the unit is included in the first draw, second draw, ..., nth draw are also mutually exclusive. Therefore the probability that the specific unit is included in the sample is, from section A2.2.2(a), the sum of the probabilities of these n mutually exclusive events, i.e. n/N, since each of the n mutually exclusive events has the same probability $1/N$.

3. In section 2.3, a simple random sample of n units out of the universe of N units has been defined as the process in which each combination of n units (numbering ${}^{N}C_n$ in sampling without replacement and N^n in sampling with replacement) has the same chance of selection as every other combination: a characteristic of simple random sampling is that the probability

that a universe unit will be selected at any given draw is the same as that on the first draw, namely $1/N$ (see note 1). The above proofs have been derived from this characteristic and the theorems on probability. Alternative proofs which are derived from the definition of simple random sampling, if required, will be found in the advanced theoretical textbooks.

Corollaries

$$E(N\,\overline{y}) = \sum^{N} Y_i \tag{A2.40}$$

in srs with and without replacement.

$$\text{Var}(N\,\overline{y}) = N^2\,\frac{\sigma^2}{n} \tag{A2.41}$$

with srs with replacement.

$$\text{Var}(N\,\overline{y}) = N^2\,\frac{\sigma^2}{n}\,\frac{N-n}{N-1} \tag{A2.42}$$

in srs without replacement.

Proof. Equation (A2.40) follows from equations (A2.30) and (A2.36) on multiplying both sides by N.

Equations (A2.41) and (A2.42) follow from (A2.31) and (A2.37) on multiplying both sides by N^2.

A2.3.3 Unbiased estimator of universe variance

Defining the per unit variance of a sample as

$$s^2 = \frac{1}{n-1}\sum^{n}(y_i - \overline{y})^2 \tag{A2.43}$$

to prove that

$$E(s^2) = \sigma^2 \tag{A2.44}$$

in srs with replacement, and

$$E(s^2) = \frac{N}{N-1}\,\sigma^2 = S^2 \tag{A2.45}$$

in srs without replacement.

Proof. The sum of squares $\sum^n (y_i - \overline{y})^2$ may be written as

$$\begin{aligned}\sum^n (y_i - \overline{y})^2 &= \sum^n [(y_i - \overline{Y}) - (\overline{y} - \overline{Y})]^2 \\ &= \sum^n (y_i - \overline{Y})^2 + n(\overline{y} - \overline{Y})^2 - 2(\overline{y} - \overline{Y})\sum^n (y_i - \overline{Y}) \\ &= \sum^n (y_i - \overline{Y})^2 + n(\overline{y} - \overline{Y})^2 - 2n(\overline{y} - \overline{Y})^2 \\ &= \sum^n (y_i - \overline{Y})^2 - n(\overline{y} - \overline{Y})^2\end{aligned}$$

as $(\overline{y} - \overline{Y})$ is a constant, and $\sum^n (y_i - \overline{Y}) = \sum^n y_i - n\overline{Y} = n\,\overline{y} - n\overline{Y} = n(\overline{y} - \overline{Y})$. Now

$$\begin{aligned}E\left[\sum^n (y_i - \overline{Y})^2\right] &= \sum^n E(y_i - \overline{Y})^2 \\ &= \sum^n \text{Var}(y_i) \qquad \text{by definition} \\ &= n\,\sigma^2 \qquad\qquad \text{(A2.46)}\end{aligned}$$

from equation (A2.33).

In ***sampling with replacement***, from equation (A2.31)

$$\begin{aligned}E[n(\overline{y} - \overline{Y})^2] &= nE(\overline{y} - \overline{Y})^2 = n\,\text{Var}(\overline{y}) \\ &= n\sigma^2/n = \sigma^2 \qquad\qquad \text{(A2.47)}\end{aligned}$$

Therefore, in sampling with replacement, taking the expectation of both sides of equation (A2.43) and using equations (A2.46) and (A2.47), we obtain

$$E(s^2) = \frac{1}{n-1}\,(n\sigma^2 - \sigma^2) = \sigma^2$$

In ***sampling without replacement***, from equation (A2.37),

$$\begin{aligned}E[n(\overline{y} - \overline{Y})^2] &= nE(\overline{y} - \overline{Y})^2 = n\,\text{Var}(\overline{y}) \\ &= n\,\frac{\sigma^2}{n}\,\frac{N-n}{N-1} = \sigma^2\,\frac{N-n}{N-1} \qquad\qquad \text{(A2.48)}\end{aligned}$$

Therefore, in sampling without replacement, taking the expectation of both sides of equation (A2.43) and using equations (A2.46) and (A4.48), we obtain

$$\begin{aligned}E(s^2) &= \frac{1}{n-1}\left[n\sigma^2 - \sigma^2\,\frac{N-n}{N-1}\right] = \frac{\sigma^2}{n-1}\,\frac{N(n-1)}{N-1} \\ &= \frac{N\sigma^2}{N-1} = S^2\end{aligned}$$

by definition of S^2 (equation (A2.38)).

A2.3.4 Unbiased estimator of $\sigma^2_{\bar{y}}$

$$E\left(\frac{s^2}{n}\right) = \sigma^2_{\bar{y}} \tag{A2.49}$$

in srs with replacement, and

$$E\left[\frac{s^2}{n}(1-f)\right] = \sigma^2_{\bar{y}} \tag{A2.50}$$

in srs without replacement where s^2 is defined in equation (A2.43).

Proof. For srs with replacement, we have seen that

$$\sigma^2_{\bar{y}} = \sigma^2/n$$

and

$$E(s^2) = \sigma^2$$

Therefore

$$E\left(\frac{s^2}{n}\right) = \frac{1}{n}\,E(s^2) = \sigma^2/n = \sigma^2_{\bar{y}}$$

For srs without replacement, we have seen that

$$\sigma^2_{\bar{y}} = \frac{S^2}{n}(1-f)$$

and $E(s^2) = S^2$. Therefore,

$$E\left[\frac{s^2}{n}(1-f)\right] = \frac{1-f}{n}\,E(s^2) = \frac{(1-f)S^2}{n} = \sigma^2_{\bar{y}}$$

Corollaries. In srs with replacement

$$E\left(N^2\,\frac{s^2}{n}\right) = \mathrm{Var}(N\,\bar{y}) \tag{A2.51}$$

In srs without replacement

$$E\left[N^2\,\frac{s^2}{n}(1-f)\right] = \mathrm{Var}(N\,\bar{y}) \tag{A2.52}$$

The results follow immediately from equations (A2.49) and (A2.50) multiplying both sides by N^2.

A2.3.5 Estimation of universe proportion P

The universe proportion P has been defined in equation (2.66) as

$$P = N'/N \tag{A2.53}$$

where N' is the number of universe units possessing a certain attribute, and N is the total number of units in the universe. The results proved above are applicable here also by defining for the universe, $Y_i = 1$ if the ith unit $(i = 1, 2, \ldots, N)$ has the attribute, and $Y_i = 0$, otherwise; and similarly for the sample. For the universe $\sum^N Y_i = N'$; $\overline{Y} = Y/N = N'/N = P$.

1. *To prove that the sample proportion p is an unbiased estimator of P.*

   The sample proportion p has been defined as

   $$p = n'/n = \sum^n y_i/n = \overline{y} \tag{A2.54}$$

   where n' is the number of sample units possessing the attribute, and n the total number of sample units; $y_i = 1$, if the ith sample units has the attribute and $y_i = 0$, otherwise.

   Taking the expectation of both sides of equation (A2.54), we obtain

   $$E(p) = E(\overline{y}) = \overline{Y} = P$$

   from equation (2.67).

2. *Sampling variance of p.* We have seen in equation (2.68) that the universe variance of Y_i is

   $$\sigma^2 = P(1 - P) \tag{A2.55}$$

   An unbiased estimator of σ^2 has been defined in equation (2.73), namely

   $$\begin{aligned} s^2 &= \frac{1}{n-1}\sum^n (y_i - \overline{y})^2 \\ &= \frac{1}{n-1}\left(\sum^n y_i^2 - n\overline{y}^2\right) = \frac{np(1-p)}{n-1} \end{aligned} \tag{A2.56}$$

   since $y_i = 1$ or 0; $\sum^n y_i^2 = n' = np$; and $\overline{y}^2 = p^2$.

   The sampling variance of p is

   $$\sigma_p^2 = \sigma_{\overline{y}}^2 = \sigma^2/n = \frac{P(1-P)}{n} \quad \text{in srswr} \tag{A2.57}$$

   $$= \frac{P(1-P)}{n}\,\frac{N-n}{N-1} \quad \text{in srswor} \tag{A2.58}$$

3. *Unbiased estimator of* σ_p^2. An unbiased estimator of σ_p^2 is

$$s_p^2 = \frac{p(1-p)}{n-1} \quad \text{in srs with replacement} \tag{A2.59}$$

$$= (1-f)\,\frac{p(1-p)}{n-1} \quad \text{in srs without replacement} \tag{A2.60}$$

Proof. Since

$$s^2 = \frac{np(1-p)}{n-1}$$

$$E(s_p^2) = E\left(\frac{s^2}{n}\right) = \frac{\sigma^2}{n} = \frac{P(1-P)}{n} = \sigma_p^2 \text{ in srswr}$$

and

$$E(s_p^2) = E\left[(1-f)\,\frac{s^2}{n}\right] = \frac{\sigma^2}{n}\,\frac{N-n}{N-1}$$

$$= \frac{P(1-P)}{n}\,\frac{N-n}{N-1} = \sigma_p^2 \text{ in srswor}$$

A2.3.6 Variance of the ratio of random variables

(a) If y and x are two random variables with expectations Y and X, variances σ_y^2 and σ_x^2, and covariance $\sigma_{yx} = \rho\sigma_y\sigma_x$ (where ρ is the correlation coefficient), the sampling variance of the ratio of two random variables $r = y/x$ is given approximately by

$$\sigma_r^2 = \frac{1}{X^2}\,(\sigma_y^2 + R^2\sigma_x^2 - 2R\sigma_{yx}) \tag{A2.61}$$

This is proved below for a simple random sample.

(b) In simple random sampling, if variables y_i and x_i are measured on each unit of a simple random sample of size n out of N universe units, the estimator $r = \overline{y}/\overline{x} = y/x$ is biased for the ratio $R = Y/X$ (where Y and X are the respective universe totals of the variables and $y = N\,\overline{y}$ and $x = N\,\overline{x}$ are the corresponding unbiased sample estimators). We shall derive the expected value, the bias and the sampling variance and the variance estimator of the estimator r.

(i) *Expected value of* r. Let $e = (y-Y)/Y$ and $e' = (x-X)/X$. Then, since $E(y) = Y$, and $E(x) = X$. $E(e) = 0 = E(e')$; $\text{Var}(e) = \text{Var}(y)/Y^2$,

$\text{Var}(e') = \text{Var}(x)/X^2$, and $\text{Cov}(e, e') = \text{Cov}(y, x)/YX$. As $y = Y(1 + e)$, and $x = X(1 + e')$, we have

$$\begin{aligned} r &= y/x = Y(1+e)/X(1+e') \\ &= R(1+e)(1+e')^{-1} \end{aligned} \tag{A2.62}$$

If we assume that $|e'| < 1$, i.e. x lies between 0 and $2X$, which is likely to happen when the sample size is large, the term $(1 + e')^{-1}$ in expression (A2.62) may be expanded by Taylor's theorem. With this assumption and taking the expectation of both sides of (A2.62), we get

$$E(r) = R\,E(1 + e - e' + e'^2 - ee' + \cdots) \tag{A2.63}$$

If we further assume that terms involving second and higher powers of e and e' are negligible, then we have, to this order of approximation, $E(r) = R$ since $E(e) = 0 = E(e')$.

(ii) *Bias of r*. If we assume that terms involving third and higher powers of e and e' are negligible, we have

$$\begin{aligned} E(r) &= R[1 + 0 - 0 + \text{Var}(e') - \text{Cov}(e, e')] \\ &= R\left[1 + \frac{\text{Var}(x)}{X^2} - \frac{\text{Cov}(y, x)}{YX}\right] \end{aligned} \tag{A2.64}$$

for $\text{Var}(e') = E(e'^2)$, and $\text{Cov}(e, e') = E(e, e')$, as $E(e) = 0 = E(e')$. Therefore, the bias of r is approximately

$$\begin{aligned} E(r) - R &= R\left[\frac{\text{Var}(x)}{X^2} - \frac{\text{Cov}(y, x)}{YX}\right] \\ &= R(CV_x^2 - \rho CV_y CV_x) \\ &= \frac{1}{X^2}\,[R\text{Var}(x) - \text{Cov}(y, x)] \end{aligned} \tag{A2.65}$$

where ρ is the correlation coefficient between the two variables, and CV_y and CV_x are the coefficients of variation of y and x respectively.

To the order of approximation assumed, the bias will be zero, if

$$R = \text{Cov}(y, x)/\text{Var}(x)$$

and this will occur when the regression line of y on x is a straight line passing through the origin (0,0).

The bias of the estimator r, given in expression (A2.65), can be estimated by replacing X, R, $\text{Var}(x)$ and $\text{Cov}(y, x)$ by their respective sample estimators. The estimator of the bias is itself subject to bias. If the distributions are not skewed, the coefficients of variation will not be very large, and $|\rho| \leq 1$, in which case the bias, given by the above expression, will be small. An upper bound of the magnitude of the relative bias $\left|\frac{E(r)-R}{R}\right|$ is given by $CV_y CV_x$.

(iii) *Sampling variance of r.* As the estimator r is biased, its mean square error is, by definition,

$$MSE(r) = E(r-R)^2 = R^2 E[(e-e')^2(1+e')^{-2}]$$

using equation (A2.62) and simplifying. If we assume as before that $|e'| < 1$ and that terms involving second and higher powers of e and e' are negligible, then r is considered to be unbiased to this order of approximation, and the variance and the mean square error become the same. In this case,

$$\begin{aligned} MSE(r) &= \text{Var}(r) = R^2[E(e-e')^2] \\ &= R^2\left[\frac{\text{Var}(y)}{Y^2} + \frac{\text{Var}(x)}{X^2} - \frac{2\,\text{Cov}(y,x)}{YX}\right] \\ &= R^2(CV_y^2 + CV_x^2 - 2\rho\, CV_y CV_x) \\ &= \frac{1}{X^2}\,[\text{Var}(y) + R^2\,\text{Var}(x) - 2R\,\text{Cov}(y,x)] \qquad \text{(A2.66)} \end{aligned}$$

The variance estimator of r is obtained on replacing X, R, Var(y), Var(x) and Cov(y, x) by their respective sample estimators. Thus, the variance estimator is given by equation (2.49) for simple random sampling with replacement and by equation (2.53) for sampling without replacement. The variance estimators are generally biased.

Note: The bias in the ratio estimator $y_R = Xy/x$ is X times that in r.

A2.3.7 Varying probability sampling

With the notations used in section 5.3, to prove that in varying probability sampling with replacement,

$$E\left(y_0^* = \frac{1}{n}\sum^n y_i^* = \frac{1}{n}\sum^n \frac{y_i}{\pi_i}\right) = Y \qquad \text{(A2.67)}$$

$$\text{Var}(y_0^*) = \sigma_{y_0^*}^2 = \frac{1}{n}\sum^N \left(\frac{Y_i}{\pi_i} - Y\right)^2 \pi_i \qquad \text{(A2.68)}$$

$$E\left[s_{y_0^*}^2 = \sum^n (y_i^* - y_0^*)^2/n(n-1)\right] = \sigma_{y_0^*}^2 \qquad \text{(A2.69)}$$

Proof. In sampling with replacement at each draw

$$E(y_i^*) = E\left(\frac{y_i}{\pi_i}\right) = \sum^N \frac{Y_i}{\pi_i}\pi_i = Y$$

so that

$$E(y_0^*) = \frac{1}{n}\sum^n E(y_i^*) = \frac{1}{n}nY = Y$$

To prove (A2.68),

$$\mathrm{Var}(y_0^*) = \sigma^2_{y_0^*} = \frac{1}{n^2}\mathrm{Var}\left(\sum^n y_i^*\right) = \frac{1}{n^2}\sum^n \mathrm{Var}(y_i^*)$$

as the y_i^*'s are independent, since sampling is with replacement, from (A2.24). Now

$$\begin{aligned}\mathrm{Var}(y_i^*) &= \sigma^2_{y_i^*} = E(y_i^* - Ey_i^*)^2 = E(y_i^* - Y)^2\\ &= \sum^N (Y_i/\pi_i - Y)^2\pi_i = \sum^N Y_i^2/\pi_i - Y^2 \qquad \text{(A2.70)}\end{aligned}$$

since $\sum^N \pi_i = 1$. Thus,

$$\begin{aligned}\sigma^2_{y_0^*} &= \frac{1}{n^2}\sum^n \mathrm{Var}(y_i^*) = \frac{1}{n^2}n\sigma^2_{y_i^*} = \sigma^2_{y_i^*}/n\\ &= \frac{1}{n}\sum^N\left(\frac{Y_i}{\pi_i} - Y\right)^2\pi_i = \frac{1}{n}\left(\sum^N \frac{Y_i^2}{\pi_i} - Y^2\right)\end{aligned}$$

To prove (A2.69), we find the per unit variance of a varying probability sample as

$$s^2_{y_i^*} = \frac{1}{n-1}\sum^n (y_i^* - y_0^*)^2 \qquad \text{(A2.71)}$$

The sum of squares $\sum^n (y_i^* - y_0^*)^2$ may be written as (see section A2.3.3)

$$\sum^n (y_i^* - y_0^*)^2 = \sum^n (y_i^* - Y)^2 - n(y_0^* - Y)^2$$

Now,

$$E\left[\sum^n (y_i^* - Y)^2\right] = \sum^n E(y_i^* - Y)^2 = \sum^n \mathrm{Var}(y_i^*) = n\sigma^2_{y_i^*} \qquad \text{(A2.72)}$$

from equation (A2.70). Also,

$$E[n(y_0^* - Y)^2] = n\,E(y_0^* - Y)^2 = n\,\text{Var}(y_0^*) = n\sigma_{y_i^*}^2/n = \sigma_{y_i^*}^2 \quad \text{(A2.73)}$$

from equation (A2.68).

Therefore, taking the expectation of both sides of equation (A2.71) and using equations (A2.72) and (A2.73), we obtain

$$E(s_{y_i^*}^2) = \frac{1}{n-1}\,(n\sigma_{y_i^*}^2 - \sigma_{y_i^*}^2) = \sigma_{y_i^*}^2 \quad \text{(A2.74)}$$

i.e. $s_{y_i^*}^2$ is an unbiased estimator of $\sigma_{y_i^*}^2$. Therefore $s_{y_0^*}^2 = s_{y_i^*}^2/n$ is an unbiased estimator of $\sigma_{y_0^*}^2 = \sigma_{y_i^*}^2/n$.

A2.3.8 Cluster sampling

With the notations of section 6.2.1, to prove that in simple random sample of clusters

(a)

$$\sigma^2 = \sigma_w^2 + \sigma_b^2 \quad (6.5)$$

or,

Total variance (per unit) = Within-cluster variances + Between-cluster variance

(b)

$$\sigma_b^2 = \frac{\sigma^2}{M_0}\,[1 + (M_0 - 1)\rho_c] \quad (6.9)$$

Proof. (a) By definition $\sigma^2 = \frac{1}{NM_0}\sum_i^N \sum_j^{M_0}(Y_{ij} - \overline{Y})^2$ so that (6.5)

$$\begin{aligned} NM_0\sigma^2 &= \sum_i\sum_j(Y_{ij} - \overline{Y})^2 = \sum_i\sum_j(Y_{ij} - \overline{Y}_i + \overline{Y}_i - \overline{Y})^2 \\ &= \sum_i\Big[\sum_j(Y_{ij} - \overline{Y}_i)^2 + M_0(\overline{Y}_i - \overline{Y})^2 \\ &\quad + 2(\overline{Y}_i - \overline{Y})\sum_j(Y_{ij} - \overline{Y}_i)\Big] \\ &= \sum_i\sum_j(Y_{ij} - \overline{Y}_i)^2 + M_0\sum_i(\overline{Y}_i - \overline{Y})^2 \end{aligned}$$

as the third term within the square brackets is zero. Thus

$$\sigma^2 = \frac{1}{NM_0}\sum_i\sum_j(Y_{ij} - \overline{Y}_i)^2 + \frac{1}{N}\sum_j(\overline{Y}_i - \overline{Y})^2 = \sigma_w^2 + \sigma_b^2$$

by definitions of σ_w^2 (equation 6.6) and σ_b^2 (equation 6.7).

(b) By definition, $\sigma_b^2 = \sum_i(\overline{Y}_i - \overline{Y})^2/N$. Now, (6.7)

$$\begin{aligned}
(\overline{Y}_i - \overline{Y}) &= \left(\frac{1}{M_0}\sum_j Y_{ij} - \overline{Y}\right) \\
&= \frac{1}{M_0}\left(\sum_j Y_{ij} - M_0\overline{Y}\right) = \frac{1}{M_0}\left(\sum_j Y_{ij} - \sum_j \overline{Y}\right) \\
&= \frac{1}{M_0}\sum_j (Y_{ij} - \overline{Y})
\end{aligned}$$

So,

$$\begin{aligned}
\sigma_b^2 &= \frac{1}{NM_0^2}\sum_i\left[\sum_j (Y_{ij} - \overline{Y})\right] \\
&= \frac{1}{NM_0^2}\left[\sum_i\sum_j (Y_{ij} - \overline{Y})^2 + \sum_i \sum_{j \neq j'} (Y_{ij} - \overline{Y})(Y_{ij'} - \overline{Y})\right] \\
&= \frac{1}{NM_0^2}\left[NM_0\sigma^2 + NM_0(M_0 - 1)\sigma^2\rho_c\right]
\end{aligned}$$

by definition of σ^2 (equation 6.5) and ρ_c (equation 6.8). This, when simplified, gives equation (6.9).

A2.3.9 Stratified simple random sampling

With the notations used in section 10.2, to prove that

(a)

$$E\left(y = \sum^L y_{h0}^*\right) = \sum^L Y_h = Y \tag{A2.75}$$

(b)

$$\sigma_y^2 = \sum^L \sigma_{y_{h0}^*}^2 \tag{A2.76}$$

where

$$\sigma_{y_{h0}^*}^2 = N_h^2 \frac{\sigma_h^2}{n_h} \quad \text{in srswr} \tag{A2.77}$$

$$
\begin{aligned}
&= N_h^2 \frac{\sigma_h^2}{n_h} \frac{N_h - n_h}{N_h - 1} \\
&= N_h^2 \frac{S_h^2}{n_h} (1 - f_h) \qquad \text{in srswor} \qquad \text{(A2.78)}
\end{aligned}
$$

and

$$f_h = n_h/N_h$$

(c)

$$E\left(s_y^2 = \sum^L s_{y_{h0}^*}^2\right) = \sum^L \sigma_{y_{h0}^*}^2 = \sigma_y^2 \qquad \text{(A2.79)}$$

Proof. (a) From equation (A2.40), we know that y_{h0}^* is an unbiased estimator of Y_h in every stratum, so that

$$E(y) = E\left(\sum^L y_{h0}^*\right) = \sum^L E(y_{h0}^*) = \sum^L Y_h = Y$$

(b) For each stratum, from equation (A2.41) in sampling with replacement, we have

$$\sigma_{y_{h0}^*}^2 = N_h^2 \frac{\sigma_h^2}{n_h}$$

Then,

$$
\begin{aligned}
\sigma_y^2 &= \mathrm{Var}(y) = \mathrm{Var}\left(\sum^L y_{h0}^*\right) \\
&= \sum^L \mathrm{Var}(y_{h0}^*) + \sum_{h \neq h'}^L \mathrm{Cov}(y_{h0}^*, y_{h'0}^*) \\
&= \sum^L \mathrm{Var}(y_{h0}^*) = \sum^L \sigma_{y_{h0}^*}^2
\end{aligned}
$$

as the covariance terms vanish because samples are drawn independently in the different strata.

The proof for sampling without replacement follows similarly, noting that for each stratum from equation (A2.42)

$$\sigma_{y_{h0}^*}^2 = N_h^2 \frac{S_h^2}{n_h} (1 - f_h)$$

Corollary. The sampling variance of the unbiased estimator of the overall mean is

$$\mathrm{Var}(y/N) = \sigma_y^2/N^2 \qquad \text{(A2.80)}$$

σ_y^2 being defined by equations (A2.76)-(A2.78).

(c) For the hth stratum, from equation (A2.51) in srs with replacement,

$$\begin{aligned} s^2_{y^*_{ho}} &= \frac{1}{n_h(n_h-1)}\sum^{n_h}(y^*_{hi}-y^*_{h0})^2 \\ &= \frac{N_h^2}{n_h(n_h-1)}\sum^{n_h}(y_{hi}-\bar{y}_h)^2 \end{aligned} \qquad \text{(A2.81)}$$

is an unbiased estimator of $\sigma^2_{y^*_{ho}} = N_h^2\sigma_h^2/n_h$. Hence $s_y^2 = \sum^L s^2_{y^*_{ho}}$ is an unbiased estimator of $\sigma_y^2 = \sum^L \sigma^2_{y^*_{ho}}$.

Similarly in srs without replacement, from equation (A2.52), $[N_h^2 s_h^2/n_h (n_h-1)](1-f_h)$ is an unbiased estimator of $\sigma^2_{y^*_{ho}} = N_h^2\, s_h^2/n(1-f_h)$. Hence the result (A2.79) also for srs without replacement.

Corollary. An unbiased variance estimator of the variance of the overall mean σ_y^2/N^2 is

$$s_y^2/N^2 \qquad \text{(A2.82)}$$

Note: The theorems for stratified varying probability sampling can be proved on similar lines as for stratified srs.

A2.3.10 Optimum allocation in stratified single-stage sampling

To prove the result given in section 12.3.2, we use the well-known Cauchy-Schwarz inequality, which states that if a_h and b_h are two sets of positive numbers, then

$$\left(\sum a_h^2\right)\left(\sum b_h^2\right) \geq \left(\sum a_h b_h\right)^2 \qquad \text{(A2.83)}$$

there being equality only if b_h is proportional to a_h, i.e. if b_h/a_h is a constant for all h.

Now, the optimum allocation is obtained on minimizing the variance function V, given by expression (12.1), for a given total cost C, obtained using expression (12.5), or *vice versa*; this is equivalent to minimizing the product VC. Taking only those components of V and C that depend on n_h, we have to minimize

$$\left(\sum N_h^2 V_h/n_h\right)\left(\sum n_h c_h\right) \qquad \text{(A2.84)}$$

From the Cauchy-Schwarz inequality (A2.83), by putting $a_h = N_h\sqrt{(V_h/n_h)}$ and $b_h = \sqrt{(n_h c_h)}$, we see that

$$\left(\sum N_h^2 V_h/n_h\right)\left(\sum n_h c_h\right) \geq \left[\sum N_h\sqrt{(V_h c_h)}\right]^2 \tag{A2.85}$$

The minimum value of expression (A2.84), i.e. equality in (A2.85), occurs only when

$$\frac{b_h}{a_h} = \frac{n_h\sqrt{c_h}}{N_h\sqrt{V_h}} = \text{constant} = k, \text{ say,} \tag{A2.86}$$

or

$$n_h = kN_h\sqrt{(V_h/c_h)} \tag{A2.87}$$

i.e. n_h should be proportional to $N_h\sqrt{(V_h/c_h)}$.

Summing both sides of (A2.87), we get

$$n = \sum n_h = k\sum N_h\sqrt{(V_h/c_h)} \tag{A2.88}$$

which gives the value of k; substituting this in (A2.87), we get

$$n_h = \frac{nN_h\sqrt{(V_h/c_h)}}{\sum N_h\sqrt{(V_h/c_h)}} \tag{A2.89}$$

The above expression for n_h presumes the value of n, which would depend on whether the total cost or the variance is specified. If the total cost is fixed at C', then we substitute the n_h values obtained from equation (A2.89) in the cost function (12.5), and solve for n, which gives

$$n = (C' - c_0)\frac{\sum N_h\sqrt{(V_h/c_h)}}{\sum N_h\sqrt{(V_h c_h)}} \tag{A2.90}$$

Substituting this value of n in equation (A2.89), we get equation (12.6).

Similarly, if the variance is fixed at V' then we substitute the optimum n_h in the variance function (12.1), whence

$$n = \left[\sum N_h\sqrt{(V_h c_h)}\right]\left[\sum N_h\sqrt{(V_h/c_h)}\right]\Big/ V' \tag{A2.91}$$

substituting this value of n in equation (A2.89), we get equation (12.7).

Notes

1. See notes 2 and 3 to section 12.3.2.
2. The Neyman allocation follows from the above formulation as indicated in section 12.3.3. A direct proof may also be derived noting that $c_h = \bar{c}$, a constant, and $n = (C - c_0)/\bar{c}$.
3. The proofs for stratified srs without replacement (note 3 to section 12.3.2, note to section 12.6.1 and note 4 to section 12.6.2) can be derived similarly.

APPENDIX III

Statistical Tables

Table A3.1: Random numbers

03 47 43 73 86	36 96 47 36 61	46 98 63 71 62	33 26 16 80 45	60 11 14 10 9
97 74 24 67 62	42 81 14 57 20	42 53 32 37 32	27 07 36 07 51	24 51 79 89 7
16 76 62 27 66	56 50 26 71 07	32 90 79 78 53	13 55 38 58 59	88 97 54 14 1
12 56 85 99 26	96 96 68 27 31	05 03 72 93 15	57 12 10 14 21	88 26 49 81 7
55 59 56 35 64	38 54 82 46 22	31 62 43 09 90	06 18 44 32 53	23 83 01 30 3
16 22 77 94 39	49 54 43 54 82	17 37 93 23 78	87 35 20 96 43	84 26 34 91 6
84 42 17 53 31	57 24 55 06 88	77 04 74 47 67	21 76 33 50 25	83 92 12 06 7
63 01 63 78 59	16 95 55 67 19	98 10 50 71 75	12 86 73 58 07	44 39 52 38 7
33 21 12 34 29	78 64 56 07 82	52 42 07 44 38	15 51 00 13 42	99 66 02 79 5
57 60 86 32 44	09 47 27 96 54	49 17 46 09 62	90 52 84 77 27	08 02 73 43 2
18 18 07 92 46	44 17 16 58 09	79 83 86 19 62	06 76 50 03 10	55 23 64 05 0
26 62 38 97 75	84 16 07 44 99	83 11 46 32 24	20 14 85 88 45	10 93 72 88 7
23 42 40 64 74	82 97 77 77 81	07 45 32 14 08	32 98 94 07 72	93 85 79 10 7
52 36 28 19 95	50 92 26 11 97	00 56 76 31 38	80 22 02 53 53	86 60 42 04 5
37 85 94 35 12	83 39 50 08 30	42 34 07 96 88	54 42 06 87 98	35 85 29 48 3
70 29 17 12 13	40 33 20 38 26	13 89 51 03 74	17 76 37 13 04	07 74 21 19 3
56 62 18 37 35	96 83 50 87 75	97 12 25 93 47	70 33 24 03 54	97 77 46 44 8
99 49 57 22 77	88 42 95 45 72	16 64 36 16 00	04 43 18 66 79	94 77 24 21 9
16 08 15 04 72	33 27 14 34 09	45 59 34 68 49	12 72 07 34 45	99 27 72 95 1
31 16 93 32 43	50 27 89 87 19	20 15 37 00 49	52 85 66 60 44	38 68 88 11 8
68 34 30 13 70	55 74 30 77 40	44 22 78 84 26	04 33 46 09 52	68 07 97 06 5
74 57 25 65 76	59 29 97 68 60	71 91 38 67 54	13 58 18 24 76	15 54 55 95 5
27 42 37 86 53	48 55 90 65 72	96 57 69 36 10	96 46 92 42 45	97 60 49 04 9
00 39 68 29 61	66 37 32 20 30	77 84 57 03 29	10 45 65 04 26	11 04 96 67 2
29 94 98 94 24	68 49 69 10 82	53 75 91 93 30	34 25 20 57 27	40 48 73 51 9
16 90 82 66 59	83 62 64 11 12	67 19 00 71 74	60 47 21 29 68	02 02 37 03 3
11 27 94 75 06	06 09 19 74 66	02 94 37 34 02	76 70 90 30 86	38 45 94 30 3
35 24 10 16 20	33 32 51 26 38	79 78 45 04 91	16 92 53 56 16	02 75 50 95 9
38 23 16 86 38	42 38 97 01 50	87 75 66 81 41	40 01 74 91 62	48 51 84 08 3
31 96 25 91 47	96 44 33 49 13	34 86 82 53 91	00 52 43 48 85	27 55 26 89 6
56 67 40 67 14	64 05 71 95 86	11 05 65 09 68	76 83 20 37 90	57 16 00 11 6
14 90 84 45 11	75 73 88 05 90	52 27 41 14 86	22 98 12 22 08	07 52 74 95 8
68 05 51 18 00	33 96 02 75 19	07 60 62 93 55	59 33 82 43 90	49 37 38 44 5
20 46 78 73 90	97 51 40 14 02	04 02 33 31 08	39 54 16 49 36	47 95 93 13 3
64 19 58 97 79	15 06 15 93 20	01 90 10 75 06	40 78 78 89 62	02 67 74 17 3
05 26 93 70 60	22 35 85 15 13	92 03 51 59 77	59 56 78 06 83	52 91 05 70 7
07 97 10 88 23	09 98 42 99 64	61 71 62 99 15	06 51 29 16 93	58 05 77 09 5
68 71 86 85 85	54 87 66 47 54	73 32 08 11 12	44 95 92 63 16	29 56 24 29 4
26 99 61 65 53	58 37 78 80 70	42 10 50 67 42	32 17 55 85 74	94 44 67 16 9
14 65 52 68 75	87 59 36 22 41	26 78 63 06 55	13 08 27 01 50	15 29 39 39 4
17 53 77 58 71	71 41 61 50 72	12 41 94 96 26	44 95 27 36 99	02 96 74 30 8
90 26 59 21 19	23 52 23 33 12	96 93 02 18 39	07 02 18 36 07	25 99 32 70 2
41 23 52 55 99	31 04 49 69 96	10 47 48 45 88	13 41 43 89 20	97 17 14 49 1
60 20 50 81 69	31 99 73 68 68	35 81 33 03 76	24 30 12 48 60	18 99 10 72 3
91 25 38 05 90	94 58 28 41 36	45 37 59 03 09	90 35 57 29 12	82 62 54 65 6
34 50 57 74 37	98 80 33 00 91	09 77 93 19 82	74 94 80 04 04	45 07 31 66 4
85 22 04 39 43	73 81 53 94 79	33 62 46 86 28	08 31 54 46 31	53 94 13 38 4
09 79 13 77 48	73 82 97 22 21	05 03 27 24 83	72 89 44 05 60	35 80 39 94 8
88 75 80 18 14	22 95 75 42 49	39 32 82 22 49	02 48 07 70 37	16 04 61 67 8
90 96 23 70 00	39 00 03 06 90	55 85 78 38 36	94 37 30 69 32	90 89 00 76 3

Table A3.1 continued

53 74 23 99 07	61 32 28 69 84	94 62 67 86 24	98 33 41 19 95	47 53 53 38 09
63 38 06 86 54	99 00 65 26 94	02 82 90 23 07	79 62 67 80 60	75 91 12 81 19
35 30 58 21 46	06 72 17 10 94	25 21 31 75 96	49 28 24 00 49	55 65 79 78 07
53 43 36 82 69	65 51 18 37 88	61 38 44 12 45	32 92 85 88 65	54 34 81 85 35
98 25 37 55 26	01 91 82 81 46	74 71 12 94 97	24 02 71 37 07	03 92 18 66 75
92 63 21 17 69	71 50 80 89 56	38 15 70 11 48	43 40 45 86 98	00 83 26 91 03
64 55 22 21 82	48 22 28 06 00	61 54 13 43 91	82 78 12 23 29	06 66 24 12 27
35 07 26 13 89	01 10 07 82 04	59 63 69 36 03	69 11 15 83 80	13 29 54 19 28
58 54 16 24 15	51 54 44 82 00	62 61 65 04 69	38 18 65 18 97	85 72 13 49 21
84 85 27 84 87	61 48 64 56 26	90 18 48 13 26	37 70 15 42 57	65 65 80 39 07
03 92 18 27 46	57 99 16 96 56	30 33 72 85 22	84 64 38 56 98	99 01 30 98 64
62 95 30 27 59	37 75 41 66 48	86 97 80 61 45	21 53 04 01 63	45 76 08 64 27
98 45 93 15 22	60 21 75 46 91	98 77 27 85 42	28 88 61 08 84	69 62 03 42 73
07 08 55 18 40	45 44 75 13 90	24 94 96 61 02	57 55 66 83 15	73 42 37 11 61
91 85 89 95 66	51 10 19 34 88	15 84 97 19 75	12 76 39 43 78	64 63 91 08 25
72 84 71 14 35	19 11 58 49 26	50 11 17 17 76	86 31 57 20 18	95 60 78 46 75
38 78 28 16 84	13 52 53 94 53	75 45 69 30 96	73 89 65 70 31	99 17 43 48 76
45 17 75 65 57	28 40 19 72 12	25 12 74 75 67	60 40 60 81 19	24 62 01 61 16
96 76 28 12 54	22 01 11 94 25	71 96 16 16 88	68 64 36 74 45	19 59 50 88 92
43 31 67 72 30	24 02 94 08 63	38 32 36 66 02	69 36 38 25 39	48 03 45 15 22
50 44 66 44 21	66 06 58 05 62	68 15 54 35 02	42 35 48 96 32	14 52 41 52 48
22 66 22 15 86	26 63 75 41 99	58 42 36 72 24	58 37 52 18 51	03 37 18 39 11
96 24 40 14 51	23 22 30 88 57	95 67 47 29 83	94 69 40 06 07	18 16 36 78 86
31 73 91 61 19	60 20 72 93 48	98 57 07 23 69	65 95 39 69 58	56 80 30 19 44
78 60 73 99 84	43 89 94 36 45	56 69 47 07 41	90 22 91 07 12	78 35 34 08 72
34 37 90 61 56	70 10 23 98 05	85 11 34 76 60	76 48 45 34 60	01 64 18 39 96
36 67 10 08 23	98 93 35 08 86	99 29 76 29 81	33 34 91 58 93	63 14 52 32 52
97 28 59 07 48	89 64 58 89 75	83 85 62 27 89	30 14 78 56 27	86 63 50 80 02
10 15 83 87 60	79 24 31 66 56	21 48 24 06 93	91 98 94 05 49	01 47 59 38 00
55 19 68 97 65	03 73 52 16 56	00 53 55 90 27	33 42 29 38 87	22 13 88 83 34
53 81 29 13 39	35 01 20 71 34	62 33 74 82 14	53 73 19 09 03	56 54 29 56 93
61 86 32 68 92	33 98 74 66 99	40 14 71 94 58	45 94 19 38 81	14 44 99 81 07
35 91 70 29 13	80 03 54 07 27	96 94 78 32 66	50 95 52 74 33	13 80 55 62 54
37 71 67 95 13	20 02 44 95 94	64 85 04 05 72	01 32 90 76 14	53 89 74 60 41
93 66 13 83 27	92 79 64 64 72	28 54 96 53 84	48 14 52 98 94	56 07 93 89 30
02 96 08 45 65	13 05 00 41 84	93 07 54 72 59	21 45 57 09 77	19 48 56 27 44
49 83 43 48 35	82 88 33 69 96	72 36 04 19 76	47 45 15 18 60	82 11 08 95 97
34 60 71 62 46	40 80 81 30 37	34 39 23 05 38	25 15 35 71 30	88 12 57 21 77
18 17 30 88 71	44 91 14 88 47	89 23 30 63 15	56 34 20 47 89	99 82 93 24 98
79 69 10 61 78	71 32 76 95 62	87 00 22 58 40	92 54 01 75 25	43 11 71 99 31
75 93 36 57 83	56 20 14 82 11	74 21 97 90 65	96 42 68 63 86	74 54 13 26 94
38 30 92 29 03	06 28 81 39 38	62 25 06 84 63	61 29 08 93 67	04 32 92 08 00
61 29 50 10 34	31 57 75 95 80	51 97 02 74 77	76 15 48 49 44	18 55 63 77 09
21 31 38 86 24	37 79 81 53 74	73 24 16 10 33	52 83 90 94 96	70 47 14 54 36
29 01 23 87 88	58 02 39 37 67	42 10 14 20 92	16 55 23 42 45	54 76 09 11 06
95 33 95 22 00	18 74 72 00 18	38 79 58 69 32	81 76 80 26 92	82 80 84 25 39
90 84 60 79 80	24 36 59 87 38	82 07 53 89 35	96 35 23 79 18	05 98 90 07 35
46 40 62 98 82	54 97 20 56 95	15 74 80 08 32	16 46 70 50 80	67 72 16 42 79
20 31 89 03 43	38 46 82 68 72	32 14 82 99 70	80 60 47 18 97	63 49 30 21 30
71 59 73 05 50	08 22 23 71 77	91 01 93 20 49	82 96 59 26 91	66 39 67 08 60

Table A3.2: Selected values of the t-distribution

Degrees of freedom	Probability (α)					
	0.5	0.2	0.1	0.05	0.01	0.001
1	1.000	3.078	6.314	12.706	63.657	636.619
2	0.816	1.886	2.920	4.303	9.925	31.598
3	0.765	1.638	2.353	3.182	5.841	12.924
4	0.741	1.533	2.132	2.776	4.604	8.610
5	0.727	1.476	2.015	2.571	4.032	6.869
6	0.718	1.440	1.943	2.447	3.707	5.959
7	0.711	1.415	1.895	2.365	3.499	5.408
8	0.706	1.397	1.860	2.306	3.355	5.041
9	0.703	1.383	1.833	2.262	3.250	4.781
10	0.700	1.372	1.812	2.228	3.169	4.587
11	0.697	1.363	1.796	2.201	3.106	4.437
12	0.695	1.356	1.782	2.179	3.055	4.318
13	0.694	1.350	1.771	2.160	3.012	4.221
14	0.692	1.345	1.761	2.145	2.977	4.140
15	0.691	1.341	1.753	2.131	2.947	4.073
16	0.690	1.337	1.746	2.120	2.921	4.015
17	0.689	1.333	1.740	2.110	2.898	3.965
18	0.688	1.330	1.734	2.101	2.878	3.922
19	0.688	1.328	1.729	2.093	2.861	3.883
20	0.687	1.325	1.725	2.086	2.845	3.850
25	0.684	1.316	1.708	2.060	2.787	3.725
30	0.683	1.310	1.697	2.042	2.750	3.646
40	0.681	1.303	1.684	2.021	2.704	3.551
60	0.679	1.296	1.671	2.000	2.660	3.460
120	0.677	1.289	1.658	1.980	2.617	3.373
∞ (normal)	0.674	1.282	1.645	1.960	2.576	3.291

Table A3.3: Values of Coefficient of Variation (CV) per unit in the universe for different values of the universe proportion P

P	*CV (%)*	*P*	*CV (%)*	*P*	*CV (%)*
0.01	995	0.25	173	0.65	73
0.02	700	0.30	153	0.70	65
0.03	569	0.35	136	0.75	58
0.04	490	0.40	127	0.80	50
0.05	436	0.45	111	0.85	41
0.10	300	0.50	100	0.90	33
0.15	238	0.55	90	0.95	23
0.20	200	0.60	82	0.99	10

$CV = \sqrt{[(1-P)/P]}$

Table A3.4: Sample size (n) required to ensure desired coefficient of variation of sample estimator (e) in sampling with replacement

Desired CV of sample estimator (e) (%)	*Value of universe CV per unit (%)*												
	5	*10*	*20*	*30*	*40*	*50*	*60*	*70*	*80*	*90*	*100*	*150*	*200*
25	1	1	1	1	3	4	6	8	10	13	16	36	64
20	1	1	1	2	4	6	9	12	16	20	25	56	100
15	1	1	2	4	7	11	16	22	28	36	44	100	178
10	1	1	4	9	16	25	36	49	64	81	100	225	400
5	1	4	16	36	64	100	144	196	256	324	400	900	1600
4	2	5	25	56	100	156	225	306	400	506	625	1406	2500
3	3	11	44	108	178	278	400	544	711	900	1111	2500	4444
2.5	4	16	64	144	256	400	576	784	1024	1296	1600	3600	6400
2	5	25	100	225	400	625	900	1225	1600	2025	2500	5625	10 000
1	25	100	400	900	1600	2500	3600	4900	6400	8100	10 000	22 500	40 000

$n = (\text{Universe CV per unit}/\text{Desired CV of sample estimator})^2$

APPENDIX IV

Hypothetical Universe of 600 Households in 30 Villages in 3 States

Table A4.1

State	Village	Area (km^2)	Previous census population	Number of households	Size of households	Total population
I	1	8.7	69	17	7 5 5 4 6 2 3 5 5 6 5 4 4 4 5 3 3	76
	2	10.6	82	18	6 5 4 5 4 5 6 5 3 5 4 4 5 3 3 5 6 4	82
	3	15.0	110	26	6 6 3 5 3 4 5 5 4 4 4 3 7 5 4 6 2 5 5 6 1 5 5 4 6 3	116
	4	6.2	80	18	6 3 6 3 6 3 4 5 4 4 4 5 6 3 5 1 3 5	76
	5	9.6	92	24	5 4 6 5 4 5 6 5 4 4 7 6 6 5 4 4 5 6 3 4 3 3 5 3	112
	6	7.3	65	17	3 4 4 6 5 7 3 5 4 5 4 6 4 5 3 3 6	77
	7	4.5	72	20	6 4 4 5 4 5 6 4 3 5 4 6 5 5 2 4 5 4 3 4	88
	8	10.6	108	24	5 3 3 7 4 4 6 6 4 5 3 7 6 4 5 6 3 5 1 3 5 6 4 4	109
	9	5.4	106	24	5 3 5 3 4 6 5 4 6 5 6 3 6 5 6 6 3 5 4 4 5 4 6 2	111
	10	3.5	80	22	4 4 5 5 4 4 5 4 3 5 5 3 4 5 4 3 4 3 5 4 4 1	88

(continued)

Table A4.1 continued

State	Village	Area (km^2)	Previous census population	Number of households	Size of households	Total population
II	1	5.8	72	15	8 4 4 6 5 5 5 7 3 5 6 6 6 4 4	78
	2	11.4	102	22	9 6 5 5 4 3 5 5 3 10 4 4 6 5 4 7 4 5 3 6 4 5	112
	3	5.8	73	17	6 5 5 3 6 5 3 7 3 5 5 6 6 5 3 4 3	80
	4	7.8	84	19	6 4 6 4 5 4 4 5 7 6 4 6 4 5 6 5 3 4 5	93
	5	6.5	98	20	8 4 3 5 6 3 7 5 5 5 6 6 7 3 6 6 7 2 5 6	105
	6	9.0	84	19	4 3 5 6 5 6 5 5 5 3 5 4 5 5 3 9 4 5 6	93
	7	7.3	85	19	7 5 5 3 5 4 4 5 8 4 4 6 5 7 4 5 5 5 4	95
	8	7.0	102	23	4 5 5 4 6 5 5 3 4 4 8 5 4 4 5 6 4 5 4 6 7 5 6	114
	9	10.5	122	25	8 4 4 5 5 4 7 5 6 4 5 5 8 5 3 6 3 4 6 5 4 7 3 5 6	127
	10	11.1	102	23	7 4 5 6 5 4 6 4 4 7 5 2 5 4 4 9 6 5 6 2 4 4 5	113
	11	6.3	86	18	4 5 4 4 7 5 5 7 5 9 6 5 7 4 3 5 4 5	94
III	1	10.0	78	15	8 6 7 4 6 5 5 2 7 7 7 5 6 4 4	83
	2	14.2	112	21	9 4 5 8 6 3 6 5 5 5 6 2 8 5 8 4 5 7 5 7 8	121
	3	8.2	97	18	8 8 4 8 3 6 6 6 4 6 6 8 7 6 4 5 6 4	105
	4	12.5	117	21	7 5 6 7 8 4 5 6 6 6 6 5 7 5 9 8 6 6 6 4 7	129
	5	6.5	106	20	7 3 5 6 6 7 5 6 5 5 4 10 7 6 5 8 6 4 6 3	114

(continued)

Table A4.1 continued

State	Village	Area (km^2)	Previous census population	Number of households	Size of households	Total population
III	6	10.0	115	21	7 4 6 7 10 4 5 5 6 7 6 3 8 7 6 4 6 5 5 3 7	121
	7	7.0	110	21	7 6 4 8 1 6 6 7 4 6 6 7 6 7 3 6 6 7 2 5 6	116
	8	12.5	104	17	3 8 5 7 6 10 7 8 4 8 7 6 6 5 7 6 6	109
	9	10.2	103	16	6 5 7 6 4 5 8 4 7 8 7 6 10 7 6 4 5	105

APPENDIX V

Case Study: Indian National Sample Survey, 1964-5

A5.1 Introduction

Along with the Current Population Survey (with probability sampling starting in 1943) in the U.S. and the Family Expenditure Survey (continuing since 1957) in the U.K., the National Sample Survey in India has a long tradition of undertaking probability-based household surveys on a continuing basis.

As a case study, the main features of the planning, execution and analysis of the Indian National Sample Survey (NSS) for the period July 1964-June 1965 are given in this chapter, based on *Technical Paper on Sample Design, The National Sample Survey, Nineteenth Round, July 1964-June 1965* by A.S. Roy and A. Bhattacharyya (1968).

At the instance of P.C. Mahalanobis, Honorary Statistical Adviser to the Cabinet of the Government of India, the National Sample Survey was started in 1950 with the object of obtaining comprehensive and continuing information on economic, social, demographic, and agricultural characteristics through sample surveys on a country-wide basis. The information collected is utilized for planning, research and other purposes by the Central and State Governments, the Planning Commission and other interested organizations. The NSS is a continuing, multi-subject, integrated survey and is conducted in the form of successive "rounds"; each round covers several topics of current interest in a specific survey period. The scope, period, sample design and program of each round are fixed by taking into account the requirements of its users and the resources available for that period. Since 1958-9, the survey period has been made one complete year coinciding approximately with the agricultural year.

A5.2 Objectives of the survey

The survey for 1964-5 was designed to provide information on the following topics: population, current and historical fertility and current mortality; employment and unemployment; indebtedness of rural labor households; land utilization, acreage and production of the major cereal crops; rural retail prices; and integrated socioeconomic activities of the households. The survey covered the whole of India, but excluded a few specific areas, the latter accounting for less than 0.5 per cent of the total estimated population. Hotel residents and persons in boarding houses were included, but inmates of hospitals, nursing homes, jails etc. were not covered by the survey: persons without fixed abode were included only for the demographic inquiry. The results were required for the nineteen state and union territories and separately for the rural and the urban sectors, but the crop survey was designed to provide estimates for all the major cereal crops taken together for rural India as a whole. The total cost of the field survey was about Rs. 8 million ($1 US = Rs. 7.50 and £1 Sterling = Rs. 18.00 at the then official exchange rates) and an equivalent sum is estimated to have been spent on the tabulation, analysis and preparation of reports.

A5.3 Budget and cost control

From the early planning stages, preliminary budgets were prepared on the basis of the experiences of a few of the preceding rounds and were examined on the basis of the time records of the staff engaged in field enumeration and processing.

A5.4 Administrative organization

A Program Advisory Committee consisting of the representatives of the Planning Commission, Central Ministries, State Governments, the Central Statistical Organization, the NSS Directorate and the Indian Statistical Institute advised the Department of Statistics under the Cabinet Secretariat of the Government of India on the overall planning, subject coverage, methodology, tabulation program, fixation of priorities, and other related matters of the NSS. The final decisions on these points were taken by the Department of Statistics.

The Directorate of the National Sample Survey under the Department of Statistics is responsible for the field work and the Indian Statistical Institute for the technical work, including the planning of the survey, processing, analysis, and the preparation of reports and studies (in 1970, the departments of the Indian Statistical Institute responsible for the technical work of the NSS were transferred to the NSS Directorate of the Government of India). Field work is done by full time and mostly permanent enumerators and supervisory staff, a large number of whom have Bachelor of Arts, Science or Commerce degrees. The technical work is undertaken by full time, qualified personnel. Thus the benefits of a permanent survey organization are realized.

Fourteen out of the fifteen states participated in the survey on a matching basis. They surveyed an additional matching sample, selected in an identical manner, used the same concepts and definitions, followed the same survey procedure, but processed and analyzed their own data. The sample surveyed by the Central Government agency is known as the "central" sample and that surveyed by the State Government agencies the "state sample". The central and state samples were "linked" members, described later, in a system of interpenetrating sub-samples. The data of the central and the state samples are sometimes pooled together to provide more efficient estimates than either sample.

A5.5 Coordination with other bodies

The interests of government departments and other institutions were ascertained and accommodated to the extent possible, as advised by the Program Advisory Committee. Close coordination with the Office of the Registrar General, revenue departments etc. was maintained for obtaining and updating the frames for the strata and the first-stage units (generally villages in the rural sector and urban blocks in the urban sector, the maps for which were obtained from the two agencies mentioned).

A5.6 Questionnaire preparation

For reasons mentioned in section 26.12.1, schedules rather than questionnaires were used. Information was collected with required details; the time reference periods for the data depended on the items. The list of schedules canvassed is given later.

A5.7 Methods of data collection and supervision

For the socioeconomic inquiries, the data were collected by interviewing a sample of households, and the data on crop acreage and yield rates were obtained by direct physical observation of the acreage and the actual harvesting of crops (known as crop-cutting experiments). Information was also obtained for a sub-sample of households on areas under different crops by the interview method with a view to comparing its utility with that of the detailed crop schedule based on physical observation: this comparison has not yet been made.

There was, on an average, one supervisor for every four enumerators. The supervisors also undertook field checks of the enumerators' work.

A5.8 Survey design

A5.8.1 Type of survey

The survey was a multi-subject inquiry on the topics mentioned in section A5.2; for a sub-sample of households, the integrated survey approach was taken for consumption and productive enterprises (section 26.13.1).

A5.8.2 Choice of sample design

The sample design was stratified multi-stage and the design was similar but completely independent for the rural and urban areas. In the rural sector, groups of villages formed the strata, 353 in number; in the urban sector, towns and cities in a state were generally grouped into two strata according to their 1961 population, there being a total of 37 urban strata from the whole of India.

For the household inquiries, the sampling plan was stratified two-stage; within each stratum, the first-stage units were villages in the rural sector, and urban blocks in the urban sector, and households formed the second-stage units. The first-stage units were selected circular systematically with probability proportional to size, the measure of size being related to the 1961 Census population, and the second-stage unites were selected linear systematically with equal probability. The sample design was self-weighting for the household inquiries, at the state level for the rural sector and at the stratum level for the urban sector. For crop-area surveys, the sampling plan was stratified two-stage, and for yield-rate surveys it was four-stage, the successive stages being villages, clusters of plots, crop plots and circular cuts. The same sample villages were used for both the households and the crop inquiries.

The first-stage units were selected in the form of four independent, interpenetrating sub-samples. The first-stage units were selected at the Indian Statistical Institute and the households, clusters of plots etc., selected in the field by the enumerators.

Experiences of the past years provided information on the allocation of the all-India sample sizes (first-stage units) to the different strata, and the allocation of the second-stage units for the different types of inquiries. The total sample size is in particular the product of the total number of enumerators and the number of villages and urban blocks that can be surveyed by an enumerator during the whole survey period. With the work load of an enumerator being more or less fixed, the demand for a big sample in order to provide estimates at the state and lower levels had to be met by increasing the number of enumerators. The total number of enumerators in the nineteenth round was 752 for the central sample (with a reserve of 10 per cent), of whom 706 were to survey both the rural and the urban areas and 46 in the urban areas only. The saving arising from the multi-subject nature of the survey was also utilized for having a large number of sample fsu's. The total number of first-stage units for the central sample was set at 8,472 villages and 4,572 urban blocks.

For the reasons mentioned in section 26.13.4, the survey period was the agricultural year, divided into six equal periods of two months each, called sub-rounds, and one-sixth of the sample surveyed in each sub-round. This ensured firstly the employment of a smaller number of skilled and well-trained enumerators, and secondly the representation of all the four seasons so that the seasonal fluctuations were taken into account.

For the household inquiries, sampling was the most intensive for the population schedule, with overall sampling fractions of 0.2 per cent in the rural sector and 0.4 per cent in the urban for the central sample, and least intensive for the integrated schedules 17 and 17(suppl.), with overall sampling fractions of 0.01 per cent in the rural sector and 0.02 per cent in the urban sector.

The time required for enumerating the different schedules was obtained on the basis of the records in the previous rounds and, for the schedules which were canvassed for the first time, on the basis of a try-out. In large-scale surveys, journey time accounts for an appreciable portion of the total time and depends on the size of the area an enumerator covers, the number of sample first-stage units and the general transport facilities. In the present survey, an enumerator's area of operation in the rural sector was the stratum in which he was posted. Most of these strata were less than 4,000 square miles in area and this was considered to be a manageable size for an enumerator. When a stratum exceeded 4,000 square miles in area, that stratum was sub-divided into two or more parts known as investigation zones and the sample villages to be surveyed by an enumerator were selected from one of these zones. An enumerator was allotted twelve villages in his stratum and a similar procedure was followed for the urban blocks, which achieved some savings in the journey time, but only marginal savings could be achieved in the enumeration time. For the socioeconomic inquiries, the preparation of sampling frames for selecting households was simplified and a schedule was canvassed in a sub-sample of households selected for another schedule. For crop surveys, a sample of clusters of plots instead of a direct sample of individual plots was chosen: this was because the time taken for identifying the plots is comparatively large in relation to the time taken for recording the land utilization of such plots. Also, the same sample of clusters of plots was surveyed in all the crop seasons instead of surveying fresh clusters during each season. Finally, when the sample village or the urban block was big, the enumerators were allowed to confine the survey only to part of a sample village or the urban block so that their individual work load remained within limits. For household inquiries, each sample village and urban-block was visited and surveyed only once during the whole round. The crop survey was conducted in each of the four seasons - autumn, winter, spring and summer. The crop-area survey was conducted in all the twenty-four sample villages of the rural sector while the crop-yield survey was taken up only in one-fourth of the villages. The price inquiry was conducted in a fixed set of sample villages which has been continuing since the sixteenth round, July 1960-June 1961. These sample villages were 491 in number. The time standards for enumerating different schedules and the average work load of an enumerator within a sample village or an urban block are shown in Table A5.1.

A5.8.3 Sample design in the rural sector

Some details of the sample design in the rural sector are given in this section. The sampling frame used for the first-stage units, namely villages, was the Primary Census Abstract (PCA) of the 1961 population census. The rural PCA provided a complete list of all the 1961 census villages with their identification particulars, as well as supplementary information, for example on population, area, number of houses and households, number of literates, number of workers in different occupational groups such as agriculture, industry etc. The abstract was available in the form of booklets (manuscript copy), each booklet giving the list of villages for one *tehsil/taluk/thana* of a state.

The villages were selected with probability proportional to size, the size of a village being related to its 1961 Census population. Each village was assigned a size in such a way that the size became a simple indicator of the village population and also an integer. The strict probability proportional to population sampling was not favored because it would have involved heavy computational work at the selection stage, would have been unnecessary for those items which are not strongly related to population, and would have made it extremely difficult to achieve a self-weighting design. The average population of villages having populations between 0-499, 500-999, 1000-1999, 2000-2999, and so on were computed for each of these population classes separately for each state and union territory. The size 1 was assigned to villages with population less than 500. The size of a village

Table A5.1: Average time requirements for different schedules and journey between the average work-load within a sample village/urban block: Indian National Sample Survey, 19th round, July 1964-June 1965

Schedule no.	Description	Sector	Time standard	Sample size
0.1	General schedule (for listing)	Rural	2 1/2 days/village	120 hh/village
0.2	General schedule (for listing)	Urban	3 1/2 days/block	160 hh/block
3.01	Rural retail (monthly)	Rural	1 1/2 days/village	419 villages
4.1	Investigator's time record for socioeconomic inquiry	Rural & urban	–	–
4.2	Investigators' time record for crop surveys and price inquiry	Rural	–	–
5.0	Land utilization survey	Rural	30 plots/day	60 plots per crop-cutting village; 20 plots in others
5.1	Crop-cutting experiments	Rural	2 plots/day	6 cuts per crop-cutting village
5.2	Driage experiment	Rural	–	5 cuts per crop-cutting village
10	Urban labor force	Urban	8 hh/day	6.7 hh/block
10.1	Employment, unemployment and indebtedness of rural households	Rural	8 hh/day	8 hh/village

(continued)

Table A5.1 (continued)

Schedule no.	Description	Sector	Time standard	Sample size
12	Population, births and deaths	Rural & Urban	25 hh/day	20 hh/village or block
16	Integrated household schedule (detailed)	Rural & Urban	1 hh/day	2 hh/village or block
17	Integrated household schedule (abridged)	Rural & Urban	2 hh/day	1 hh/village or block
17(S)	Integrated household schedule (abridged)-land utilization	Rural	1 hh/day	1 hh/village
–	Field progress reports	Rural & Urban	–	–
–	Journey to a village/block	Rural & Urban	– –	– –

Source: Roy and Bhattacharyya (1968).

Notes:

1. Schedules 5.0, 5.1, 5.2, and 17 (S) were taken up in each of the four crop seasons.
2. For Schedule 5.1, the sample size may go up to 12 cuts, depending upon the nature of sowing of pure and mixed crops.
3. "hh" denotes household(s); "S" denotes Supplementary.

in any other population size class was taken as the ratio of the average population on the class to the average population of villages in the first class, rounded off to a suitable integer. The total size of a *tehsil* was obtained by cumulating the sizes of all the villages contained in that *tehsil*. Similarly, the total size of a region was obtained by cumulating the sizes of all *tehsils* contained in that region (there were forty-eight regions, formed by grouping contiguous districts within a state mainly on the basis of information on topography, crop pattern, and population density: the ultimate rural strata were formed within these regions).

The all-India central sample of 8,472 villages was allocated to the different states (and union territories) on a joint consideration of the rural populations, area under food crops, and the enumerator strength. This allocation was modified to ensure a minimum sample size of 360 villages in each state, which was made a multiple of 24 in order to facilitate having four sub-samples of six villages each per stratum. The allocation to any rural stratum was always 24 sample villages in order to achieve uniform work load for the enumerators all over India.

The total number of sample villages allocated to a state (or union territory) was divided by 24 to get the total number of strata to be formed in the state (or union territory). The number was allocated to the regions in that state in proportion to the region sizes. The proportional allocations were adjusted to obtain an integral number of strata in all the regions. After deciding the number of strata, the next step was to determine the stratum sizes for a state. The strata were made of equal size within a state: the first consideration was that it would justify having equal allocations of sample size per stratum, and the second that it would considerably help to implement the self-weighting design. With these ideas in view, the total size of the state, that is the sum of the sizes of all the villages in that state, was divided by the proposed number of strata (say L) to yield the average stratum size (say Z') for that state. This average size was rounded off and made a multiple of 12 in order to attain an integral interval at the selection stage as 12 villages (central and state samples taken together) were selected systematically in each sub-sample of a rural stratum. The formation of strata had the following requirements to satisfy:

1. Each stratum should comprise a geographically compact area.

2. The stratum boundaries should not cut across the regions since estimates were required to be obtained separately for each region; (region-wise estimates can be built up even if the regions contain some part-strata, but such estimates are liable to large sampling errors due to the "randomness" of sample size on which they are based.)

3. The stratum boundaries should not cut across the state blocks which were compact groups of districts in a big state. Each state block was an administrative unit of the NSS directorate and there would have arisen some administrative difficulties of travelling allowance, inspection, supervision, etc. if the enumerator under the charge of one state block had to survey sample villages falling in another state block.

4. Different parts of a stratum should be similar with respect to population densities and altitude.

5. There should be good transport and communication facilities among different parts of a stratum so that an enumerator posted in that stratum could travel throughout it without much difficulty.

6. The original size (say Z') of a stratum (i.e. sum of the sizes of villages in it) should not differ by more than 10 per cent from the planned average figure Z, because otherwise the large adjustment to be done in order to equalize this Z' to Z would lower the sampling efficiency of the pps design.

In all, 353 compact strata were formed by grouping contiguous *tehsils* such that the first, second, and the sixth conditions were completely satisfied and conditions three, four and five satisfied to the extent possible. Finally, the original stratum size Z' was made equal to Z by slightly increasing or decreasing the sizes of the villages belonging to the stratum.

In each stratum, four independent sub-samples of 12 villages each were selected circular systematically with probability proportional to the size, the size being defined earlier. Out of these 12 villages in a sub-sample, the six villages with odd orders of selection constituted the central sample and the rest, with even orders of selection, the state sample. The linking of central and state samples ensured a better spread of samples over the whole stratum and consequently a better estimate when the central and the state sample data are pooled together.

The interval for systematic selection for either the central or the state sample was obtained as $I = Z/6$; the interval I, an integer, was the same for all the strata in a state since stratum sizes were equalized during the formation of the strata. Four independent random starts were used for selecting the four sub-samples of the 12 villages each.

Because of pps selection, larger villages occurred more frequently in the sample. This resulted in too heavy a work load for some enumerators. The planned work load involved the listing of 120 households per sample village on an average, i.e. a total of 1,440 households per year per enumerator. To reduce the work load of household listing in large villages, the hamlets were grouped in such a manner as to contain approximately the same population and the survey confined to one hamlet group, selected at random with equal probability. The number of hamlet groups to be formed in a village was specified by the Indian Statistical Institute, but the formation and selection of the hamlet groups were done by the enumerators on reaching the sample villages.

For selection of households for the socioeconomic inquiries, the households were not stratified according to the available information for this would have complicated the computation of the estimates and would have rendered difficult the task of making the design self-weighting. The same objectives of stratification were to a large extent achieved by arranging the households in the manner described in section 4.3.3. From the sampling frame thus prepared, a sample of 22 households on an average was selected linear systematically with the interval and a random start prescribed for this purpose. For schedule 16, a sub-sample of 2 households on an average was selected systematically with interval 11 and a specified random start for this schedule; for schedule 17, from the sample households selected for schedule 16, a sub-sample of 1 household on an average was selected linear systematically with interval 2 and a random start 1 or 2, and the next household in the frame was surveyed for schedule 17 (the actual sample household for schedule 17 thus did not belong to the total of 22 households referred to earlier); in the remaining 20 households, schedule 12 was canvassed.

The following methods were used for selecting the sample of clusters or plots in the sample villages:

1. Suppose the sample village was cadastrally surveyed and a village map showing the location and boundaries of the plots was available without much trouble. Then clusters of size 10 (i.e. of 10 plots each) were formed by combining plots with major survey numbers 1-10, 11-20, 21-30, etc., and similarly clusters of size 5 were groups of major survey numbers, 1-5, 6-10, 11-15, 16-20, etc. Then 4 clusters of 5 plots each were selected in villages meant for land utilization only and 6 clusters of 10 plots each were selected in sample villages meant for both acreage and yield-rate surveys. The actual procedure of selection was as follows:

   Suppose N is the highest survey number in a village/hamlet-group; n the number of clusters to be selected; c the cluster size = 5 or 10; and N' is N increased as little as possible to become a multiple of cluster size c.

   Then n plots, known as "basic plots", were selected circular systematically with interval I and random start R where $I = (N'/n)$ and $I \leq R \leq N$. Next, clusters were formed around the selected basic plots, e.g. if the basic plot with survey number 472 was selected, the corresponding sample cluster was taken as a group of plots with survey numbers 471-480 and 471-475 for $c = 10$ and 5 respectively.

2. Suppose the village map was not available but a list of all plots in the village was available. In this case, all the plots in the village were first given a continuous sampling serial number. The selection of basic plots and formation of clusters around them were exactly the same as in method 1 with the only difference that survey numbers were replaced by sampling serial numbers.

3. In case neither the village map nor the list of plots was available, plots were selected indirectly through a sample of households. A sample of 4 to 6 households was selected systematically depending on whether

the village was meant for just acreage survey or both acreage and yield-rate surveys. All the plots possessed by these sample households within a 5 mile radius of the sample village were taken up for survey.

Self-weighting sample. The sample was made self-weighting with respect to the unbiased estimator of the state total of a study variable of the household inquiries in the following manner.

The following notations are used for a state

- h: subscript for a stratum
- i: subscript for a village
- j: subscript for a household
- L: total number of strata
- n_h: number of sample villages in any sub-sample of the hth stratum ($n_h = 6$)
- Z_h: total size of the hth stratum ($Z_h = Z_0$, as the strata were of equal size)
- Z_{hi}: size of the ith sample village in the hth stratum
- D_{hi}: number of hamlet-groups formed in the hith sample village ($D_{hi} = 1$, if the whole village was surveyed)
- I_{hi}: interval for selecting the combined sample in the hith sample village
- m_{hi}: number of households in the combined sample in the hith sample village
- y_{hij}: value of the study variable for the jth sample household in the hith sample village ($j = 1, 2, \ldots, m_{hi}$)
- Y: total value of the study variable for the state

From estimating equations (21.15), (21.17) and (21.18), an unbiased estimator of Y, based on any sub-sample is given by

$$\begin{aligned} y &= \sum_{h=1}^{L} \frac{1}{6} \sum_{i=1}^{6} \frac{Z_0}{Z_{hi}} D_{hi} \; I_{hi} \sum_{j=1}^{m_{hi}} y_{hij} \\ &= \sum_{h}^{L} \sum_{i}^{6} \sum_{j}^{m_{hi}} w_{hij} y_{hij} \end{aligned} \qquad \text{(A5.1)}$$

where

$$w_{hij} = \frac{Z_0}{6\,Z_{hi}} \; D_{hi} I_{hi}$$

is the multiplier for y_{hij}. The objective is to make w_{hij} a constant, w_0, within a state. Now $Z_h = \sum Z_{hi}$ (sum over i) $= Z_0$ is the same for all the strata within a state, and the values of Z_{hi} and D_{hi} are already determined by the village populations. So the only item that can be properly chosen so as to equalize all the w_{hij} values is I_{hi}; i.e.

$$I_{hi} = \frac{6\,w_0}{Z_0}\frac{Z_{hi}}{D_{hi}} \tag{A5.2}$$

which determines the interval for selecting the combined sample in a sample village, when the value of the multiplier w_0 is fixed.

The unbiased estimator y then becomes

$$y = w_0 \sum_h^L \sum_i^6 \sum_j^{m_{hi}} y_{hij} \tag{A5.3}$$

The next problem is to find an appropriate w_0 that will ensure the desired sample size in terms of households. From equation (A5.3) putting $y_{hij} = 1$, the estimated total number of households in the state is

$$\begin{aligned} m &= \sum^L \sum^6 m_{hi} \\ &= w_0 \times \text{number of sample } hh \text{ per } ss \text{ in the state} \end{aligned} \tag{A5.4}$$

(hh and ss denoting households and sub-sample respectively).

To be more precise about the value of w_0, we can replace y by its true value Y, and write

$$\begin{aligned} w_0 &= \text{(total number of } hh \text{ in the state)/(number of sample } hh \text{ per sub-sample in the state)} \\ &= \text{reciprocal of the overall sampling fraction} \end{aligned} \tag{A5.5}$$

Now, the number of sample households per sub-sample in the state $= nm_0$, where n is the number of sample villages (in any sub-sample) for the state and m_0 the average number of households planned to be selected per sample village. In this particular case, $n = 6L$ and $m_0 = 22$. The true total number of households at the mid-point of the survey period, i.e. on 1 January 1965, is of course unknown; it is replaced by y obtained by first projecting the rural population of the state from the 1961 Census (and the values of the previous NSS rounds) and then dividing that by the average household size (obtained also from the previous NSS rounds). Thus w_0 finally works out as

$$w_0 = y'/nm_0 = y'/132\,L \tag{A5.6}$$

First, w_0 is to be obtained from equation (A5.6) for each state, and then the values of I_{hi} are determined from equation (A5.2) for a sample village, noting that the factor $6w_0/Z_0$ is the same for all the sample villages in a state.

When, as in most cases, I_{hi} as computed was a fraction, it was rounded off in the randomized manner, as described in note 3 to section 12.3. The rounded off (integer) values of I_{hi} were given to the enumerators.

The intervals obtained by equation (A5.2) ensured that the total sample size (number of households) in the state would be near the desired value $nm_0 = 132\,L$, but they do not ensure anything for the individual sample villages. Some enumerators were allotted too many big villages, and as each enumerator had to survey 12 sample villages, the total sample size for them was much larger than the planned figure of 12 × 22 households. In such cases, to provide relief to the enumerators, they were allowed to survey a smaller number of sample households than was strictly required for a self-weighting design: the shortfall was made good at the scrutiny stage by repeating some of the filled-in schedules.

Once the design was made self-weighting for the combined sample, it became automatically so for individual household inquiries, for constant fractions of the combined sample households were surveyed for each of these inquiries in all the sample villages.

Estimating procedures. With the same notation as for the rural sector, the unbiased estimators for the state totals of study variables were as follows.

Schedule 0.1 (complete enumeration of the households in the sample village or the selected hamlet-group):

$$\begin{aligned} y &= \sum_{h=1}^{L} \frac{1}{6} \sum_{i=1}^{6} \frac{Z_0}{Z_{hi}} D_{hi} y_{hi} \\ &= I_0 \sum_{h}^{L} \sum_{i}^{6} D_{hi} \frac{y_{hi}}{Z_{hi}} \end{aligned} \tag{A5.7}$$

where $I_0 = Z_0/6$.

Schedules 10.1, 12, 16, and 17:

$$y = w_0 \sum_{h}^{L} \sum_{i}^{6} \sum_{j}^{m_{ij}} y_{hij} \tag{A5.8}$$

where w_0 is the constant multiplier for the corresponding schedule.

Information on fertility history was collected for one-fifth of the sample households for Schedule 12; the constant multiplier for such information was therefore $5w_0$, where w_0 was the constant multiplier for Schedule 12.

Schedule 3.01: An estimator of the average price P of a commodity was

$$p = \sum_{i=1}^{n'} y_i/n' \tag{A5.9}$$

where n' is the number of reporting villages in the state and the sub-sample considered.

In the urban sector, as there were 2 strata in each state (i.e. $L = 2$), the unbiased estimators took the following form:

Schedule 0.2:

$$y = \sum_{h=1}^{2} I_h \sum_{i=1}^{n_h} D_{hi} \frac{y_{hi}}{Z_{hi}} \tag{A5.10}$$

where $I_h = Z_h/n_h$.

Schedules 10, 12, 16, and 17:

$$y = \sum_{h=1}^{2} w_h \sum_{i=1}^{n_h} \sum_{j=1}^{m_{hi}} y_{hij} \tag{A5.11}$$

where w_h is the constant multiplier for a schedule in the hth stratum.

For **crop surveys**, an unbiased estimator of the acreage under a particular cereal crop in a season from a sub-sample in a state was

$$a = \sum^{L} a_h \tag{A5.12}$$

where

$$a_j = I \sum_{i=1}^{L} \frac{D_{hi}}{Z_{hi}} \frac{M'_{hi}}{m_{hi}} \sum_{j=1}^{m_{hi}} a_{hij} \tag{A5.13}$$

where M'_{hi} is the adjusted highest survey number/sampling serial number (method 1 or 2) or total number of households (method 3) in the ith sample village of the hth stratum, m_{hi} is the number of sample plots or households planned for survey, and a_{hij} is the area under the specified cereal crop for the jth sample plot or in all plots of the jth sample household in the ith sample village.

An estimator of the **yield rate** (Schedule 5.1) for a cereal crop in a season from sub-sample 1 or 2 is given by

$$x = \frac{\sum^L a_h \bar{y}_j}{\sum^L a_h} \tag{A5.14}$$

where a_h is given by equation (A5.13), and $\bar{y}_h$ is the simple average of the yield rates over the crop-cuts taken for the crop in the hth stratum; the summation $\sum$ extends over all strata reporting crop-cutting experiments. The yield rates were obtained on the basis of the sample cuts of radius 4′; the yields of all the plants inside the circle together with half the yield of the border plants of the 4′ circle constituted the yield of a cut.

While calculating the production estimates, the estimate of yield rate for sub-sample 1 was multiplied by the corresponding area estimate, pooled over sub-samples 1 and 3, and the estimate of yield rate for sub-sample 2 was multiplied by the corresponding area estimate, pooled over sub-samples 2 and 4. Thus two independent estimates y_1 and y_2 of total production were obtained. For obtaining the dry weights, these two production estimates were multiplied by the driage factors based on sub-samples 1 and 2 respectively. The driage factor was obtained on the basis of the sample cuts of radius 2′3″, including the full border. If d and g were the weights of dry crop and green crop respectively, the driage factor was given by $\sum d/\sum g$, where $\sum$ denotes summation over all cuts taken on that crop in all strata. For paddy crop, the production figure related to "clean rice" and for this the figure for dry paddy was multiplied by 0.662.

Substitution of casualties. For socioeconomic inquiries, all casualty households, villages and blocks were substituted. Similarly for crop survey, all casualty plots/clusters/households/villages were substituted by the previously surveyed corresponding units in the same sub-sample and stratum.

Combined estimators. If y_l (l = 1, 2, 3, 4) is the lth sub-sample estimator of the state total of a study variable, the combined estimator was

$$y_0 = \frac{1}{4}\,(y_1 + y_2 + y_3 + y_4) = \frac{1}{4}\sum_{h=1}^{L}(y_{h1} + y_{h2} + y_{h3} + y_{h4}) \tag{A5.15}$$

where y_{hl} is the lth sub-sample estimator of the hth stratum total.

The ratio $R = Y/X$ was estimated by $r_l = y_l/x_l$ from the lth sub-sample and the combined estimator of R was

$$r = \frac{y_0}{x_0} = \frac{y_1 + y_2 + y_3 + y_4}{x_1 + x_2 + x_3 + x_4} \tag{A5.16}$$

The combined estimator of the area under a crop was similarly the mean of the four sub-sample estimates. The combined production estimator y_0 for a season is the mean of the two production estimators y_1 and y_2. Whenever season-wise total crop estimates were obtained, the estimate for the year was the sum of the season-wise estimates.

Variance estimators (see also section 25.6.5). Two unbiased variance estimators of y_0 are

$$s_{y_0}^2 = \frac{1}{12} \sum_{h=1}^{L} \sum_{i=1}^{4} (y_{hi} - \bar{y}_h)^2 \tag{A5.17}$$

and

$$s_{y_0}'^2 = \frac{1}{12} \sum_{l}^{4} (y_l - y_0)^2 \tag{A5.18}$$

where

$$\bar{y}_h = \frac{1}{4} \sum_{l}^{4} y_{hl}$$

$s_{y_0}'^2$ is easier to compute but less efficient than $s_{y_0}^2$. Unbiased variance estimators of the acreage estimator a in equation (A5.12) are obtained similarly.

Two variance estimators of r are

$$s_r^2 = \frac{1}{12} \frac{1}{x^2} \sum_{h}^{L} \sum_{l}^{4} \left[(y_{hl} - \bar{y}_h)^2 + r^2 (x_{hi} - \bar{x}_h)^2 - 2r(y_{hl} - \bar{y}_h)(x_{hi} - \bar{x}_h) \right]$$

and

$$s_r'^2 = \frac{1}{12} \sum_{l}^{4} (r_l - r)^2$$

the latter being easier to compute but less efficient than the former.

A variance estimator of the combined production estimator y_0 is

$$s_{y_0}^2 = \frac{1}{4} (y_1 - y_2)^2$$

A5.9 Pilot inquiries and pre-tests

The previous 18 rounds, covering the period 1950-64, acted as pilot inquiries, the results of which were used to design the sample for the 19th round effectively, including information on the variability and the cost. In addition, special tryouts were organized for new schedules introduced in this survey.

A5.10 Selection and training of enumerators

The type of enumerators has been mentioned in paragraph A5.4. The training of enumerators was done in two phases, the supervisors being trained by the technical staff and in their turn training the enumerators (section 26.15).

A5.11 Processing of data

The schedules were first scrutinized by the field supervisors and then by the technical staff of the Indian Statistical Institute. Queries of a technical nature from the field were answered by the technical staff.

The edited information was put into punch cards and the final tabulation done by mechanical processing. Electronic processing is also resorted to for special computations and analysis.

A5.12 Some results

Some results of the survey obtained from the central sample are given in Table A5.2. The coefficients of variation of the estimates are seen to be reasonably small.

Further reading

Full details of the survey methods of the Indian National Sample Survey, 1964-65, are given by Roy and Bhattacharyya; Murthy (1967) gives in Chapter 15 the survey methods of the Indian National Sample Survey, 1958-9, and in Chapter 16 those of the **Family Living Survey in urban areas of India, 1958-8**, the latter from Chinappa (1963).

The following are references to some readily available case studies:

Agricultural surveys: (a) in India conducted by the Indian Statistical Institute: Mahalanobis (1940, 1944, 1946a, 1968); (b) in India conducted by the Indian Council of Agricultural Research: Panse and Sukhatme, and Sukhatme and Panse; (c) Surveys of Fertilizer Practice in England and Wales: Yates.

Auditing and accounting: Trueblood and Cyert.

Demographic surveys: (a) in Greece: Deming (1950), Chapter 12; (b) in India: Som *et al.* (1961), Mahalanobis (1966), and Office of the Registrar-General, *Sample Registration of Births and Deaths in India: Rural 1965-68*; (c) in Pakistan: Pakistan Institute of Development Economics, *Report of the Population Growth Estimation Experiment*; (d) in Trinidad and Tobago: Harewood; (e) in the U.S.: the U.S. Bureau of the Census (1963b), *Technical Paper No. 7* (summarized in Hansen *et al.* (1953), Vol. I, Chapter 12, and in Kish (1965), section 10.4).

Demographic and health surveys: reports by Demographic and Health Surveys, Macro International (see Appendix VI).

Employment and unemployment surveys of Greece, 1962: Raj, Appendix I.

Family planning surveys: Ascádi, Klinger and Szabady (1969, 1970); and reports of surveys mentioned in Appendix VI.

Fertility surveys: reports by World Fertility Survey (see Appendix VI).

Health surveys: Som, De and Das (1961) and Das (1969) for mobidity studies in India through the Indian NSS; Harris (1949) for the Canadian Sickness Survey, 1950-51; Krupka *et al.* (1968) for the sampling studies on trachoma in Southern Morocco; and U.S. NCHS (1963) for the U.S. National Health Survey.

Inventory surveys: Deming (1950), Chapter 11, and (1960), Chapters 6-8. Chapter 11.

Marketing research survey in Great Britain, 1952: Moser and Kalton, Chapter 8.

Road/rail traffic surveys: Dillman (1978); Frey; Kish *et al.* (1961); Rosander.

Sample surveys in Sweden : Dalenius (1957).

Sexuality surveys: Laumann *et al.*

Social surveys in Great Britain: Gray and Corlett; Moser and Kalton.

Surveys of retail stores and annual survey of manufactures in the U.S.: Hansen *et al.* (1953), Vol. I, Chapter 12.

Telephone survey methods: Lavrakas (1993).

Hansen *et al.* (1953) also provide case studies of the following in the U.S.: the Current Population Survey; some variances and covariances for cluster sample of persons and dwelling units; and the sample verification and quality control in the 1950 Population Census.

Slonim in Chapter 17 gives a simple account of a large number of sample surveys. Sudman in Chapter 1 gives brief descriptions of the following sample surveys in the U.S.A.: The Current Population Survey, Equality of Educational Opportunity (The Coleman Report), Supplemental Studies for the National Advisory Commission on Civil Disorders (The Kerner Commission), the Education of Catholic Americans, the Gallup Poll, Public Perception of the Illinois Legislature, the National Labor Relations Board (NLRB) Election Study, the Detroit Area Study, and Unfunded Doctoral Dissertation Research.

Table A5.2: Some estimates from the Indian National Sample Survey, 19th round, July 1964-June 1965

Item		Estimate	CV(%)	No. of villages/ urban blocks	No. of house-holds
1. Rural India:	(a) Birth rate	37.0/1000	0.75	8472	169,440
	(b) Death rate	13.0/1000	1.49	8472	169,440
2. Urban India:	(a) Birth rate	31.9/1000	1.13	4572	91,440
	(b) Death rate	8.0/1000	1.69	4572	91,440
3. Total monthly consumer expenditure in the urban sector in some states:					
(a) Andhra Pradesh		Rs. 208 million	6.20	384	768
(b) Gujerat		Rs. 163 million	2.40	192	384
(c) Rajasthan		Rs. 111 million	6.68	216	432
(d) West Bengal		Rs. 329 million	6.28	432	564

Source: Courtesy of Indian Statistical Institute, National Sample Survey Department, 1970.

APPENDIX VI

Multi-country Survey Programs

A6.1 Introduction

An important development in sample survey operations since the early 1970s has been the launching of coordinated multi-country survey programs, sponsored by the United Nations and its agencies or by donor government agencies, primarily the United States Agency for International Development (USAID), or by both. In a multi-country survey program, a number of major issues arise that do not arise in individual country programs; it is thus more than the sum of its component country programs.

In section A6.2 of this appendix we shall briefly review these issues and shall illustrate them from the major multi-country survey programs in socioeconomic and health areas. The survey programs themselves are summarized in Section A6.3.

A6.2 Major issues in multi-country survey programs

A6.2.1 Principal objectives

Two sets of principal objectives in multi-country survey programs are often cited:

(a) To assist countries in acquiring survey-based information that would permit them to describe and interpret the topics under study in a coordinated program with inter-country comparability.

(b) To help in building national capabilities in household surveys through technology-transfer.

To attain the first objectives, survey programs should be designed to ensure a continuing flow of up-to-date and accurate data to be used together with data from other sources to meet national needs. The data requirement could, for example, relate to the current rate of population growth, agricultural production (yields of different crops), employment and unemployment rates, etc.

To attain the second objective, the survey programs should be so designed as to build up national capacities for the different aspects of survey sampling, of which training is an important element; and training, by its very nature, takes a little time to produce results.

Some survey programs tend to emphasize the attainment of one objective over the other. For example, in one developing country that participated in the National Household Capability Program (NHSCP, sponsored by the U.N. and its agencies), two U.N. agencies funded two separate surveys, conducted one after the other, the donor agency for the second survey declined to continue the data processing expert recruited by the donor agency for the first survey but fielded its own expert; moreover, the computer equipment brought in by the second donor was incompatible with the previous set: thus, the input of one agency did not mesh with that of the other, resulting in a less significant contribution towards the enhancement of the data processing capabilities of the recipient country (deGraft-Johnson, 1993, para. 17).

In other international survey programs too, the objective of augmenting national survey capability is often ignored when it comes into conflict with the competing objective of "getting out the data"; "for example, a number of countries had their survey results [of inter-country survey programs other than NHSCP] processed outside the country because of technical difficulties within the countries" (*op. cit.*, para. 18). In the World Fertility Survey (WFS), country reports (including the first results and the sampling errors of major items), were prepared in the participating countries, but detailed country analysis and inter-country comparative analysis, based on computer data tapes of the survey results, were made by the WFS secretariat, the U.N. Secretariat, and a number of universities and research institutions, mostly in developed countries: such a procedure does not fully contribute to the capacity-building in the developing countries.

It also became clear that unless multi-country survey programs are made product-oriented (in terms of obtaining results and analysis in a standardized time-frame), the objective of strengthening the process (of surveys themselves) cannot be achieved easily. Capacity-building will not flourish unless data are delivered in a timely manner; a survey that is well designed and well executed and which furnishes the required data and analysis as input to national planning can well be self-sustaining. Nothing succeeds like success.

Technical cooperation with United Nations and its agencies and other bilateral and multilateral organizations has resulted in the **transfer of**

technology of survey practices to co-operating countries that were not otherwise readily available to them. Of special interest is the transfer of Personal Computer (PC) technology. Most of the countries of the third world do not have the critical mass, in terms of human and physical resources, to produce PCs or computer software. However, among these countries, those that received technical assistance from the U.N. and its agencies or bilateral organizations have reaped benefits not generally enjoyed by others that did not participate in multi-country survey programs and have, in the short run at least, leapfrogged a developing country such as India that has both a tradition of survey technology and are producing computers – mainframe, mini, and PCs. Countries like Viet Nam and Bhutan have, by now, established computer support centers that are assembling PCs or developing software, and others, such as Algeria, Ethiopia, Ghana, Uganda, and Mexico, are now providing experts to other developing countries.

A6.2.2 Funding

External funding for the survey program for a country follows two routes:

(1) Funding is guaranteed by the funding agencies for the multi-country survey program, the headquarters of which then invites some selected countries to participate in the programs: this was the case with WFS, DHS (Demographic and Health Surveys), and the World Bank's LSMS (Living Standards Measurement Study).

(2) Sponsoring agencies provide core funding for a multi-country program's headquarters but the extent and timing of external funding for national surveys are not guaranteed; the headquarters of the program approaches a number of countries to participate in its programs, and then jointly submit a country project proposal for funding to potential donor agencies: this was the case with the U.N.'s NHSCP and the World Bank's SDA (Social Dimensions of Adjustment) Program.

Absent a guarantee from donor agencies for national surveys, the second approach inevitably results in delays and other problems in implementing survey programs. The problems are compounded by the data needs of a particular country not meshing with a donor organization's mandate, priority and interests.

A6.2.3 Management

Different survey programs adopt different systems of in-country project management. In the NHSCP, it was observed that in general, projects with externally recruited Chief Technical Advisers did much better than those managed by nationals – a prescription that runs counter to the objective of national capability building (deGraft-Johnson, 1993, para. 92).

In order to get the program a higher degree of priority within national offices and to attract qualified staff, incentive payments are sometimes made to them, but that leads to internal disparities. A better solution would be to give training in project management to the national staff, and to associate them and the technical staff in publications. Cases are not uncommon where computing and printing equipment was acquired but no one had been trained to operate it. This calls for installing **project management software**, such as TimeLine/DOS (by Symantec Corporation) and Project/WIN (by Microsoft Corporation).

A6.2.4 Inter-country data comparability

In survey programs sponsored or supported by the United Nations family of agencies and by other multilateral or bilateral organizations, emphasis is given rightly on the importance of producing data and estimates that are comparable across a wide range of countries. This should be also of interest to the participating countries themselves.

At the same time, it should be axiomatic that data should be collected and analytical findings interpreted in a culture-specific context – a truism that is generally self-evident in nationally sponsored survey programs, but one that tends to be less focused in an externally sponsored multi-country survey program with set questionnaires.

Programs, such as NHSCP, have flexible frameworks to meet national needs – their first priority – and cannot always provide detailed inter-country comparability of data, that other programs, such as WFS and DHS, with generally fixed frameworks do.

But some surveys in the developed countries do not always adopt full standardization: to wit, the "Labour Force Surveys in the European Community are standardized in terms of the variables generated, sample size, and survey timing; but not as concerns the design of questionnaires and other aspects of data collection methodology. ... Moreover, several of the Family Budget Surveys [of the European Community] suffer from the serious problems of non-probability sampling and high rates of non-response" (Verma, 1992).

A survey program with a rigid framework cannot adapt the survey procedures to different cultural settings. To take an example, in parts of the Indian sub-continent, young branches of the "neem" (*Melia Azadirachta*) tree, that are astringent and believed to have medicinal properties (toothpastes made of "neem" extracts are sold in specialty pharmacies in the U.S.), are used by many to clean their teeth: one cannot conclude that just because they are not using "modern" toothbrush and toothpaste that they are not taking care of oral hygiene. To cite another example, some surveys include a topic on pre-natal care of mothers, specifically asking questions on whether any vitamin or milk was taken; when the answers are in the negative, the conclusion is drawn that there was no pre-natal care. But in several segments of population in Afghanistan, Bangladesh, Ethiopia, and India (and no doubt in many other countries), pregnant women in rural and small urban areas follow traditional practices of ingesting items such as dried yogurt, roasted chick peas (*kurut* and *nakhod* respectively in the local language in Afghanistan), and special red clay that are rich in calcium or protein: the body knows what is good for it. But questions on such indigenous pre-natal care are seldom asked in multi-country surveys with set questionnaires.

The suggestion is that while maintaining a standard set of questions on any topic, local variations should be allowed to supplement the set questions.

It may also be observed, as an aside, that the tradition of indigenous pre-natal and post-natal care of mothers in developing countries is being swamped by the spreading waves of urbanization and industrialization (and extended family norms giving way to nuclear family norms), without an increase, *pari passu*, in attendant community health care – a price that has *par force* to be paid for progress until such a time that a health network covers the entire population. The argument, raised in connection with the environmental damages resulting from some capital-intensive projects for which the World Bank provides loans, therefore, remains moot: "the fundamental problem of modernity may be that development pursued as an absolute goal is nihilistic" (Rich, 1994).

A6.2.5 Sampling and non-sampling errors

For over 45 years, the United Nations has been exhorting countries to make the evaluation of survey results an integral part of their survey programs (see section 26.18.2 of this book). The computation of sampling errors and study of non-sampling errors constitute such an evaluation; by now, the computation of sampling errors has been standardized (aided more recently by computer software: see Chapter 27) and great strides have been made in the study of non-sampling errors (see Chapter 25).

But not all multi-country (or, for that matter, national) survey programs, including the U.N.-sponsored NHSCP and GCHS (Gulf Child Health Survey) programs and World Bank-sponsored programs, follow the U.N. dictum, an irony that compounds itself by the fact that the NHSCP itself had published a study on non-sampling errors over twelve years ago

(United Nations, 1982b) and has recently published another study on the computation of sampling variance (U.N., 1993a); however, UN-supported Pan Arab Child Health Program has plans to calculate sampling errors for the major items (see Section A6.3.8 later in the appendix).

To their credit, WFS and DHS with their strong central management had made mandatory the procedures for computing sampling variance (for the major items) and indications of non-sampling errors in surveys conducted under their auspices. Some studies of the sampling and non-sampling errors have also been undertaken for the Contraceptive Prevalence Surveys (see Section A6.3.2).

However, of the 20 European countries that had executed their own WFS-type of fertility surveys, without financial or technical assistance from the WFS secretariat, a number used non-probability samples, and even among those that had adopted probability sampling, none computed sampling errors (Kish, 1994).

A6.2.6 Coordination among multi-country programs

Co-ordination among different multi-country programs, even those sponsored by the U.N. and its agencies, is essential but cannot, unfortunately, always be taken as granted.

In the cooperating countries, the implementation of a survey project becomes the responsibility of the concerned line ministries, e.g., the ministry of health for a health survey or a survey on children (if sponsored by UNICEF, the United Nations Children's Fund), or the ministry of planning (under which national statistical offices are generally located) for a household or demographic survey. Generally, a coordinating machinery is available, being located in the planning office to minimize overlap and to fill in gaps left by different survey programs. Such coordinating units may not always be administratively and technically strong enough to ensure the desired objectives, or to take a not so trivial a case, to ensure compatibility between computer equipment provided by different external agencies (see section A6.2.1 above).

Ad hoc committees of potential data users and of technical personnel both from the government organizations and the non- or quasi-governmental sectors (such as the private sector, universities, and research institutions) have proved successful in many a country to achieve the needed in-country coordination.

Coordination becomes difficult when programs such as World Bank's SDA promote special purpose surveys, using small samples ostensibly to provide indicators of current change at national level (a procedure that cannot be technically supported), while other programs such as U.N.'s NHSCP promotes multi-purpose surveys, with large samples to provide national and sub-national reliable estimates.

A6.2.7 Technical cooperation among developing countries

Multi-country survey programs should seek to encourage technical cooperation among developing countries. As examples, NHSCP sponsored, in India, the training and exchange experiences of statisticians from developing countries, and DHS arranges regional training courses and seminars.

A6.3 Some multi-country survey programs

A6.3.1 World Fertility Survey

The World Fertility Survey (WFS) was sponsored by the United Nations Population Fund (UNFPA) and the U.S. Agency for International Development (USAID), with additional contribution provided, among others, by the governments of the U.K., France, Japan, the Netherlands, and the International Development Research Center of Canada. It was executed by the International Statistical Institute in collaboration with the International Union for the Scientific Study of Population and the U.N.; the surveys were implemented in participating countries by relevant government agencies. WFS secretariat was based in London, U.K. WFS was conceptualized in 1971 and was concluded in 1984.

Under WFS auspices, fertility surveys were conducted in 42 developing countries/territories in all the world regions. External assistance was $250,000 per country on an average.

In the participating countries, standardized core and optional questionnaire modules were used that permitted international comparison of data. The household schedules covered topics such as age, sex and marital status of household members and the individual questionnaire (for women) included detailed information on maternity and marriage histories, contraceptive knowledge and use, and fertility regulation.

The sample design was stratified multi-stage. The strata were geographical sub-divisions of a country. Area units, for which a sampling frame was available, and whose boundaries were reasonably well defined and whose individual population sizes were as small as possible were termed "basic area units" (bau's). In each selected ultimate area units, all the occupied dwellings (or households) were listed, from which a sample of enough dwellings (or households) were selected so as to yield an expected sample of between 20 and 50 "eligible" women (eligibility being determined by age and marital status); the probability of selection at that stage was such as to give a self-weighting design. The sample comprised, on an average, 10,000 households from whom household information was obtained and 5,000 "eligible" women for whom detailed fertility particulars were recorded.

For processing and analyzing survey data, WFS adapted available computer software and evolved new ones, including "CLUSTERS", for computing sampling variances. In case where there was only one first-stage unit (fsu) in a stratum, as in the Kenyan Fertility Survey, some strata were collapsed to provide two or more fsu's in the re-constituted stratum (see section 24.8 of this book).

For additional information on WFS, copies of publications and data files, contact International Statistical Institute, P.O. Box 950, 2270 AZ Voorburg, Netherlands.

A6.3.2 Contraceptive Prevalence/Family Planning Surveys

Contraceptive Prevalence Surveys (CPS) were sponsored by the USAID to provide the rapid feedback necessary to evaluate and improve family planning information and service delivery programs. The project was awarded in 1977 to Westinghouse Public Applied Systems; it was concluded in 1985. CPS surveys were in operation in 43 countries/territories in all the world regions.

Essentially, the surveys were stratified, multi-stage household samples of women of childbearing age in a national or state population. A sample of households was first selected, using an existing sampling frame, where available. The number of households sampled ranged from 3,000 in the Pernambuco State in Brazil to 20,000 in South Korea. Approximately equal numbers of households were selected in two or three residence strata – urban areas and rural areas, or the capital city, other urban areas, and rural areas. Roughly two-thirds of the sample households included at least one woman of child-bearing age, out of whom one respondent was selected for interview in each household.

Response rate was quite high, averaging 92 percent. The data were entered on computer tapes. As the questionnaire was short and relatively simple, usually taking 20 minutes to be completed, preliminary results could be produced within four to six months of the completion of the field work (*Population Reports*, 1981). Sampling errors were computed following standard methods.

In addition to CPS, another USAID-supported study, previously called Maternal-Child Health/Family Planning Surveys and now called **Family Planning Surveys** (FPS), also started in 1975 to collect data on the use and source of contraceptives, and information to plan and evaluate family planning programs. This continuing study is executed by the United States Centers for Disease Control. As of 31 December 1992, FPS surveys were, or are being, conducted in 16 countries in Africa, Latin America, the Caribbean, Western Asia, and Europe (*Population Reports*, 1981, 1985, and 1992).

Data collected In Brazil, El Salvador, and Paraguay revealed that the non-respondents were younger and had fewer children for their age than respondents, who were more likely to be at home and thus available for interview, perhaps because they had no outside jobs or had children to care for: CPS results may then slightly overestimate fertility and underestimate contraceptive use (*Population Reports*, Series M, No. 5, May-June 1981, citing Anderson, 1979).

For more information on CPS and FPS, their publications and data files, contact: Division of Reproductive Health, Center for Health Promotion and Education, Centers for Disease Control, Atlanta, Georgia, 30333, USA. CPS and FPS data are also archived by DHS (see the next section).

A6.3.3 Demographic and Health Surveys

The Demographic and Health Surveys (DHS) is a worldwide program of surveys that was started in 1984 intended to obtain data on reproduction and fertility preferences, contraception, infant mortality and morbidity and health (particularly of infants, children, and mothers) and related issues. Between 1985 and 1992, forty-eight DHS surveys had been conducted in 48 countries in the different world regions.

DHS is sponsored by the USAID, executed by Macro International (previously, Westinghouse's Institute for Resource Development), and implemented in participating countries by government agencies. The external assistance is US$200,000-500,000 per survey.

DHS uses stratified, multi-stage sample designs. Suitably chosen geographical subdivisions of a country constitute the strata; Census Enumeration Districts (ED's) comprise the first-stage units and "segments" of standard size, 500 census persons, the second- stage units. Every census ED, usually containing 1,000-2,000 persons, is assigned a measure of size equal to the number of standard segments it contains, by dividing its census population by 500 and rounding to the nearest whole number. A sample of EDs is then selected with probability proportional to this measure of size. In the selected EDs, a mapping operation is carried out to create the designated number of segments and one of these is selected with equal probability. In the selected segments, all dwellings or households are listed, and a fixed fraction of them are selected by systematic sampling. In the selected households, a roster is completed to identify women aged 15-49 and all of these are interviewed. The DHS program is designed to provide samples of 5,000 to 6,000 women aged 15-49 years (For details and discussion of issues, see DHS *Sampling Manual* and the articles by Aliaga and Verma (1991) and by Than Le (1993)).

For data processing, including computing sampling variances, DHS developed the software package, "ISSA" (see Section 27.10.3 of this book). An experiment was conducted in the DHS of Guatemala for interactive processing of data (Ochoa *et al.*).

For copies of DHS publications and data files, contact Macro International/Demographic and Health Surveys, 11785 Beltsville Drive, Suite 300, Calverton, Maryland, 20705-3119, U.S.A. phone: 01-301-572-0000; fax: 01-301-572-0999.

A6.3.4 National Household Survey Capability Program

The National Household Survey Capability Program (NHSCP) is sponsored and executed by the U.N. and implemented in participating countries by respective government agencies, with external funding provided by U.N. Development Program (UNDP), UNFPA, World Bank, UNICEF, and a number of donor countries, such as the USA, Canada, and Sweden; technical assistance is also being provided by countries such as India for training in household surveys.

The objectives of NHSCP are to assist countries through a systematic program to collect, process, analyze and disseminate integrated data on the household and its members for use in development plans, policies, administration and research; and to help in building national capabilities in household surveys.

NHSCP survey designs are country-specific and flexible in subject coverage and sample design. In general, the sample designs are stratified (with geographical stratification, based on administrative units in a country) and multi-stage, the sample households being selected from a listing of all the households in the selected penultimate-stage unit. The sample size varied in accordance with the measurement objectives of the survey; it ranged from 2,000 to 40,000, the modal sample size being 20,000 households.

The topics of study include agriculture, household budgets, income and expenditure, demography, family planning, health, nutrition, migration, labor force, and energy consumption. Samples for selected topics, in particular countries, are often chosen as sub-samples from an existing master sample of households.

By the end of 1992, NHSCP had supported the development of household survey programs in 50 countries in the different world regions.

For processing the survey data, mainframes, or increasingly personal computers, are used, adapting available software such as CONCOR, REDATAM, DBase, and ISSA (see chapter 27). A study on non-sampling errors in household surveys had been published in 1982 (UN, 1982b) and another study made on the computation of sampling errors (UN, 1993a), but computation of sampling errors and indications of non-sampling errors are not mandatory for the program.

For copies of publications and other information, contact Director, Statistical Division, United Nations, New York, New York, 10017, U.S.A.; phone: 01-212-963-4996; fax: 01-212-963-9851.

A6.3.5 The Living Standards Measurement Study

The Living Standards Measurement Study (LSMS) was initiated by the World Bank as an effort to capture the many dimensions of living conditions using integrated, multi-topic, household surveys, with a strong experimental and research orientation.

Between 1985 and 1992, LSMS surveys were conducted in eleven countries in Africa, Asia, Latin America, and the Caribbean.

One special feature of LSMS was the decentralization of data processing. Data processing units (with microcomputers) were set up in the field and batches of completed questionnaires were sent to these units for editing. Questionnaires with errors were then sent back to the field for checking and corrections. Soon after enumeration, fully edited responses in machine-readable form were thus available for final tabulations. This editing was, however, restricted only for sections of the questionnaire covered in the first interviewing round.

For copies of publications and other information, contact World Bank, 1818 H Street N.W., Washington D.C., 20233, U.S.A.

A6.3.6 The Social Dimensions of Adjustment Program

The Social Dimension of Adjustment (SDA) Program was launched in 1987 by the UNDP Regional Program for Africa, the African Development Bank, and the World Bank in collaboration with other multilateral and bilateral agencies, with the objective of strengthening the capacity of governments in the Sub-Saharan African region to integrate social dimensions in the design of their structural adjustment programs.

The standard SDA survey uses a stratified two-stage sample design. Census enumeration areas (ea's) constitute the first stage units, ea's with very large ea's being split and very small ea's being combined. The resulting ea's (clusters) are then stratified and a sample of clusters is selected systematically with pps within the strata, all the households in the selected clusters listed, and finally a fixed number of sample households selected in each selected cluster. The normal sample size is 3,000-4,000 households.

The survey data are processed using either commercial software, such as SAS-PC and SPSS, or software specifically developed, such as IMPS or ISSA (see Chapter 27).

Since the inception of the SDA program, 30 countries have received or have requested, assistance to establish national SDA programs, and of these, 20 now have on-going programs. For copies of publications and other information, contact World Bank, 1818 H Street N.W., Washington D.C., 20233, U.S.A.

A6.3.7 Gulf Child Health Survey

The Gulf Child Health Survey (GCHS) is a research program implemented by the Council of Health Ministers of the Gulf Co-operation Council States, sponsored by the Arab Gulf Program for the United Nations Development Organizations (AGFUND), and with the collaboration of UNFPA, UNICEF, and WHO.

GCHS was initiated in 1986 to provide detailed information on the bio-demographic determinants of maternal and child health in the region that are required for the evaluation of on-going programs as well as for the formulation of new health policies and programs. GCHS surveys were conducted in all the seven member-countries of the Gulf Co-operation Council. GCHS used a stratified multi-stage sampling design, with households as the ultimate stage units and geographical areas as the other stage units.

The survey included a basic household survey, a maternal care survey and a child health survey.

For further information on GCHS and copies of publications and information, contact Director-General, Executive Board, Council of Health Ministers of the Gulf Co-operation Council, Post Box 7431, Riyadh 11462, Saudi Arabia.

A6.3.8 Pan-Arab Project for Child Development

The Pan-Arab Project for Child Development (PAPCHILD) was initiated in 1987 with the long-rage objective of undertaking comprehensive surveys covering aspects of household socio-economic and demographic characteristics, maternal care (including maternal history), child care, and community characteristics for 14 Arab counties including six least-developed countries. PAPCHILD is sponsored by the League of Arab States and supported by the AGFUND, UNFPA, UNICEF, WHO, and the UN. The project is based at the headquarters of the League of Arab States in Cairo, Egypt. In its first phase, the project covered six Arab countries; phase II of the program, through 1996, is now in operation, with surveys planned, or now under way, in three other Arab countries. The average external assistance for a PAPCHILD survey was US$380,000.

PAPCHILD survey is carried out using a stratified two-stage sample of households. The prototypical plan contains 300 first-stage units (fsu's) and 6,000 households. The 300 fsu's are "standard segments" derived from census enumeration districts as the sampling frame. In some countries, resources permitting, the second stage of sampling may be expanded to 18,000 households. Within a selected fsu, a sample of 20 households (second-stage units, ssu's) will be selected for interviewing (60 households for countries opting for 18,000 total sample households). Individual interviews will relate to all ever-married women aged under 54 and all children (under 5), irrespective of whether their mothers were household members.

Survey data are processed and analyzed, adapting available computer software, such as ISSA (see Section 27.10.3). Sampling errors have been, or are being, computed for major estimates.

For copies of publications and information, contact: Project Manager, Pan-Arab Project for Child Development, League of Arab States, 22A Taha Hussein Street, Cairo, Egypt; phone 202-340-4306; fax: 202-340-1422.

A6.3.9 Expanded Program on Immunization Surveys

In 1977 the World Health Organization's Expanded Program on Immunization (EPI) established the objective of immunizing all children throughout the world against six major childhood diseases – diphtheria, whooping cough, tetanus, measles, polio, and tuberculosis. An international effort was led by UNICEF and WHO to support ministries of health of governments all over the world, so that their national immunization services reach 80 per cent of the world's children against these diseases by 1990, starting from immunization coverage of 15 per cent in 1980 (*Population Reports*, 1992). Towards that end, WHO has developed a methodology for estimating, by use of relatively quick and inexpensive sample surveys, immunization levels of children in target areas.

The sample for the EPI surveys comprises 30 sample cluster with seven children in each cluster, yielding a sample size of 210 children (Levy and Lemeshow (1991), Section 14.1; also see Lemeshow, Hosner and Klar (1989) and Lemeshow and Robinson (1985)).

In each selected area (health service district, village, town, or city), a pps sample of 30 clusters is taken at the first stage, but the procedure of selection of the second-stage sample of seven children differed from probability sampling. In the EPI surveys, a "starting household" was first selected; after collecting the required information on eligible subjects in that household, another household was proceeded to whose front door was phys-

Table A6.1: Estimates and standard errors from some multi-country survey programs

Item	Country	Year	Survey[a]	Sample Size[b]	Estimate	S.E.[c]	P.E.[d]
Children ever born (per woman)	Kenya	1977-8	WFS	8,100	3.896	0.060	1.3
	Nepal	1981	CPS	5,880	3.28	0.072	2.2
Proportion of children fully immunized	Cameroon	1991	DHS	644	0.406	0.030	7.5
	Jordan	1990	DHS	6,462	0.880	0.009	1.0
Proportion of currently pregnant women	Colombia	1986	DHS	5,331	0.097	0.006	6.1
	Mauritania	1990-1	PAP	5,319	0.130	0.010	7.7
Proportion of women ever-used contraceptives	Korea, South	1974	WFS	5,420	0.566	0.009	1.9

Source: WFS, CPS, DHS, and PAPCHILD Country Reports.

a CPS = Contraceptive Prevalence Survey; DHS = Demographic and Health Surveys.

PAP = Pan Arab Child Health Program; WFS = World Fertility Survey.

b Number of "eligible" (generally, currently married and aged 15-49 years) interviewed.

c S.E. = Standard Error.

d P.E. = Percentage Error = 100 (Standard Error)/(Estimate). Termed Relative Error in DHS publications.

ically closest to the starting household. This procedure of visiting the next closest households and collecting information on all eligible individuals in those households continued until the required seven subjects are studied (For details, see Lemeshow and Stroh, 1988).

About 4,500 EPI surveys had been conducted in 122 countries all throughout the world between 1978 and 1992.

For obtaining estimates, relating both to children and adults, and their sampling variances from EPI surveys, a software, COSAS (COvearge Survey Analysis System), which is now in version 4.3, was developed (Desvé, Havreng, and Brenner, 1991).

Absent the total number of children and the number of persons in the clusters, EPI surveys cannot be said to have adopted strictly probability samples in selecting the second-stage sample of 7 children in each of the 30 selected cluster. However, a study with computer simulation models has led to the conclusion that "the method appears quite useful when used for the target areas as a whole, but could provide highly unacceptable estimates if used for particular clusters or subgroups" (Levy and Lemeshow, *ibid.*).

For further information and copies of publications, contact: EPI Project, World Health Organization, CH 1211 Geneva-27, Switzerland.

A6.3.10 Some estimates

Estimates relating to fertility, immunization and family planning from the World Fertility Survey, the Contraceptive Prevalence Survey, the Demographic and Health Surveys and the Pan Arab Child Health Program, are given in Table A6.1 for Cameroon, Colombia, Jordan, Kenya, Mauritania, Nepal, and South Korea.

Further reading

For the World Fertility Survey: see its *Final Report* and national reports. For the World Bank sponsored Social Dimensions of Adjustment Surveys in Sub-Saharan Africa, see Grootaert and Marchant (1991) and Delaine *et al.* (1992). For other survey programs, see the references cited in the text. On the medicinal properties of the "neem" (*Melia Azadirachta*) tree, referred to in Section A6.2.4, see "From the Ancient Neem Tree, a New Insecticide," *New York Times*, 5 June 1994, p. 49.

References

This list comprises publications listed under "Further Reading" at the end of each chapter and others cited in the text. For the abbreviations of the names of journals, organizations and programs, see the *List of Abbreviations*.

ABERNATHY, J.R., B.G. GREENBERG and D.G. HORVITZ (1970). "Estimates of induced abortion in urban North Carolina," *Demography*, 7, 19.

ACKOFF, R.L. (1953). *Design of Social Research.* University of Chicago Press, Chicago.

ACSADI, GEORGES, ANDREAS KLINGER, and EGON SZABADY (1969). *Survey Techniques in Fertility and Family Planning Research: Experience in Hungary.* Central Statistical Office, Demographic Research Institute, Budapest.

ACSADI, GEORGES, ANDREAS KLINGER, and EGON SZABADY (1970). *Family Planning in Hungary: Main Results of the 1966 in Fertility and Family Planning (TCS) Study.* Central Statistical Office, Demographic Research Institute, Budapest.

ADLAKA, ARJUN L., HEATHER BOOTH and JOAN W. LINGNER (1977). "The Dual Record System: Sampling Design–POPLAB Experience," Laboratories for Population Statistics, *Scientific Report Series No.* 30. The University of North Carolina at Chapel Hill, North Carolina.

ALDER, JAMES E. (1985). "Focus Groups," *1985 Test Census: Preliminary Research and Evaluation Memorandum No. 19.* U.S. Bureau of the Census, Washington D.C.

ALIAGA, ALFREDO and VIJAY VERMA (1991). "An Analysis of Sampling Errors in the DHS," *DHS World Conference, Washington, D.C., 5-7 August 1991, Proceeding, Vol. I,* IRD/Macro International Inc., Columbia, Maryland, 513-537.

ANDERSON, J. E. (1979). "Fertility estimation: a comparison of results of Contraceptive Prevalence Surveys in Paraguay, São Paulo State (Brazil), and El Salvador," *1979 Proceedings of the Social Statistics Section*, ASA, 532-537.

APPEL, MARTIN V. (1992). "Touchtone Data Entry: Initial technical assessment," CASIC Committee on Technology Testing, U.S. Bureau of the Census, Washington, D.C.

ARBIS, G. and M. COLI. (1991). "The case of GIS in spatial and space-time statistical surveys," *ISI/IASS Booklet, Invited Papers,* Cairo, 9-17 September 1991, 464-491.

ARDILLY, PASCAL (1994). *Les Techniques de Sondage.* Technip, Paris.

AZORIN, F. and J.L. SANCHEZ-CRESPO (1986). *Metodos y Aplicaciones del Muestro.* Allianza Universidad Textos, Madrid.

BAILAR, BARBARA and PAUL P. BIEMER (1984). "Some methods of evaluating nonsampling error in household censuses and surveys." In *W.G. Cochran's Impact on Statistics,* edited by Poduri S.R.S. Rao and Joseph Sedransk, Wiley, New York.

BAILEY, N.T.J. (1951). "On estimating the size of mobile population from recapture data," *Biometrika,* 38, 293-306.

BAKER, REGINALD P. and NORMAN M. BRADBURN (1992). "CAPI: Impacts on data quality and survey costs." In *Proceedings of the 1991 Public Health Conference on Records and Statistics.* NCHS, Washington, D.C., 459-464.

BANGLADESH: BUREAU OF STATISTICS (1991). *Report of the Child Nutritional Status Survey 1989-1990.* Dhaka.

BARNETT, VIC (1991). *Sample Survey Principles and Methods.* Edward Arnold, London and Wiley, New York.

BEHMOIRAS, J.P. (1965). *La situation démographique au Tchad: Résultats provisoires de l'enquête démographique 1964.* République du Tchad, Commissariat général au plan, Société d'Etudes pour le Développment Economique et Sociale, Paris.

BELLHOUSE, D.R. (1988a). "A brief history of random sampling methods." In *Handbook of Sampling, Vol. 6, Sampling,* edited by P.R. Krishnaiah and C.R. Rao. Elseiver, Amsterdam, 1-14.

BELLHOUSE, D.R. (1988b). "Systematic sampling." In *Handbook of Statistics, Vol. 6, Sampling,* edited by P.R. Krishnaiah and C.R. Rao. Elseiver, Amsterdam, 125-145.

BENJAMIN, BERNARD (1955). "Quality of response in census taking," *Population Studies,* 8, 288-293.

BENJAMIN, BERNARD (1968). *Health and Vital Statistics.* Allen and Unwin, London.

BETHLEHEM, JELKE G., LON HOFMAN and MAARTEN SCHUERHOFF (1989). "Computer assisted coding at the NCBS," NCBS, Voorburg, The Netherlands, BPA no. 4090-89-M3, 17 March 1989.

BETHLEHEM, JELKE G. and HUNDEPOOL, ANCO J. (1992). "Integrated statistical information processing on microcomputers," IUSSP/NIDI Expert Meeting on Demographic Software and Computing. NIDI, The Hague, 29 June-3 July 1992.

BHATTACHARYA, BHAIRAB C. (1992). "Cattle AIDS," *Proc. of the Zoological Society*, Calcutta, 45, 1-6.

BIEMER, PAUL P., ROBERT M. GROVES, LARS E. LYBERG, NANCY A. MATHIOWETZ and SEYMOUR SUDMAN (eds.) (1991). *Measurement Errors in Surveys*. Wiley, New York.

BIEMER, PAUL P. and S. LYNNE STOKES (1991). "Approaches to the modelling of measurement errors." In *Measurement Errors in Surveys* (eds. Biemer *et al.*). Wiley, New York.

BINDER, A. and M.A. HUDIROGLOU (1988). "Sampling in time." In *Handbook of Statistics, Vol. 6, Sampling,* edited by P.R. Krishnaiah and C.R. Rao. Elseiver, Amsterdam.

BLALOK, HUBERT M. (1979). *Social Statistics*. McGraw-Hill, New York.

BLANC, ROBERT (1962). *Manuel de Recherche démographiques en pays sous-développés*. INSEE, Paris. (Translated at the UN Economic Commission for Africa, *Manual of Demographic Research in Under-developed Countries*, E/CN.14/ASPP/L.14.).

BLANKENSHIP, ALBERT B. (1978). *Consumer and Opinion Research*. Ayer, Salem, New Hampshire.

BOCHTLER, ERWIN and ODED LÖWENBEIN (1991). "Combining house-based and establishment-based frames in marketing research surveys the single source approach in Germany," *ISI/IASS Booklet, Invited Papers, Cairo, 9-17 September 1991*, 126-132.

BOSWELL, M.T., K.P. BURNHAM and G.P. PATIL (1988). "Role and use of composite sampling and capture-recapture sampling in ecological studies." In *Handbook of Statistics, Vol. 6, Sampling,* edited by P.R. Krishnaiah and C.R. Rao. Elseiver, Amsterdam.

BOURKE, PATRICK D. and TORE DALENIUS (1974). "A Note on inadmissible estimates in randomized inquiries," *Technical Report 72*. Institute of Statistics, University of Stockholm, Stockholm.

BRADBURN, NORMAN M., MARTIN R. FRANKEL, REGINALD P. BAKER and MICHAEL R. PERGAMIT (1992). "A comparison of computer-assisted personal interview (CAPI) with paper-and-pencil (PAPI) interviews in the National Longitudinal Study on Youth," NLS Discussion Paper 92-2. Bureau of Labor Statistics, U.S. Department of Labor, Washington, D.C.

BREWER, K.R.W. and M. HANIF (1983). *Sampling with Unequal Probabilities*. Springer-Verlag, New York.

BRICK, J.M. (1992). "Questions and answers," *The Survey Statistician*, 28, 23-24.

BROOKES, B.C. and W.F.L. DICK (1969). *Introduction to Statistical Methods* (2nd ed.). Heinemann, London.

BULLOUGH, VERN L. (1994). *Science in the Bedroom: A History of Sex Research*. Basic Books. New York.

BUSH, CAROLEE (1985). "Observation of Mississippi Test Census Focus Groups, December 2 and 3, 1985, in Meridian and Philadelphia." Internal memorandum, U.S. Bureau of the Census, Washington D.C.

CALDER, BOBBY J. (1977). "Focus groups and the nature of qualitative marketing research," *JMR,* 14(3), 353-364.

CAMEROUN: DIRECTION NATIONALE DU DEUXIEME RECENSEMENT GENERAL DE LA POPULATION ET DE L'HABITAT and DHS (1992). *Enquête Démographique et de Santé, Cameroun, 1991.* Yaoundé and Columbia, Maryland.

CANTOR, DAVID C. and GUILLERMO ROJAS (1992). "Future prospects for survey processing in developing countries: Technologies for data collection," IUSSP/NIDI Expert Meeting on Demographic Software and Computing. NIDI, The Hague, 29 June-3 July 1992.

CANTRIL, H. (1944). *Gauging Public Opinion.* Princeton University Press, Princeton.

CARLSON, BEVERLEY A. (1987). "Core indicators for the Interagency Food and Nutrition Programme." UNICEF, New York.

CARLSON, BEVERLEY A. and JOHN JAWORSKI (1994). *Surveying Nutritional Status of Young Children.* Package contains text, video-cassette and audio-cassette. UNICEF, New York.

CASLEY, D.J. and D.A. LURY (1982). *Data Collection in Developing Countries.* Clarendon Press, Oxford.

CHAKRAVARTI, I.M., R.G. LAHA and J. ROY (1967). *Handbook of Methods of Applied Statistics, Vol. II, Planning of Surveys and Experiments.* Wiley, New York.

CHAMIE, MARY (1989). "Survey design strategies for the study of disability," *WHSQ,* 42 (3), 122-140.

CHANDRA SEKAR, C. and W.E. DEMING (1949). "On a method of estimating birth and death rates and the extent of registration," *JASA,* 44, 101-115.

CHATTERJEE, S. (1967). "A note on optimum stratification." *Skand. Akt.,* 50, 40-54.

CHAUDHURI, ARIJIT and RAHUL MUKHERJEE (1988). *Randomized Response: Theory and Practice.* Marcel Dekker, New York.

CHAUDHURI, ARIJIT and HORST STENGER (1992). *Survey Sampling: Theory and Methods.* Marcel Dekker, New York.

CHINAPPA, NANJAMMA (1963). "Technical paper on sample designs of Working Class and Middle Class Family Living Surveys." *Sankhyā (B),* 25, 359-418; reprinted as chapter 16, "Family Living Surveys" in Murthy (1967).

CLAUSEN, J.A. and R.N. FORD (1947). "Controlling bias in mail questionnaire," *JASA,* 42, 497-511.

COALE, ANSLEY J. (1955). "The population of the United States in 1950 classified by age, sex, and color a revision of census figures." *JASA*, 50, 16-54.

COCHRAN, WILLIAM G. (1977). *Sampling Techniques* (3rd ed.). Wiley, New York.

COCHRAN, WILLIAM G., F. MOSTELLER, and J. W. TUKEY (1953). "Statistical Problems of the Kinsey Report", *JASA*, 48, 673-716.

COEFFIC, NICOLE (1993). "En France, une enquête originale de measure du degré d'exhaustivité du recensement de 1990," *ISI/IASS Booklet, Invited Papers, Florence, 25 August-2 September 1993*, 151-182.

COLE, ROBERT A. and DAVID G. NOVICK (1993). "Rapid prototyping of spoken language system: The year 2000 census project," U.S. Bureau of the Census, Washington, D.C. (mimeographed).

COLLINS, MARTIN (1991). Sampling issues in U.K. Telephone Surveys," *ISI/IASS Booklet, Invited Papers, Cairo, 9-17 September 1991*, 146-163.

COX, KEITH K., J.B. HIGGINBOTHAM and J. BURDON (1976). "Applications of focus group interviews in marketing," *Journal of Marketing*, 40 (1), 77-80.

CREIGHTON, K., S. MATCHETT, and C. LANDMAN (1994). "Building Integrated Systems of CASIC Technologies at the U.S. Bureau of the Census (Draft)," U.S. Bureau of the Census, March 23, 1994 (mimeographed).

CROFT, TREVOR (1992). "The Integrated System for Survey Analysis (ISSA): Design principles behind the development of ISSA," IUSSP/NIDI Expert Group Meeting on Demographic Software and Micro-Computing, The Hague, 29 June-3 July 1992.

DALENIUS, TORÉ (1957). *Sampling in Sweden: Contributions to the Method and Theories on Sample Survey Practice.* Almqvist and Wiksell, Stockholm.

DALENIUS, TORÉ (1985). *Elements of Survey Sampling.* SAREEC, 8-105 25 Stockholm.

DALENIUS, TORÉ (1988). "A first course in survey sampling." In *Handbook of Statistics, Vol. 6, Sampling,* edited by P.R. Krishnaiah and C.R. Rao. Elseiver, Amsterdam.

DAS, N.C. (1969). *Special Study on Morbidity.* NSS No. 119. The Cabinet Secretariat, Government of India, New Delhi.

DASGUPTA, AJIT (1958). "Determination of fertility level and trend in defective registration areas," *BISI*, 36, 127-136.

DATAQUEST (1991). "The year of uncertainty." New Delhi, July 1991.

DAVID, H.A. (ed.)(1978). *Contributions to Survey Sampling and Applied Statistics: Papers in Honor of H.O. Hartley.* Academic Press, New York.

DEAN, A.G. *et al.*, (1990). *EPI INFO, Version 5.0: A Word Processing, Database and Statistics Program for Epidemiology on Microcomputers.* USD Inc., Stone Mountain, Georgia.

DEGRAFT-JOHNSON, K.T. (1993). "Review of the NHSCP, 1979-1992." UN, Department of Economic and Social Development, Doc. No. 93-09635, New York.

DEKKER, ARIJ L. (1991). "New or uncommon computer methods in population census data processing," *ISI/IASS Booklet, Invited Papers,* Cairo, 9-17 September 1991, 392-405.

DELAINE, GHISLAINE, LIONEL DEMERY, JEAN-LUC DUBOIS, BRANKO GRDJIC, CHRISTIAAN GROOTAERT, CHRISTOPHER HILL, TIMOTHY MARCHANT, ANDREW McKAY, JEFFERY ROUND and CHRISTOPHER SCOTT (1992). *The Social Dimensions of Adjustment Integrated Survey,* SDA Working Paper No. 14. World Bank, Washington D.C.

DEMING, W. EDWARDS (1950). *Some Theory of Sampling.* Reprint (1984), Dover, New York.

DEMING, W. EDWARDS (1960). *Sample Design in Business Research.* Wiley, New York. Reprint (1990).

DEMING, W. EDWARDS (1963). "On some of the contributions of interpenetrating networks of samples." In *Contributions to Statistics, Presented to Professor P.C. Mahalanobis on the Occasion of His 70th Birthday,* edited by C.R. Rao. Pergamon Press, London, and Statistical Publishing Society, Calcutta.

DEMING, W. EDWARDS (1969). "Boundaries of statistical inference." In *New Developments in Survey Sampling,* edited by Norman L. Johnson and Harry Smith, Jr. Wiley, New York.

DEMING, W. EDWARDS and G.J. GLASSER (1959). "On the problem of matching data by samples." *JASA,* 54, 403-415.

DEROO, MARC and DUSSAIX, ANNE-MARIE (1980). *Pratique et Analyse des Enquêtes par Sondages.* Presses Universitaires de France, Paris.

DESABIE, M.J. (1962). *Théorie et Pratique des Sondages, Tomes I et II.* INSEE, Paris.

DESVÉ, GILLES, JEAN FRANÇOIS HAVRENG and ERIC BRENNER (1991). *COSAS 4.3 (COverage Survey Analysis System): Programme for Analysis of Immunization Coverage Surveys.* EPICENTRE, Paris, and WHO, Geneva.

DHS (1987). *Sampling Manual.* IRD, Columbia, Maryland.

DILLMAN, DON A. (1978). *Mail and Telephone Surveys: The Total Design Method.* Wiley, New York.

DROITCOUR, JUDITH, ROBERT A. CASPAR, MICHAEL E. HUBBARD, TERESA L. PARSLEY, WENDY VISSCHER and TRENA M. EZZATI (1991). "The item count technique as a method of indirect questioning: a review of its development and a case study application." In *Measurement Error in Surveys,* edited by P. Biemer *et al.*, Wiley, New York, 185-210.

DURBIN, J. (1959). "A note on the estimation of Quenouille's method of bias reduction in the estimation of ratios," *Biometrika,* 46, 477-480.

DUSSAIX, ANNE-MARIE and JEAN-MARIE GROSBRAS (1993). *Les Sondages: Principes et Méthodes,* Collection "Que sais-je," no. 701. Presses Universitaires de France, Paris.

DUTKA, SALOMON and LESTER R. FRANKEL (1991). "Role of data banks in marketing research," *ISI/IASS Booklet, Invited Papers, Cairo, 9-17 September 1991,* 112-125.

EFRON, BRADLEY (1979). "Bootstrap methods: another look at the jackknife," *AMS,* 7, 1-26.

EFRON, BRADLEY (1982). *The Jackknife, the Bootstrap, and Other Resampling Plans.* Society for Industrial and Applied Mathematics, Philadelphia.

EFRON, BRADLEY (1989). "Bootstrap and other resampling methods," *IMSB,* 18, 406-408.

EFRON, BRADLEY and G. GONG (1983). "A leisurely look at the Bootstrap, Jackknife and cross validation," *American Statistician,* 37, 36-38.

EFRON, BRADLEY and ROBERT J. TIBSHIRANI (1993). *An Introduction to the Bootstrap.* Chapman & Hall, New York.

EL-BADRY, M.A. (1956). "A sampling procedure for mailed questionnaire," *JASA,* 51, 209-227.

ELLIS, CARLOS (1991). "Population census data processing in Latin America: new developments," *ISI/IASS Booklet, Invited Papers, Cairo, 9-17 September 1991,* 435-446.

ERICSON, W.A. (1969). "Subjective Bayesian models in sampling finite populations: stratification." In *New Developments in Survey Sampling,* edited by Norman L. Johnson and Harry Smith, Jr. Wiley, New York.

ERICSON, W.A. (1988). "Bayesian Inference in finite populations." In *Handbook of Statistics, Vol. 6, Sampling,* edited by P.R. Krishnaiah and C.R. Rao. Elseiver, Amsterdam.

ESPANA, EDUARDO GARCIA (1978). *Design of the General Population Survey.* Ministerio de Planification del Desarrolo, Instituto Nacional de Estadistica, Madrid.

FAO (1969). *Improvement of Agricultural Census and Survey Reports.* African Commission on Agricultural Statistics, Fourth Session, Algiers (AGS:AF/4/69/4), Rome.

FAO (1984). *Guidelines on Selected Nutrition Indicators.* Bangkok.

FAO (1989). *Sampling Methods for Agricultural Surveys.* FAO Statistical Development Series, Rome.

FELLEGI, IVAN P. (1964). "Response variance and its estimation," *JASA,* 59, 1016-1041.

FELLEGI, IVAN P. and ALAN R. SUNTER (1969). "A theory for record linkage," *JASA,* 64, 1183-1210.

FELLER, W. *An Introduction to Probability Theory and its Applications.* Vol. I (3rd ed., 1968), Vol. II (2nd ed., 1971). Wiley, New York.

FERBER, ROBERT, PAUL SHEATSLEY, ANTHONY TURNER and JOSEPH WAKSBERG (undated). *What Is a Survey?* ASA, Washington, D.C.

FERN, EDWARD F. (1982). "The use of focus groups in ideas generation: the effects of groups size, acquaintanceship, and moderator on response quantity and quality," *JMR,* 19(1), 1-13.

FICHNER, R.R., K.M. SULLIVAN, F.L. TROWBRIDGE and B.A. CARLSON (1989). "Report of the Technical Meeting on Software for Nutritional Surveillance." *Food and Nutrition Bulletin,* 11(2), 57-61. UN University, Tokyo.

FIENBERG, S.E. and J.M. TANUR (1983). "Large-scale social surveys: perspectives, problems, and prospects," *Behavioral Science,* 28, 135-153.

FISHER, R.A. and F. YATES (1973). *Statistical Tables for the Biological, Agricultural and Medical Research* (6th ed.). Longman, Harlow, U.K.

FOLCH-LYON, EVELYN and JOHN F. FROST (1981). "Conducting Focus Group Sessions," *SFP,* 12 (12), 443-449.

FOLCH-LYON, EVELYN, LUIS DE LA MACORRA and S. BRUCE SHEARER (1981). "Focus Group and Survey Research on Family Planning in Mexico," *SFP,* 12 (12), 409-432.

FOREMAN, E.K. (1991). *Survey Sampling Principles.* Marcel Dekker, New York.

FORSYTH, BARBARA H. and JUDITH T. LESSLER (1991). "Cognitive laboratory methods: a taxonomy." In *Measurement Error in Surveys,* edited by Biemer *et al.*, Wiley, New York, 393-418.

FOWLER, F. (1988). *Survey Research Methods.* Applied Social Research Methods Series, Vol. I. Sage Publications, Beverly Hills.

FRANCE: MINISTERE DE LA FRANCE D'OUTRE-MER, SERVICE DES STATISTIQUES; HAUT COMMISSARIATE DE L'AFRIQUE OCCIDENTALE FRANCAISE; DIRECTION DE LA STATISTIQUE GENERALE DE L'AOF ET SERVICE STATISTIQUE DE LA GUINEE (1956). *Etude démographique par sondage, Guineé, 1954-55, Iere Partie – Technique d'enquete.* Mission Démographique de Guineé, Paris.

FRANKEL, L.R. (1969). "The role of accuracy and precision of response in sample surveys." In *New Developments in Survey Sampling,* edited by Norman L. Johnson and Harry Smith, Jr. Wiley, New York.

FREEMAN, MICHAEL (1986). "Los Angeles Post Census Focus Groups." Internal note, Bureau of, Washington D.C.

FREY, JAMES H. *Survey Research by Telephone. Sage Publications* (2nd ed.). Newbury Park, California.

FURRIE, ADELE D. (1989). *Comparison of the Results from the 1986 Census and the Health and Activity Limitation Survey for Persons with Disabilities*

Residing in Households. Statistics, Canada, Ottawa.

GAGE, T.J. (1978). "Theories differ on use of focus groups," *Advertising Age*, S-19, 20-22.

GALLUP, GEORGE (1948). *A Guide to Public Opinion Polls.* Princeton University Press, Princeton.

GANGULY, AMALENDU N. and RANJAN K. SOM (1958). "A Note on the Estimated Variances of Vital Rates in the National Sample Survey, Seventh Round," (Unpublished), Indian Statistical Institute, Calcutta (mimeographed).

GERLAND, PATRICK (1992). "Software development: Past, present and future trends and tools." IUSSP/NIDI Expert Group Meeting on Demographic Software and Micro-Computing, The Hague, 29 June-3 July 1992.

GINI, C. and L. GALVANI (1929), "Di una applicazione de metodo rapprsentativo all'ultimo censimento Italiano della poplazione," *Annali di Statistica*, 6, 1-107.

GLASS, DAVID V. and EUGENE GREBENIK (1954). *The Trend and Pattern of Fertility in Great Britain: A Report on the Family Census of 1946, Part I (Report).* H.M.S.O., London.

GODAMBE, V.P. (1955). "A unified theory of sampling from finite populations," *JRSS(B)*, 17, 269-278.

GODAMBE, V.P. and *D.A. SPROTT* (eds.) (1971). *Foundations of Statistical Inference.* Holt, Rinehart and Winston, Toronto.

GODAMBE, V.P. and M.E. THOMPSON (1988). "On single stage unequal probability sampling." In *Handbook of Statistics, Vol. 6, Sampling,* edited by P.R. Krishnaiah and C.R. Rao. Elsevier, Amsterdam, 111-124.

GOLDSTEIN, RICHARD (1990). "A review of Resampling software for MS-DOS computers," *Proceedings of the Statistical Computing Section,* ASA, 42, 86.

GOODMAN, R. and L. KISH (1950). "Controlled selection a technique in probability sampling," *JASA*, 45, 350-372.

GOURIEROUX, C., A.M. DUSSAIX, J.C. DEVILLE and J.M. GROSBARS (1987). Contribution dans *Les Sondages* par Jean-Jacques Droesbeke, Bernhard Fichet and Philippe Tassi (éditeurs). Economica, Paris.

GRAY, HENRY L. and W.R. SCHUCANY (1972). *The Generalized Jackknife Statistics.* Marcel Dekker, New York.

GRAY, P.G. and T. CORLETT (1950). "Sampling for the Social Survey." *JRSS(A)*, 113, 150-206.

GRIK, D.C., K. PARKER and G.M.B. HATEGIKAMANA (1987). "Integrating quantitative and qualitative survey techniques." *Community Health Education*, 7(3), 181-200.

GROOTAERT, CHRISTAAN and TIMOTHY MARCHANT (1991). *The Social Dimensions of Adjustment Priority Survey,* SDA Working Paper No. 14. World Bank, Washington D.C.

GROSBRAS, JEAN-MARIE (1987). *Méthodes Statistiques des Sondages.* Economica, Paris.

GROVES, ROBERT M. (1989). *Survey Errors and Survey Costs.* Wiley, New York.

GROVES, ROBERT M., PAUL P. BIEMER, LARS E. LYBERG, JAMES T. MESSLEY, WILLIAM T. NICHOLS II, and JOSEPH WAKSBERG (eds.) (1989). *Telephone Survey.* Wiley, New York.

HAJEK, J. (1981). *Sampling from a Finite Population.* Marcel Dekker, New York.

HAJEK, JAROSLAV and VAVCLAC DUPAC (1967). *Probability in Science and Engineering.* Academic Press, New York.

HALDANE, J.B.S. (1945). "On a method of estimating frequencies," *Biometrika,* 33, 222-225.

HAMMERSLEY, J.M. (1953). "Capture-recapture analysis." *Biometrika,* 40, 265-278.

HANSEN, MORRIS H., TORE DALENIUS and BENJAMIN J. TEPPING (1985). "The Development of Sample Surveys of Finite Populations." In *A Celebration of Statistics* edited by A.C. Atkinson and S.E. Fienberg. Springer-Verlag, New York, 327-354.

HANSEN, MORRIS H. and WILLIAM N. HURWITZ (1943). "On the theory of sampling from finite population," *AMS,* 14, 333-362

HANSEN, MORRIS H. and WILLIAM N. HURWITZ (1946). "The problem of nonresponse in sample surveys," *JASA,* 41, 517-529.

HANSEN, MORRIS H., WILLIAM N. HURWITZ and M.A. BERSHAD (1961). "Measurement errors in censuses and surveys," *BISI,* 38, 359-374.

HANSEN, MORRIS H., WILLIAM N. HURWITZ and T.J. JABINE (1964). "The use of imperfect lists for probability sampling at the United States Bureau of the Census," *BISI,* 40, 497-517.

HANSEN, MORRIS H., WILLIAM N. HURWITZ and WILLIAM G. MADOW (1953). *Sample Survey Methods and Theory: Vol. I, Methods and Applications; Vol. II, Theory.* Wiley, New York. Reprint (1993).

HANSEN, MORRIS H., WILLIAM N. HURWITZ, ELI S. MARKS and W. PARKER MAULDIN (1951). "Response errors in surveys," *JASA,* 46, 147-190.

HANSEN, MORRIS H., WILLIAM N. HURWITZ and L. PRITZKER (1964). "The estimation and interpretation of gross differences and the simple response variance." In *Contribution to Statistics,* edited by C.R. Rao, Pergamon Press, Oxford.

HANSEN, MORRIS H. and L. PRITZKER (1956). "The Post-Enumeration Survey of the 1950 Census of Population: some results, evaluations and implications." Paper presented at the Annual Meeting of the Population Association of America, Ann Arbor, Michigan, 19 May 1956. Mimeographed.

HANSEN, MORRIS H. and BENJAMIN J. TEPPING (1969). "Progress and problems in survey methods and theory illustrated by the United States Bureau of the Census." In *New Developments in Survey Sampling,* edited by Norman L. Johnson and Harry Smith, Jr. Wiley, New York.

HAREWOOD, JOHN (1968). *Continuous Sample Survey of Population, General Reports, Rounds 1-8.* Continuous Sample Survey of Population, Publication no. 11, Central Statistical Office, Trinidad and Tobago, Port of Spain.

HARRIS, F.F. (1945). "The use of sampling methods for ascertaining total morbidity in the Canadian Sickness Survey, 1950-51," *WHO Bulletin,* 11, 25-50.

HARTLEY, H.O. (1946). Discussions on "A review of recent statistical developments in sampling and sampling surveys," by Frank Yates, *JRSS (A),* 109, 19-43.

HEDAYAT, A.S. and BIKAS K. SINHA (1991). *Design and Inference in Finite Population Sampling.* Wiley, New York.

HENDRICKS, WALTER A. (1949). "Adjustment for bias by non-response in mailed surveys," *Agricultural Economics Research,* 1, 52-56.

HENDRICKS, WALTER A. (1956). *The Mathematical Theory of Sampling.* Scarecrow, New Brunswick.

HERBERGER, L. (1971). "Organization and functioning of the Micro-Census in the Federal Republic of Germany," *Population Data and Use of Computers with Special Reference to Population Research,* German Foundation for Developing Countries, Berlin, and Federal Statistical Office, Wiesbaden. Mimeographed.

HESS, I., D.C. RIEDEL and T.B. FITZPATRICK (1981). *Probability Sampling of Hospitals and Patients.* University of Michigan, Ann Arbor.

HIGGINBOTHAM, J.R. and K.K. COX (eds.) (1979). *Focus Group Interview: A Reader.* American Marketing Association, Chicago.

HITE, SHERE (1979). *The Hite Report on Female Sexuality.* Knopf, New York.

HITE, SHERE (1981). *The Hite Report on Male Sexuality.* Knopf, New York.

HOGAN, HOWARD (1993). "Planning for census correction: the 1990 United States experience." *ISI/IASS Booklet, Invited Papers, Florence, 25 August-2 September 1993,* 133-149.

HORVITZ, D.G., B.G. GREENBERG and J.R. ABERNATHY (1975). "Recent developments in randomized response designs." In *A Survey of Statistical Design and Linear Models,* edited by J.R. Srivastava. American Elseiver, New York, 271-285.

HORVITZ, D.G., B.V. SHAH and W.R. SIMMONS (1967). "The unrelated question randomized response model," *Proceedings of the ASA, Social Statistics Section*, 65-72.

HOSNI, E. (1975). "Analyse de l'effet de rétrospection (Méthode SOM)," *As-Soukan: étude de centre de recherches et d'études démographiques.* Direction de la statistiques, Rabat, Maroc, 3, 17-27.

I-CHENG, C., L.P. CHOW and R.V. RIDER (1972). "The randomized response technique as used in the Taiwan outcome of pregnancy study," *SFP*, 3, 265.

IDE, MITSURU and SHIGERU KAWASAKI (1991). "Some new attempts in the data processing of the 1990 round of population censuses of Japan," *ISI/IASS Booklet, Invited Papers*, Cairo, 9-17 September 1991, 424-434.

ILO (1990). *Surveys of Economically Active Population, Employment, Unemployment and Underemployment.* ILO Manual on Concepts and Methods. Geneva.

IMPULSE RESEARCH CORP. (1993). *Impulse Survey of Focus Facilities.* Los Angeles.

INDIA: OFFICE OF THE REGISTRAR-GENERAL (1955). *Sample Census of Births and Deaths in 1953-54, Uttar Pradesh.* Ministry of Home Affairs, New Delhi.

INDIA: OFFICE OF THE REGISTRAR-GENERAL (1970). *Sample Registration of Births and Deaths in India: Rural 1965-68.* Ministry of Home Affairs, New Delhi.

ISI (1964). "Proceedings of the 34th Session," *BISI*, 40(1).

IUSSP (1959). *Problems in African Demography: A Colloquium.* Paris.

JANUS, SAMUEL S. and CYNTHIA L. JANUS (1993). *The Janus Report on Sexual Behavior.* Wiley, New York.

JEMIAI, HEDI and ATEF KHALIFA (1993). "Monitoring goals for children and women using the PAPCHILD surveys." Presented at the 49th session of the ISI, 25 August-2 September 1993, Florence.

JESSEN, R.J. (1978). *Statistical Survey Techniques.* Wiley, New York.

JOHNSON, NORMAN L. and HARRY SMITH, Jr. (eds.) (1969). *New Developments in Survey Sampling,* Wiley, New York.

JOHNSON, P.O. (1949). *Statistical Methods in Research.* Prentice-Hall, Englewood, N.J.

JONES, D.C. (1949). *Social Surveys.* Hutchinson, London.

KALTON, GRAHAM (1983). *Introduction to Survey Sampling.* Sage Publications, Beverly Hills, California.

KALTON, GRAHAM (1983). *Compensating for Missing Survey Data.* Institute for Social Research, Ann Arbor.

KALTON, GRAHAM and DALLAS W. ANDERSON (1986). "Sampling rare population," *JRSS (A)*, 148, 65-82.

KASPRZYK, D., G. DUNCAN, G. KALTON, and M.P. SINGH (1989). *Panel Surveys*. Wiley, New York.

KENDALL, MAURICE G. and WILLIAM R. BUCKLAND (1970). *A Dictionary of Statistical Terms* (3rd ed.). Oliver and Boyd, Edinburgh.

KENDALL, MAURICE G. and B. BABBINGTON SMITH (1954). *Tables of Random Sampling Numbers*. Cambridge University Press, Cambridge and New York.

KENYA: CENTRAL BUREAU OF STATISTICS and WFS (1980). "Annex II: Sample errors," *Kenya Fertility Survey 1977-1978, First Report, Vol. I.* Nairobi and London.

KINSEY, ALFRED C., WARDELL B. POMEROY and CLYDE E. MARTIN (1948). *Sexual Behavior in the Human Male*. W.B. Saunders, Philadelphia,

KINSEY, ALFRED C., WARDELL B. POMEROY, CLYDE E. MARTIN and PAUL H. GEBHHARD (1953). *Sexual Behavior in the Human Female*. W.B. Saunders, Philadelphia.

KISER, C.V. (1934). "Pitfalls in sampling for population study." *JASA*, 29, 250-256.

KISH, LESLIE J. (1965). *Survey Sampling*. Wiley, New York. Reprint (1995).

KISH, LESLIE J. (1971). "Special aspects of demographic samples," *International Population Conference, London, 1969, Tome I.* IUSSP, Liège, Belgium.

KISH, LESLIE J. (1987). *Statistical Design for Research*. Wiley, New York.

KISH, LESLIE J. (1994). "Multipopulation survey designs: five types with seven shared aspect," *ISR*, 62(2), 167-186.

KISH, LESLIE J., W. LOVEJOY and P. RACKOW (1961). "A multistage probability sample for continuous traffic surveys," Proceedings of the Social Statistics Section, ASA.

KNODEL, JOHN (1990). "Focus group discussions for social science research: a practical guide emphasis on the topic of aging," *Comparative Studies of the Elderly in Asia*. University of Michigan, Population Studies Center, Michigan.

KNOOP, H. (1966). "Some demographic characteristics of a suburban squatting community of Leopoldville, a preliminary analysis," *Cahiers Economi-ques et Sociaux*. Institut de Recherches Economiques et Sociales, Université Lovanium, Kinshasa, Congo, IV (2).

KOOP, J.C. (1960). "On theoretical questions underlying the technique of replicated or interpenetrating samples," Proc. Soc. Stat. Sec., *ASA*, 196.

KOOP, J.C. (1988). "The technique of replication or interpenetrating samples." In *Handbook of Statistics, Vol. 6, Sampling*, edited by P.R. Krishnaiah and C.R. Rao. Elseiver, Amsterdam.

KRISHNAIAH, P.R. and C.R. RAO (eds.) (1988). *Handbook of Statistics, Vol. 6, Sampling*. Elseiver, Amsterdam.

KROTKI, KAROL J. (1966). "The problem of estimating vital rates in Pakistan," *World Population Conference, 1965, Vol. III.* UN, Sales No. 66.XIII.7.

KROTKI, KAROL J. (ed.) (1978). *Developments in Dual System Estimation of Population Size and Growth.* The University of Alberta Press, Edmonton, Alberta.

KRUPKA, K., B. NIZETO and J.N. REINHARDS (1968). "Sampling studies on the epidemiology and control of trachoma in Southern Morocco," *Bulletin of the WHO,* 39.

LAHIRI, D.B. (1951). "A method of sample selection providing unbiased ratio estimates," *BISI,* 33, 133-140.

LAHIRI, D.B. (1958). "Observations on the use of interpenetrating samples in India," *BISI,* 36, 144-152.

LAPLANTE, M. (1988). *Data on Disability from the National Health Interview Survey, 1983-1985: An InfoUse Report.* National Institute on Disability and Rehabilitation Research, U.S. Department of Education, Washington D.C.

LAUMANN, EDWARD O., JOHN H. GAGNON, ROBERT T. MICHAEL and STUART MICHAELS (1994). *The Social Origins of Sexuality: Sexual Practices in the United States.* University of Chicago Press, Chicago.

LAURIAT, P. and A. CHINTAKANANDA (1965). "Technique to measure population growth: Survey of Population Change in Thailand," *World Population Conference, 1965, Belgrade.* UN (B6/V/E/507).

LAVRAKAS, PAUL J. (1993). *Telephone Survey Methods: Sampling, Selection, and Supervision* (2nd ed.). Sage Publications, Newbury Park, California.

LEMESHOW, S., D.W. HOSMER Jr. and J. KLAR (1989). *Adequacy of Sample Size in Health Studies.* Wiley, New York.

LEMESHOW, STANLEY and DAVID ROBINSON (1985). "Surveys to measure programme coverage and impact: A review of the methodology used by the Expanded Programme on Immunization," *WHSQ,* 38 (1), 65-75.

LEMESHOW, S. and G. STROH, Jr. (1988). *Sampling Techniques for Evaluating Health Parameters in Developing Countries.* National Academy Press, Washington, D.C.

LEPAGE, RAOUL and LYNNE BILLARD (eds.) (1992). *Exploring the Limits of Bootstrap.* Wiley, New York.

LESLIE, P.H. (1952). "The estimation of population parameters from data obtained by the capture-recapture method, II. The estimation of total numbers," *Biometrika,* 39, 363-388.

LESSLER, JUDITH T. and WILLIAM D. KALSBEEK (1992). *Nonsampling Error in Surveys.* Wiley, New York.

LEVY, PAUL S. and STANLEY LEMESHOW (1991). *Sampling of Populations: Methods and Applications.* Wiley, New York.

LEWONTIN, R.C. (1995). "Sex, Lies, and Social Sciences," *New York Review of Books*, 42(7), 24-29. (Review of *Sex in the Bedroom...* by Bullough, *The Social Organization of Sexuality...* by Laumann *et al.*, and *Sex in America...* by Michael *et al.*)

LICHTY, TOM (1994). *Desktop Publishing with Word for Windows* (2nd ed.). Ventana Press, Chapel Hill, North Carolina.

LINCOLN, F.C. (1930). "Calculating waterfowl abundance on the basis of banding returns," *Circ. U.S. Department of Agriculture*, No. 118.

LINDLEY, D.V. (1965). *Introduction to Probability and Statistics from a Bayesian Viewpoint, Pts. 1 and 2.* Cambridge University Press, Cambridge, U.K.

LINDLEY, D.V. (1972). *Bayesian Statistics: A review.* Society for Industrial and Applied Mathematics, Philadelphia.

LORD, M.G. (1994). "What that survey didn't say," Op-Ed, *New York Times*, October 25, 1994.

LYBERG, LARS and DANIEL KASPRZYK (1991). "Data collection methods and measurement error: An overview." In *Measurement Error in Surveys*, edited by P. Biemer *et al.*, Wiley, New York, 237-258.

MACURA, M. and V. BALABAN (1961). "Yugoslav experience in evaluation of population censuses and sampling." *BISI*, 38, 375-399.

MADOW W.G., H. NISSELSON and I. OLKIN (eds.) (1983). *Incomplete Data in Sample Surveys, Vol. 1: Report and Case Studies.* Academic Press, New York.

MADOW W.G., I. OLKIN and D.B. RUBIN (eds.) (1983). *Incomplete Data in Sample Surveys, Vol. 2: Theory and Bibliographies.* Academic Press, New York.

MADOW W.G., H. NISSELSON and I. OLKIN (eds.) (1983). *Incomplete Data in Sample Surveys, Vol. 3: Proceedings of the Symposium.* Academic Press, New York.

MAGUIRE, D.J., M.F. GOODCHILD and D.W. RHIND (eds.) (1991). *Geographical Information Systems: Principles and Applications.* Longmans, London.

MAHALANOBIS, P.C. (1940). "A sample survey of the acreage under jute in Bengal." *Sankhyā*, 4, 511-530.

MAHALANOBIS, P.C. (1944). "On large-scale sample surveys," *Philosophical Transactions of the Royal Society*, B, 231, 329-451.

MAHALANOBIS, P.C. (1946a). "Recent experience in statistical sampling in the Indian Statistical Institute." *JRSS (A)*, 108, 326-378.

MAHALANOBIS, P.C. (1946b). "Use of small-size plots in sample surveys for crop yields," *Nature*, 158, 798-799.

MAHALANOBIS, P.C. (1958). "Recent experience in statistical sampling in the Indian Statistical Institute." *JRSS (A)*, 108, 326-378. Reprint with a

Preface, *Sankhyā*, 20, 1-68, and (1961), Asia Publishing House, London, and Statistical Publishing Society, Calcutta.

MAHALANOBIS, P.C. (1960). "A method of fractile graphical analysis." *Econometrika*, 28, 325-351. Reprinted in *Sankhyā, A*, 23, 325-358.

MAHALANOBIS, P.C. (1966). "Some concepts of sample surveys in demographic investigations," *World Population Conference, Vol. III.* United Nations, Sales No. 66.XIII.7, 246-250.

MAHALANOBIS, P.C. (1968). *Sample Census of Area Under Jute in Bengal.* Statistical Publishing Society, Calcutta.

MAHALANOBIS, P.C. and J.M. SENGUPTA (1951). "On the size of sample cuts in crop cutting experiments in India," *BISI*, 33(2), 359-403.

MAMMON, E. (1992). *When Does Bootstrap Work? Asymptomatic Results and Simulations.* Springer-Verlag, New York.

MARKET DYNAMICS (undated). *Respondent Motivation: Los Angeles Focus Groups: Final Report.*

MARKS, ELI S., WILLIAM SELTZER and KAROL J. KROTKI (1974). *Population Growth Estimation: A Handbook of Vital Statistics Measurement.* Population Council, New York.

MARTIN, ELIZABETH (1993). "Response errors in survey measurements of facts," *ISI/IASS Booklet, Invited Papers, Florence, 25 August-2 September 1993*, 17-34.

MASTERS, WILLIAM H. and VIRGINIA E. JOHNSON (1966). *Human Sexual Response.* Little, Brown: Boston.

MAURITANIA: NATIONAL STATISTICAL OFFICE and LEAGUE OF ARAB STATES (1992). *Mauritania: Maternal and Child Health Survey (1990-1991), Principal Report.* Nouakchot and Cairo.

McDANIEL, C. (1979). "Focus groups – their role in the marketing research process," *Akron Business and Economic Review*, 10(4), 14-19.

MERTON, ROBERT K. (1987). "The focussed interview and focus groups: Continuities and discontinuities," *POQ*, 51, 550-566.

MERTON, ROBERT K. and PATRICIA L. KENDALL (1946). "The focused interview and focus groups: Continuities and discontinuities," *American Journal of Sociology*, 51, 541-557.

MERTON, ROBERT K., MARJORIE FISKE and PATRICIA L. KENDALL (1956). *The Focussed Interview: A Manual of Problems and Procedures* (2nd ed., 1990). Free Press, New York and Collier Macmillan, London.

MICHAEL, ROBERT T., JOHN H. GAGNON, EDWARD O. LAUMANN and GINA KOLATA (1994). *Sex in America: A Definitive Study.* Little Brown, Boston.

MILLER, J. (1984). "A new survey technique for studying deviant behavior." Ph.D. Dissertation, Sociology Department, The George Washington University, Washington, D.C.

MILLER, R.G. (1974). "The Jackknife – a review." *Biometrika*, 61, 1-17.

MONROE, JOHN and A.L. FINKNER (1959). *Handbook of Area Sampling.* Chilton Book Division, Philadelphia and New York.

MORGAN, DAVID L. (1988). *Focus Groups as Qualitative Research.* Sage Publications, Beverly Hills, California.

MORIN, HERVÉ (1993). *Théorie de l'Echantillonnage.* Presse de l'Université Laval, Quebec.

MOSER, CLAUS A. (1949). "The use of sampling in Great Britain," *JASA,* 44, 231-259.

MOSER, CLAUS A. and GRAHAM KALTON (1971). *Survey Methods in Social Investigation* (2nd ed.). Heinemann, London.

MURTHY, M.N. (1964). "On Mahalanobis' contribution to sample survey theory and methods." In *Contributions to Statistics, Presented to Professor P.C. Mahalanobis on the Occasion of His 70th Birthday,* edited by C.R. Rao. Pergamon Press, London, and Statistical Publishing Society, Calcutta, 283-316.

MURTHY, M.N. (1967). *Sampling Theory and Methods.* Statistical Publishing Society, Calcutta.

MURTHY, M.N. and N.S. NANJAMMA (1959). "Almost unbiased ratio estimates based on interpenetrating sub-sample estimates," *Sankhyā,* 21, 381-392.

MURTHY, M.N. and T.J. RAO (1988). "Systematic sampling with illustrative examples." In *Handbook of Statistics, Vol. 6, Sampling,* edited by P.R. Krishnaiah and C.R. Rao. Elseiver, Amsterdam, 147-186.

MURTHY, M.N. and A.S. ROY (1970). "A problem of integration of surveys – a case study," *JASA,* 65, 123-135.

MYERS, ROBERT J. (1976). "The Dual Record System: An Overview of Experience in Five Countries," Laboratories for Population Statistics, *Scientific Report Series No. 27.* The University of North Carolina at Chapel Hill, North Carolina.

NADOT, ROBERT (1966). *Afrique Noire, Madagascar, Comores – démographie comparée, 3 – Fécondité: Niveau.* INSEE, Paris.

NAG, MONI and M. BADRUD DUZA (1989). "Application of Focus Group Discussion Technique in Understanding Determinants of Contraceptive Use: A Case Study in Matlab, Bangladesh," *International Population Conference, New Delhi 1989,* 3, 367-378. IUSSP, Liège,

NAMBOODIRI, KRISHNAN (ed.) (1978). *Survey Sampling and Measurement.* Academic Press, New York.

NETER, J. and WAKSBERG, J. (1965). *Response Errors in Collection of Expenditure Data by Household Interviews: An Experimental Study.* Technical Paper No. 11, U.S. Bureau of the Census, Washington D.C.

NETHERLANDS, CENTRAL BUREAU OF STATISTICS (1992). *Netherlands Health Interview Survey 1981-1991.* The Hague.

NISSELSON, H. and T.D. WOOLSEY (1959). "Some problems of the household interview design for the National Health Survey," *JASA*, 54, 88-101.

OCHOA, LUIS HERNADEDZ, JULIO ORTUZAR and JOSEPH L. WIL-LARD (undated). "Procesamiento Interactivo de Encuesta Complejas con Microcomputadora Portailes: Un experimento en Guatemeala." (Mimeographed). DHS, Columbia, Maryland.

OLSEN, RANDALL J. (1991). "The effects of computer assisted interviews on data quality." Center for Human Resource Research, Ohio State University, Columbia, Ohio.

PAKISTAN: INSTITUTE OF DEVELOPMENT ECONOMICS (1968). *Report of the Population Growth Estimation Experiment 1961-1963.* Karachi.

PAKISTAN: NATIONAL INSTITUTE OF POPULATION STUDIES and IRD/ MACRO INTERNATIONAL (1992). *Pakistan DHS, 1990/1991.* Islamabad and Columbia, Maryland.

PANSE, V.G. (1954). *Estimation of Crop Yields.* FAO, Rome.

PANSE, V.G. (1966). *Some Problems of Agricultural Census Taking.* FAO, Rome.

PANSE, V.G. and P.V. SUKHATME (1948). "Crop surveys in India," *Journal of the Indian Society of Agricultural Statistics,* 1, 34-58.3, 96-168.

PAPPAIOANOU, MARGUERITE, TIMOTHY J. DONDERO, Jr., LYLE R. PETERSEN, IDA M. ONORATO, CAROLYN D. SANCHEZ and JAMES W. CURRAN (1990a). "The series of HIV Seroprevalence Surveys: Objectives, methods, and uses of sentinel surveillance for HIV in the United States," *PHR,* 105(2), 113-118.

PAPPAIOANOU, MARGUERITE, J. RICHARD GEORGE, W. HARRY HANNOX, MARTA OWINN, TIMOTHY J. DONDERO, Jr., GEORGE F. GRADY, RODNEY HOFF, ANNE D. WILLOUGHBY, AUDREY WRIGHT, ANTONIA C. NOVELLO and JAMES W. CURRAN (1990b). "HIV Seroprevalence Surveys of Childbearing Women – Objectives, methods, and uses of data," *PHR,* 105(2), 147-152.

PARKER, ROGER C. (1994a). *Desktop Publishing with WordPerfect 6.* Ventana Press, Chapel Hill, North Carolina.

PARKER, ROGER C. (1994b). *Desktop Publishing with WordPerfect 6 for Windows.* Ventana Press, Chapel Hill, North Carolina.

PARTEN, M.B. (1950). *Surveys, Polls and Samples: Practical Procedures.* Harpers, New York.

PATHAK, P.K. (1988). "Simple random sampling." In *Handbook of Statistics, Vol. 6, Sampling,* edited by P.R. Krishnaiah and C.R. Rao. Elseiver, Amsterdam.

PAYNE, S.L. (1951). *The Art of Asking Questions.* Princeton University Press, Princeton, and Oxford University Press, London.

POLITZ, ALFRED and *SIMMONS, WILLARD (1949 and 1950)*. "An attempt to get the 'not at home' into the sample without callbacks," *JASA*, 44, 9-31, and 45, 136-137.

POLLOCK, K.H., J.D. NICHOLS, J.E. HINES and C. BROWNIE (1990). "Statistical inference for capture-recapture experiments," *Wildlife Monographs*, 107, 1-97.

PONIKWOSKI, C.H., K.R. COPELAND and S.A. MEILY (1989). "Applications for Touchstone Recognition Technology in Establishment Surveys," *Proceedings of the Fourth Annual Research Conference, 1988*, U.S. Bureau of the Census. Washington, D.C.

POPULATION COUNCIL (1970). *A Manual for Surveys of Fertility and Family Planning: Knowledge, Attitudes, and Practice.* New York.

POPULATION REPORTS (1981). "Special topics: Contraceptive Prevalence Surveys: a new source of family planning data," *Population Reports, Series M, No. 5.* Johns Hopkins University, Population Information Program.

POPULATION REPORTS (1985). "Special topics: fertility and family planning surveys: an update," *Population Reports, Series M, No. 8.* Johns Hopkins University, Population Information Program.

POPULATION REPORTS (1992). "Special topics: the reproductive revolution: new survey findings," *Population Reports, Series M, No. 11.* Johns Hopkins University, Population Information Program.

PUBLIC HEALTH REPORTS (1990). Special issue, 105 (2).

QUENOUILLE, M.H. (1956). "Notes on bias in estimation," *Biometrika*, 43, 353-360.

RAJ, DES (1968). *Sampling Theory.* McGraw-Hill, New York.

RAMACHANDRAN, K.V. (1959). "On the Studentized smallest Chi-squares," *JASA*, 53 (284), 868-872.

RAMSEY, F.L., C.E. GATES, G.P. PATIL and C. TAILLIE (1988). "On transect sampling to assess wildlife populations and marine resources." In *Handbook of Statistics, Vol. 6, Sampling,* edited by P.R. Krishnaiah and C.R. Rao. Elseiver, Amsterdam, 515-532.

RAND CORPORATION (1955). *A Million Random Digits.* Free Press, Glencoe, Illinois.

RAO, C.R. (1973). *Advanced Statistical Methods in Biometric Research* (2nd ed.). Hafner, New York.

RAO, C.R. (1993). "Statistics must have a purpose: The Mahalanobis dictum," *ISI/IASS Booklet, Invited Papers, Florence, 25 August-2 September 1993,* 1-16.

RAO, J.N.K. (1988). "Variance estimation in sample surveys." In *Handbook of Statistics, Vol. 6, Sampling,* edited by P.R. Krishnaiah and C.R. Rao. Elseiver, Amsterdam.

RAO, J.N.K. and C.F.J. WU (1988). "Resampling inferences with complex survey data," *JASA*, 83, 231-241.

RAO, P.S.R.S. (1988). "Ratio and regression estimators." In *Handbook of Statistics, Vol. 6, Sampling,* edited by P.R. Krishnaiah and C.R. Rao. Elseiver, Amsterdam, 440-468.

RAO, T.J. (1966). "On certain unbiased ratio estimators," *Annals of Inst. Stat. Math.*, 18, 117-121.

RESEARCH TRIANGLE INSTITUTE (1989). *SUDAAN: Professional Software for SUrvey DAta ANalysis.* Research Triangle Institute, Research Triangle Park, North Carolina.

REYNOLDS, F.D. and D.R. JOHNSON (1978). "Validity of focus group findings," *Journal of Advertising Research,* 18(3), 21-24.

RICE, S.C., R.A. WRIGHT and B. ROWE (1989). "Development of computer assisted personal interview for the National Health Interview Survey," *Proceedings of the Section on Survey Research Methods*, ASA, 1988, 397-400.

RICH, BRUCE (1994). *Mortgaging the Earth.* Beacon Press and Earthscan (paperback).

RIES, P. and S. BROWN (1991). "Disability and Health: Characteristics of Persons by Limitation of Activity and Assessed Health Status, United States, 1984-1988," *Advanced Data from Vital and Health Statistics of the NCHS, 197.* Department of Health and Human Services, Public Health Service, NCHS. Washington D.C.

ROBINSON, PAUL (1994). "The Way We Do the Things We Do," *New York Times Book Review*, October 9, 1994. (Review of *The Social Organization of Sexuality...* by Laumann *et al.* (1994), and *Sex in America...* by Michael *et al.* (1994).)

ROTSCHILD, B.B. and L.B. WILSON (1989). "Nationwide Food Consumption Survey 1987: a landmark personal interview survey using laptop computers," *Proceedings of the Fourth Annual Research Conference, 1988,* U.S. Bureau of the Census. Washington, D.C., 347-356.

ROSANDER, A.C. (1977). *Case Studies in Sample Design.* Marcel Dekker, New York.

ROWE, ERROL, TOM PETKUNAS, MARTY APPEL and GLORIA CABLE (1994). "Using Optical Character Recognition to Process M3 Survey Faxed Images," Casic Technologies Interchange: U.S. Census Bureau 1994 Annual Research Conference, U.S. Bureau of the Census, Washington, D.C.

ROY, A.S. and A. BHATTACHARYYA (1968). *Technical Paper on Sample Design, Nineteen Round, July 1964-June 1965.* NSS Report No. 125. The Cabinet Secretariat, Government of India, New Delhi.

ROYER, JACQUES (1958). *Handbook on Agricultural Sample Surveys in Africa, I. Principles and Examples.* FAO and Commission for Technical Co-operation in Africa South of the Sahara, Government of France, Rome.

RUBIN, DONALD R. (1987). *Multiple Imputation for Nonresponse in Surveys.* Wiley, New York.

SALANT, PRISCILA and DON A. DILLMAN (1994). *How to Conduct Your Own Survey.* Wiley, New York.

SAMPFORD, M.R. (1962). *An Introduction to Sampling Theory with Applications to Agriculture.* Oliver and Boyd, Edinburgh.

SAMPSON, PETER (1986). "Qualitative Research and Motivation Research." In *Consumer Market Research Handbook,* edited by Robert Worcester and John Downham. McGraw-Hill, London.

SANCHEZ-CRESPO, J. L. (1979). *Curso Intensive de Muestro en Poblaciones Finitas.* Instituto Nacional de Estatdistica, Madrid.

SANDERSON, F.H. (1954). *Methods of Crop Forecasting.* Harvard University Press, Cambridge, Massachusetts.

SANWAL, MUKUL (1988). "Micro computers and development: organization and management issues in local-level computing," *Economic and Political Weekly,* Bombay, August 27, 1988, M-121-M131.

SARIS, W.E. (1989). "Technological revolution in data collection," Quality and Quantity, 23, 333-349.

SCHEAFFER, RICHARD L., WILLIAM MENDENHALL and LYMAN OTT (1990). *Elementary Survey Sampling* (4th ed.). Duxbury Press, Belmont, California.

SCHEARER, R. BRUCE (1981). "The value of focus group research for social action programs," *SFP,* 12 (12), 407-408.

SCHERR, MARVIN G. (1980). "The use of focus group interviews to improve the design of an administrative form: A case study at the Social Security Administration." U.S. Department of Health & Human Services, Social Security Administration, Baltimore.

SCHMITT, S.A. (1969). *Measuring Uncertainty: An Elementary Introduction.* Addison-Wesley, Reading, Massachusetts.

SCHREUDER, HANS T., TIMOTHY G. GREGOIRE and GEOFFREY WOOD (1993). *Sampling Methods for Multiresource Forest Inventory.* Wiley, New York.

SCHUCANY, W., H. GRAY and O. OWEN (1971). "On bias reduction in estimation," *JASA,* 66, 524-533.

SCOPP, THOMAS S. and STEVAN W. TORNELL (1991). "The 1990 Census experience with industry and occupation coding." Presented at the Southern Demographic Association Annual Meeting, Jacksonville, Florida, October 11, 1991.

SCOTT, C. (1967). "Sampling for demographic and morbidity surveys in Africa," *RISI,* 35, 154-171.

SCOTT, C. (1968). "Vital rate surveys in Tropical Africa: some new data affecting sample design." In *The Population of Tropical Africa,* edited by J.C. Caldwell and C. Okonjo. Longmans, London.

SEN, A.R. (1971). "Some recent developments in waterfowl sample survey techniques," *JRSS(C)*, 20, 139-147.

SEN, A.R. (1972). "Some nonsampling errors in the Canadian Waterfowl Mail Survey," *Journal of Wildlife Management*, 36, 951-954.

SEN, A.R. (1973). "Response errors in Canadian Waterfowl Surveys," *Journal of Wildlife Management*, 37, 485-491.

SEN, A.R. (1991). "Review of sampling techniques for estimation of marine fish catch and effort in North America with particular reference to Hawaii," *ISI/IASS Booklet, Invited Papers*, Cairo, 9-17 September 1991, 506-529.

SIKES, O.J. (1993). "Appropriate action to narrow the KAP-gap." In *Family Planning: Meeting Challenges: Promoting Choices, the Proceedings of the IPPF Family Planning Congress, New Delhi, November 1962*, edited by Pramila Senanyake and Ronald L. Kleinman. Parthenon Publishing, Pearl River, New York, U.S.A.

SINGH, DAROGA and F.S. CHAUDHARY (1986). *Theory and Analysis of Sample Survey Designs.* Wiley Eastern, New Delhi.

SLAVSON, S.R. (1979). *Dynamics of Group Psychotherapy.* Jason Aronson, New York.

SLONIM, MORRIS J. (1967). *Sampling* (3rd paperback printing). Simon and Schuster, New York.

SMITH, T.M.F. (1976). *Statistical Sampling for Accountants.* Haymarket Publishing, London.

SMITH, T.M.F. (1984). "Sample surveys, present position and potential developments: Some personal views (with discussion)," *JRSS(A)*, 147, 208-221.

SNEDECOR, GEORGE W. and WILLIAM G. COCHRAN (1967). *Statistical Methods* (3rd ed.). Iowa State University Press, Ames, Iowa.

SOGUNRO, B.O. (1991). "Nonresponse in developing countries – evolving an adequate approach to handle nonresponse," *ISI/IASS Booklet, Invited Papers*, Cairo, 9-17 September 1991, 242-262.

SOM, RANJAN K. (1958-9). "On sample design in opinion and marketing research," *POQ*, 32, 564-566.

SOM, RANJAN K. (1959). "Self-weighting sample design with an equal number of ultimate stage units in each of the selected penultimate stage units," *Calcutta Statistical Association Bulletin*, 9, 59-66.

SOM, RANJAN K. (1965). "Use of interpenetrating samples in demographic studies," *Sankhyā (B)*, 27(3&4), 329-342.

SOM, RANJAN K. (1973). *Recall Lapse in Demographic Enquiries.* Asia Publishing House, Bombay.

SOM, RANJAN K. (1993). "Use of microcomputers in national development." In *Probability and Statistics*, edited by S.K. Basu and B.K. Sinha, Narosa Publishing, New Delhi, 287-295.

SOM, RANJAN K. (1994). "Role of Prof. Mahalanobis in the United Nations and the International Statistical Institute." In a volume under publication by the Indian Statistical Institute.

SOM, RANJAN K., AJOY K. DE and NITAI C. DAS (1961). *Report on Morbidity.* NSS Report No. 49. The Cabinet Secretariat, Government of India, New Delhi.

SOM, RANJAN K., AJOY K. DE, NITAI C. DAS, B. TRIVIKRAMAN PILLAI, HIRALAL MUKHERJEE and S.M. UMAKANTA SARMA (1961). *Preliminary Estimates of Birth and Death Rates and of the Rate of Growth of Population, Fourteenth Round, July 1958-July 1959.* NSS Report No. 48. The Cabinet Secretariat, Government of India, New Delhi.

SPEIZER, HOWARD and DOUG DOUGHERTY (1991). "Automating data transmission and case management functions for a nationwide CAPI study." In *Proceedings of the 1991 Annual Research Conference.* U.S. Bureau of the Census, Washington, D.C., 389-397.

STATISTICS CANADA (1986). *Report of the Canadian Health and Disability Survey 1983-1984.* Social Trends Analyses Directorate, Ottawa, Canada.

STATISTICS CANADA (1991). *Health and Activity Limitation Survey – 1991: User's Guide.* Ottawa, Canada.

STATISTICS CANADA (1992). *Report of the International Workshop on the Development and Dissemination of Statistics on Persons with Disabilities, October 13-16, 1992, Ottawa, Canada.* Statistics, Canada, Ottawa, and the UN Statistical Division, New York.

STATISTICS CANADA (1994). *Health and Activity Limitation Survey – 1991. Making Disability Statistics Accessible: A Workshop on the 1991 Health and Activity Limitation Survey (HALS).* Ottawa, Canada.

STEPHAN, F.F. and McCARTHY, P.J. (1958). *Sampling Opinions.* Wiley, New York.

STUART, ALAN (1987). *Ideas of Sampling.* Edwin Arnold, London (3rd ed.).

STYCOS, MAYONE J. (1981). "A critique of focus group and survey research: the machismo case." *SFP,* 12 (12), 450-456.

SUDMAN, SEYMOUR (1976). *Applied Sampling.* Academic, San Diego.

SUDMAN, SEYMOUR and NORMAN M. BRADBURY (1974). *Response Effects in Surveys: A Review and Synthesis.* Aldine, Chicago.

SUDMAN, S. and NORMAN M. BRADBURN (1982). *Asking Questions: A Practical Guide to Questionnaire Design.* Jossey-Bass Publishers, San Francisco.

SUDMAN, SEYMOUR, MONROE J. SIRKEN and CHARLES D. COWAN (1988). "Sampling rare and elusive populations," *Science,* 240, 991-996.

SUKHATME, P.V. (1946a). "Bias in the use of small-size plots in sample surveys for yield," *Current Science,* 15, 119-120; *Nature,* 157, 630.

SUKHATME, P.V. (1946b). "Size of sampling unit in yield surveys," *Nature,* 158, 345.

SUKHATME, P.V. (1947a). "Use of small size plots in yield surveys," *Nature,* 160, 542.

SUKHATME, P.V. (1947b). "The problem of plot size in large-scale yield surveys," *JASA,* 42, 297-310, 460.

SUKHATME, P.V. and V.G. PANSE (1951). "Crop surveys in India II," *Journal of the Indian Society of Agricultural Statistics*, 3, 96-168.

SUKHATME, P.V. and G.R. SETH (1952). "Non-sampling errors in surveys," *Journal of the Indian Society of Agricultural Statistics*, 4, 5-41.

SUKHATME, PANDURANG V. and BALKRISHNA V. SUKHATME (1970b). "On some methodological aspects of sample surveys of agriculture in developing countries." In *New Developments in Survey Sampling*, edited by Norman L. Johnson and Harry Smith, Jr. Wiley, New York.

SUKHATME, PANDURANG V., BALKRISHNA V. SUKHATME, SHASHIKALA SUKHATME and C. ASOK (1984). *Sampling Theory of Surveys with Applications* (3rd ed.). Iowa State University Press, Ames, Iowa, and Indian Society of Agricultural Statistics, New Delhi; first edition (1953) by Pandurang V. Sukhatme, FAO, Rome, and Indian Society of Agricultural Statistics, New Delhi; second edition (1970a) by Pandurang V. Sukhatme and Balkrishna V. Sukhatme, FAO, Rome, and Asia Publishing House, Bombay.

SULLIVAN, K.M., R.R. FICHNER, J. GORSTEIN and A.G. DEAN (1990). "The use and availability of Anthropometry software." *Food and Nutrition Bulletin*, 12(2), 116-119. UN University, Tokyo.

SULLIVAN, K.M. and J. GORSTEIN (1990). *ANTHRO Version 1.01: Software for Calculating Pediatric Anthropometry*. Division of Nutrition, Centers for Disease Control, Atlanta, Georgia, U.S.A., and Nutrition Unit, WHO, Geneva.

SURVEY RESEARCH CENTER (1983). *General Interview Techniques: A Self-Instructional Workbook for Telephone and Personal Interview Training*. University of Michigan, Ann Arbor.

SUYONO, HARYONO, NANCY PIET, FARQUHAR STIRLING and JOHN ROSS (1981). "Family planning attitudes in urban Indonesia: Findings from focus group research," *SFP*, 12 (12), 433-442.

SZAMETIAT, K. and SCHAFFER, K.A. (1964). "Imperfect frames in statistics and the consequences for their use in sampling," *BISI*, 40, 517-538.

SZYBILLO, G.J. and R. BERGER (1979). "What advertising agencies think of focus groups," *Journal of Advertising Research*, 19(3), 29-33.

TAVRIS, CAROL and SUSAN SADD (1975). *The Redbook Report on Female Sexuality*. Delacorte, New York.

THANH LE (1993). "Sampling Practice in the DHS," *Proceedings of the 49th session of the ISI, Florence, Italy, 1993*. (Abstract).

THIONET, P. (1953). *Applications des Methodes de Sondage*. INSEE, Paris.

THIONET, P. (1958). *La Théorie des sondages*. INSEE, Paris.

THOMPSON, STEVEN K. (1992). *Sampling*. Wiley, New York.

TIPPET, L.H.C. (1952). *The Methods of Statistics* (4th ed.). Williams and Moorgate, London.

TORO, VIVIAN and KATHLEEN CHAMBERLAIN (1989). "Recent work with microcomputers for census processing in developing countries," *JOS*, Statistics, Sweden, 5(1), 69-91,

TRUEBLOOD, R.M. and R.M. CYERT (1952). *Sampling Techniques in Accountancy.* Prentice Hall, Englewood, New Jersey.

TUKEY, J. (1958). "Bias and Confidence in Not Quite Large Samples," *AMS*, 29, 614.

UN (1947). Report of the Sub-Commission on Statistical Sampling to the Statistical Commission. New York. Reprint (1948), *Sankhyā* (8), 393-402.

UN (1961). *The Mysore Population Study.* Sales No.: 61.XIII.3. New York.

UN (1964a). *Recommendations for the Preparation of Sample Survey Reports (Provisional Issue).* Sales No.: 64.XVII.7. New York.

UN (1964b). *Handbook of Household Surveys.* Sales No.: 64.XVII.13. New York.

UN (1971a). *Handbook of Population and Housing Census Methods, Part VI, Sampling in Connection with Population and Housing Censuses.* Sales No.: E70.XIII.9. New York.

UN (1971b). *Methodology of Demographic Sample Surveys.* Sales No.: E71.XVII. 11. New York.

UN (1972). *A Short Manual on Sampling, Vol. I, Elements of Sample Survey Theory.* Sales No.: E72.XIII.5. New York.

UN (1982a). *NHSCP: Survey Data Processing: A Review of Issues and Procedures.* UN Document No. DP/UN/INT-81-041/1, New York.

UN (1982b). *NHSCP: Non-sampling Errors in Household Surveys: Sources, Assessment and Control.* UN Document No. DP/UN/INT-81-041/2, New York.

UN (1982c). *Sample Surveys of Current Interest: Fourteenth Report.* Sales No.: E82.XIII.9. New York.

UN (1984a). *Improving Concepts and Methods for Statistics and Indicators on the Status of Women.* Sales No. E.84.XVII.3, New York.

UN (1984b). *Sampling Frames and Sample Designs for integrated Household Survey Programmes (Preliminary Version).* Document No. DP/UN/INT-81-014/5, New York.

UN (1986). *Development of Statistics of Disabled Persons: Case Studies.* Sales No. 86.XVII.17. New York.

UN (1988a). *Development of Statistical Concepts and Methods on Disability for Household Surveys.* Sales No. E.88.XVII.4 New York.

UN (1988b). *Improving Statistics and Indicators on Women using Household Surveys.* Sales No. E.88.XVII.11, New York.

UN (1989). *The Use of Microcomputers for Census Data Processing.* Document No. UNFPA/INT-88-P09/1, New York.

UN (1990a). *Supplementary Principles and Recommendations for Population and Housing Censuses.* Sales No. E.90.XVII.9. New York.

UN (1990b). *Methods of Measuring Women's Participation and Production in the Informal Sector.* Sales No. E.90.XVII.16. New York.

UN (1990c). *Manual on Population Census Data Processing Using Microcomputers.* Sales No. E.90.XVII.19, New York.

UN (1990d). *Assessing the Nutritional Status of Young Children.* NHSCP. Document No. DP/UN/INT-88-X01/8E, New York.

UN (1992). *WISTAT: Women's Indicators and Statistics Spreadsheet Database for Microcomputers (Version 2) – User's Guide and Reference Manual.* Sales No. E.92.XVII.11. New York.

UN (1993a). *Sampling Errors in Household Surveys.* NHSCP, Document No. DP/UN/INT-88-P80-15E, New York.

UN (1993b). *Sampling Rare and Elusive Populations.* NHSCP, Document No. INT-92-P80-16E, New York.

UN (1993c). *Methods of Measuring Women's Economic Activity.* Sales No. E.93.XVII.6. New York.

UN (1993d). *Rural Women in Development Model.* Department of Economic and Social Development, New York.

UN (1993e). *Urban Women in Development Model.* Department of Economic and Social Development, New York.

UN (1994). *Manual for the Development of Statistical Information for Disability Programmes and Policies* (Draft). U.N. Statistical Division, Document No. ESA/STAT/AC.4/INF.3. New York.

UN ECONOMIC COMMISSION FOR AFRICA (1971). *Manual on Demographic Sample Surveys in Africa* (Draft). Addis Ababa. Mimeographed.

UN ECONOMIC COMMISSION FOR ASIA AND THE PACIFIC (1988). *Statistical Software Packages for Processing Censuses and Surveys: A Brief Guide to Present Availability and Expected Future Developments.* No. ST/ESCAP/682, Bangkok.

U.S. AGENCY FOR INTERNATIONAL DEVELOPMENT (1981). *Framework for Preparing Census Reports on Women's Status and Roles in National Development.* Office of Population, Washington, D.C.

U.S. BUREAU OF THE CENSUS (1960). *The Accuracy of Census Statistics with and without Sampling.* Technical Paper No. 2. Washington, D.C.

U.S. BUREAU OF THE CENSUS (1963a). *The Current Population Re-interview Survey – Some Notes and Discussions.* Technical Paper No. 6. Washington, D.C.

U.S. BUREAU OF THE CENSUS (1963b). *The Current Population Survey – A Report on Methodology.* Technical Paper No. 7. Washington, D.C.

U.S. BUREAU OF THE CENSUS (1968). *Sampling Lectures.* Washington, D.C.

U.S. BUREAU OF THE CENSUS (1993a). *Protocol for Pretesting Demographic Surveys at the Census Bureau.* Report of the Pretesting Committee. Washington, D.C.

U.S. BUREAU OF THE CENSUS (1993b). *Recent HIV Seroprevalence Levels by Country: December 1993.* Center for International Research, Research Note, No. 11, Washington, D.C.

U.S. BUREAU OF THE CENSUS (1993c). *Trends and Patterns of HIV/AIDS Infection in Selected Developing Counties: Country Profile, December 1993.* Center for International Research, Research Note, No. 12, Washington, D.C.

U.S. BUREAU OF THE CENSUS (1993d). *HIV/AIDS Surveillance Data Base: Software User's Guide.* Center for International Research, Washington, D.C.

U.S. NCHS (1963). *Origin, Progress, and Operations of the U.S. National Health Survey.* Vital and Health Statistics Series 1, No. 1. U.S. Department of Health, Education and Welfare, Washington, D.C.

U.S. NCHS. *Vital and Health Statistics, Series I, Programs and Collection Procedures; Series II, Data Evaluation and Methods Research.* Vital and Health Statistics Series 1, No. 1. U.S. Department of Health, Education and Welfare, Washington, D.C.

U.S. NCHS (1970a). *Estimation and Sampling Variance in the Health Interview Survey,* Vital and Health Statistics Series 2, No. 38. U.S. Department of Health, Education and Welfare, Washington, D.C.

U.S. NCHS (1970b). *Development of the Design of NCHS Hospital Discharge Survey,* Estimation and Sampling Variance in the Health Interview Survey, Vital and Health Statistics Series 2, No. 39. U.S. Department of Health, Education and Welfare, Washington, D.C.

U.S. NCHS (1976). *Sample Design and Estimation Procedures for a National Health Examination Survey of Children,* Vital and Health Statistics Series 2, No. 43. U.S. Department of Health, Education and Welfare, Washington, D.C.

U.S. OFFICE OF MANAGEMENT AND BUDGET (1983). "Approaches to developing questions." *Statistical Policy Working Paper 10,* Statistical Policy Office, Office of Information and Regulatory Affairs.

U.S. OFFICE OF MANAGEMENT AND BUDGET (1990). "Computer-Assisted Survey Information Collection," *Statistical Policy Working Paper 19,* National Technical Information Service, Springfield, Virginia.

VAN BASTELAER, A., F. KERSEMAJERS and D. SIKKEL (1988). "Data collection with hand-held computers: Contributions to questionnaire design," *JOS,* 4, 141-154.

VANCE, L.L. (1950). *Scientific Method of Auditing: Applications of Statistical Sampling Theory to Auditing Procedure.* University of California Press, Berkeley, and Cambridge University Press, London.

VANCE, L.L. and J. NETER (1950). *Statistical Sampling for Auditors and Accountants.* Wiley, New York.

VELU, R. and G.M. NAIDU (1988). "A review of current survey sampling methods in marketing research (Telephone, Mail Intercept, and Panel Surveys)." In *Handbook of Statistics, Vol. 6, Sampling,* edited by P.R. Krishnaiah and C.R. Rao. Elseiver, Amsterdam, 533-554.

VERMA, V. (1992). "Household surveys in Europe: Some issues in comparative methodologies." *Seminar: International Comparison of Survey Methodologies,* Athens.

WAKSBERG, J. (1978). "Sampling methods for random digit dialing," *JASA,* 73, 40-46.

WARNER, S. (1965). "Randomized response: a survey technique for eliminating evasive answer bias," *JASA,* 60, 63-69.

WARWICK, DONALD P. and *CHARLES A. LININGER (1975). The Sample Survey: Theory and Practice.* McGraw-Hill, New York.

WAY, PETER O. and KAREN A. STANECKI (1993). *An Epidemiological Review of HIV/AIDS in Sub-Saharan Africa.* Center for International Research, U.S. Bureau of the Census, Washington, D.C.

WEBER, A.A. (1967). *Les Méthods de Sondage.* Bureau regional de l'Europe, Organisation Mondiale de la Santé, Copenhagen. Mimeographed.

WELBURN, ARTHUR J. (1984). *Practical Statistical Sampling for Auditing.* Marcel Dekker, New York.

WERKING, G.S. and R.L. CLAYTON (1991). "Enhancing data quality through the use of a mixed mode collection," *Survey Methodology,* Statistics Canada, 17 (1), 3-14.

WFS (1975). *Manual on Sample Design,* Basic Documentation No. 3. London.

WFS (undated). *Final Report.* London.

WHO (1966). *Sampling Methods in Morbidity Surveys and Public Health Investigations.* Technical Report Series No. 336, Geneva.

WHO (1980). *International Classification of Impairments, Disabilities and Handicaps – A Manual of Classification Relating to the Consequences of Disease.* Geneva.

WHO Eastern Mediterranean Region/South-East Asian Region Meeting on the Prevention of Neonatal Tetanus, Lahore, (1982). *EMRO Technical Publication No. 7,* Alexandria, Egypt, and SEARO *Technical Publication No. 3,* New Delhi.

WILLEMAN, THOMAS R. (1994). "Bootstrap on a shoestring: Resampling using spreadsheets," *American Statistician,* 48 (1), 40-42.

YATES, FRANK (1946). "A review of recent statistical developments in sampling and sampling surveys," *JRSS (A),* 109, 19-43.

YATES, FRANK (1981). *Sampling Methods for Censuses and Surveys* (4th ed.). Edwin Arnold, London and Oxford University Press, New York.

YERUSHALMY, J. and J. NEYMAN (1947). *PHR*, 67.

YULE, G. UDNY and MAURICE G. KENDALL (1950). *An Introduction to the Theory of Statistics* (14th ed.). Griffin, London, and Hafner, New York.

ZARKOVICH, S.S. (1962). "Agricultural statistics and multisubject household surveys," *Monthly Bulletin of Economics and Statistics*, 11, 1-5.

ZARKOVICH, S.S. (1963). *Sampling Methods and Censuses, Vol. II, Quality of Statistical Data* (Draft, mimeographed). Printed in 1966, FAO, Rome.

ZARKOVICH, S.S. (1965a). *Sampling Methods and Censuses*. FAO, Rome.

ZARKOVICH, S.S. (1965b). *Estimation of Areas in Agricultural Statistics*. FAO, Rome.

ZONGMING, SHAO and ZHANG WEIMIN (1993). "The evaluation of the coverage of China's 1990 Population Census," *ISI/IASS Booklet, Invited Papers, Florence, 25 August-2 September 1993*.

Answers to Exercises

Chapter 2 (page 66)

1. Following Example 2.1, $y = 2.91 \pm 0.235$[1] acres per plot; $y_0^* = 291.0 \pm 24.5$ acres; CV for both, 8.42%.

2. Following Example 2.3, (a) $940,597 \pm 19,128$; the 95% probability limits are 903,106 and 978,088; (b) 12.49 ± 0.254 cattle per farm; the 95% probability limits are 11.99 and 12.99.

5. Using the method of section 2.13, the number of households possessing radios is 40 ± 13; the number of persons in these households is 175 ± 60.

6. Follow the methods of section 2.13, and noting that the finite sampling correction is $(1 - f)$, where $f = n/N$ ($N = 175$ and $n = 60$), the estimated incidence of HIV seroconversion is 0.15 with standard error of 0.038.

8. For (a) using the methods of section 2.9 and considering the sample as having been drawn with replacement, (i) $25,902 \pm 1797$; (ii) 6093 ± 650; (iii) 0.2331 ± 0.01872. Here $n = 43$, $N = 325$. For (b) use equation (2.74) of section 2.13, which gives the standard error of the proportion absent as 0.007224. Here n has to be taken as 3427.

9. Considering the sample as with replacement and following Example 2.2, (a) 2460 ± 175; (b) $\$47,880 \pm 1215$; (c) $\$18,143 \pm 1164$; (d) 3.73 ± 0.2657; (e) $\$27.49 \pm \$1.76$; (f) $\$7.38 \pm \$0.534$; (g) $\$19.46 \pm \$1.476$; (h) 0.3789 ± 0.0758.

Chapter 3 (page 78)

1. $y_R^* = 962,096 \pm 14,218$, CV 1.48%; $y_0^* = 943,609 \pm 19,188$, CV 2.03%.

2. $961,348 \pm 12,012$, CV 1.25%.

3. $959,620 \pm 14,081$, CV 1.47%.

4. $957,579 \pm 11,349$, CV 1.19%.

[1] The sign $\pm$ after an estimate indicates the estimated standard error of the estimate.

Chapter 4 (page 89)

1. Following the method of Example 4.1, the estimated total area under wheat is 9855 ± 238 acres, CV 2.4%.

Chapter 5 (page 112)

1. Using the methods of section 5.3, $y_0^* = 351,664$ acres, CV 3.84%.
2. Using the methods of section 5.5, $\bar{p} = 0.2045 \pm 0.011085$, $y_0^* = 1178 \pm 63.84$. For details of computations, see the reference.
3. For (a), use the methods of section 5.3, and for (b) section 5.6. (a) $19,943 \pm 1242$ acres; CV 6.23%; (b) $19,453 \pm 946$ acres, CV 4.86%.

Chapter 6 (page 123)

1. Following the method of note 4 in section 6.4 and taking M_0 = number of persons per *kraal* = 3427/43 = 79.7, the estimated intraclass correlation coefficient is 0.085.
2. Relative cost efficiencies: 100, 94, 86, 92; optimum size is 1 ft row.

Chapter 7 (page 139)

2. Using equation (7.8), and noting that the permissible margin of error $d = 0.1\,P$, where P is the universe proportion, the number of wells is 44 for $P = 0.9$ to 400 for $P = 0.5$. For details, see the reference.
3. See Exercise 2, using $P = 0.5$, $n = 400$ persons.
4. In Example 2.2, the CV of the estimator obtained from 20 sample households was 0.0863. Using equation (7.22), the required sample size is 60 households.
5. Assuming normality and using Table 7.2 (last line), the estimated s.d. is 15, whereas the true s.d. is 16.1.

Chapter 10 (page 183)

1. As students are often asked to analyze the data of a stratified srs in the above form, the required computations are given in Table 1. The estimated standard errors of the stratum means will be obtained on dividing column (14) by N_h.
2. For (a) use equations (10.5(d)) and (10.15); $y = 1,353,572$ households, CV 9.12%. For (b) use equations (10.55) and (10.15); gain 407%.
3. From equation (10.51), average household size = 3022/598.8 = 4.95; from equation (10.52), its standard error is 0.1306 and CV 2.64%. Note that in this case, the use of separate ratio estimates for the estimate of a ratio has not led to an improvement.

Table 1: Computations for Chapter 10, Exercise 1

Stratum	$\bar{y}_h = \sum_{i=1}^{n_h} \frac{y_{hi}}{n_h}$	$y^{\bullet}_{h0} = N_h\bar{y}_h$	$\left(\sum_{j=1}^{n_h} y_{hi}\right)^2$	$\frac{\left(\sum_{j=1}^{n_h} y_{hi}\right)^2}{n_h}$	SSy_{hi}	$n_h(n_h-1)$	$\frac{SSy_{hi}}{n_h(n_h-1)}$	$s^2_{y^{\bullet}_{h0}} = N_h^2 \cdot$col.(12)	$s_{y^{\bullet}_{h0}}$	CV of $y^{\bullet}_{h0}$ (%)
(1)	(6)	(7)	(8)	(9)	(10)	(11)	(12)	(13)	(14)	(15)
I	4.05	2572	383 161	2504	3075	23,256	0.132210	53,310	231	9.0
II	10.31	5878	2,024,929	14,673	9580	18,906	0.506696	164,626	406	6.9
III	15.29	7263	3,090,564	26,874	7208	13,110	0.549773	124,043	352	4.8
IV	23.16	7017	2,859,481	39,171	12,248	5256	2.330295	213,942	462	6.6
V	28.71	2565	363,609	17,315	990	420	2.357833	18,676	137	5.2
All strata	12.21† (y/N)	25,293 (y)						574,597 (s^2_y)	758† s_y	3.0† CV_y (%)

† Not additive.

Table 2: Computations for Chapter10, Exercise 3

Stratum (acres)	No. of farms in stratum N_h	No. of farms sampled n_h	$\sum_{i=1}^{n_h} y_{hi}$	$\sum_{i=1}^{n_h} y_{hi}^2$	r_h	$p_h = r_h/n_h$
1-5	435	25	0	0	0	0
6-20	519	26	3	9	1	0.0385
21-50	357	16	6	36	1	0.0625
51-150	519	17	159	3969	8	0.4706
151-300	400	26	762	38,510	20	0.7692
300	266	15	1371	164,737	15	1
All strata	2496 (N)	125 (n)	2301	207,231	45 (h)	0.3600 (p)†

† Not additive.

4. For convenience and checking, Table 2 is provided with additional columns (4)-(7); y_{hi} is the wheat acreage of the ith farm ($i = 1, 2, \ldots, n_h$) in the hth stratum ($h = 1, 2, \ldots, 6$), r_h is the number and p_h the proportion of farms growing wheat in the hth stratum.

   For (i), using the formulae of sections 2.9 and 2.12, the estimated total area of wheat $= (2496 \times 2301/125) = 45,946 \pm 8142$ acres, CV 17.7%. Estimated number of farms growing wheat $N_p = 2496 \times 45/125 = 899$, with estimated standard error of $N\sqrt{[p(1-p)/(n-1)]} = 109$, CV 12.1%.

   For (ii), using the formulae of section 10.7(1), $y = \sum N_h \overline{y}_h = 41,106 \pm 4444$ acres, CV 10.8%; using the results of section 10.3, the estimated number of farms growing wheat $N_p = \sum N_h p_h = 860$, with estimated standard error $\sqrt{[\sum N_h^2 p_h(1-p_y)/(n_h-1)]} = 79$, CV 10.2%. Note the gain in efficiency in stratification after sampling.

6. Follow the methods of sections 10.4 and 10.5, noting that the finite sampling correction $(1 - f_h) = 0.5$, where $f_h = n_h/N_h = 0.5$, are not small and have, therefore, to be applied to estimate variances. Note, further, that since one hospital is selected in a sample county, there would be no need to assign a subscript for the sample hospital. Let x denote hospital beds and y AIDS admissions, then $x_{hi}^* = N_h M_{hi} x_{hi}$ and $y_{hi}^* = N_h M_{hi} y_{hi}$, where M_{hi} is the number of hospitals in the ith selected county and x_{hi} and y_{hi} are respectively the number of beds and the number of AIDS admissions in the

one selected hospital in the ith selected county. Arrange the computation as in the tables in Example 10.2. The following are the results: (a) unbiased estimate of the total number of AIDS admissions: 1096, s.e. 308; proportion of AIDS admissions: 0.4769, s.e. 0.06299; (b) separate ratio estimate of AIDS admissions: 1170, s.e. 506; combined ratio estimate: 1192, s.e. 157.

Chapter 11 (page 204)

1. Following the method of section 11.2, $y = 38,040 \pm 3027$ acres.

Chapter 12 (page 221)

1. Use equations (12.6) and (12.27); the results are given in Table 12.6.
2. Using the Neyman allocation (equation (12.10)), with $V_h = P_h(1 - P_h)$ and $n = 2000$, the allocations are 1222, 167 and 611 persons.

Chapter 15 (page 268)

1. (a) Using the methods of section 15.2.6(b-i), note 2, the average weight of a tablet is 1.08 ± 0.035 g; (b) using the methods of section 15.4, note 3, the proportion of sub-standard tablets is 0.058 ± 0.020; (c) using the method of section 15.2.5, note 2, the ratio is 0.170 ± 0.0020.
2. (a) Using equations (15.13) and (15.14), the estimated total number of cattle in the area is $28,820 \pm 3427$, with CV 11.98%. (b) (i) Following the method of section 15.2.6(b-ii), the average number of cattle per farm is, from equations (15.45) and (15.46), 12.4595 ± 1.0940, with CV 8.40%; (ii) following the methods of section 15.2.6(b-iii), the average is 13.0508 ± 1.195, with CV 9.16%; and (iii) following the methods of section 15.2.6(b-i), the average is 13.9090 ± 1.654, with CV 11.89%.
3. Following the methods of section 15.2.3, the estimated total number of beetles is 24.594 ± 2706. For details of computation, see the reference in the exercise.

Chapter 16 (page 287)

1. Following the method of section 16.3.1, the estimated total number of cattle is $28,421 \pm 2899$, with CV 10.20%.
2. Following section 16.3.2, the estimated total number of cattle is 24,188 with CV 9.79%. For details of computation, see the reference.

Chapter 17 (page 299)

1. Using equation (17.11), $m_0 = 2.9$ or 3. From the cost function, $n = (C - c_0)/(c_1 + m_0 c_2) = 336$.

2. Following the same method as for Example 17.2, the estimated variance of the total income $= 3180 + 653,049 = 3,833,529$.

3. (a) Following the same methods as for Example 17.1, $\hat{V}_1 = 0.0788$, $\hat{V}_2 = 2.1374$, and the unbiased variance estimate of the mean $= 0.000788 + 0.0021374 = 0.0029254$.

   (b) From equation (17.3), expected variance $= 0.0843766$.

Chapter 20 (page 353)

1. Although the sample design is stratified three-stage srs with villages, fields and plots selected at different stages, we can use the notations of a stratified two-stage design as only one plot is selected from each field. Denote by y_{hij} the yield of the jth sample field in the ith sample villages of the hth stratum ($h = 1, 2, \ldots, 10$; $i = 1, 2$; $j = 1, 2$). Then if $a_0 = \frac{1}{160}$ acre is the area of plot selected in each sample field, the unbiased estimator of the total yield of the hth stratum is $y^*_{h0} = \frac{1}{2}(y^*_{h1} + y^*_{h2})$ with an unbiased variance estimator $\frac{1}{4}(y^*_{h1} - y^*_{h2})^2$, where $y^*_{hi} = \frac{1}{2} A_h(y_{hi1} - y_{hi2})/a_0$, and A_h is the total area of the hth stratum. The estimated total yield is $27,716 \pm 2187$ metric tonnes (1 metric tonne $= 2204.6$ lb).

2. Following the same methods as for Example 20.1, the estimated average number of adults per household in the three states combined is 2.72 ± 0.097.

3. See section 20.2.10(b) and 10.5. The ratio estimate of the average household size is $y_{RS}/X_{RS} = 3017.67/598.78 = 5.04 \pm 0.0306$, with the estimated CV of 0.61%, a very considerable improvement over the CV of the ratio of the unbiased estimators, 3.22%, in Example 10.2.

Chapter 21 (page 376)

1. Following the method of Example 21.1, the results are: (a) total expenditure on cereals, Rs. $18,370,605 \pm 2,230,232$; (b) per capita expenditure on cereals, Rs. 9.79 ± 1.498; (c) average household size, 5.73 ± 0.442.

2. Following Example 21.1, the estimated average number of adults per household is 2.67 ± 0.078.

Chapter 24 (page 399)

1. Using equations (24.26) and (24.29), $N^* = 1087 \pm 159$.

Index of Names and Organizations

This list comprises names of persons, countries, governments, institutions, and organizations referred to in the book. Programs such as the Current Population Survey of the U.S. are listed in the Index of Subjects. Co-authors have been listed only once, unless they had more than one publication, singly or jointly.

Index of Subjects